Signals and Communication Technology

This series is devoted to fundamentals and applications of modern methods of signal processing and cutting-edge communication technologies. The main topics are information and signal theory, acoustical signal processing, image processing and multimedia systems, mobile and wireless communications, and computer and communication networks. Volumes in the series address researchers in academia and industrial R&D departments. The series is application-oriented. The level of presentation of each individual volume, however, depends on the subject and can range from practical to scientific.

Indexing: All books in "Signals and Communication Technology" are indexed by Scopus and zbMATH

For general information about this book series, comments or suggestions, please contact Mary James at mary.james@springer.com or Ramesh Nath Premnath at ramesh.premnath@springer.com.

Yu-Chu Tian · Jing Gao

Network Analysis and Architecture

Springer

Yu-Chu Tian
School of Computer Science
Queensland University of Technology
Brisbane, QLD, Australia

Jing Gao
College of Computer and Information Engineering
Inner Mongolia Agricultural University
Hohhot, China

ISSN 1860-4862 ISSN 1860-4870 (electronic)
Signals and Communication Technology
ISBN 978-981-99-5647-0 ISBN 978-981-99-5648-7 (eBook)
https://doi.org/10.1007/978-981-99-5648-7

This Springer imprint is published by the registered company Springer Nature Singapore Pte Ltd.
The registered company address is: 152 Beach Road, #21-01/04 Gateway East, Singapore 189721, Singapore

To our families and friends

Preface

Computer networks and the Internet have become integral parts of the fundamental infrastructure in modern industries and societies. Building a new network, upgrading an existing network, or planning for using a public network requires a profound understanding of the concepts, principles, approaches, and processes involved in advanced network planning. This book helps develop such a deep understanding. The knowledge and skills acquired from this book are relevant to computer networks, data communication, cybersecurity, and other related disciplines.

There have been numerous books on computer networks and network data communication, ranging from introductory to more advanced levels. The number of such books continues to increase. However, from our teaching experience over the last two decades, we have noticed a dearth of network books that provide detailed discussions on systematic approaches and best practices for high-level network planning in a structured process. As modern service-based networking becomes increasingly complex, integrating advanced network technologies, mechanisms, and policies into network architecture to serve business goals and meet network requirements poses a significant challenge. Topics relevant to these aspects are scattered throughout the networking literature. In our teaching and learning of advanced network subjects, specifically network planning, we have had to gather references from various sources and compile them into modular teaching materials. This has inspired us to write a dedicated book on network analysis and architecture design for network planning.

Therefore, this book distinguishes itself from existing network books by describing systematic approaches and best practices for network planning in a structured process. It introduces high-level network architecture and component-based architecture and discusses how advanced network technologies, such as security, are integrated into network architecture. The book can be used as a textbook for senior undergraduate students or postgraduate students. It is also valuable as a reference book for network practitioners seeking to develop or enhance their skills in network analysis and architecture design, particularly for large-scale computer networks with complex network service requirements. In our case, we have used the materials from this book in teaching postgraduate students specializing in computer science and electrical engineering.

As a textbook, this book compiles materials from various references, including books, international standards such as Request for Comments (RFCs) from the Internet Engineering Task Force (IETF), research articles, and network products. Therefore, the majority of these materials are not our original contributions, although the book does incorporate some research and development from our own group. We will cite references whenever possible throughout the book to acknowledge the original contributors of these materials that are not authored by us. However, to avoid distracting readers from the main theme of the topics covered in this book, we will not provide citations for every sentence. In fact, it would not be practical to do so. It is safe to assume that all materials discussed in this book are contributed by others unless explicitly indicated that a contribution is from our own group.

Using This Book

This book covers many advanced topics of computer networks, particularly focusing on network analysis and architecture. When used as a textbook for senior undergraduate students or postgraduate students, it can easily fit into a one-semester advanced networks subject. For example, in Australian universities, there are typically 13 teaching weeks each semester. The 13 chapters of the book can be taught within these 13 teaching weeks. We teach Chaps. 1 and 2 in a single module, and each of the remaining chapters in separate modules, reserving week 13 as a revision week.

If there are additional teaching weeks available, the following options can be considered:

- Divide Chap. 6 (Network Addressing Architecture) into two modules, with one module dedicated to IPv4 addressing and the other module focusing on IPv6 addressing.
- Teach Chap. 10 (Network Security and Privacy Architecture) in two modules to allow for comprehensive discussions of security architecture.
- Split Chap. 12 (Virtualization and Cloud) into two modules, with one module focusing on virtualization and the other module dedicated to cloud architecture.
- Extend Chap. 13 (Building TCP/IP Socket Applications) to two teaching modules to provide insightful practice in developing practical network communication systems.

If there are fewer teaching weeks, several options are available, by combining multiple chapters into a single module and/or setting some chapters aside. For example,

- Combine Chap. 1 (Introduction) and Chap. 2 (Systematic Approaches) into a single teaching module as we have done in our teaching practice.
- Teach Chap. 11 (Data Centers) and Chap. 12 (Virtualization and Cloud) in a single module.
- Use Chap. 13 (Building TCP/IP Socket Applications) for setting up assignment projects.

Contacting us

We have compiled a set of questions and exercises for each chapter in this book to aid students in better understanding the content. Professors who are using this book as a textbook are more than welcome to reach out for access to specific sections or the complete question bank.

We encourage professors, instructors, tutors and teaching assistants, and even students to create supplementary modules of materials that complement the existing content of this book. If you have materials in any format that you believe suitable for potential inclusion in a future edition of this book, we would love to hear from you. If any part of the materials that you provide is included in a future edition, we will give you clear and explicit acknowledgment for your valuable contribution.

We also encourage readers of this book to share any comments or insights with us. Whether you spot typographical errors, grammatical inaccuracies, or inappropriate expressions, or if you identify any misconstrued descriptions or interpretations, we would welcome your observations. You may also discuss with us about the assertions made in the book, and let us know what resonates and what may require more investigations. You may further tell us what you think should be incorporated or excluded in a future edition.

We sincerely hope that this book is useful to you.

Brisbane, Australia — Yu-Chu Tian
Hohhot, China — Jing Gao

Acknowledgements

Prior to its publication, the majority of the drafts of this book have served as lecture notes for several years, benefiting hundreds of postgraduate students. We are grateful to these students and their tutors for their valuable comments and suggestions, which have contributed to enhancing the content and presentation of the book.

We would like to acknowledge the editors and coordinators of this book from the publisher for the professional management of the whole process of the publication of this book. We extend special thanks to Mr. Stephen Yeung, Mr. Ramesh Nath Premnath, Mr. Karthik Raj Selvaraj, and Ms. Manopriya Saravanan from Springer. It has been enjoyable to work with them and the publisher.

The contents presented in this book are based on numerous contributions from the networking community. We are sincerely grateful to all the authors who have made these valuable contributions, and the organizations that have led the research and development of many of these areas. For instance, many RFCs from the IETF serve as the primary sources of references for various networking concepts, mechanisms, and technologies. We would like to express our deep gratitude to the authors and organizations involved in these contributions from the networking community. An exhaustive list of these contributors would be too lengthy to include within the limited space here. All the contributions from the networking community have been appropriately cited and/or discussed, alongside our own understanding and practical experience. Consequently, we bear full responsibility for any errors, incorrect or inappropriate interpretations, and/or inconsistent descriptions that may arise. Any feedback and suggestions for further improvement of this book are most welcome.

In this book, we have also included some original contributions from our own group. We would like to thank our colleagues and students who have worked and collaborated with us to create these contributions. Working with all of you in an exciting team environment filled with energy, inspiration, and enthusiasm has been a truly enjoyable experience.

Many of our own contributions cited in this book have received financial support from various funding agencies through research grants. We are particularly grateful to the Australian Research Council (ARC) for its support through the Discovery Projects scheme and Linkage Projects scheme under several grants, such as DP220100580, DP170103305, and DP160102571. We would also like to acknowledge the Australian Innovative Manufacturing Cooperative Research Centre (IMCRC), the Australian CRC for Spatial Information (CRCSI), and other funding agencies and organizations that have supported our research and development endeavors.

Contents

Part I Network Analysis

1 Introduction ... 3
1.1 Motivation for Network Planning ... 4
1.2 Deliverables from Network Planning ... 6
1.3 Strategic, Tactical and Operational Planning ... 7
1.4 Structured Network Planning Processes ... 9
1.4.1 Zoline's Network Planning Activities ... 9
1.4.2 McCabe's Three-Phase Network Planning ... 10
1.4.3 Oppenheimer's Structured Process ... 10
1.4.4 A General Process for Network Planning ... 12
1.5 Network Planning as an Art ... 12
1.6 Support from Customers and Executives ... 14
1.7 Main Objectives and Contents ... 14
1.8 Book Organization ... 15
References ... 17

2 Systematic Approaches ... 19
2.1 Systems Approach ... 20
2.1.1 System, Subsystems, and Environment ... 20
2.1.2 Holism and Emergent Behavior ... 21
2.1.3 Satisfaction and Trade-offs ... 23
2.1.4 Solving Problems yet to be Defined ... 24
2.1.5 Black, Gray, and White Boxes ... 25
2.2 Waterfall Model ... 26
2.2.1 Standard Waterfall Model ... 26
2.2.2 Waterfall Software Development ... 28
2.2.3 Waterfall Networking ... 28
2.3 A Generic Network Analysis Model ... 29
2.3.1 Model Architecture ... 30
2.3.2 Model Components ... 30
2.3.3 Model Interface ... 31

2.4 Top-Down Methodology 32
2.4.1 Descriptions of the Methodology 32
2.4.2 Use of the Methodology 33
2.4.3 Advantages and Disadvantages 34
2.5 Service-Based Networking 35
2.5.1 What are Network Services 35
2.5.2 Service Request and Service Offering 37
2.5.3 Resource Sharing Among Best-Effort Services 40
2.5.4 Service Performance Metrics 40
2.5.5 Multi-dimensional View of Service Performance 43
2.6 Summary 44
References 46

3 Requirements Analysis 47
3.1 Concepts of Requirements Analysis 47
3.2 Business Goals and Constraints 48
3.2.1 Understanding Core Business 49
3.2.2 Understanding Organizational Structure 51
3.2.3 Gathering Physical Location Information 51
3.2.4 Identifying Key Applications 51
3.2.5 Understanding Future Changes 52
3.2.6 Defining Project Scope 53
3.3 User Requirements 54
3.4 Application Requirements 57
3.4.1 Application Category 57
3.4.2 Performance Characteristics of Applications 58
3.4.3 Application Group 59
3.4.4 Top N Applications 59
3.4.5 Application Map 60
3.5 Device Requirements 61
3.5.1 Device Categories 62
3.5.2 Performance Characteristics of Devices 63
3.5.3 Location Dependency of Devices 63
3.6 Network Requirements 64
3.6.1 Extending Existing Networks 65
3.6.2 Addressing and Routing Requirements 66
3.6.3 Performance and Management Requirements 67
3.6.4 Security Requirements 68
3.7 Characterizing Existing Networks 69
3.7.1 Characterizing Network Architecture 69
3.7.2 Characterizing Addressing and Routing 70
3.7.3 Characterizing Performance and Management 71
3.7.4 Characterizing Security 72
3.7.5 Analyzing Protocols in Use 72

3.8 Requirements Specifications and Trade-Offs 73
3.8.1 Requirements Specifications 73
3.8.2 Technical Goals and Trade-Offs 73
3.9 Summary 77
References 78

4 Traffic Flow Analysis 79
4.1 Traffic Flows 80
4.1.1 Concepts of Traffic Flows 80
4.1.2 Individual and Composite Flows 81
4.1.3 Critical Flows 83
4.1.4 Flow Sources and Sinks 84
4.1.5 Flow Boundary 85
4.1.6 Identifying Traffic Flows 86
4.2 QoS Requirements 87
4.2.1 The Concept of QoS 87
4.2.2 The Importance of QoS 88
4.2.3 The Objective of QoS 89
4.2.4 QoS Requirements of Flows 90
4.2.5 Flow Prioritization 91
4.3 Traffic Flow Models 92
4.3.1 Peer-to-Peer Flow Model 92
4.3.2 Client-Server Flow Model 93
4.3.3 Hierarchical Client-Server Flow Model 95
4.3.4 Distributed-Computing Flow Model 96
4.4 Traffic Flow Measurement 98
4.4.1 Flow Measurement Architecture 98
4.4.2 Interactions Between Measurement Components 99
4.4.3 Multiple Managers, Meters, and Meter Readers 100
4.4.4 Granularity of Flow Measurements 101
4.4.5 Meter Structure 102
4.5 Traffic Load and Behavior 103
4.5.1 Characterizing Traffic Load 103
4.5.2 Examples of Traffic Load Analysis 105
4.5.3 Characterizing Traffic Behavior 109
4.5.4 Bandwidth Usage Pattern 112
4.5.5 Weekday Pattern 114
4.6 Flow Specification 114
4.6.1 Flowspec from Flow Analysis 115
4.6.2 Flowspec for QoS Management 117
4.6.3 Flowspec for Traffic Routing 118
4.7 Summary 118
References 120

Part II Network Architecture

5 Network Architectural Models ... 123
5.1 Hierarchical Network Architecture ... 124
5.1.1 The Core/Distribution/Access Architecture ... 124
5.1.2 Mesh Topology for Core Layer ... 126
5.1.3 Hierarchical Redundancy for Distribution Layer ... 126
5.1.4 The LAN/MAN/WAN Architecture ... 127
5.2 Enterprise Edge Architecture ... 129
5.2.1 Redundant WAN Segments ... 129
5.2.2 Multihomed Internet Connectivity ... 130
5.2.3 Secure VPN Links ... 130
5.3 Flow-based Architectural Models ... 131
5.3.1 Peer-to-Peer Architectural Model ... 132
5.3.2 Client-Server Architectural Model ... 133
5.3.3 Hierarchical Client-Server Architectural Model ... 133
5.3.4 Distributed Computing Architectural Model ... 134
5.4 Functional Architectural Models ... 135
5.4.1 Application-Driven Architectural Model ... 136
5.4.2 End-to-End Service Architectural Model ... 137
5.4.3 Intranet-Extranet Architectural Model ... 137
5.4.4 Service-Provider Architectural Model ... 138
5.5 Component-Based Architectural Models ... 139
5.5.1 Component-Based Architecture Design ... 139
5.5.2 Addressing Component Architecture ... 141
5.5.3 Routing Component Architecture ... 141
5.5.4 Network Management Component Architecture ... 142
5.5.5 Performance Component Architecture ... 143
5.5.6 Security Component Architecture ... 144
5.6 Redundancy Architectural Models ... 145
5.6.1 Router Redundancy ... 146
5.6.2 Workstation-to-Router Redundancy ... 146
5.6.3 Server Redundancy ... 151
5.6.4 Route and Media Redundancy ... 156
5.7 Integration of Various Architectural Models ... 157
5.8 Summary ... 159
References ... 160

6 Network Addressing Architecture ... 161
6.1 Review of IPv4 Address Representation ... 161
6.1.1 IPv4 Address Structure ... 162
6.1.2 Binary and Decimal Representations ... 162
6.1.3 Static, Dynamic, and Automatic IP Addressing ... 164

6.2 IPv4 Addressing Mechanisms ... 164
6.2.1 Classful IPv4 Addressing ... 165
6.2.2 Private Addressing ... 167
6.2.3 Classless IPv4 Addressing ... 168
6.2.4 Fixed-Length Subnetting ... 169
6.2.5 Variable-Length Subnetting ... 170
6.2.6 Supernetting ... 173
6.3 IPv4 Addressing Strategies ... 177
6.3.1 Effective Addressing Strategies ... 177
6.3.2 Hierarchical Address Assignment ... 178
6.4 IPv6 Addressing ... 184
6.4.1 IPv6 Notation ... 184
6.4.2 IPv6 Site Prefix, Subnet ID, and Interface ID ... 185
6.4.3 IPv6 Anycast Addresses ... 186
6.4.4 IPv6 Unicast Addresses ... 186
6.4.5 IPv6 Multicast Addresses ... 188
6.4.6 Multicast Flooding ... 191
6.4.7 Assignment of IPv6 Addresses to Interfaces ... 192
6.4.8 IPv6 Header Structure ... 193
6.4.9 Key Benefits of IPv6 Addressing ... 197
6.5 IPv6 Autoconfiguration ... 198
6.5.1 IPv6 Stateless Autoconfiguration ... 199
6.5.2 IPv6 Stateful Autoconfiguration ... 202
6.6 Built-in Security in IPv6 ... 203
6.6.1 IPsec in IPv6 and IPv4 ... 203
6.6.2 Extension Headers for Security and Privacy ... 204
6.7 Built-in True QoS in IPv6 ... 208
6.7.1 Traffic Class ... 208
6.7.2 Flow Labeling ... 209
6.8 Coexistence of IPv6 and IPv4 ... 211
6.8.1 Dual Stack ... 211
6.8.2 IPv6-over-IPv4 Tunneling ... 213
6.9 IPv6 Network Planning ... 214
6.9.1 Checklist for IPv6 Planning ... 214
6.9.2 Preparing Network Services ... 215
6.9.3 Planning for Tunnels ... 216
6.9.4 Security Considerations ... 216
6.10 Summary ... 217
References ... 218

7 Network Routing Architecture ... 221
7.1 Categories of Routing Protocols ... 221
7.1.1 Path-Vector Routing Protocol ... 223
7.1.2 Distance-Vector Routing Protocols ... 224
7.1.3 Link-State Routing Protocols ... 229

7.1.4 Classful and Classless Routing Protocols 237
7.1.5 Comparisons of Routing Protocols 238
7.2 Routing Architecture and Strategies 238
7.2.1 Architectural Considerations 238
7.2.2 Choosing Routing Protocols 240
7.2.3 Route Redistribution 242
7.2.4 Router Filtering 243
7.3 Software Defined Networks 244
7.3.1 Motivations Behind SDN 244
7.3.2 Layered Architecture of SDN 246
7.3.3 SDN Interfaces 247
7.3.4 SDN in a Service Provider Environment 248
7.3.5 Key Features of SDN 249
7.4 Routing in Publish-Subscribe Networks 250
7.4.1 Wide Area Networks in Smart Grid 250
7.4.2 Constrained Optimization of Multicast Routing 252
7.4.3 Algorithm Design for Problem Solving 254
7.4.4 An Illustrative Example 258
7.4.5 Multiple Multicast Trees with Shared Links 261
7.5 Routing in Large-Scale Wireless Networks 263
7.5.1 Classification of Wireless Routing 263
7.5.2 Proactive and Reactive Routing 265
7.5.3 Reactive Routing Protocols 266
7.5.4 Proactive Routing Protocols 268
7.6 Summary 271
References 272

8 Network Performance Architecture 275
8.1 Quality-of-Service Control 276
8.2 Resource and Traffic Control 277
8.2.1 Prioritization 277
8.2.2 Traffic Management 283
8.2.3 Queuing and Scheduling 286
8.2.4 Frame Preemption 288
8.3 Network Policies 290
8.3.1 Policies and Their Benefits 290
8.3.2 Types of Network Policies 292
8.3.3 Simple Yet Effective Performance Policies 292
8.3.4 Implementing Network Policies 294
8.4 Differentiated Services (DiffServ) 294
8.4.1 DiffServ Architecture 295
8.4.2 DiffServ Domain and Region 296
8.4.3 Traffic Classification and Conditioning 297
8.4.4 Per-hop Behavior (PHB) 299
8.4.5 DSCP to Service Class Mapping 299

8.5 Integrated Services IntServ 300
8.5.1 Assumptions for IntServ Architecture 300
8.5.2 IntServ Service Model 303
8.5.3 Overall Architecture 305
8.5.4 Controlled-Load Service and Guaranteed QoS 307
8.5.5 Reference Implementation Framework 309
8.5.6 RSVP Protocol 310
8.6 Service-Level Agreements (SLAs) 315
8.7 Summary 316
References 319

9 Network Management Architecture 321
9.1 Concepts of Network Management 322
9.1.1 Network Management Hierarchy 322
9.1.2 Network Management Framework 323
9.1.3 Network Management Questions 324
9.2 Functional Areas of Network Management 324
9.3 Network Management Protocols 326
9.3.1 SNMP 326
9.3.2 Syslog Protocol 334
9.3.3 IPFIX and PSAMP 334
9.3.4 NETCONF for Configuration Management 335
9.3.5 CMIP 336
9.3.6 Protocols and Mechanisms with Specific Focus 338
9.4 Network Management Data Models 339
9.4.1 Generic Infrastructure Data Models 340
9.4.2 Management Infrastructure Data Models 340
9.4.3 Data Models at Specific Layers 340
9.4.4 An FCAPS View of Management Data Models 341
9.4.5 Model-Based Network Management 343
9.5 Network Management Mechanisms 344
9.5.1 Characterizing Network Devices for Management 344
9.5.2 Instrumentation 345
9.5.3 Event Notification and Trend Analysis 346
9.5.4 Configuration of Network Parameters 347
9.6 Management Architectural Models 348
9.6.1 In-Band and Out-of-Band Network Management 349
9.6.2 Centralized and Distributed Network Management 351
9.6.3 Hierarchical Network Management and MoM 354
9.6.4 Consideration of Network Management Traffic 355
9.7 Summary 356
References 358

10 Network Security and Privacy Architecture 361
10.1 Overall Recommendations 361
10.2 Architectural Considerations for Security 363
10.2.1 Objectives 363
10.2.2 Network and Service Configuration for Security 364
10.2.3 NSA's Guide for Secure Network Architecture 364
10.3 Security and Privacy Plan 366
10.3.1 Basic Approach 367
10.3.2 Identifying What to Protect 367
10.3.3 Clarifying Security Threats 368
10.3.4 Developing a Security and Privacy Plan 368
10.4 Security Policies and Procedures 370
10.4.1 Characteristics of Security Policies 370
10.4.2 Components of Security Policies 371
10.5 Security and Privacy Mechanisms 372
10.5.1 Security Awareness 372
10.5.2 Physical Security 372
10.5.3 Authentication, Authorization, and Accounting 373
10.5.4 Firewalls 375
10.5.5 Intrusion Detection and Prevention 379
10.5.6 Encryption and Decryption 382
10.6 Securing Internet Connections 387
10.7 Securing Network Services and Management 390
10.7.1 Securing End-User Hosts and Applications 390
10.7.2 Securing Sever Farms 391
10.7.3 Hardening Routers and Switches 392
10.7.4 Securing Wireless Networks 393
10.7.5 Securing Network Management 394
10.8 Remote Access and VPNs 394
10.8.1 Layer-2 and Layer-3 Tunneling Protocols 395
10.8.2 Site-to-Site, Remote-Access, and Client-to-Site VPNs 398
10.8.3 Choosing and Hardening VPNs 399
10.9 Summary 401
References 402

Part III Network Infrastructure

11 Data Centers 405
11.1 Data Centers Around the World 406
11.2 Functions and Categories of Data Centers 408
11.2.1 Edge Data Centers 408
11.2.2 Cloud Data Centers 408
11.2.3 Enterprise Data Centers 409

11.2.4 Managed Data Centers ... 410
11.2.5 Colocation Data Centers ... 410
11.3 Standards for Building Data Centers ... 410
11.3.1 ANSI/TIA-942 ... 411
11.3.2 Telcodia's GR-3160 ... 412
11.3.3 EN 50600 and ISO/IEC 22237 ... 413
11.4 Tiered Reliability of Data Centers ... 416
11.4.1 ANSI/TIA-942 Data Center Levels ... 417
11.4.2 EN50600 and ISO/IEC 22237 Data Center Classes ... 418
11.4.3 Uptime Institute Data Center Tiers ... 418
11.4.4 The Choice of a Data Center Tier ... 421
11.5 Site Space, Cabling, and Environments ... 422
11.5.1 Site Space and Layout ... 422
11.5.2 Cabling Infrastructure ... 424
11.5.3 Environmental Considerations ... 425
11.6 Considerations of Data Center Locations ... 425
11.6.1 Safety and Security of Physical Locations ... 426
11.6.2 Power Supply ... 426
11.6.3 Environment Temperatures ... 427
11.6.4 Access to Communication Networks ... 428
11.6.5 Proximity to End Users and Skilled Labor ... 429
11.6.6 Other Considerations ... 429
11.7 Network Architecture ... 430
11.7.1 Three-layer Data Center Architecture ... 430
11.7.2 Data Center Design Models ... 431
11.7.3 HPC Cluster Types ... 433
11.7.4 Data Center Network Virtualization ... 436
11.8 Data Center Security ... 441
11.8.1 Physical Security ... 441
11.8.2 Network Security ... 442
11.9 Summary ... 443
References ... 444

12 Virtualization and Cloud ... 447
12.1 Virtualization and Virtual Networking ... 447
12.1.1 How Virtualization Works ... 448
12.1.2 Types of Virtualization ... 450
12.1.3 Internetworking of VMs ... 453
12.1.4 Advantages and Limitations of Virtualization ... 458
12.2 Network Functions Virtualization (NFV) ... 459
12.2.1 NFV Objectives and Standards ... 460
12.2.2 NFV Use Cases ... 461
12.2.3 NFV Architecture Framework ... 470
12.2.4 NFV Framework Requirements ... 472
12.2.5 Cloud-native Network Functions (CNFs) ... 473

12.2.6 Open-Source Network Virtualization and Orchestration 474
12.2.7 NFV Challenges 475
12.3 Cloud Computing and Its Characteristics 477
12.3.1 Cloud Computing and Cloud Services 477
12.3.2 Essential Characteristics of Cloud Computing 478
12.4 Deployment Models of Cloud Computing 480
12.4.1 Public Cloud 480
12.4.2 Private Cloud 481
12.4.3 Community Cloud 481
12.4.4 Hybrid Cloud 482
12.5 Service Models of Cloud Computing 483
12.5.1 IaaS (Infrastructure as a Service) 484
12.5.2 PaaS (Platform as a Service) 485
12.5.3 SaaS (Software as a Service) 486
12.5.4 The Use of Cloud Service Models 487
12.6 Cloud Computing Reference Architecture 489
12.6.1 Conceptual Reference Model 489
12.6.2 Cloud Provider 491
12.6.3 Cloud Consumer 492
12.6.4 Cloud Broker 492
12.6.5 Cloud Auditor 493
12.6.6 Cloud Carrier 494
12.7 Cloud Security 494
12.7.1 Cloud Security Responsibilities 494
12.7.2 Cloud Security Challenges 495
12.7.3 Cloud Security Methods 495
12.8 Summary 497
References 498

13 Building TCP/IP Socket Applications 501
13.1 Why Socket Programming 502
13.2 Example Client-Server Systems 504
13.3 Socket APIs for TCP/IP Communications 505
13.3.1 Socket()—Open a Socket for Communications 507
13.3.2 Bind()—Bind the Socket to an Address 509
13.3.3 Listen()—Set the Number of Pending Connections 511
13.3.4 Connect() from Client and Accept() from Server 511
13.3.5 send()/recv()-Send/Receive Data 512
13.3.6 close()—Close Socket 514
13.4 Example Server-Client Programs 514
13.4.1 System Specifications and Requirements 514
13.4.2 System Design 515
13.4.3 System Implementation 515

13.5 IPv6 Sockets 519
13.5.1 Changes in IPv6 Sockets from IPv4 520
13.5.2 Example IPv6 Socket Programs 521
13.6 Keyboard Input Processing 525
13.6.1 Simple Keyboard Input Processing 525
13.6.2 Multithreading Keyboard Input Processing 527
13.7 Socket Programming in Windows 531
13.7.1 Comparisons Between Winsock and Linux Sockets 531
13.7.2 Example Code for Winsock Server and Client 531
13.7.3 Compiling C Programs in Command Prompt 535
13.7.4 Further Discussions on Windows Programming 537
13.8 Summary 539
References 540

Index 541

About the Authors

Prof. Yu-Chu Tian is a professor of computer science at the School of Computer Science, Queensland University of Technology, Brisbane QLD 4000, Australia. He received the Ph.D. degree in computer and software engineering from the University of Sydney, Sydney NSW 2006, Australia, in 2009, and the Ph.D. degree in industrial automation from Zhejiang University, Hangzhou 310027, China, in 1993. He has previously worked at Hong Kong University of Technology, Hong Kong, China; Curtin University, Perth WA 6102, Australia; and the University of Maryland at College Park, MD 20742, USA. His research interests encompass a wide range of areas, including big data computing, cloud computing, computer networks, smart grid communications and control, optimization and machine learning, networked control systems, and cyber-physical security. Email: y.tian@qut.edu.au.

Prof. Jing Gao is a professor and the Dean of the College of Computer and Information Engineering, Inner Mongolia Agriculture University, Hohhot 010018, China. She also serves as the director of the Inner Mongolia Autonomous Region Key Laboratory of Big Data Research and Application for Agricultural and Animal Husbandry, Hohhot 010018, China. She received the Ph.D. degree in computer science and technology from Beihang University, Beijing 100191, China, in 2009. She has previously worked as a visiting professor at the School of Computer Science, Queensland University of Technology, Brisbane QLD 4000, Australia. Her research interests include computer networks, big data analytics and computing, knowledge discovery, and agricultural intelligent systems. Email: gaojin@imau.edu.cn.

Acronyms

3GPP	3rd Generation Partnership Project
AAA	Authentication, Authorization, and Accounting
AAAA	Authentication, Authorization, Accounting, and Allocation
ABR	Area Border Router
ACL	Access Control List
ACSE	Association Control Service Element
AES	Advanced Encryption Standard
AH	Authentication Header
AIFS	Arbitration Inter-Frame Space
AIFSN	AIFS Number
ANSI	American National Standards Institute
AODV	Ad-hoc On-demand Distance Vector
AOMDV	Ad-hoc On-demand Multipath Distance Vector
API	Application Programming Interface
APIPA	Automatic Private IP Addressing
APNIC	Asia Pacific Network Information Centre
AQM	Active Queue Management
ARP	Address Resolution Protocol
AS	Autonomous System
ASBR	AS Boundary Router
ATM	Asynchronous Transfer Mode
AVF	Active Virtual Forwarder
AVG	Active Virtual Gateway
BC	Betweenness Centrality
BCBT	Betweenness Centrality to Bandwidth ratio Tree
BER	Bit Error Rate
BGP	Border Gateway Protocol
BNG	Broadband Network Gateway
BPDU	Bridge Protocol Data Unit
BR	Backbone Router
CA	Certificate Authority

CBQ	Class-based Queuing
CDN	Content Delivery Network
CE	Congestion Experienced
CENELEC	European Committee for Electrotechnical Standardization
CIA	Confidentiality, Integrity, and Availability
CIDR	Classless Inter-Domain Routing
CIM	Common Information Model
CISA	Cybersecurity and Infrastructure Security Agency
CMIP	Common Management Information Protocol
CMIS	Common Management Information Service
CMISE	CMIS Element
CMOT	CMIP over TCP/IP
CNF	Cloud-native Network Function
CoS	Class of Service
CPE	Customer Premises Equipment
CRC	Cyclic Redundancy Check
CW	Contention Window
DAL	Device and resource Abstraction Layer
DCF	Distributed Coordination Function
DDoS	Distributed DoS
DEI	Drop Eligible Indicator
DHCP	Dynamic Host Configuration Protocol
DiffServ	Differentiated Service
DMTF	Distributed Management Task Force
DMZ	DeMilitarized Zone
DNS	Domain Name System
DoS	Denial of Service
DSCP	Differentiated Services field CodePoint
DSDV	Destination Sequence Distance Vector
DSR	Dynamic Source Routing
DTLS	Datagram Transport Layer Security
DUAL	Diffused update algorithm
eBGP	External BGP
ECC	Elliptical Curve Cryptography
ECN	Explicit Congestion Notification
ECT	ECN-Capable Transport
EDA	Equipment Distribution Area
EDCA	Enhanced Distributed Channel Access
EGP	Exterior Gateway Protocol
EIGRP	Enhanced Interior Gateway Routing Protocol
EM	Element Management
EMS	Element Management System
EPC	Evolved Packet Core
EPO	Emergency Power Off
ERP	Enterprise Resource Planning

ESP	Encapsulating Security Payload
ETSI	European Telecommunications Standards Institute
EUI	Extended Unique Identifier
FCAPS	Fault, Configuration, Accounting, Performance, and Security
FIB	Forwarding Information Base
FIFO	First In First Out
FQDN	Fully Qualified Domain Name
FTP	File Transfer Protocol
FTTK	Fiber To The Kerb
FTTP	Fiber To The Premises
GLBP	Gateway Load Balancing Protocol
GPS	Global Positioning System
GUI	Graphical User Interface
HAN	Home Area Network
HCCA	HCF Controlled Channel Access
HCF	Hybrid Coordination Function
HDA	Horizontal Distribution Area
HPC	High Performance Computing
HSRP	Hot Standby Router Protocol
HSS	Home Subscriber Server
HTTP	HyperText Transfer Protocol
IaaS	Infrastructure as a Service
IAM	Identity and access management
IANA	Internet Assigned Numbers Authority
iBGP	Internal BGP
ICMP	Internet Control Message Protocol
ICV	Integrity Check Value
IDE	Integrated Development Environment
IDPS	Intrusion Detection and Prevention Systems
IDS	Intrusion Detection System
IET	Interspersing Express Traffic
IETF	Internet Engineering Task Force
IFG	Inter-Frame Gap
IGMP	Internet Group Management Protocol
IGP	Interior Gateway Protocol
IGRP	Interior Gateway Routing Protocol
IHL	Internet Header Length
IKE	Internet Key Exchange
IKEv2	Internet Key Exchange Protocol Version 2
IMAP	Internet Message Access Protocol
IMS	IP Multimedia Subsystem
IntServ	Integrated Service
IoT	Internet of Things
IP	Internet Protocol
IPFIX	IP Flow Information eXport

IPS	Intrusion Prevention System
IPsec	Internet Protocol Security
IR	Internal Router
IRDP	ICMP Router Discovery Protocol
IS-IS	Intermediate System to Intermediate System
ISDN	Integrated Services Digital Network
ISP	Internet Service Provider
KPI	Key performance indicators
KVM	Kernel-based Virtual Machine
L2F	Layer 2 Forwarding
L2TP	Layer 2 Tunneling Protocol
LAC	L2TP Access Concentrator
LAN	Local Area Network
LEACH	Low Energy Adaptive Clustering Hierarchy
LLC	Logical Link Control
LNS	L2TP Network Server
LPP	Lightweight Presentation Protocol
LSA	link state advertisement
MAC	Media Access Control
MAN	Metropolitan Area Network
MANO	Management and Orchestration
MDA	Main Distribution Area
MIB	Management Information Base
MLD	Multicast Listener Discovery
MME	Mobility Management Entity
MPLS	Multi-Protocol Label Switching
MPP	Management Plane Protection
MTBF	Mean Time Between Failure
MTTR	Mean Time to Repair
MTU	Maximum Transmission Unit
NaaS	Network as a Service
NACM	NETCONF Access Control Model
NAN	Neighborhood Area Network
NAPT	Network Address Port Translation
NAS	Network Access Server
NETCONF	Network Configuration Protocol
NFV	Network Function Virtualization
NFVI	Network Functions Virtualization Infrastructure
NFVIaaS	Network Functions Virtualization Infrastructure as a Service
NIC	Network Interface Card
NIST	National Institute of Standards and Technology
NLRI	Network Layer Reachability Information
NMD	Network Management Data
NMS	Network Management System
NSA	Network Security Agency

NSAL	Network Services Abstraction Layer
NTP	Network Time Protocol
NVE	Network Virtualization Edge
NVO3	Network Virtualization over Layer 3
OAM	Operation, Administration, and Maintenance
OLSR	Optimized Link State Routing
OLT	Optical Line Termination
ONT	Optical Network Terminal
ONU	Optical Network Unit
OSA	Open System Adapters
OSI	Open System Interconnection
OSPF	Open Shortest Path First
OSS/BSS	Operations Support Systems and Business Support Systems
PaaS	Platform as a Service
PCF	Point Coordination Function
PCP	Priority Code Point
PDC	Phasor Data Concentrator
PE	Provider Edge
PGW	Packet Gateway
PHB	Per-Hop Behavior
PKI	Public Key Infrastructure
PM	Physical Machine
PMU	Phasor Measurement Unit
PoP	Point of Presence
POP3	Post Office Protocol version 3
POS	Point of Sale
PPP	Point-to-Point Protocol
PPTP	Point-to-Point Tunneling Protocol
PSAMP	Packet Sampling
PSN	Publish-Subscribe Network
QoS	Quality of Service
RA	Router Advertisement
RADIUS	Remote Authentication Dial-In User Service
RED	Random Early Detect
RESTCONF	REpresentational State Transfer CONFiguration
RFC	Request for Comment
RGW	Residential Gateway
RIB	Routing Information Base
RIP	Routing Information Protocol
RIR	Regional Internet Registry
RMA	Reliability, Maintainability, and Availability
RMON	Remore MONitoring
ROLL	Routing Over Low powered and Lossy networks
ROSE	Remote Operations Service Element
RP	Rendezvous Point

RS	Router Solicitation
RSA	Rivest–Shamir–Adleman
RSPEC	Request SPECification
RSTP	Rapid Spanning Tree Protocol
RSVP	Resource Reservation Protocol
RTE	RouTe Entry
RTFM	Real-Time Flow Measurement
RTP	Real-time Transport Protocol
RTT	Round Trip Time
SA	Security Association
SaaS	Software as a Service
SAN	Storage Area Network
SCTP	Stream Control Transmission Protocol
SDN	Software Defined Networking
SGW	Serving Gateway
SLA	Service Level Agreement
SLAAC	StateLess Address AutoConfiguration
SLM	Service Level Management
SLS	Service Level Specification
SMS	Subscriber Management System
SMTP	Simple Mail Transfer Protocol
SNAP	SubNetwork Access Protocol
SNMP	Simple Network Management Protocol
SPF	Sender Policy Framework
SPI	Security Parameters Index
SPIN	Sensor Protocols for Information via Negotiation
SPT	Shortest Path Tree
SSH	Secure Shell
SSID	Service Set IDentifier
SSL	Secure Sockets Layer
SSM	Source-specific Multicast
STB	Set Top Box
STP	Spanning Tree Protocol
TCA	Traffic Conditioning Agreement
TCI	Tag Control Information
TCS	Traffic Conditioning Specification
TFTP	Trivial File Transfer Protocol
TIA	Telecommunications Industry Association
TLS	Transport Layer Security
ToS	Type of Service
TPID	Tag Protocol Identification
TSI	Tenant System Interface
TSN	Time-Sensitive Networking
TSPEC	Traffic SPECification
TTL	Time to Live

TXOP	Transmit Opportunity
UI	User Interface
UPS	Uninterruptible Power Supply
VAP	Virtual Access Point
vCDN	Virtualization of Content Delivery Network
VDI	Virtual Desktop Infrastructure
vE-CPE	Virtualization of the CPE
VID	VLAN Identifier
VLAN	Virtual Local Area Network
VM	Virtual Machine
VNA	Network Virtualization Authority
VNF	Virtualized Network Function
VNFaaS	Virtual Network Function as a Service
VNI	Virtual Network Instance
vNIC	Virtual NIC
VNPaaS	Virtual Network Platform as a Service
VoIP	Voice over IP
vPE	Virtualization of the PE
VPN	Virtual Private Network
vRGW	Virtualized RGW
VRRP	Virtual Router Redundancy Protocol
vSTB	Virtualized STB
vSwitch	Virtual Switch
WAC	Wide Area Control
WAN	Wide Area Network
WBEM	Web-Based Enterprise Management
WEP	Wired Equivalent Privacy
WLAN	Wireless LAN
WMI	Windows Management Instrumentation
WPA	Wi-Fi Protected Access
WRED	Weighted RED
ZDA	Zone Distribution Area

List of Figures

Fig. 1.1	Logical network planning for a large-scale network across multiple geographically remote locations with network services from a private data center and a public cloud	5
Fig. 1.2	Three-phase network planning [2, pp. 9–12]	11
Fig. 1.3	Organization of book chapters	16
Fig. 2.1	System and environment	21
Fig. 2.2	Three perspectives of a network considered as a system	22
Fig. 2.3	A complex system initially treated as a black box	25
Fig. 2.4	The waterfall model	27
Fig. 2.5	The waterfall model for networking	28
Fig. 2.6	OSI's layered architecture and a generic network analysis model	30
Fig. 2.7	Top-down versus bottom-up approaches	32
Fig. 2.8	Service requests and service offerings [3, p. 35]	37
Fig. 2.9	Three-dimensional view of delay, throughput, and RMA [3, p. 51]	44
Fig. 3.1	Requirements analysis in which overlap exists in the technical analysis components. Traffic flow analysis will be discussed in a separate chapter	48
Fig. 3.2	User requirements	55
Fig. 3.3	Application requirements	57
Fig. 3.4	Application map	61
Fig. 3.5	Device requirements	62
Fig. 3.6	Network requirements	65
Fig. 3.7	A simplified requirements map. A complete requirements map has detailed information about the location dependencies of comprehensive requirements and a large number of important applications	75

Fig. 4.1 A traffic flow with attributes applied end-to-end and over network ... 82
Fig. 4.2 A composite flow aggregated from individual flows ... 83
Fig. 4.3 Representations for flow sources and sinks ... 84
Fig. 4.4 Examples of flow sources and sinks ... 84
Fig. 4.5 Flow boundaries in a network with three sites ... 86
Fig. 4.6 Peer-to-peer flow model in which all flows are individual flows ... 93
Fig. 4.7 Client-server flow model with asymmetric traffic flows [6, p. 184]. The traffic of responses is much bigger than that of requests. Thus, the server is more likely a flow source whereas the clients are more likely flow sinks ... 94
Fig. 4.8 Hierarchical client-server flow model [6, p. 186], also known as cooperative computing flow model, with two or more hierarchical tiers of servers acting as both flow sources and sinks. The distributed clients act more likely as flow sinks ... 95
Fig. 4.9 Distributed-computing flow model with a task manager and multiple computing nodes [6, p. 189]. Depending on the application scenarios, the task manager can be either or both of flow sources and sinks, and so does each of the computing nodes. Direct interactions among the computing nodes may or may not exist for information exchange ... 97
Fig. 4.10 The real-time flow measurement architecture [1, p. 6] ... 99
Fig. 4.11 Multiple managers, meters, and meter readers ... 101
Fig. 4.12 Meter structure [1, p. 18] ... 102
Fig. 4.13 Access to a local database ... 106
Fig. 4.14 An example of traffic flow calculation ... 107
Fig. 4.15 Flow distribution for access to remote databases ... 108
Fig. 4.16 Calculation results of traffic flows from a single pair of query and response for the example in Fig. 4.14 ... 108
Fig. 4.17 Weekly bandwidth usage pattern normalized as a percentage of the maximum bandwidth usage over a week ... 113
Fig. 4.18 Applications of flowspec rules to flowspecs ... 117
Fig. 5.1 Cisco's core/distribution/access architecture ... 125
Fig. 5.2 Full- and partial-mesh architectural models for core layer ... 126
Fig. 5.3 Hierarchical redundant architecture for distribution layer ... 127
Fig. 5.4 LAN/MAN/WAN architecture ... 128
Fig. 5.5 Multihomed Internet connectivity ... 131
Fig. 5.6 Using VPN to connect to an enterprise network ... 132
Fig. 5.7 Peer-to-peer architectural model ... 133
Fig. 5.8 Client-server architectural model ... 133
Fig. 5.9 Hierarchical client-server architectural model ... 134

Fig. 5.10 Distributed-computing architectural model ... 135
Fig. 5.11 An example of application-driven networking with distributed application components everywhere and localized/remote datasets ... 136
Fig. 5.12 An example of end-to-end architecture model that considers all components along the path of the traffic flow ... 137
Fig. 5.13 Intranet/extranet architectural model ... 138
Fig. 5.14 Service-provider architectural model ... 139
Fig. 5.15 Tiered-performance architectural models ... 143
Fig. 5.16 The physical and logical views of HSRP in a local area network 192.168.10.0/24 with the default gateway 192.168.10.1/24. All hosts have IP addresses on this network 192.168.10.0/24 ... 148
Fig. 5.17 The format of HSRP datagram [6, p. 5] ... 149
Fig. 5.18 Examples of MX records and CNAME records looked up by using `nslookup` ... 155
Fig. 5.19 Integration of network architectural models ... 157
Fig. 5.20 Network architecture with integrated components ... 158
Fig. 6.1 Relationship between binary and decimal representations ... 162
Fig. 6.2 An IPV4 address (Table 6.1): from binary to decimal format ... 163
Fig. 6.3 Five classes of IPv4 addresses each begins with a Class ID followed by a network ID and then a host ID ... 165
Fig. 6.4 Network portion and host portion in IPv4 addresses in Classes A, B, and C ... 166
Fig. 6.5 Subnetting that uses one or more significant bits from the original host portion to form subnets ... 168
Fig. 6.6 Fixed-length subnetting of a /16 address for equal-size subnets ... 171
Fig. 6.7 Variable-length subnetting of a /16 address for variable-size subnets ... 172
Fig. 6.8 Illustration of supernetting and subnetting ... 173
Fig. 6.9 Aggregation of two addresses ... 174
Fig. 6.10 Three-level hierarchical IP addressing for the requirements shown in Table 6.6 ... 179
Fig. 6.11 Hierarchical allocation of a Class B IP address 129.80.0.0/16 for the requirements shown in Table 6.6 ... 180
Fig. 6.12 Block diagram of three-level hierarchical IP addressing for the requirements shown in Table 6.6 ... 181
Fig. 6.13 Global IPv6 deployment (https://pulse.internetsociety.org/technologies, accessed 6 Aug 2022) ... 184
Fig. 6.14 IPv6 site prefix, subnet ID, and interface ID ... 186

Fig. 6.15 IPv6 unicast addresses. The global ID in unique local addresses is random with a high probability of global uniqueness [12] ... 187
Fig. 6.16 General format of IPv6 multicast addresses [10, 15] ... 189
Fig. 6.17 Unicast-prefix-based IPv6 multicast addresses specified in RFC 7371 [14], which updates RFC 3306 [17] and RFC 3956 [18] ... 191
Fig. 6.18 Comparison between IPv4 and IPv6 headers ... 194
Fig. 6.19 Examples of IPv6 packets with and without extension packets ... 196
Fig. 6.20 AH format [33, p. 4] ... 205
Fig. 6.21 Top-level format of an ESP packet [34, p. 5] ... 205
Fig. 6.22 Transport mode and tunnel mode of IPsec ... 206
Fig. 6.23 AH in an IP packet using TCP ... 207
Fig. 6.24 ESP in an IP packet using TCP ... 207
Fig. 6.25 AH and ESP in an IP packet using TCP ... 208
Fig. 6.26 Demonstration of dual stack in Windows by using command `ipconfig` in a command window. Use command `ifconfig` in MacOS and other Linux systems ... 212
Fig. 6.27 Dual stack configured in a network ... 212
Fig. 6.28 Encapsulation of IPv6 within IPv4 ... 213
Fig. 6.29 Router-to-router tunneling over IPv4 ... 214
Fig. 7.1 Classification of routing protocols ... 222
Fig. 7.2 IGPs and EGP in networks ... 222
Fig. 7.3 Growth of the BGP table—1994 to present as of 15 Sep. 2021 [3] ... 223
Fig. 7.4 eBGP and iBGP ... 224
Fig. 7.5 Updating routing tables in distance vector routing protocols ... 225
Fig. 7.6 Dijkstra's algorithm for a graph with source node n_0 ... 231
Fig. 7.7 An OSPF network with a backbone connected with multiple routing areas via ABRs ... 235
Fig. 7.8 Interconnection of IS-IS networks, in which the string of the L2 system in Area 1 and all L1-L2 systems in the other three areas form the backbone ... 236
Fig. 7.9 Conceptual interconnection with ISPs via border routers (firewall and other security components are omitted here) ... 239
Fig. 7.10 Routing in multiple levels ... 240
Fig. 7.11 Route redistribution through router R2 ... 243
Fig. 7.12 Networking in a cloud data center ... 245
Fig. 7.13 SDN architecture ... 247
Fig. 7.14 Communication networks in smart grid ... 251
Fig. 7.15 A typical WAC system with WAN support [17, 19] ... 251

Fig. 7.16 Logical diagram of a PSN with a multicast tree from Pi to multiple subscribers [17, 19] 252
Fig. 7.17 The prim algorithm for the minimum spanning tree problem from source n_0 to multicast nodes n_1, n_2, n_3, and n_5 256
Fig. 7.18 Minimum bandwidth and minimum weight spanning trees from source n_0 to multicast nodes n_1, n_2, n_3, and n_5. Assume that the total bandwidth of the tree is bounded by $B^c = 30$ 260
Fig. 7.19 Demonstration of cluster-based LEACH 264
Fig. 7.20 Illustration of SPIN. Node 1 advertises a new data and Node 3 requests the new data 265
Fig. 7.21 An illustrative diagram of AODV 266
Fig. 7.22 An illustrative diagram of AOMDV that finds multiple paths 267
Fig. 7.23 Demonstration of the DSR protocol 268
Fig. 7.24 OLSR messages 269
Fig. 7.25 Illustration of OLSR topology 269
Fig. 7.26 The DSDV routing protocol 270
Fig. 7.27 Babel's topology dissemination through communications between neighbors 270
Fig. 8.1 IEEE 802.1p Ethernet frame header 278
Fig. 8.2 The Structure of the one-octet DS field consisting of 6-bit Differentiated Services Codepoint (DSCP) and 2-bit Explicit Congestion Notification (ECN) in the IP header 281
Fig. 8.3 Multiple FIFO queues each with a different priority level 286
Fig. 8.4 Frame preemption 289
Fig. 8.5 DiffServ architecture 296
Fig. 8.6 A logical view of DiffServ classifier and conditioner [8, p. 16] 298
Fig. 8.7 IntServ architecture for end-to-end QoS 306
Fig. 8.8 IntServ architecture in a router 309
Fig. 8.9 IntServ architecture in a host, which generates data in this example 310
Fig. 8.10 RSVP operation from end to end 313
Fig. 9.1 Hierarchical network management 323
Fig. 9.2 FCAPS model for network management [3, 4] 325
Fig. 9.3 SNMPv3 with security 333
Fig. 9.4 The CMOT protocol suite [36, p. 5] 338
Fig. 9.5 End-to-end and per-link/per-network/per-element management of network devices 345
Fig. 9.6 Trend analysis for VoIP latency over 24 h and network traffic over a week 348

Fig. 9.7 In-band and out-of-band network management. In in-band management, management traffic flows follow the same paths as the traffic flows of user's applications. In comparison, in out-of-band management, separate and dedicated paths are used for management traffic flows 350
Fig. 9.8 Configuration of in-band and out-of-band management by using Cisco's *Management Plane Protection Commands* [43] 351
Fig. 9.9 Centralized network management, in which a centralized NMS manages all network domains across the Internet 352
Fig. 9.10 Distributed network management with distributed local EMS nodes or distributed local monitoring nodes 353
Fig. 9.11 Hierarchical network management with two-, three-, and four-tire models 354
Fig. 9.12 Manager of managers (MoM), in which each manager has its own database NMD 355
Fig. 10.1 The use of an ACL in two types of security policies 377
Fig. 10.2 Encryption and decryption 383
Fig. 10.3 Public and private keys are used in encryption and decryption 384
Fig. 10.4 Encryption with digital signature for authentication 384
Fig. 10.5 DMZ secure topology 388
Fig. 10.6 Public servers and resources on a DMZ network 389
Fig. 10.7 Critical servers separated with front- and back-end components 390
Fig. 10.8 L2TP topological reference models [18, pp. 8–9] 396
Fig. 11.1 Illustration of top 10 countries with the most data centers [2] 407
Fig. 11.2 EN 50600 Structure 414
Fig. 11.3 Key functional areas in a TIA-942 compliant data center [4] 423
Fig. 11.4 Hot and cold airflows in data center cooling 426
Fig. 11.5 Facebook's Luleå data center (*Source*: https://m.facebook.com/LuleaDataCenter, accessed on 10 Apr. 2022) 428
Fig. 11.6 Cisco's three-layer data center architecture [11] 431
Fig. 11.7 Segregation between tiers in the multi-tier model 433
Fig. 11.8 Logical view of a server cluster 434
Fig. 11.9 Physical view of a server cluster data center 434
Fig. 11.10 Generic reference model for data center network virtualization overlays [16, p. 9] 437
Fig. 11.11 Generic NVE reference model [16, p. 11] 438
Fig. 11.12 NVE reference model from the IETF RFC 8014 [18, p. 9] 440

Fig. 12.1 Traditional versus virtual computing models for three applications. In the virtual computing model, depending on the type of hypervisor, each PM may or may not need an OS ... 449
Fig. 12.2 Comparison between Type 1 (bare-metal) and Type 2 (hosted) hypervisors ... 450
Fig. 12.3 Virtualization with VMs and containers ... 453
Fig. 12.4 Switch and bridge of VM connection ... 454
Fig. 12.5 Internetworking of VMs ... 455
Fig. 12.6 Three modes of vSwitch configuration: host-only mode, NAT mode, and bridge mode ... 456
Fig. 12.7 Network settings in the Oracle VirtualBox ... 457
Fig. 12.8 NFVIaaS multi-tenant support of both cloud computing apps and NFVs from different administration domains [6, p. 12] ... 463
Fig. 12.9 Example of three-party enterprises sharing a service provider's infrastructure [6, p. 22] ... 465
Fig. 12.10 An example of EPC virtualization [6, p. 29] ... 467
Fig. 12.11 An example of EPC virtualization, in which both STB and RGW are virtualized for Home A, RGW is virtualized for Home B, and no NFV for Home C ... 469
Fig. 12.12 High-level NFV architectural framework [8, p. 10] ... 471
Fig. 12.13 NFV reference architectural model in detail [8, p. 14] ... 472
Fig. 12.14 A simple scenario of cloud computing accessible from anywhere and any platforms in an on-demand and pay-as-you-go manner ... 478
Fig. 12.15 A logical view of public cloud [21, p. 15] ... 481
Fig. 12.16 Logical diagrams of private cloud [21, p. 13] ... 481
Fig. 12.17 Logical diagrams of community cloud [21, p. 14] ... 482
Fig. 12.18 A logical diagram of hybrid cloud consisting of multiple distinct cloud infrastructures [21, p. 15] ... 483
Fig. 12.19 Service models of cloud computing ... 484
Fig. 12.20 A screenshot of web-based Office 365 interface showing all applications available to the author from cloud Office 365 ... 488
Fig. 12.21 The use of cloud service models from the perspective of cloud service customers ... 488
Fig. 12.22 Cloud computing reference architecture [19, p. 3] ... 490
Fig. 12.23 Interactions among the five actors in cloud computing [19, p. 4] ... 491
Fig. 13.1 Layered architecture of computer networks ... 503
Fig. 13.2 TCP/IP communications between two hosts ... 504
Fig. 13.3 A client-server network system with communicating hosts as clients ... 504

Fig. 13.4 The communication requirements and logical flows of the server and client in client-server network systems. The dotted arrows represent information flows between the server and client. Other application tasks that need to execute forever or periodically are normally integrated with the Block of Send/Receive data in both the server and client 506
Fig. 13.5 The operations and APIs of the server and client in client-server network systems. The dotted arrows indicate logical information flows between the server and client. The bind() and listen() operations on the server are not needed on the client 507
Fig. 13.6 A simple client-server network system 514
Fig. 13.7 Timelines of TCP/IP communications and other tasks 516
Fig. 13.8 Screenshots of server and client operations on Mac OS 520

List of Tables

Table 3.1	Examples of application groups [2, pp. 73–75]	59
Table 3.2	An example of requirements specifications	74
Table 4.1	Flow characteristics of applications [7, p. 96]	87
Table 4.2	Protocol overhead measured in the number of bytes	105
Table 4.3	Traffic load with database access and synchronization for the example shown in Fig. 4.14	109
Table 4.4	Some protocols that use TCP or UDP	112
Table 4.5	Application usage from 27 Mar to 3 Apr 2011 [17]	114
Table 4.6	Three types of flow specifications	115
Table 5.1	Comparisons of LAN, MAN, and WAN	128
Table 5.2	Network components, functions, capabilities, and mechanisms	140
Table 6.1	Decimal and binary representations of IPv4 address 192.168.1.10	162
Table 6.2	IPv4 address classes illustrated in Fig. 6.3	166
Table 6.3	Calculating the first address of a subnet	169
Table 6.4	Fixed length subnetting of 129.80.0.0/16 for equal-size subnets	169
Table 6.5	Supernetting examples	176
Table 6.6	Network hierarchy and requirements for IP address allocation	178
Table 6.7	Feasible address allocation for the requirements specified in Table 6.6. It can be further refined to save address resources	182
Table 6.8	Refined address allocation to AS1 129.80.0.0/17 with subnets of appropriate sizes meeting the requirements in Table 6.6	182
Table 6.9	Refined address allocation to AS2 129.80.128.0/17	183
Table 6.10	Pre-defined IPv6 multicast addresses in RFC 4291 [10, pp. 16–17]	190

Table 6.11 IPv6 Extension Headers 196
Table 6.12 Checklist for IPv6 planning [41, Chap. 4] 215
Table 7.1 Comparisons of routing protocols 238
Table 7.2 Default seed metrics for route distribution 243
Table 7.3 Comparisons between SPIN and LEACH 265
Table 8.1 IEEE-recommended use of the eight PCP levels 279
Table 8.2 Four access categories (ACs) 279
Table 8.3 Default EDCA settings of Contention Window (CW), AIFNSN, and Transmit Opportunity (TXOP) 280
Table 8.4 Applications of ACs in medical and industrial control systems 281
Table 8.5 Three pools of DSCP codepoints specified in RFC 2474 [2, pp. 14–15] and RFC 8436 [4, p. 4] ('x' takes a value of either 0 or 1) 282
Table 8.6 Mapping DSCP 'xxx000' to IP Precedence 282
Table 8.7 ECN codepoints [6, p. 7] 283
Table 8.8 Types of network policies 293
Table 8.9 Seven steps of policy implementation [10] 295
Table 8.10 Ten user/subscriber service classes defined and recommended in RFC 4594 [11, p. 15] 298
Table 8.11 Assured Forwarding behavior group defined in RFC 2597 [14] 300
Table 8.12 DSCP to service class mapping (sr+bs: single rate with burst size token bucket policer) recommended in RFC 4594 [11, pp. 19–20] 301
Table 8.13 IntServ RSVP and related RFCs 311
Table 8.14 Key components of an SLA 317
Table 8.15 Examples of SLA performance metrics 318
Table 9.1 SNMP commands 328
Table 9.2 Ten groups of managed objects in MIB-II [10] 330
Table 9.3 Ten groups of managed objects from RMON1 330
Table 9.4 Additional groups of managed objects from RMON2 331
Table 9.5 Advanced NETCONF requirements [30] 336
Table 9.6 CMIP management operation services 337
Table 10.1 Common sources of firewall rules [7, p. 5] 376
Table 10.2 Comparisons between layer-2 and layer-3 tunneling 398
Table 10.3 Comparisons between client-to-site and remote access VPNs 400
Table 11.1 Top 10 countries with the most data centers [2] 406
Table 11.2 Four classes/levels/tiers of data centers 416
Table 11.3 Availability classes of data centers from EN50600/ISO/IEC 22237 (Class 1: single path; Classes 2 through 4: multi-path) 419

Table 11.4 Summary of tier requirements [9] 419
Table 11.5 Performance confirmation tests for data centers [9] 421
Table 11.6 Top 10 supercomputers as of June 2023 [14] 435
Table 12.1 ETSI GS NFV specifications [7] 461
Table 12.2 Actors in cloud computing [19, p. 4] 490
Table 13.1 Programmer's view of socket APIs 508
Table 13.2 The prototype of `socket ( )` 509
Table 13.3 The prototype of `bind ( )` 510
Table 13.4 The prototype of `listen ( )` 511
Table 13.5 The prototypes of `connect ( )` and `accept ( )` 511
Table 13.6 The prototypes of `send ( )` and `recv ( )` 512
Table 13.7 The prototype of `close ( )` 514
Table 13.8 Summary of sockets components in Windows 531

Part I
Network Analysis

This part consists of four chapters:

- Chapter 1: Introduction.
- Chapter 2: Systematic Approaches.
- Chapter 3: Requirements Analysis.
- Chapter 4: Traffic Flow Analysis.

This part is devoted to network analysis. It will discuss basic concepts of computer network planning through structured processes and systematic approaches. A comprehensive requirements analysis will be conducted for network planning to align with business goals under various constraints. This will be followed by detailed discussions of traffic flow analysis with the focus on Quality of Service (QoS) requirements of traffic flows. The outcomes from this part serve as a foundation for detailed network planning.

Chapter 1
Introduction

The main theme of this book is network analysis and architecture design for network planning. While there are various types of networks, this book specifically focuses on computer networks for data communication. Therefore, it uses the term *networks* to specifically refer to *computer networks* unless specified explicitly otherwise.

Computer networks interconnect computing and network devices for data communication and network services, typically utilizing shared network resources. They have become an integral part of fundamental infrastructure in modern industries and societies. Organizations reply on computer networks to gather, exchange, store, and analyze information for business intelligence, which assists in, and supports, business decisions. The worldwide system of interconnected computer networks forms the Internet.

With the increasing complexity of modern computer networks and network services, it becomes a challenging task to build a new network, upgrade an existing network, or use a third-party or public network for an organization, especially when dealing with a large-scale network. It requires a deep understanding of the concepts, principles, mechanisms, process, and methodology of network planning. This book will help develop such a profound understanding of network analysis and architecture design, which are pivotal aspects of high-level network planning.

Network planning involves strategic planning of building a new network, upgrading an existing network, or utilizing a third-party or public network for an organization before the network is implemented. It serves business goals within various constraints and ensures that the network meets current and future requirements in a cost-effective way. Additionally, network planning encompasses determining how to provide network connectivity and scalability, as well as how to provision and secure network functions and services at the expected levels of Quality of Service (QoS).

This introductory chapter will begin with a discussion of the motivation behind network planning. It will then clarify what is involved in network planning. This is followed by a brief explanation of how to plan a network. After that, the main objectives, contents, and organization of this book will be presented.

Y.-C. Tian and J. Gao, *Network Analysis and Architecture*, Signals and Communication Technology, https://doi.org/10.1007/978-981-99-5648-7_1

1.1 Motivation for Network Planning

In traditional computer networking, network planning focused primarily on capacity planning due to the fact that only a limited number of network technologies and services were available. Throwing more bandwidth was a practically useful solution to a network problem. This motivated over-provisioning of bandwidth capacity in network planning and operation. The excessive bandwidth was not a real waste because it could be utilized later to accommodate the rapid growth in the demand to the network. Capacity planning, along with the development and application of simple rules for network, made a good sense in traditional networking. One such rule was the 80/20 rule, which assumed that 80% of network traffic was local while 20% was remote. These simple rules helped guide capacity planning decisions and optimize network performance.

However, the era of solving network problems by simply throwing more bandwidth is over. Several reasons contribute to this change, for instance:

- The availability of a much wider range of network technologies and services than before with increased complexity, leading to the need for comprehensive analysis and detailed design,
- The rapid growth of the number of users and network devices, resulting in complex interactions among network components,
- The increasing demand for predictable and guaranteed network services, necessitating specific QoS designs,
- The increasing importance of privacy and security in networking, requiring well-planned security and privacy measures,
- The widespread deployment of cloud and other network services through third-party infrastructure, demanding Service Level Agreement (SLA) and enhanced security requirements, and
- Elevated expectations for network connectivity, scalability, and performance derived from network architecture, design, implementation, and deployment.

These scenarios cannot be effectively addressed through the over-provisioning of bandwidth alone. For example, excessive bandwidth does not necessarily mean any performance guarantee for critical services if no QoS mechanisms are in place. This is because network services are provisioned as best-effort services by default in Internet Protocol (IP) networks. As another example, the use of third-party network infrastructure, such as Wide Area Network (WAN) links and routers, poses inherent security risks. How to secure an enterprise network that relies on third-party network infrastructure is a challenging task that cannot be solved solely by throwing more bandwidth. It requires a detailed design for security and privacy.

A typical example of larges-scale networks is shown in Fig. 1.1. This network interconnects hundreds or thousands of network devices across multiple geographically remote locations. It provides network services from both a private data center and a public cloud. Its WAN connections rely on third-party fibre optic infrastructure. This means that the network services from the public cloud and the private data

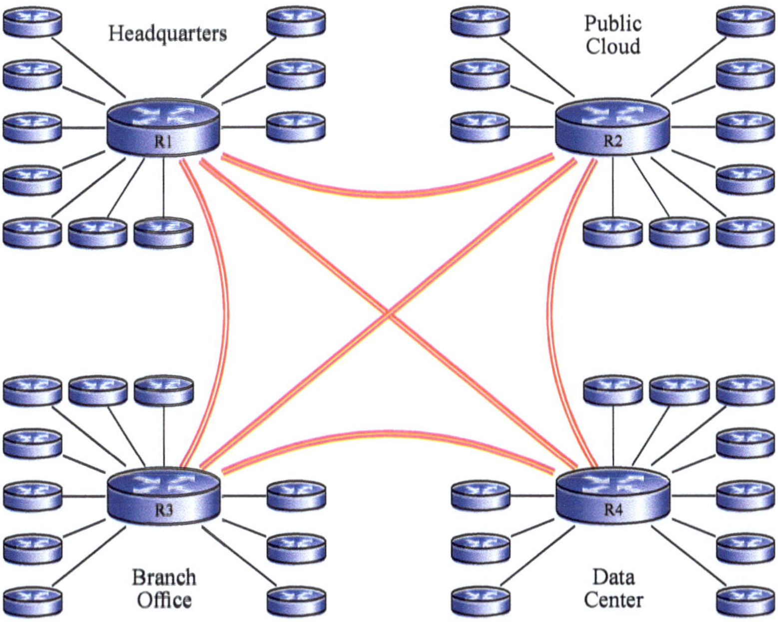

Fig. 1.1 Logical network planning for a large-scale network across multiple geographically remote locations with network services from a private data center and a public cloud

center must be provisioned over the third-party WAN links to the whole network. The network must be able to provide connectivity, scalability, performance, security, and network services that meet current and future requirements.

However, many questions need to be answered before the network becomes truly useful. For example, what current and future requirements are? What SLAs with the third-party service providers should be developed? What security requirements are and how they are enforced over the third-part infrastructure? What trade-offs should be developed between scalability and connectivity, performance and costs, and new technologies and complexity? All these and other aspects demand a comprehensive analysis of network behavior and requirements, followed by systematical planning of the network architecture to align with the identified requirements.

As depicted in the example presented in Fig. 1.1, network planning needs to address multiple factors, requirements, and/or objectives, such as:

- Complexity of network topology and behavior,
- Network connectivity versus hierarchy,
- Network scalability,
- Bandwidth capacity and allocation,
- Network services and QoS,
- Network management,
- Confidentiality, Integrity, and Availability (CIA), and
- Costs.

These factors, requirements, and objectives often compete with each other. For example, improving QoS may result in higher costs, optimizing resource utilization may lead to more complex topology and/or protocols, enhancing security may introduce additional overhead. Therefore, it is important to identify and establish trade-offs through network planning for a satisfactory solution to the network planning problem.

A computer network should be able to deal with short-term, medium-term, and long-term dynamic changes in both the network itself and its requirements. An example of such changes is the growth in the number of users, network devices, and/or network services, leading to a requirement of well-planned network augmentation. Network augmentation may involve modifying the network topology, introducing additional QoS mechanisms, designing new Virtual Local Area Networks (VLANs), adjusting server placement, or adopting additional protocols. To address network augmentation effectively, it is important to engage in network planning in advance to adequately prepare for the upcoming network changes.

Overall, network planning is essential to a computer networking project. It offers a number of benefits. For example,

- It clarifies business goals and constraints,
- It defines network planning problems,
- It identifies technical requirements and trade-offs, network services, and QoS levels,
- It gives a top-level view of network architecture,
- It presents a detailed design for the implementation of network services, and
- It provides component-based architecture, in which security and QoS are at the heart.

Good network planning fosters a comprehensive understanding of the challenges a network must address and how the network is used and managed. It enhances network performance, accounts for future growth, and ensures that the network conforms to security and QoS requirements.

1.2 Deliverables from Network Planning

Ultimately, network planning aims to plan the network to serve business goals within various constraints. These goals and constraints are transformed into current and future requirements that should be fulfilled. To fulfill these requirements, logical network architecture and physical network connections are developed and planned incorporating with various mechanisms, protocols, and technologies. From this understanding, the following is a list of main deliverables in network planning:

- Clarification and documentation of business goals and constraints relevant to network planning,

- Identification and documentation of current and future technical requirements and trade-offs,
- A top-level view of network architecture, showing the main components, functions, services, technologies, interconnections, and possibly locations,
- An architectural view of each significant network component with regard to its sub-components, functions, services, mechanisms, technologies, protocols, connections, performance, security, and other aspects. Example components include
  - addressing,
  - routing,
  - performance,
  - management,
  - security,
  - data center,
  - cloud, and
  - others
- A physical view of the network and its components to drive further physical network design and implementation.

It can be observed from the above discussions that requirements analysis plays an important role in network planning. Network planning would not be possible without requirements analysis. The results obtained from requirements analysis serve as the input for the development of network architecture.

After the overall network architecture is discussed, architectural models will be delved in this book for a few significant network components. These components include addressing, routing, performance, management, security, data center, and cloud. This is not an exhaustive list of network components. There may be other components that are critical for specific network and thus require careful planning. For instance, in satellite network communication, space-ground connectivity is particularly important and thus can be considered as a separate space-ground connectivity component in the planning of component-based architecture.

It is worth mentioning that the physical view of the network and its components serves as the input for, and thus drives, further physical network design, implementation, and deployment. Physical network design encompasses tasks such as selecting hardware devices and vendors, designing structured cabling systems, planning server rooms, creating floor plans, and establishing physical connections. However, as the main focus of this book is on network analysis and high-level architecture, physical network design will not be discussed in the book.

1.3 Strategic, Tactical and Operational Planning

As in other planning projects, a network planning project can be considered from the perspectives of strategic, tactical, and operational planning. Strategic planning deals with long-term requirements and sets up strategic goals for a given period of time.

Tactical planning breaks down long-term strategic planning into short-term objectives and actions. Operational planning considers the requirements of current network operation and translates strategic goals into technical ones. Network planning should investigate all strategic, tactical, and operational requirements through clearly defined long-term, short-term, and current targets and actions.

More specifically, strategic planning involves the development of strategies to enable networks to focus on common objectives for a given period of time, typically three to five years. It provides a big picture of the network in the long term to cast a vision, and thus requires mission processing and a high-level thinking of the entire business. Some examples of strategic objectives are listed below:

- In the next three years, move the majority of office network services to cloud or replace them with cloud-based Software as a Service (SaaS), such as word processing, spreadsheet, and email.
- In the next four years, replace current circuit-switching voice services with Voice over IP (VoIP).
- In the next five years, decommission current private data center and move to a public data center infrastructure.

These strategic targets will have a significant impact on the planning of current and short-term networking. Tactical and operational planning should fulfill the requirements of the strategic targets.

Tactical planning aims to achieve specific and short-term objectives, which are derived from strategic planning. It presents short-term steps and actions that should be taken to accomplish strategic objectives described in the strategic planning phase. The tenure of a tactical plan is typically short, usually one year or shorter. Here are some examples of tactical objectives and actions:

- In the next three months, initiate testing of cloud-based SaaS for selected services in a specific network segment.
- In the next six months, expand the testing of SaaS to more services in multiple network segments.
- In the next nine months, replace local DNS servers with cloud-based DNS servers from a public cloud and then decommission local DNS servers.
- In the next 12 months, replace local mail servers with cloud-based servers from a public cloud and then decommission local mail servers.

Operational planning focuses on current and immediate operational issues and requirements of the network. It makes decisions based on detailed information specific to network segments, functions, services, technologies, and/or protocols, providing an opportunity to use network resources effectively and efficiently. The following are some examples of operational planning:

- Segment a Local Area Network (LAN) into two to reduce the impact of broadcast traffic on latency performance.
- Create a Virtual Local Area Network (VLAN) for a new work group with staff members sitting in two different buildings.

- Implement stricter access control measures for a specific network segment.
- Provide one-hour security awareness training to users every month.

As operational planning turns strategic goals into technical requirements, it should align with strategic planning and tactical planning.

In addition, contingency planning is also necessary for mitigating risks and uncertainties. It can be incorporated into strategic, tactical, and operational planning as a backup plan in case the original planning fails. For example, consider a scenario where the mail servers are mitigated to the public cloud, but they are unable to send emails from the network domain of the organization due to a mismatch in the original Sender Policy Framework (SPF) settings. Should any contingency planning be developed in advance before the actual migration of the email servers? This is a real case that we have investigated for a company. An SPF record is a text record attached to the Domain Name System (DNS) to help validate messages that are sent from the specified domain.

1.4 Structured Network Planning Processes

Network planning should not be conducted in an ad-hoc manner. In particular, traditional networking, which focuses on connectivity and capacity planning through a set of pragmatic rules, is no longer suitable for network planning. It needs to be extended to service-based networking in modern networks. Service-based networking considers users and applications in addition to devices and the network, which are considered in traditional networking. To address the complexity of large-scale networks, network planning for modern service-based networking should be carried out by following a structured process and using systematic approaches. Various effective approaches for network planning will be discussed later in Chap. 2. The process of network planning will now be explored below from different perspectives.

1.4.1 Zoline's Network Planning Activities

Overall, network planning converts networking visions and ideas into meaningful actions and results [1]. It transforms business communication objectives and networking needs into networking requirements, budgets, and project plans. The process of network planning consists of the following main steps and activities [1]:

(1) Document customer's business problem and understand what is really needed.
(2) Abstract, formulate, and document a conceptual solution to the business problem, leading to some results that can be visualized and discussed.
(3) Define a conceptual solution in terms of requirements, such as functional, operational, administrative, and performance aspects to satisfy the customer's needs and address the business problem.

(4) Research and select appropriate product technologies for deploying the solution.
(5) Create a realistic budget for solution deployment by considering both one-time and recurring expenses.
(6) Develop a project plan for designing and implementing the deployable solution.

This book will not discuss budget issues. Also, as mentioned earlier. it will not discuss physical network design. Instead, the book will concentrate on high-level architectural models of larges-scale networks.

1.4.2 McCabe's Three-Phase Network Planning

Network planning is also considered as a process of three main sequential phases: network analysis, network architecture, and network design [2, pp. 9-12]. The output from each phase is the product of that phase. It also serves as the input to the next phase. These three phases are briefly described below:

(1) Network analysis for requirements analysis and traffic analysis
    - Input: problem statement, initial conditions, workflow data, and existing policies.
    - Output: requirements specifications, sets of services, requirements boundaries, service location information, traffic flow specifications, and architecture and design trusts.
(2) Network architecture through development, selection, and evaluation
    - Output: network topology, network technologies, equipment requirements, strategic locations, components, and architectural boundaries.
(3) Network design for vendor/hardware selection and evaluations, network layout, and other detailed network implementation and deployment
    - Output: vendor selections, equipment selections, configuration details, network blueprints, component plans, and design boundaries.

For each of the above three main phases, a model or multiple models are developed and simulated to evaluate the output of that phase. Risk assessment is also conducted for the output of each phase. Figure 1.2 depicts the three-phase process of network planning presented in the book by McCabe [2, pp. 9-12].

1.4.3 Oppenheimer's Structured Process

It is recommended that a structured process be followed for network planning and design [3, p. 5]. This will allow for a more accurate representation of customer needs

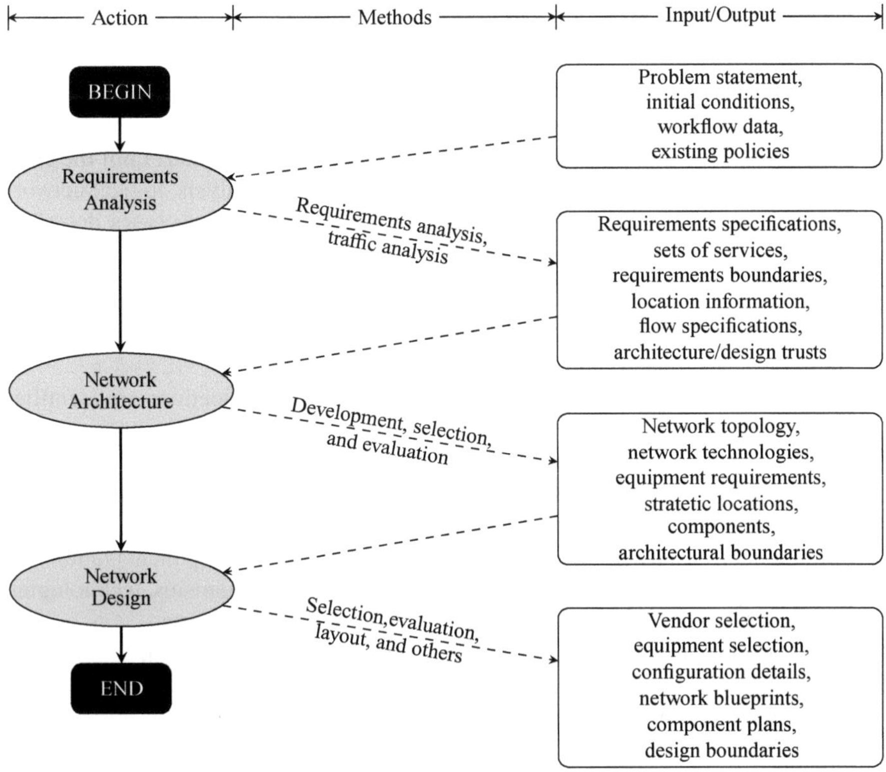

Fig. 1.2 Three-phase network planning [2, pp. 9–12]

and ensure the manageability of the networking project. The structure process, as recommended by Oppenheimer, exhibits the following characteristics:

- A top-down design sequence, which begins with gathering and analyzing requirements,
- The use of multiple techniques and models to characterize networks, determine requirements, and propose a structure for future systems,
- A focus on data flow, data types, and processes for accessing to, or changing, the data,
- An understanding of the location and needs of data access and processing, and
- The development of a logical model ahead of a physical model.

From these characteristics, four main phases are identified for the process of network design in the system development life cycle [3, p. 6]:

(1) Requirements analysis,
(2) Logical design,
(3) Physical design, and
(4) Testing, validation, and documentation of the design.

1.4.4 A General Process for Network Planning

While the process of network planning is described from different perspectives, it has become a common understanding that network planning should follow a structured process with sequential phases or steps. Each phase should not start until the completion of its preceding phase. The phases of requirements analysis, logical network architecture, and physical network design are common to all processes described above. The general process of network planning is outlined below with multiple sequential phases:

(1) Business goals analysis: This phase involves understanding and clarifying business goals and constraints.
(2) Requirements analysis: In this phase, requirements specifications are identified and formalized, which take possible trade-offs into consideration.
(3) Top-level network architecture: This phase provides a top-level view of network components, topology, functions, services, technologies. interconnections, policies, and possible locations.
(4) Component-based network architecture: Each significant component is addressed in this phase with the focus on its functions, services, mechanisms, technologies, protocols, connections, performance, security, and other aspects.
(5) Network design: This phase primarily focuses on physical design.
(6) Implementation and deployment.
(7) Evaluation, testing, and verification.
(8) Operation, Administration, and Maintenance (OAM).

This general process will be formalized later in Chap. 2 in the waterfall model as one of the systematic approaches for network planning. This book will mainly focus on the first four phases with an emphasis on network analysis and architecture planning. As mentioned earlier, it will not cover physical network design.

1.5 Network Planning as an Art

Network planning can begin with current requirements and then incorporate additional enhancements and/or technologies for future network growth. Alternatively, it can start with a strategic perspective on future targets and then narrow down to current operational requirements. The ways chosen by different planners for network planning can vary significantly. This makes network planning more of an art than a science or technology. From this perspective, network planning requires a deep understanding of the insights into computer networks with regard to [2, p. 3],

- Individual rules on evaluating and choosing network technologies,
- Ideas about how network technologies, services, and protocols work together and interact,

- Experience in determining what works and what does not work, and
- Selecting network topological models, often based on arbitrary factors.

Such a profound understanding of computer networks requires much knowledge that can only be developed through experience [4].

As an art, network planning largely relies on the expertise and experience of the planner. There is no standard solution to a network planning problem. This explains the observation that no two networks are exactly the same in the real world. Different network planers will likely propose different plans and designs, all of which can function effectively. The solutions are not differentiated solely for their correctness or incorrectness. However, a solution may be better than others in the sense that it provides better trade-offs among competing objectives and requirements.

Good network planning can be achieved by following best practices. For example, one such practice is to follow a structured process. Utilizing systematic approaches is also a good practice in network planning. Petryschuk has summarized top five best practices for network design [5]:

- Integrate security early on. Security is always crucial in all networks, and sometimes even more so than performance. Therefore, it is highly recommended to consider security as a priority requirement from the beginning of the network planning project.
- Know when to use top-down versus bottom-up. The top-down methodology is always recommended for planning a large-scale network from scratch. It allows us to focus on the fulfillment of business goals and constraints through the development of technical requirements and trade-offs. However, for one or more specific segments of the network, if the requirements and their relationships with business goals are already clear, the bottom-up methodology may provide a quick solution.
- Standardize everything. Some examples are hostnames (e.g., printer02.area01.lan03), IP addressing, structured cabling, and security policies.
- Plan for growth. Consider factors such as bandwidth capacity, segmentation, and IP address allocation to accommodate future expansion.
- Create and maintain network documentation. Documentation plays a vital role in network management and troubleshooting.

This book will present best practices from various perspectives for network planning.

No matter how a network is planned and who conducts the planning, it is always imperative to ensure that the needs of the business are met from the network planning. As network complexity continues to grow, comprehensive analysis and detailed technical design become increasingly important in meeting the requirements of modern service-based networking. Therefore, network planning should be approached not only as an art but also as a science and technology. In this book, it is recommended to follow a structured process and incorporate systematic approaches for network planning. This will be discussed in detail throughout the book, providing valuable insights for effective network planning.

1.6 Support from Customers and Executives

As analyzed previously, the success of a network planning project requires the planner to follow a structured process incorporating systematic approaches. It also necessitates a good understanding of customers requirements. It further calls for a high-level logical view of the network before any physical design is conducted. The logical view should be hardware- and vendor-independent.

In addition, engaging customers and executives in the network panning project is critical for the success of the project. This will ensure that the customer requirements are well clarified and specified, and the planning well aligns with the business goals and constraints. It is important to discuss the network planning project with customers and executives during each planning phase, seeking their endorsement of intermediate outputs or gathering clear suggestions for changes and amendments.

Observations indicate that large Information Technology projects that fail often exhibit the following features [1]:

- Lack of customer (end-user) involvement in planning,
- Focus on product technology and vendor selection,
- Misunderstanding of customer (end-user) business requirements,
- Understatement of project start-up (build) and recurring (run) costs,
- Poor definition of the business value of the proposed project, and/or
- Lack of interest and support from executive management.

1.7 Main Objectives and Contents

The overall aim of this book is to provide systematic approaches and best practices for network planning in a structured process. It describes how to assemble various network technologies, services, mechanisms, and policies into network architecture in a cohesive way. Specific objectives of this book include:

- Establishing a structured process for planning large-scale computer networks with increasing complexity,
- Introducing systematic approaches that are effective in, and suitable for, network planning,
- Developing insights into business goals and constrains that network planning should meet,
- Understanding technical requirements specifications and exploring possible trade-offs through comprehensive requirements analysis for network planning,
- Presenting a top-level view of network architecture that aligns with business goals and technical requirements, and
- Investigating logical architecture for important components of large-scale computer networks.

To achieve these objectives, this book is designed with the following main contents:

Part I: Network Analysis

- The concepts of network planning in a structured process.
- Systematic approaches for network planning. More specifically, the following approaches will be discussed: the systems approach, the waterfall model, a generic network analysis model, the top-down methodology, and service-based networking.
- Requirements analysis for network planning to define what network problems are, clarify business goals and constraints, and develop technical requirements specifications with possible trade-offs.
- Traffic flow analysis to identify predictable and guaranteed traffic flows, and the requirements to serve these flows.

Part II: Network Architecture

- Top-level network architecture covering network topology, functional entities, and service models.
- Component-based architecture for key components of large-scale networks with complex service and security requirements. More specifically, architectural models will be discussed for addressing, routing, performance, management, and security components.

Part III: Network Infrastructure

- Data centers with various national and international standards on topology, architecture, security, and design.
- Virtualization and cloud in relation to virtualization mechanisms, virtualized resources, virtualized network functions, cloud architecture, cloud service models, cloud security, and other related topics.
- Building practical TCP/IP network communication systems by using sockets. Comprehensive examples will be provided for different scenarios and requirements of TCP/IP communication applications.

1.8 Book Organization

The overall structure of this book is shown in Fig. 1.3. The present introductory chapter introduces basic concepts of network planning through systematic approaches in a structured process. Then, network analysis is discussed in detail for network planning. It is covered in three main chapters: Chap. 2 on systematic approaches for network analysis and architecture design, Chap. 3 on requirements analysis, and Chap. 4 on traffic flow analysis, respectively.

Next, Chap. 5 is dedicated to providing a top-level view of network architecture for large-scale networks. It is supported by component-based architectural models

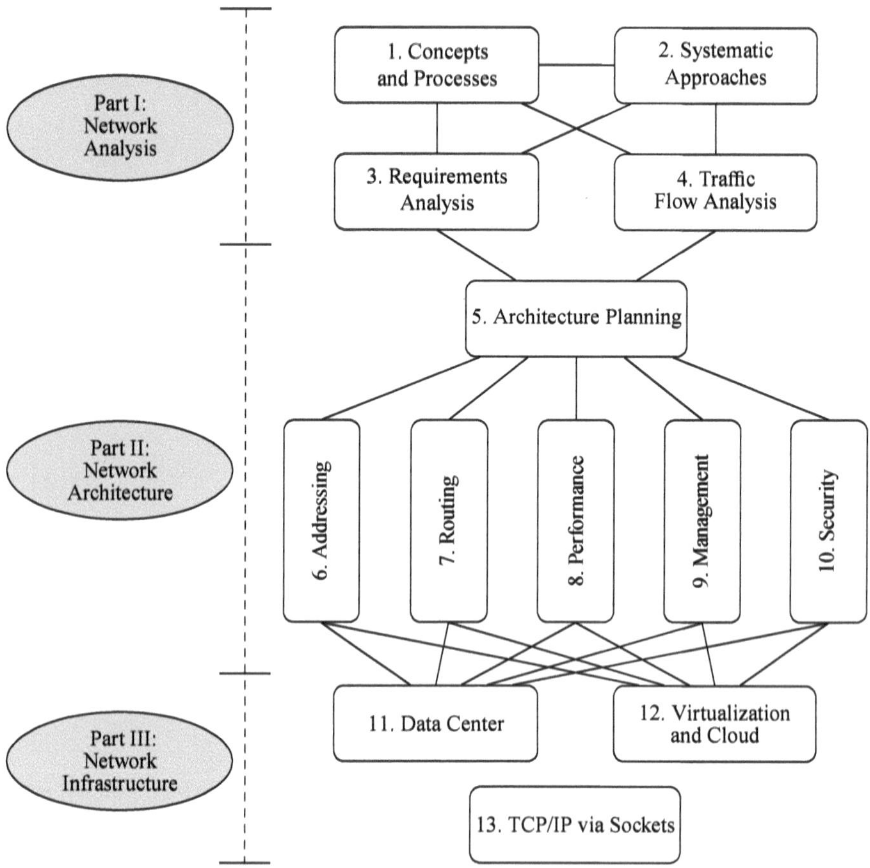

Fig. 1.3 Organization of book chapters

for key network components in Chaps. 6 through 10, which cover addressing, routing, performance, management, and security, respectively.

After that, two critical network infrastructure components are discussed in Chaps. 11 and 12 to support various network functions, services, and applications. Chapter 11 focuses on data centers, covering topics such as topology, architecture, security, and standards. Chapter 12 explores virtualization and cloud with detailed discussions on virtualization mechanisms, virtualized resources, virtualized network functions, cloud architecture, cloud service models, cloud security, and other related topics.

Moreover, Chap. 13 is devoted to building TCP/IP communication systems via sockets. It delves into the concepts, principles, and practices of socket programming. Comprehensive examples are also provided in this chapter.

References

1. Zoline, K.O.: Network planning. http://www.zoline.com/network/Planning.html (2000). Accessed 12 Nov 2022
2. McCabe, J.D.: Network Analysis, Architecture, and Design, 3rd edn. Morgan Kaufmann Publishers, Burlington, MA 01803, USA (2007). ISBN 978-0-12-370480-1
3. Oppenheimer, P.: Top-Down Network Design, 3rd edn. Cisco Press, Indianapolis, IN 46240, USA (2011). ISBN 978-1-58720-283-4
4. Smythe, C.: Internetworking: Designing the Right Architectures. Addison-Wesley, Boston, MA 02116, USA (1995). ISBN 978-0-201-56536-2
5. Petryschuk, S.: Network design and best practices. https://www.auvik.com/franklyit/blog/network-design-best-practices/ (2021). Accessed 12 Nov 2022

Chapter 2
Systematic Approaches

A computer network is a system consisting of a large number of physical and logical components that are interconnected in a complex manner. This complexity makes network planning, particularly for large-scale networks, a challenging task in terms of analysis and architecture design. To achieve a satisfactory solution to the network planning problem, it is essential to adopt systematic approaches along with a structured planning process. This chapter introduces a few systematic approaches that have been shown to be effective in, and suitable for, network planning:

(1) The systems approach, which focuses on the decomposition of the network into components, the interactions among the components, and a satisfactory solution for multi-objective network planning,
(2) The waterfall model, which emphasizes sequential phases of network planning,
(3) A generic network analysis model, which analyzes requirements from users, applications, devices, and network itself,
(4) The top-down methodology, which highlights the process from a top-level view of network objectives to high-level network architecture, intermediate-level network design, and finally a bottom-level physical implementation, and
(5) Service-based networking, which focuses on ensuring QoS by considering service requests and offerings in a configurable and manageable manner to meet the diverse requirements of users, applications, devices, and the network.

These approaches address network planning from different perspectives. They are not mutually exclusive but overlap in many aspects. For example, they all involve decomposing a network into components although they have different ways for decomposition. Therefore, these approaches can and should be used concurrently in a network planning project.

When should these systematic approaches be applied? They should be used in the following scenarios:

- When designing a new computer network,
- When upgrading an existing network, or

Y.-C. Tian and J. Gao, *Network Analysis and Architecture*, Signals and Communication Technology, https://doi.org/10.1007/978-981-99-5648-7_2

- When planning to use third-party or public networks for network services and applications.

Let us begin our discussions below with the systems approach as one of the effective systematic approaches.

2.1 Systems Approach

The systems approach is a notation of the general systems theory, which was originally developed by Ludwig Von Bertalanffy in the 1960's. It refers to the decomposition of a complex system into smaller and easy-to-understand subsystems for better understanding of the complexity of the overall system. The system approach finds applications in various areas including network planning. This section discusses the systems approach and its specific applications within the context of network planning.

2.1.1 System, Subsystems, and Environment

In the general system theory, a *system* is a unitary whole integrated from interacting and interdependent subsystems. It can be a natural, human-made, or even conceptual entity. Depending on the application context, a system can represent an organization, a software system, a management process process, or a computer network. In each of these examples, the system is treated as a whole, which is composed of inter-related and interdependent elements or subsystems. In the context of computer networking, multiple individual Local Area Networks (LANs), functional areas, network services, and other logical or physical entities work together to form a unified whole, namely the network.

Most logical or physical systems of practical relevance are open systems, meaning that they interact with their *environment*. Therefore, in order to understand or define the functions of a system as a whole, it is important to clearly delineate the boundary between the system and its environment. By doing so, the interactions between the system and its environment can be investigated and understood. This helps address questions such as:

- What inputs does the system receive from the environment?
- What outputs does the system generate and send to the environment?
- What constraints exist at the boundary of the system?

Figure 2.1 provides a graphical representation of the concepts of a system and its environment. The system consists of multiple subsystems that are interacting and interdependent. It also interacts with the environment through inputs and outputs.

The concepts of system and environment are directly applicable to computer networks as well. In the context of computer networking, it if crucial to define a

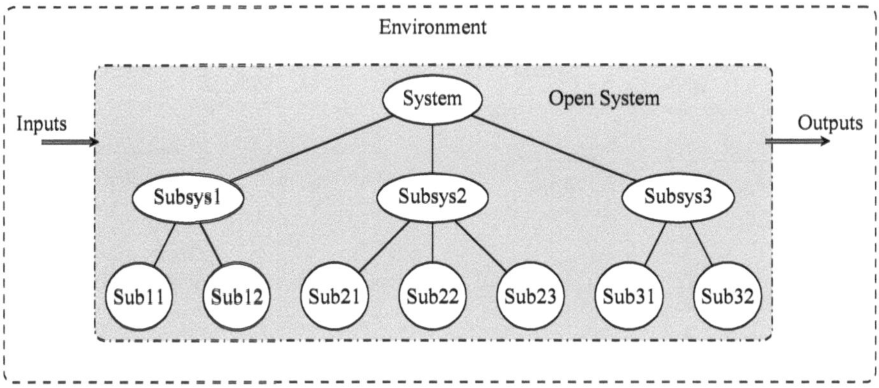

Fig. 2.1 System and environment

clear boundary for each routing area. Border routers, for instance, are installed at the boundary between an enterprise network and its external ISP. The performance of network services as specified in Service Level Agreements (SLAs) is often measured at the boundary of the network. In this scenario, the external ISP can be considered as part of the network's environment.

In the context of computer networking, Fig. 2.2 illustrates three perspectives of a network considered as a system. It shows three topological models for the same network, i.e., geographical topology, functional topology, and component-based topology, respectively. This highlights that fact that a system, such as a network in this example, can be investigated from different perspectives for a comprehensive understanding of its functions, behavior, and performance.

2.1.2 *Holism and Emergent Behavior*

While a complex system is investigated through decomposing it into smaller subsystems in the systems approach, the functions of the overall system should be considered as a whole. It is important to realize that any change, no matter how small the change is, in any subsystem of the system, can directly or indirectly impact the entire system. This change may affect the system's functions, or the level of the system's performance, either resulting in improvement or degradation. For example, a functional failure or performance degradation in any of the components shown in Fig. 2.2 can lead to the failure of the entire network, disrupt critical network services, or violate SLAs. The consideration of the impact of a change in any part of the system on the system functions, behavior, and performance is captured by the concept of *holism.*

Therefore, every subsystem or part of a subsystem, contributes to the whole, and is therefore important for the overall system. This characteristic is known as the non-summation feature of the systems approach. In order to get a holistic perspective, it is

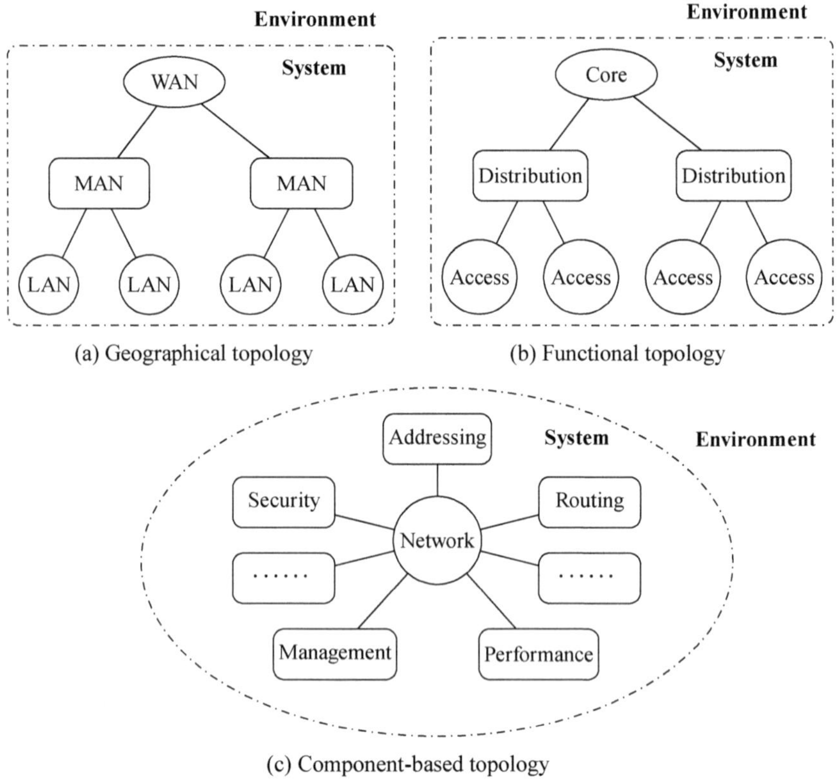

Fig. 2.2 Three perspectives of a network considered as a system

essential to have a good understanding of the behavior of each component within the overall system. In the case of a computer network, this requires a good understanding of the behavior of each network segment or component, such as LANs, Metropolitam Area Networks (MANs), Wide Area Networks (WANs), access, distribution, core, and physical and logical components shown in Fig. 2.2.

However, comprehending each subsystem individually is not sufficient to fully understand the functions, behavior, and performance of the overall system. This is because interactions between the subsystems exist, and these interactions give rise to unique system dynamics that may not exist in the individual subsystems. Having a deep understanding of each atom does not mean that we understand the entire universe. Similarly, knowing each individual from a school does not mean that we know the school's culture. For a network shown in Fig. 2.2b, a high level of traffic in each access area may not indicate an issue for that specific area, but collectively, the traffic from the access areas may cause congestion in the distribution area.

From this perspective, systems possess an important feature known as emergent behavior or synergy. This means that a system is more than the sum of its parts.

We can simply express this feature as $1 + 1 > 2$. This can be understood from two perspectives:

- Due to the interactions between components and their collective contributions, a system can exhibit some functions, behavior, and features that do not exist in its individual components. Each individual computer can function alone in isolation. But when multiple computers are networked, they can exchange information through data communication over the network.
- Well-designed system components with appropriate use of their interactions will make the system stronger than the sum of its individual components. Being stronger in this context can be interpreted as having richer dynamics, improved performance, or additional functions in our favor. Conversely, a poor design of the system may cause the system to behave worse than the sum of its parts.

The understanding of system synergy from these two perspectives highlights the importance of designing network components and segments individually and collectively in a systematic way.

2.1.3 Satisfaction and Trade-offs

A complex system, such as an enterprise network, typically encompasses multiple objectives that are expected to be optimized. When planning an enterprise network, the general goal is to minimize capital cost and running cost while maximizing network performance. In particular, it is expected to guarantee the reliability, security, and performance of mission-critical, safety-critical, and time-critical services. It is worth noting that not only are there multiple objectives, but many of these objectives are also competing or conflicting in nature. Optimizing some objectives will inevitably result in the degradation of other objectives. For example, in TCP/IP networks, maximizing the throughput of a network link often leads to an increase in communication delay. Using UDP instead of TCP can enhance throughput and bandwidth utilization, but the reliability of the communication is sacrificed.

Therefore, optimizing all objectives simultaneously is generally not possible for a complex system like an enterprise network. In such cases, it becomes important to clarify and identify primary, secondary, and even tertiary objectives. Then, the primary objectives are given the highest priority in network planning. The secondary and tertiary objectives are considered with lower priority. In practical network planning, the objectives with lower priority can be converted into constraints that must be satisfied. For example, the time delay for specific network links should be bounded within 100 ms. As a result, the overall solution is no longer an optimal one in terms of the original multiple objectives. However, it represents a *satisfactory* solution that takes into account all objectives, even though some of these objectives are competing or conflicting.

In order to achieve a satisfactory solution, it is necessary to find trade-offs among multiple objectives, subsystems, and components. One approach to achieving trade-offs is to prioritize primary objectives while converting the remaining objectives into constraints, as mentioned above. These constraints then become conditions that must be satisfied. They will be no longer optimized anymore. By doing so, the primary objectives and the functions of the overall system can be maintained without compromising their performance.

Another approach to achieving trade-offs is to use game theory for multiple objectives that are competing or conflicting with each other. This approach can yield a solution known as *Nash Equilibrium*, which represents a balanced trade-off among the objectives. For example, the concept of Nash equilibrium has been applied to scenarios such as semantic caching in mobile sensor grid database systems [1] and multi-factor multicast routing in sensor grid networks [2].

Identifying objectives, constraints, and possible trade-offs is an important aspect of system analysis. By carefully considering these factors, a network planner can make informed decisions and find a satisfactory solution that balances the multiple objectives and requirements of the network system.

2.1.4 Solving Problems yet to be Defined

Tertiary students are generally well-trained in solving given problems using specific techniques or tools. For example, when presented a linear programming problem that has a well-defined objective function and constraints, they can apply the simplex method to find a solution. However, when it comes with complex systems like enterprise network planning, the technical specifications for solving complex systems problems are not always readily available. This implies that the problem itself, which needs to be solved, is not clearly defined.

Let us consider a scenario where one of the distribution areas in the network shown in Fig. 2.2b experiences traffic congestion and significant latency. This is a problem that needs to be solved, but what is the actual underlying technical problem? Could it be attributed to:

- An inappropriate use of network resources by a user or group of users?
- An inadequate segmentation design in the access areas that interact with the distribution area?
- An improper topological design of the distribution area?
- Or other underlying issues?

Without clarifying the specific problem, it becomes challenging to address it from the technical perspective. This highlights a distinctive feature of complex system, such as network planning: we need to tackle problems that have not yet been clearly defined.

Therefore, when dealing with complex systems, the first step to solve a problem, which is typically yet to be defined, is to clarify and clearly define the problem itself. How can we accomplish this? The answer lies in conducting a comprehensive system analysis, specifically a requirements analysis. This will aid in gathering and clarifying the business goals and constraints, which serve as the foundation for developing technical specifications and trade-offs. Only after we have a complete set of technical specifications can we begin to explore and develop techniques and tools to lead a solution satisfying those specifications.

Additionally, the process of system analysis also provides insights into potential solutions or directions to pursue in the search of a satisfactory solution. For example, if system analysis reveals that the traffic congestion mentioned earlier in a distribution area depicted in Fig. 2.2b results from an inappropriate use of network resources by a group of users in an access area, it may be necessary to develop and enforce a user access policy. If network access from all access areas is functioning properly, the issue could be attributed to inadequate segmentation. In such a case, it may be necessary to break down large segments of the network into smaller ones.

2.1.5 *Black, Gray, and White Boxes*

Although a complex system may lack a clear description or complete understanding, it still needs to be managed, controlled, and operated. While understanding its components or subsystems is important as discussed above, comprehending its overall functions and behaviors is more critical. In the absence of detailed knowledge about the system and its subsystems, the system can be treated as a *black box*. By injecting an input to the system, it will produces an output. We can then observe the system output and establish a relationship between the output and input. This allows us to capture the characteristics and behaviors of the system. Furthermore, the performance and behavior of the system can be improved through a feedback loop from the output to the input. This process is illustrated in Fig. 2.3.

To investigate the traffic congestion issue discussed previously, there are various approaches to injecting input into the network. One simple approach is to apply traffic shaping to one or more access areas. Another approach is to identify the users who have generated a significant amount of traffic within a short period of time. We can then temporarily restrict the traffic from these users, as long as the applications they are using are not critical. For either of these two approaches, we can observe

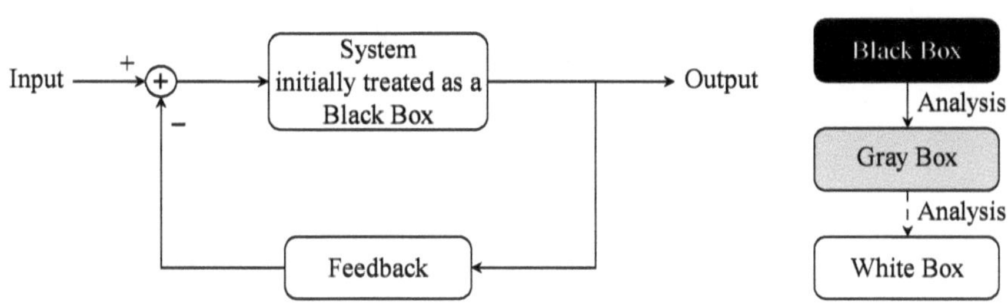

Fig. 2.3 A complex system initially treated as a black box

the behavior of the network and analyze the impact of these measures. If necessary, we can feed the information obtained from the output back to the system, further adjust the actions of traffic shaping or restriction based on the obtained information, and observe the resulting improvement or degradation in system performance. This iterative process enables us to gain a better understanding of the system dynamics and behavior, as well as uncover potential solutions for improving system performance.

Through this input-output-feedback process, we will acquire more knowledge about the system. As a result, the black box gradually becomes a *gray box*, and potentially even a *white box*. A white box signifies that the system is fully understood for the purpose of system control and operation. System analysis, including requirements analysis, in network planning is a process that assists in understanding the network from an initial black box gradually to a gray box and potentially even a while box.

2.2 Waterfall Model

The waterfall model, also referred to as the waterfall methodology, is a well-known and effective project management approach. It involves gathering, analyzing, and developing requirements at the beginning of the project, followed by a linear sequential process of well-planned phases to fulfill those requirements. More specifically, each phase in the waterfall model cascades into the next, flowing steadily down like a waterfall. The waterfall model finds applications in a wide range of areas such as engineering design, software development, and many others. It is discussed here in the context of network planning.

2.2.1 *Standard Waterfall Model*

The waterfall model typically consists of five to seven phases that proceed strictly in a linear sequential order, implying that a phase cannot commence until its previous phase is completed. The names of the waterfall phases may vary depending on the specific application scenario. In its early version defined by Winston W. Royce, the waterfall model is composed of five phases, i.e., requirements, design, implementation, verification, and maintenance. This is shown in Fig. 2.4a.

The five phases in the waterfall model depicted in Fig. 2.4a are briefly described below:

- Requirements: In this phase, all customer requirements are gathered at the beginning of the project, enabling all other phases to be planned without further interactions with the customer until the conclusion of the project.
- Design: The design phase is divided into two steps: logical design and physical design. In the logical design step, conceptual and/or theoretical solutions are

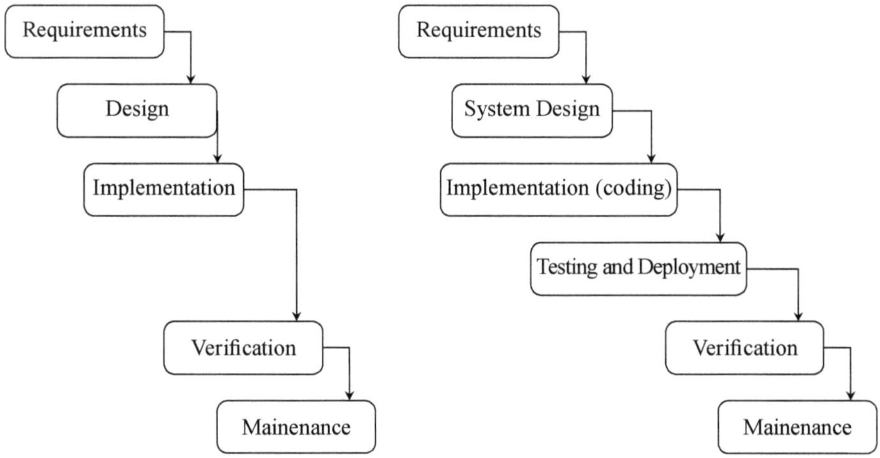

(a) Standard waterfall model (b) Waterfall software development life cycle

Fig. 2.4 The waterfall model

developed based on the gathered requirements. In the physical design step, the conceptual and/or theoretical solutions are converted into concrete specifications that can be implemented. The design phase provides a blueprint for the construction or implementation of the final product.

- Implementation: In this phase, the concrete specifications developed in the design phase are implemented. For example, in software development, this involves writing code based on the design specifications.
- Verification: Once the implementation is complete, the product undergoes thorough review and testing. Both the developer and particularly the customer examine the product to ensure that it functions as intended and meets the requirements developed in the initial phase.
- Maintenance: The customer uses the product, maintains it, discovers any bugs and deficiencies. The developer applies fixes as necessary.

These five phases provide a structured and sequential process to project management, enabling a systematic progression from requirements analysis to product delivery and maintenance.

To make the waterfall model more effective, feedback can be introduced from the verification and maintenance phases to the requirements, design, and implementation phases. This enables the refinement of the first three phases even after the complete product is delivered to the customer. In order to show the original waterfall model clearly, the feedback feature has not been visualized in Fig. 2.4a.

2.2.2 *Waterfall Software Development*

When applied to software development projects, the waterfall model is adapted to accommodate the specific features and requirements of the software development life cycle. Typically, it comprises the following phases: requirements, system design, implementation (coding), testing and deployment, verification, and maintenance. This is illustrated in Fig. 2.4b.

In practice, the results obtained from the testing, verification, and maintenance phases can be fed back to previous phases. This allows for further improvement of requirements analysis, system design, and implementation. However, to emphasize the sequential waterfall nature of the waterfall software development approach, the feedback mechanism is intentionally omitted in Fig. 2.4b.

2.2.3 *Waterfall Networking*

In the context of computer networking, the waterfall model typically comprises the following sequential phases: requirements analysis, logical network design, physical network design, implementation and deployment, evaluation/testing/verification, and Operation, Administration, and Maintenance (OAM). It is depicted in Fig. 2.5. The results from the evaluation/testing/verification and OAM phases could be fed back to previous phases for their refinement.

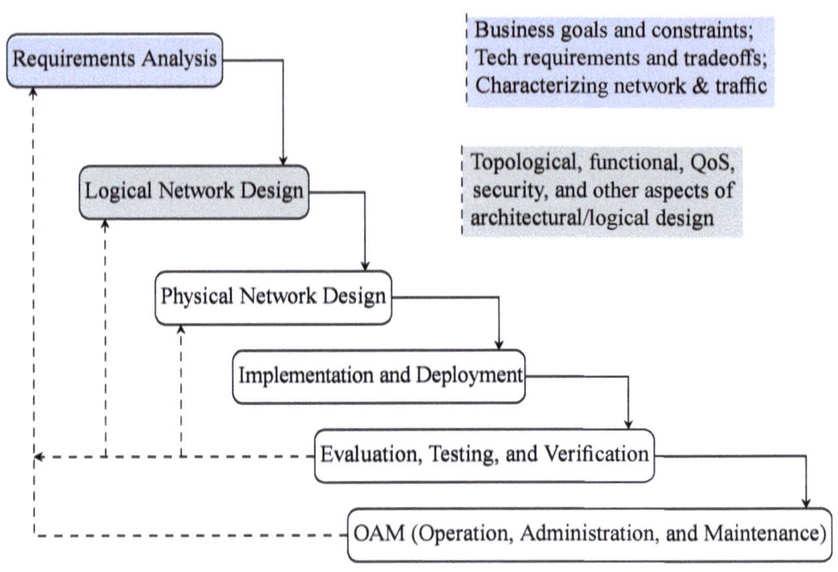

Fig. 2.5 The waterfall model for networking

For the purpose of network planning, this book will focus more on the first two phases in the waterfall networking model illustrated in Fig. 2.5, i.e., requirements analysis, and logical network design.

The requirements analysis phase in waterfall networking requires a good understanding of business goals and constraints. From this understanding, comprehensive specifications of technical requirements and potential trade-offs can be developed. Therefore, this phase consists of three main steps:

- Analyzing business goals and constraints,
- Analyzing technical requirements and exploring potential trade-offs, and
- Characterize existing network and also network traffic. This step considers current and future network traffic, and assesses its impact on the protocol behavior and Quality of Service (QoS) requirements.

The logical network design phase in the waterfall networking model deals with architectural and logical design, including topology, functionality, QoS, security, and other related aspects. More specifically, it starts with a top-level architectural deign. This is followed by component-based architecture for addressing, routing, performance, management, security, cloud, and others. For each of these components, to fulfill the requirements of the component, we need to

- Understand what this component can provide,
- Plan and design the topology, and
- Identify, select, and develop appropriate mechanisms.

It is worth noting that network planning can be evaluated before its actual implementation and deployment. The evaluation can be conducted through modelling and simulation under typical use cases. The results obtained from the evaluation can be fed back to the requirements analysis and system design phases for their further refinement. However, in order to maintain our focus on the main theme of network planning, this book will not extensively cover network modelling and simulation.

2.3 A Generic Network Analysis Model

As discussed in previous sections, both the systems approach and waterfall model emphasize the significance of system analysis with a particular focus on requirements analysis. This section introduces a generic model for network analysis from the perspective of complex systems comprising multiple components and entities. This model can be viewed as an application of system analysis in the context of network planning [3, pp. 27–31].

2.3.1 Model Architecture

Recall the OSI seven-layer architectural model consisting of application, presentation, session, transport, network, data link, and physical layers, as shown in Fig. 2.6a. Network devices are interconnected into a network, which provides users with network services through applications. Therefore, from OSI's layered architecture, four basic components, or subsystems in the jargon of the systems approach, can be abstracted. They are user, application, device, and network. When considered together, they form a generic model for network analysis as shown in Fig. 2.6b.

It is worth noting that traditional networking primarily focused on the network itself, ensuring sufficient bandwidth capacity for device connectivity to the network. Generally, it did not consider users and applications extensively. By contrast, modern networking takes into account users and applications in addition to device interconnection to the network. Particularly, it approaches network analysis in a systematic manner, links network requirements with sets of network services, and also considers interfaces between users and applications, applications and devices, and devices and the network.

2.3.2 Model Components

From the systems approach perspective, the network component depicted in Fig. 2.6b differs from the OSI network protocol layer illustrated in Fig. 2.6a. It is a subsystem with functions spanning the bottom three OSI protocol layers (i.e., the network, data link, and physical layers) in Fig. 2.6a. Its main functions include routing and

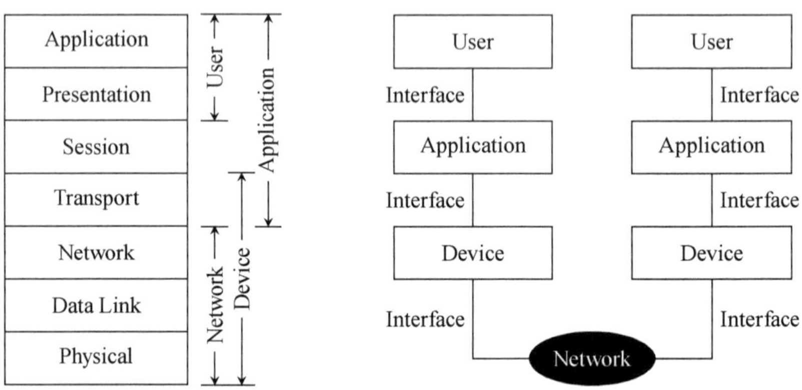

(a) OSI's layered architecture (b) A generic network analysis model [119, p. 28]

Fig. 2.6 OSI's layered architecture and a generic network analysis model end-to-end delivery of data packets at the network layer, media access control at the data link layer, and bit streaming through NICs and communication medium at the physical layer.

The device component or subsystem shown Fig. 2.6b represents an abstraction of the functionalities performed by hardware devices such as routers, switches, servers, hosts, and other networking devices. These hardware device functions are implemented within the OS and span across the bottom four OSI protocol layers (i.e., the transport, network, data link, and physical layers). In general, the device subsystem manages end-to-end routes through the use of transport protocols, end-to-end data delivery by employing network-layer protocols, media access control with support from data-link-layer protocols, and the transmission of data bits via NICs and communication medium by using physical-layer protocols.

Various network services are provisioned over the network through applications with support from protocols spanning multiple OSI layers. The application component or subsystem depicted in Fig. 2.6b is an abstraction of the main functions performed across the top four OSI protocol layers shown in Fig. 2.6a, i.e., the application, presentation, session, and transport layers. It deals with application-layer protocols and services (e.g., web service via HTTPS), presentation control (e.g., data format, encryption, and compression), session control, and the establishment of end-to-end links through transport protocols.

It is important to realize that network services and applications ultimately serve users. Therefore, the user component or subsystem illustrated in Fig. 2.6b captures the functions and requirements of users. It incorporates the top two OSI protocol layers (i.e., the application and presentation layers) to address application requirements and data format.

It is worth noting that depending on how complex a network is and what needs to be analyzed for the network, each of the four subsystems depicted in Fig. 2.6b can be further decomposed into two or more components. For example, the user subsystem can be investigated by further dividing it into specific user groups. In the case of an enterprise network within a university, user groups could include IT support staff, general administration staff, academic staff, senior executives, students, and visitors. Different user groups may have different requirements and security policies, necessitating specific considerations in network analysis and architectural planning.

2.3.3 Model Interface

In the generic model depicted in Fig. 2.6b, each of two neighboring subsystems interacts with each other through appropriate interfaces. The user subsystem interacts with the application subsystem through displays (e.g., monitors), Graphical User Interfaces (GUIs), and general User Interfaces (UIs). Between the application and devices subsystems, application-device interfaces can be various Application Progamming Interfaces (APIs), QoS configuration and management, and device monitoring and management systems. The device-network interface between the device and network subsystems could be as simple as device drivers and a standard LAN (Ethernet) interface. In more complex scenarios, it may also be coupled with QoS management, application-device APIs, or even cross-layer interactions. Nevertheless, identifying four subsystems, their boundaries, and their interfaces in Fig. 2.6b is helpful for understanding the characteristics and requirements of the overall network.

2.4 Top-Down Methodology

For the analysis and design of systems such as computer networks, either top-down or bottom-up approach could be used. These two approaches reflect different styles and processes of thinking and decision-making. A comparison between top-down and bottom-up approaches is provided in Fig. 2.7. The top-down methodology has been described in [4, pp. 3–7] as a general network design methodology.

2.4.1 Descriptions of the Methodology

The top-down approach is a recursive heuristic for problem-solving by going from the general down to the specific. It starts with a big picture of the system in terms of its functions, requirements, and constraints. It then breaks down the system into multiple

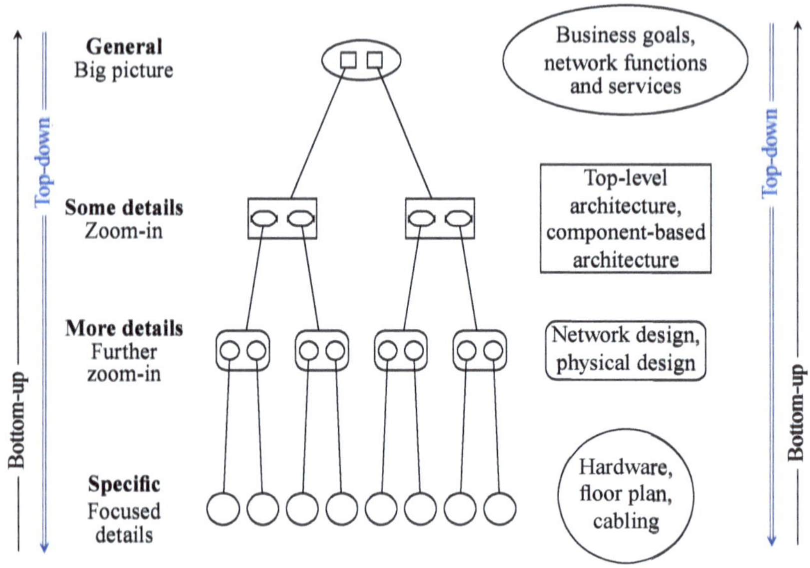

Fig. 2.7 Top-down versus bottom-up approaches

big subsystems. After that, investigate how each of these subsystems is solved and how these subsystems interact with each other. This process continues recursively, further dividing the subsystems into smaller ones until sufficient details are achieved to support the functions of the overall system. This recursive decomposition of the system into subsystems is similar to that in the systems approach. However, after the system is decomposed, either top-down or bottom-up approach can be employed. The top-down methodology emphasizes the recursive process from the top to the bottom. It also requires to consider not only the problem to be solved, but also the way of solving the problem. In general, the top-down approach is effective for large-scale and complex systems. In the context of network planning, the top-down and bottom-up approaches address network problems in different ways. The top-down approach first considers the upper layers of OSI's layered network architecture before moving to lower layers. This means that it focuses on applications, data format, session control, and transport at upper layers before delving into routing, switching, and bit streaming at lower layers.

The top-down methodology for network planning is also an interactive process. The business and technical requirements with high priority should be addressed first. They can be presented in a top-level architecture of the network. Later, more information will be gathered regarding specific technical and non-technical requirements, such as business service models, protocol behavior, connectivity and performance requirements, access control policies, and security policies. Then, the top-level architecture can be improved and refined, from which component-based architectural models can be further developed.

Different from the top-down approach, the bottom-up approach starts from the specific and moves up to the general. It focuses on dealing with individual components first and then integrates logical or physical components with clearly defined or well understood interactions to form larger subsystems. This process is repeated recursively, moving up the hierarchy to derive a solution with the expected functions and dynamics of the overall system. Overall, the bottom-up approach is useful for small-scale and simple systems. In the context of network planning, it primarily concentrates on the hosts, switches, routers, and their interconnections before considering upper-layer functions and behaviors of the overall network.

2.4.2 *Use of the Methodology*

For a small-scale system with a small number of functional or physical components and simple interactions between the components, the bottom-up approach is well-suited. This is because it is relatively easy to match local technical requirements and trade-offs of subsystems with global business goals and constraints of the overall system. Staring from the specific allows for easy progression to the general, and vice versa. For example, consider a small-scale network in a small business with only tens of networking devices. The network connectivity is straightforward, and the network scalability is not a concern. In this case, the bottom-up approach works effectively.

The devices can be interconnected with one or two Ethernet LANs using switches. Then, implement a simple Dynamic Host Configuration Protocol (DHCP) server, add a border router, configure firewalls, and install and deploy applications. It is not difficult to configure the network to satisfy the requirements and constraints of the business. In this bottom-up design process, individual components of the network are specified in detail. They are then integrated to form larger components, which are subsequently integrated to create a complete network. This is a process of decision-making about smaller components first and then deciding how to put the components together to get a complete system with the desired functions.

For the planning of a large-scale network, connectivity and scalability are among the major concerns. Meeting the requirements for availability, reliability, security, performance, and management is also challenging. Due to the interactions among all such logical or physical components, a design that appears satisfactory for individual components may not be effective for the overall system. Let us take addressing as an example, which is a logical component that spans almost all aspects of a network, e.g., connectivity, scalability, security, performance, and management. Addressing cannot be adequately tackled without a global view of the entire network. The same holds true for security, QoS, and network management. As another example, if two routing areas are designed separately with different routing protocols, each routing protocol may function perfectly within its own routing area. However, integrating these two routing areas into the network's routing system would present a significant challenge: route redistribution would be required. Hence, the top-down approach is preferable to the bottom-up approach for complex networks. It helps avoid such issues resulting from the lack of the global knowledge of the overall network.

2.4.3 *Advantages and Disadvantages*

It is observed from above discussions that the top-down approach has its advantages and disadvantages. Let us consider the advantages of the top-down approach:

- It facilitates the alignment of business goals and constraints.
- Expectations from the system are unified while functions and responsibilities are clearly defined. They all are independent of hardware devices and vendors.
- The logical correctness of primary system functions and services can be ensured. It can be confirmed before the system implementation and deployment.
- Having a big picture of the system aids in fulfilling technical requirements and trade-offs. As a result, no major logical flaws will exist in the system.
- Implementation is quick after system analysis and high-level design are completed.

However, there are certain drawbacks associated with the top-down approach. It requires a lengthy process for system analysis and architectural design before any further actions can be taken. Not all team members may possess the skills or expertise required for this process, resulting in a potential waste of resources in the initial phases of top-down network planning. Moreover, starting from high-level analysis and design may limit the creative thinking of individual team members. Additionally, during implementation, individual members may encounter difficulties in implementing the components derived from the high-level analysis and design.

The bottom-up approach turns these issues arising from the top-down approach into advantages. Starting from the bottom allows for quick progress in finding solutions for small and local components. Team members can make a full use of their respective skills and expertise to work on specific components or areas of the system. However, integrating these local solutions to form the overall solution for the entire system is not a trivial task. It requires a significant effort, fine tuning, and a lengthy process of configurations. The advantages discussed above in the the top-down approach become questionable in the bottom-up approach.

2.5 Service-Based Networking

Traditionally, network planning focused primarily on capacity planning. It was also a practice to solve network problems by throwing more bandwidth. This approach was based on a few assumptions that were later found to be inappropriate, e.g., bandwidth would be infinite, simple priority would suffice, and applications could adapt [5, pp. 4–5]. Over-provisioning bandwidth was effective in solving network problems when the number of network services was limited. However, it faced significant challenges with the growing number of network services and particularly the increasing requirements of guaranteed services. This prompted the emergence of service-based networking.

Service-based networking is a new approach to looking at computer networking from the perspective of network services, specifically in IP-based networks. It shifts the focus away from bandwidth capacity and instead emphasizes network services and service support in network planning. This is especially important for future networks with certain functions of self-configuration and self-administration, for which the network itself needs to make its own decisions. Self-configuration is a prominent feature of IPv6 networks.

This section discusses the general concepts and characteristics of network services. The mechanisms, strategies, and policies related to QoS will be investigated in later chapters, specifically in the context of architectural planning for component-based network performance and network management.

2.5.1 *What are Network Services*

There have been significant efforts to define and specify *services* in the OSI reference model. In general, network services can be understood from different perspectives. The European Telecommunications Standards Institute (ETSI) NFV Industry Specification Group (ISG) has defined a network service from its functional and behavioral specifications as a composition of network functions [6, p. 10]. Functional specifications of network services can be Internet connectivity, cloud connectivity, security, accounting, management, and many others. Behavioral specifications of network services can be the levels of performance such as reliability, availability, delay, and other performance characteristics. Overall, network services contribute to the behavior of higher-layer services, which are characterized by at least performance, dependability, and security specifications. The end-to-end network service behavior results from the combination of the behaviors of individual network functions and network infrastructure composition mechanism [6, p. 10].

In general, network services can be viewed as sets of network capacities that can be configured and managed within the network and between networks to facilitate network operation. In this sense, the configurability and manageability are important features of network services. Anything that cannot be configured or managed is not within the scope of network services for the purpose of network planning. Also, network services are strictly limited to the network or between networks. They are something that the network can deliver to the end system. Anything that are delivered by other parts of the system is not considered as part of network services. For example, a GUI provided by an application to end users is not a network service. Similarly, APIs used for interactions between applications are also not network services.

Network services must be provisioned *end-to-end*. To support a network service, multiple network devices, components, functions, and protocols need to work together to meet the service requirements and specifications. Typically, some of these elements are part of the network and can be managed and configured to support the service. However, other elements may belong to third parties and thus cannot be directly configured and managed. For example, visiting a remote web server will requires intermediate routers to route HTTP traffic over the Internet. Obviously, many of these intermediate routers can only be configured and managed by third parties. Therefore, if network services are not provisioned in an end-to-end manner, some components along the path of the network traffic may not be able to provide the required level of support for the services.

Network services can be categorized as best-effort, predictable, and guaranteed services in terms of the required level of performance. By default, network services are provisioned as *best-effort* delivery. The majority of network services in a network fall under the category of best-effort services. Best-effort network services do not provide differentiation of traffic flows from different users, applications, and devices. Typical examples include email, FTP, and HTTP services. These services share the available network resources without service differentiation or priority management. Therefore, they must adapt their traffic flows to the resource sharing environment. For example, FTP service on top of TCP uses the flow control mechanism embedded in TCP to adapt to the dynamic network environment.

Some network services in a network require a *predictable* level of QoS. This implies that the dynamics and behaviors of the corresponding network traffic are predictable, e.g., with latency shorter than 100 ms. In this example, the traffic latency is always within 100 ms although it varies over time. An application designed to use this predictable network service will function well as long as the traffic latency does not exceed this 100 ms threshold.

In a computer network, there may be a small number of mission-, safety-, and/or time-critical services that must be provisioned with *guaranteed* QoS. For example, an application may require a reservation of 100 kbps bandwidth along the path between end points. Otherwise, critical data or commands will not reach the destination within their respective deadlines, leading to functional failure of the application. As another example, when a server is down, a backup server at hot standby must take over within 3 s to prevent system crashes. Therefore, the switching over to the hot standby server within 3 s must be guaranteed.

Predictable and guaranteed network services are configured and managed differently from best-effort services. They require specific mechanisms, strategies, and policies in network planning and operation. Detailed discussions on predictable and guaranteed services will be provided later in the context of network analysis and architecture.

2.5.2 Service Request and Service Offering

Network services can be considered from the perspectives of *service requests* and *service offerings*. Briefly speaking, service offerings refer to network services that are offered by the network to the system. Service requests are requirements that are requested from the network by users, applications, or devices, and expected to be fulfilled by the network. They form part of the requirements for the network. Figure 2.8 illustrates the concepts of service requests and offerings.

Fig. 2.8 Service requests and service offerings [3, p. 35]

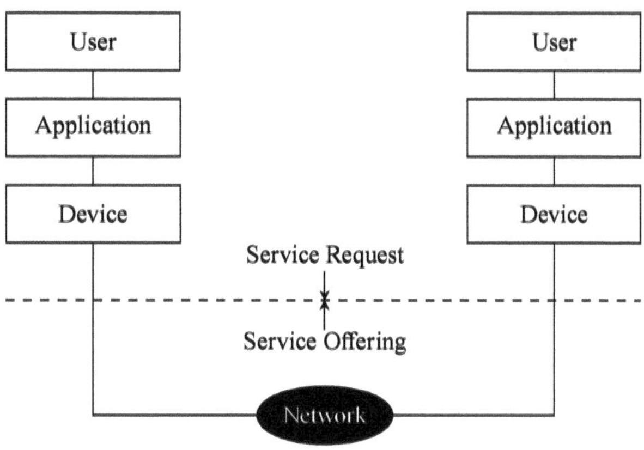

Service Offerings

Network service offerings refer to network services themselves that the network offers to the system. Well-known examples of network services offered by networks include DHCP, DNS, email, file sharing, FTP, HTTP and WWW, print, SNMP, SSH, VoIP, and many more. DHCP assigns IP addresses to hosts dynamically. DNS translates domain names to IP addresses. HTTP and WWW enable web browsing. SNMP is deployed for network management. SSH is a secure shell for remote login.

Service offerings are provisioned to meet the requirements of service requests made by users, applications, and devices. Therefore, in order to understand what service offerings should be provisioned and how they are provisioned, it is important to match service offerings with the corresponding service requests. For instance, an application requiring web service to support its functions would need the provisioning of the HTTP service.

In computer networking, service offerings are provisioned as the best-effort delivery by default, as mentioned earlier. Naturally, they will meet the requirements of best-effort service requests. The network resources that are actually available to a specific service will change dynamically over time. There may be occasions when no sufficient resources, such as bandwidth, are available to a network service for a period of time. Therefore, it is understandable that the level of performance of the service offerings is neither predictable nor guaranteed by default. For example, the FTP service may be unable to establish a connection with the remote file server due to insufficient bandwidth. In the case of a VoIP service, the quality of the VoIP service may become very poor for a few minutes or longer if insufficient bandwidth is available.

To support predictable and guaranteed service requests, simply providing service offerings is not sufficient without performance management and traffic differentiation. It is essential to develop and deploy QoS mechanisms, strategies, and policies to meet the performance requirements specified by the service requests. For example, in the case of the VoIP service mentioned above, traffic delay and jitter should be limited within a certain range, thus exhibiting predictable traffic behavior. For time-critical services that require a guaranteed level of performance, resource reservation would be essential along the path between the endpoints. In this specific example, end-to-end support is particularly important because each of the routers along the path must be able to reserve network resources to ensure the guaranteed service.

Service Requests

Unlike service offerings, which indicate network services themselves, service requests do not refer to network services. Instead, they represent the service requirements that are requested by users, applications, or devices. These requests are expected to be supported by the network through service offerings. Here are a few examples of service requests:

(1) Connecting hosts to the network: this is a requirement for hardware connectivity.
(2) Placing a specific group of users in a virtual network–this is a requirement for segmentation and management.
(3) Ensuring reliable network connection with traffic latency within 100 ms for a specific application, e.g., APPLx: this is a requirement for differentiated traffic management, which should show at least predictable behavior.
(4) Ensuring the availability of the online teaching system during teaching hours: this is an availability requirement.
(5) Establishing remote connection to the network over the Internet: this is a requirement for secure remote connection.

These examples illustrate various types of service requests that users, applications, or devices may make to the network, outlining their specific requirements and expectations.

It is seen from the aforementioned examples that some service requests correspond to one or more network services. For instance, connecting hosts to the network requires many network services, e.g., physical connectivity to layers 1 and 2, layer-3 connection, as well as DHCP service and DNS service. Fulfilling this service request requires the provisioning of multiple service offerings. In this particular example, clarifying what services should be offered by the network is relatively straightforward.

In many cases, it is necessary to conduct a detailed analysis to clarify what services should be offered by the network in order to meet the requirements of the service requests. Let us revisit the example discussed earlier: remote connection to the network over the Internet. There is no single mapping of this request to network services. Multiple options are available to meet this requirement, e.g., VPN tunneling, encryption, and other mechanisms. Depending on the architectural design, these options or mechanisms can be used independently or in combination.

The majority of service requests are addressed through best-effort service offerings. There is no differentiation of network traffic among best-effort services. As a result, all best-effort services share network resources without prioritization. The behaviors of these services are neither predictable nor guaranteed.

Some service requests may require predictable or guaranteed service offerings. A predictable service request may specify a predictable traffic behavior, e.g., end-to-end traffic latency below 100 ms. In this example, the actual latency value may be unknown in advance, but it is known to be within the 100 ms threshold. In comparison with a predictable service request, a guaranteed service request typically arises from mission-, safety-, and/or time-critical requirements. Let us consider the aforementioned example of the online teaching system that must be available during teaching hours. This implies that the availability of the online teaching system must be guaranteed for the specified time period. To meet this availability requirement, multiple network service offerings will be needed. They will be investigated later in the context of network architecture.

2.5.3 Resource Sharing Among Best-Effort Services

Network resources, such as bandwidth, are always finite for a specific network. Let us take bandwidth as an example. After a certain amount of bandwidth is reserved for predictable and guaranteed services, the remaining bandwidth will be available for best-effort services. There are basically two methods for managing the sharing of finite bandwidth:

- Serve all: This method admits all new services and share the bandwidth among existing and new services, or
- Serve with quality: This method admits new services only when the QoS can be maintained for both existing and new services.

Consider a scenario where a finite amount of bandwidth, e.g., 10 Mbps, is allocated for a web service. If there is only one session open, the session will use the entire 10 Mbps bandwidth, which is sufficient to maintain a good quality of the web service. However, if there are 100 sessions open simultaneously, each session will have an average bandwidth of 100 kbps, which may still be acceptable despite the possibility of traffic collisions and increased delays. If additional 100 sessions are introduced to the service, the available bandwidth for each session is reduced by half to an average of 50 kbps, which may cause significant performance degradation or make the system practically unusable.

However, if the number of sessions is limited to a maximum of 100, any requests to open additional sessions will be rejected. This ensures that the number of sessions does not exceed 100 at any given time, thereby maintaining an average bandwidth of 100 kbps per session. When an existing session is closed, a new session can be admitted. This will require a well-designed *admission control* mechanism.

An important part of network planning is to describe how to deal with performance management with best-effort, predictable, and guaranteed requirements. This is addressed specifically in performance-component architecture and will be discussed later in this book.

2.5.4 Service Performance Metrics

As described earlier, network services must be configurable and manageable. Therefore, they must be measurable and verifiable within the system. *Service metrics* are measurable quantities used to quantitatively describe and characterize the performance requirements of network services.

Threshold and Limit

A *threshold* is a commonly used service metric. It represents a boundary indicating whether the performance of the network conforms to a service requirement. For example, in the case of VoIP, a latency threshold of 300 ms is considered acceptable. If the latency exceeds 300 ms, call interruptions and poor voice quality may occur, indicating a violation of the service requirement. In this example, the value of 300 ms is a threshold to characterize whether the performance of the VoIP service meets the service requirement.

Another service metrics is a *limit*. It is a conforming and nonconforming region with lower and upper bounds. Within the lower bound, the performance satisfies the service requirement. Beyond the upper bound, the service requirement is violated. Between the lower and upper bounds, the performance may degrade but still remains acceptable. Let us consider a specific best-effort service: 70% of bandwidth utilization in a network segment would show satisfactory performance of the best-effort service. Between 70% and 80% of bandwidth utilization, the performance of the service is still acceptable but packet losses and collisions are increasing. Once the utilization surpasses 80%, packet dropout increases significantly, making the service performance unacceptable.

Directly Measurable Metrics

There are a number of service performance metrics that can be directly measured or quantified through measurements within a short period of time. Some key metrics are listed below in alphabetic order:

- Accuracy, which refers to the amount of error-free traffic successfully transmitted, relative to total traffic.
- Bandwidth capacity, which indicates data-carrying capacity measured in bits per second (bps).
- Bandwidth usage, which measures how much bandwidth is used for a period of time. For optimal network operation, one may expect to get as close to the maximum bandwidth as possible without overloading the network.
- Bandwidth utilization, which is the percentage of total available bandwidth capacity in use.
- Jitter, which is the variation of time delay. For example, if delay varies between 3 ms and 10 ms, the corresponding jitter is 7 ms. If an application is jitter-sensitive, the jitter must be maintained within a threshold.
- Latency, which quantifies the amount of time taken to transmit data from one point to another.
- Packet loss, which is the packet dropout during transmission from one point to another.
- Response time, which measures the amount of time taken to receive a response after sending a request for a network service.
- Round Trip Time (RTT), which is the amount of time it takes for a data packet travels to its destination plus the amount of time it takes for an acknowledgment of that packet to be received at the origin.

- Re-transmission, which refers to the number of lost or dropped packets that need to be re-transmitted to complete a successful data delivery.
- Throughput, which quantifies the rate of data successfully transmitted from one point to another. For example, the throughput over a link is measured as 300 kbps.

The metric of *accuracy* is measured differently in WANs and LANs. In WAN links, it is measured as the Bit Error Rate (BER), which is typically in the order of 10^{-11} for fiber-optic links. In LANs, the measurement of successful transmissions usually focuses on frames rather than individual bits. Therefore, a BER is not usually specified for LANs. Instead, the accuracy of data transmission in LANs can be quantified by a bad frame in a certain number of bytes, e.g., a bad frame in 10^6 bytes in a typical scenario.

Calculated Metrics

In addition to performance metrics discussed above, there are also performance metrics that are calculated based on statistical values over an extended period of time. A typical example is Reliability, Maintainability, and Availability (RMA), which are statistical indicators of performance. In computer networking, any downtime can be costly, especially for mission-critical, safety-critical, and time-critical services. Therefore, maintaining good RMA performance is an essential requirement.

The metrics of RMA cannot be measured in a short period of time. They need to be calculated based on the measurements collected over a long duration, e.g., a few months or even a year. These metrics are derived from the concepts of uptime and downtime, from which Mean Time Between Failure (MTBF) and Mean Time to Repair (MTTR) can be calculated. MTBF specifies how long the network service will last before it fails. MTTR indicates how much time is required for the failed network to be fixed.

By using MTBF and MTTR, the following equation specifies *availability*:

$$\text{Availability} = \frac{\text{MTBF}}{\text{MTBF} + \text{MTTR}}$$

This equation represents the percentage of time the network remains operational during a given period of time, i.e., uptime divided by the total period of time. Let us examine some examples to get a sense what the network availability really means in practical network operations. Consider a full year of 365 days (8, 760 h). A network that is available 99% of the time is actually out of service for 87.6 h (i.e., more than three days). The availability of 99.9% means 8.76 h of failure downtime each year. A requirement of 99.99% availability implies 0.876 h (52 min 33.6 s) of downtime per year due to failures.

In the calculation of network availability, MTBF and MTTR are mean values. Actual time before failure and actual time to repair can fluctuate around these mean values. Depending on the requirements of network services offered by the network, it may be necessary to consider the worst-case scenario for the network or its components.

Furthermore, scheduled maintenance of the network should be excluded in the calculation of the availability metric. The network may undergo planned shutdowns for maintenance purposes, e.g., once a year during Christmas or New Year. These planned shutdowns are not considered failures and therefore are not included in the calculation of MTTR.

Related to availability, *maintainability* is a statistical measure of the time required to fully restore system functions after a system failure. It is usually represented by MTTR.

Network *reliability* is related to, but different from, availability. It characterizes how long the network keeps functional without failure interruption. Practically, it can be quantified in different ways, e.g., by

- the mean service time between two failures on average, i.e., MTBF divided by the number of failures, or
- failure rate, which is the number of failures divided by the total time in service.

Mathematically, network reliability is a complex topic, which still attracts active research and development, e.g., the work presented in [7].

While the concepts of reliability and availability are similar in some aspects, they are fundamentally distinct. Quite often, they are erroneously used interchangeably. Also, reliability is sometimes represented by a percentage value, e.g., 99.99% reliability. In this case, it should not be confused with availability, but should be interpreted as the percentage of the time the network is reliable without failures or interruptions. Moreover, a network can be highly available with a short MTBF, but not practically reliable because of frequent failure interruptions. This highlights the distinction between availability, which focuses on the uptime of the network, and reliability, which considers the ability of the network to function without interruptions over an extended period of time.

2.5.5 Multi-dimensional View of Service Performance

Performance metrics discussed previously are typically coupled with each other to some extend. However, optimizing all performance metrics in a network is not a realistic task. For example, in practice, achieving a high throughput usually comes with some packet losses and increased delay. In order to quantify the requirements of service performance, it is helpful to clarify *thresholds* or *limits* for the performance metrics. By considering these individual thresholds together, a multi-dimensional view of performance requirements can be formed, which is also known as *performance envelopes* [3, pp. 50–51].

The multi-dimensional view of performance characterizes the acceptable performance in a high-dimensional space. Figure 2.9 illustrates a three-dimensional performance envelope that considers delay, throughput, and RMA performance. When the

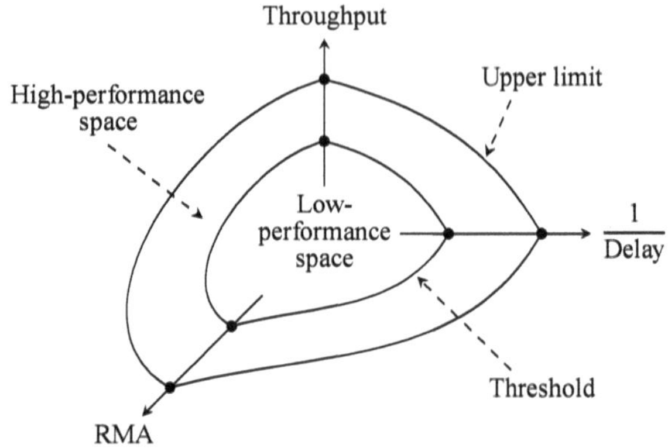

Fig. 2.9 Three-dimensional view of delay, throughput, and RMA [3, p. 51]

performance metrics fall within the defined thresholds or lower bounds of the limits in the multi-dimensional view, the service performance meets the requirements. Conversely, if the performance exceeds the thresholds or upper bounds of the limits, it no longer satisfies the requirements. When the performance lies between the lower and upper bounds, it still conforms to the requirements, but warnings may be triggered to indicate a risky state with a potential to cross the limits.

2.6 Summary

A computer network consists of a large number of entities including physical and logical network components and services. Those entities are integrated with complex interactions or relationships, leading to complex network behavior. This makes network analysis and architecture planning challenging, particularly for large-scale networks. To address these challenges, systematic approaches have been introduced in this chapter for network planning.

The systems approach focuses on the decomposition of a complex network into relatively simpler components with interactions. It shows holism and emergent behavior, leading to complex network dynamics that cannot be simply derived from the sum of the individual components. This motivates the investigations into not only the functions of the components, but also the impact of their interactions on network performance. Given the multi-objective nature of large-scale networks, it is more practical to seek trade-offs rather than an optimal solution, resulting in a satisfactory solution. More importantly, the systems approach highlights the concept of solving problems that have not yet been defined. Thus, an important step of solving the problems is to clarify what the problems are from the technical perspective. If limited information is available, the network can be treated as a black box. Then, inject inputs to the system, observe the outputs generated by the system, and develop an understanding of the system behavior.

The waterfall model emphasizes the sequential phases involved in a network planning project. Six essential phases are identified, which are requirements analysis, logical network design, physical network design, implementation and deployment, evaluation/testing/verification, and OAM. Each of these phases can be further decomposed into multiple sub-phases. With the sequential phases in the waterfall model, a phase should not commence until the successful completion of its preceding phase. The decomposition of the entire project into multiple sequential phases in the waterfall model aligns with the decomposition of a system into multiple subsystems in the systems approach.

Building upon the concept of system decomposition in network analysis, a generic network analysis model has been presented. It considers requirements from four perspectives: users, applications, devices, and the network. Traditional networking mainly focuses on the interconnection of devices into the network. In comparison, modern networking additionally considers requirements from users and applications. All requirements from users, applications, devices, and the network will be translated to technical network requirements, from which network architecture can be developed.

The top-down methodology is an effective tool to deal with complex network analysis and architectural planning, particularly for large-scale networks. It begins by developing a top-level view of the network to capture business goals, essential network functions, and critical network services. It then zooms in on specific network components for the development of top-level architectural models and component-based architecture. This is followed by detailed network design and physical design at the bottom level to cover hardware, floor plans, and structured cabling systems. The top-down methodology shares the similarities with the systems approach in terms of system decomposition and with the waterfall model in terms of sequential phases.

Unlike traditional networking with the focus on capacity planning, modern networking places greater emphasis on network services, which are configurable, manageable, and provisioned end-to-end. This shift leads to the concepts of service requests and service offerings. Service requests are requirements requested from the network, whereas service offerings are services offered by the network to the system to fulfill the requirements. The majority of network services are best-effort services. Some services are required to be predictable. A small number of services are critical and therefore must be guaranteed. To characterize and measure services quantitatively, various service performance metrics are used. From these metrics, a multi-dimensional view of service performance can be developed, which gives a basic understanding of the services in conformance or non-conformance with the requirements.

References

1. Fang, Q., Zeitouni, K., Xiong, N., Wu, Q., Camtepe, S., Tian, Y.C.: Nash equilibrium based semantic cache in mobile sensor grid database systems. IEEE Trans. Syst. Man Cybern. Syst. **47**(9), 2550–2561 (2017)
2. Fang, Q., Xiong, N., Zeitouni, K., Wu, Q., Vasilakos, A., Tian, Y.C.: Game balanced multi-factor multicast routing in sensor grid networks. Inf. Sci. **367–368**, 550–572 (2016)
3. McCabe, J.D.: Network Analysis, Architecture, and Design, 3rd edn. Morgan Kaufmann Publishers, Burlington, MA 01803, USA (2007). ISBN 978-0-12-370480-1
4. Oppenheimei, P.: Top-Down Network Design, 3rd edn. Cisco Press, Indianapolis, IN 46240, USA (2011). ISBN 978-1-58720-283-4
5. Braden, R., Clark, D., Shenker, S.: Integrated services architecture. RFC 1633, RFC Editor (1994). https://doi.org/10.17487/RFC1633
6. ETSI: Network functions virtualisation (NFV); terminology for main concepts in NFV. ETSI GS NFV 003 V1.4.1, ETSI NFV ISG. https://www.etsi.org/deliver/etsi_gs/NFV/001_099/003/01.04.01_60/gs_nfv003v010401p.pdf (2018). Accessed 3 Jun 2022
7. Chaturvedi, S.K.: Network Reliability: Measures and Evaluation. Wiley (2016). ISBN 978-1-119-22400-6

Chapter 3
Requirements Analysis

Requirements analysis is the first phase of systematic network planning in the systems approach, waterfall model, and top-down methodology, which are discussed previously in the last chapter. Subsequent phases will not commence until this phase is completed. The primary objective of requirements analysis is to clarify and define the network planning problems that need to be solved but have not been clearly specified. This chapter will discuss what and how requirements are developed in a systematic manner.

3.1 Concepts of Requirements Analysis

Ultimately, a computer network should be designed to support business goals subject to business constraints. From the technical perspective, network planning requires detailed specifications of technical network requirements. Therefore, requirements analysis plays a pivotal role in developing network requirements specifications, which will further drive network planning to support business goals. As understood from the systems approach, waterfall model, and top-down methodology, requirements analysis helps develop a comprehensive understanding of technical objectives in line with business goals, the selection of network technologies and services, network connectivity choices, and potential technical trade-offs.

How is requirements analysis conducted? After business goals and constraints are clarified, specific requirements are analyzed based on the generic network analysis model, which is discussed in the previous chapter, for users, applications, devices, and the network itself. Eventually, these requirements are translated into detailed specifications of technical network requirements for network planning. This process enables the network analyst to gain a clear picture of the characteristics and behavior of the network. A good understanding of the requirements form users, applications, devices, and the network itself is essential for the success of a network planning

Y.-C. Tian and J. Gao, *Network Analysis and Architecture*, Signals and Communication Technology, https://doi.org/10.1007/978-981-99-5648-7_3

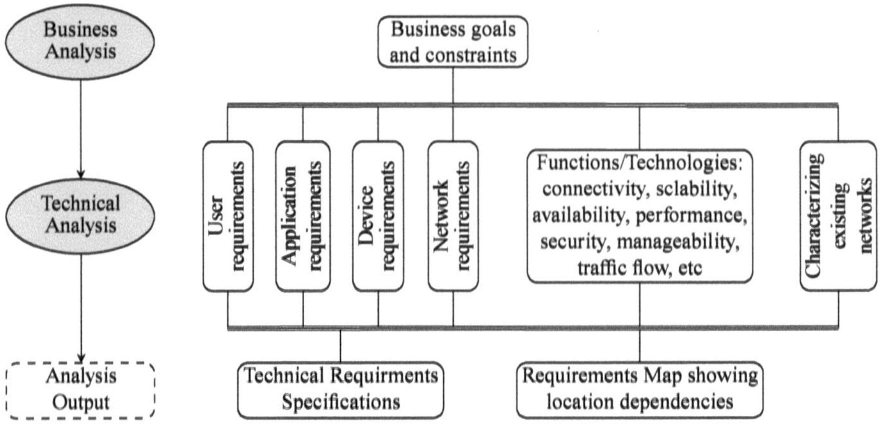

Fig. 3.1 Requirements analysis in which overlap exists in the technical analysis components. Traffic flow analysis will be discussed in a separate chapter

project. The resulting network planned from these requirements will effectively support its users, applications, and devices.

From the perspectives of network functions and technologies, requirements analysis considers connectivity, scalability, availability, performance, security, manageability, and other related aspects. More specifically, it develops performance thresholds, determines the nature of services that the network must deliver, and decides where the services must be delivered for a new or existing network.

Requirements analysis produces two types of documents:

- Network requirements specifications, which are integrated from all types of technical requirements, and
- Requirements map, which describes the location dependencies of various requirements.

The requirements map is an extension of the application map, which shows the location dependencies of applications.

Figure 3.1 illustrates a block diagram of requirements analysis. It shows the main components, processes, and their dependencies. The discussions of requirements analysis in this chapter will be structured around this diagram.

3.2 Business Goals and Constraints

As mentioned earlier, to support business goals within various constraints, requirements analysis for network planning begins with clarifying and understanding business goals and constraints. This has been briefly discussed in the book by Oppenheimer [1, pp. 8–23]. Overall, business goals and constraints should be clarified in conjunction with strategic and tactical network planning. As discussed previously in Sect. 1.3 of Chap. 1, strategic planning addresses long-term requirements and objectives, while tactical planning deals with short-term requirements and actions. Both of them are related to business goals and constraints.

The importance of clarifying business goals and constraints is often overlooked in network planing. This oversight is likely due to the fact that network analysts and planners are predominantly trained in computer science or engineering, and therefore tend to focus more on technical problems such as technical requirements regarding bandwidth capacity, network security, Quality of Service (QoS), and network management. However, it is not a feasible or justifiable assumption that technical objectives and trade-offs for a network are not related to business goals and constraints. According to the waterfall networking model discussed in Sect. 2.2 of Chap. 2, the step of identifying technical requirements and trade-offs should not commence until the step of clarifying business goals and constraints has been completed. Consequently, the overall requirements analysis for network planning begins with the analysis of business goals and constraints. This ensures that the resulting network planning well aligns with business objectives, thereby making business sense.

This section will clarify business goals and constraints from the following aspects:

- Understanding the core business,
- Understanding the organizational structure,
- Gathering physical location information,
- Identifying key applications for the core business,
- Understanding future changes, and
- Defining the project scope within constraints.

For all these aspects, it is important to work closely with customers and their executive managers. A good network planning would not be possible without their involvement and support, as discussed previously in Sect. 1.6 of Chap. 1.

3.2.1 Understanding Core Business

The core business of an organization determines, to a large extent, how a computer network should be planned, what critical network services should be provisioned over the network, and what network policies should be implemented on the network. Below are a few examples that illustrate this concept.

A higher-education institution, such as a university, has the core business of teaching and learning, research and development, and providing services to various industries. Therefore,

(1) Support for teaching and learning is not only essential but also mission-critical. The system for teaching and learning should be highly reliable and available with an acceptable level of latency, especially during teaching hours. It should also be accessible not only on-campus over the enterprise network, but also off-campus

over the Internet from anywhere in the world. The system should further provide network services integrated with text, voice, video, and multimedia communications. Additionally, the system for teaching and learning should interact with many other systems, such as those for student management, enrollment management, class allocation, timetabling, and grade center. Due to the involvement of private information in these systems, strict security and privacy polices must be in place.

(2) Support for research and development is also essential and must be highly reliable. It should be able to manage grant applications, and research projects, and host research data and results. Since some projects and generated data may be sensitive, a high level of security is also required.

For a research organization, teaching and learning may not be part of its core business. However, its requirements to support research are similar to those of a higher-education institution. Depending on the size of the organization, network services can be provisioned on-premises or off-premises. For small organizations, third-party cloud services can be considered as a cost-effective alternative without sacrificing the required level of QoS. In the real world, there are many research organizations that use cloud services to support their core business. Examples of such cloud services commonly employed by various organizations include cloud-based mail services, cloud or public Domain Name System (DNS) services, and other Software as a Service (SaaS) applications.

A financial organization manages highly-sensitive financial data or databases. The storage, operation, and transmission of these data must maintain a very high level of reliability and security. Redundant database servers may be in place with one serving as the primary server and the others in hot standby mode. Given that such an organization serves a large customer base across multiple cities, states, and counties, network services over the Internet are an essential requirement. Any financial transactions conducted over the Internet must adhere to strict security policies, employ strong encryption, and ensure guaranteed Quality of Service (QoS) management.

A company for online sales will have different requirements from those of an higher-education, research, or financial organization discussed above. Web or web-based services would become mission-critical, which serve as front-end interfaces to customers. These services must be highly reliable and available, likely operating seven days a week and 24 h a day. An online payment system will be integrated into these web-based services. The back-end financial databases supporting the sales operations must also maintain a high level of reliability, availability, and security.

Overall, understanding the core business of an organization provides valuable insights into its products, services, as well as internal and external relationships. From this understanding, basic ideas can be developed regarding the key requirements and top priorities for the network planning task. These ideas will be further supported later through the development of detailed technical requirements specifications, architectural topology, and various mechanisms and policies.

3.2.2 Understanding Organizational Structure

Typically, an organization has a hierarchical structure. The organizational structure will affect network planning in at least two ways:

- It provides information about who the decision-makers are for the network planning project and who holds the authority to approve the network planning proposal. As discussed previously in Sect. 1.6 of Chap. 1, obtaining support from customers and executive managers is critical for the success of a network planning project.
- It offers insights into the potential hierarchical topology of the network being planned. While the network hierarchy may not necessarily need to align precisely with the organizational hierarchy, it is natural to develop a network topology that, to some extent, mirrors the organizational structure. This is particularly helpful in developing network segmentation, Virtual Local Area Networks (VLANs), network management, and security and privacy policies.

These considerations highlight the significance of understanding the organizational structure when conducting network planning.

3.2.3 Gathering Physical Location Information

An organization may be located on a single site with one or multiple buildings. It may also have multiple sites spread across different cities, states, or countries. The physical location information of the organization largely determines network physical topology, network connectivity, and network policies. For example, consider an organization with headquarters and branch offices located in two different cities (see Fig. 1.1). In this case, the headquarters and branch offices must be interconnected through WAN links, which are typically provided by a third-party service provider. For multiple buildings within a single site, some applications may be used across several buildings, while others may be exclusive to a specific building. This information provides valuable insights into network connectivity and application dependencies across multiple buildings.

Overall, the physical location information of the organization helps the development of a complete application map, which illustrates the location dependencies of key applications. The application map will be further extended to a requirements map, which is one of the two main deliverables derived from the requirements analysis process.

3.2.4 Identifying Key Applications

There are numerous applications that are commonly found across various networks. Examples of such applications include web services, mail services, File Transfer

Protocol (FTP), Secure Shell (SSH), print services, remote desktop, and many others. By default, these network services are provisioned as best-effort services unless designed differently to meet specific QoS requirements.

However, every organization has its own unique core business, which may share similarities with other organizations but often diverges in specific aspects. Consequently, each organization relies on its own set of key applications to support its core business. These key applications drive network planning from the perspective of service-based networking.

As a good practice, it is advisable to identify and list top N key applications that are critical for the organization. This helps in prioritizing resource allocation and implementing effective QoS management strategies. Depending on the scale and complexity of the network, the value N for a specific network planning project may vary. For example, it could be 5, 10, or another suitable value. A more detailed analysis of application requirements will be conducted later in Sect. 3.4.

3.2.5 Understanding Future Changes

Potential future changes to an organization or network have a significant impact on the development of network requirements and architectural design. They drive both strategic and tactical network planning, as discussed previously in Sect. 1.3 of Chap. 1.

Future changes can manifest in various aspects. They may come from one or more aspects listed below:

- Core business,
- Scales of the organization or its divisions and departments,
- Organizational structure,
- Physical locations,
- QoS requirements and policies,
- Security policies, and
- Other relevant aspects.

After these changes are clarified, they should be taken into consideration in network planning. This proactive approach enables future network growth and ensures strategic alignment of the network with future requirements.

For example, suppose a company has decided to gradually cease its operation in a city and establish a new branch office in a different city within a span of three years. This means that minimal additional network resources should be invested in the branch office soon to be closed. Instead, a strategic plan should be developed in network planning for the upcoming opening of the new branch office. In addition, tactical actions should be meticulously planned to facilitate the design, implementation, and deployment of the network over the next few years.

With the rapid development of network technologies, enterprise networks have been undergoing significant changes over time. Technology-driven changes are also important for strategic network planning. A few examples are presented below:

- A typical example is the escalating demand for wireless coverage with easy accessibility and enhanced security. Some wireless mechanisms and services that are currently considered optional may become essential and widely adopted in the coming years. For some wireless network services, low-latency and seamless switching of a mobile device between access points may become imperative.
- Another example is the growing need for secure extranet services. Enterprise networks are expected to serve not only internal employees but also an increasing number of external customers and business partners.
- Traditionally, network services have been primarily provisioned and accessed on-premise. However, there is an increasing requirement for off-premise access from anywhere in the world. This necessitates the establishment of more secure enterprise edge and VPN connections over the Internet. In some cases, current VPN servers may need to be upgraded to higher-speed alternatives.

It is worth emphasizing repeatedly that network security has become a critical challenge and will continue to be so in the future. The emergence of new types of cyberattacks that have never appeared before poses a constant threat. Planning the network to maximize resilience against both existing and new threats is a topic of great significance in both theory and practice.

3.2.6 Defining Project Scope

A network planning project should have a clearly defined scope. However, the scope is often unclear initially and thus needs to be defined. In order to define the project scop, a few questions must be answered. For example, where are the boundaries of the project? What is not part of the project? What aspects must, or should, be addressed?

In some cases, a project may focus on upgrading specific segments of an existing network. In such a scenario, other segments are not part of the project and thus should remain untouched. The project boundaries would be defined by the routers that separate these segments from others. If any settings of the routers need to be modified, make sure to assess whether these changes will affect the interactions with other parts of the network.

In some other cases, a project may focus on migrating an existing local private data center to a third-party cloud data center. Various options exist, for example,

- Use third-party Infrastructure as a Service (IaaS) to host network services, or
- Use third-party SaaS offered by the same or different cloud service provider.

Through a high-level analysis, the scope and boundaries of this specific project can be clarified.

A project may be initiated to enhance the security of an existing network with no intention of investing additional hardware devices. This would require security enhancement focusing on software-based protocols, policies, and services. For example, enterprise edge security could be enhanced with additional security mechanisms and policies. Better Authentication, Authorization, and Accounting (AAA) management could be considered as an option for the security enhancement.

The process of defining the project scope and boundaries also helps in clarifying business constraints associated with the project. For instance,

- The aforementioned example of security enhancement imposes a constraint that no additional hardware devices should be considered.
- In the example of migrating an existing local private data center to a third-party cloud data center, it indicates that services will no longer be provisioned from the existing local private data center.
- The upgrading of specific network segments implies that no other parts of the network should be modified in general.

In addition, a network planning project is also subject to other business constraints. The most notable constraints include budget, skilled manpower, and time frame:

- **Budget:** A network planning project must align with the customer's budget. It should consider not only capital investment but also operational costs. Some projects have a strict budget that must be met. However, in many cases, the budget is actually a flexible requirement within a certain range. If a higher budget is proposed with convincing justification, it may be acceptable to the customer.
- **Skilled manpower:** The availability of skilled manpower is a critical business constraint that significantly impacts network planning. Insufficient in-house expertise can pose challenges in managing, operating, and upgrading the network. In such cases, opting for cloud network services from a third party might be a preferable choice over local provisioning of services.
- **Time frame:** Time frame is a vital business constraint that influences network planning decisions. Considering what is practically feasible within the given time frame, network planners will be able to determine what network mechanisms and technologies can be chosen. For example, a requirement for rapid deployment of a new network service may necessitate the selection of cloud-based SaaS. To ensure alignment with the given time frame, a well-designed schedule with key milestones and tangible deliverables for each milestone can serve as an effective tool for project management.

3.3 User Requirements

Let us analyze various requirements based on the generic network analysis model introduced in Sect. 2.3 of Chap. 2 (Fig. 2.6). The generic model highlights four main components: users, applications, devices, and the network itself. Each of these components is mapped to multiple layers of the OSI seven-layer architectural model.

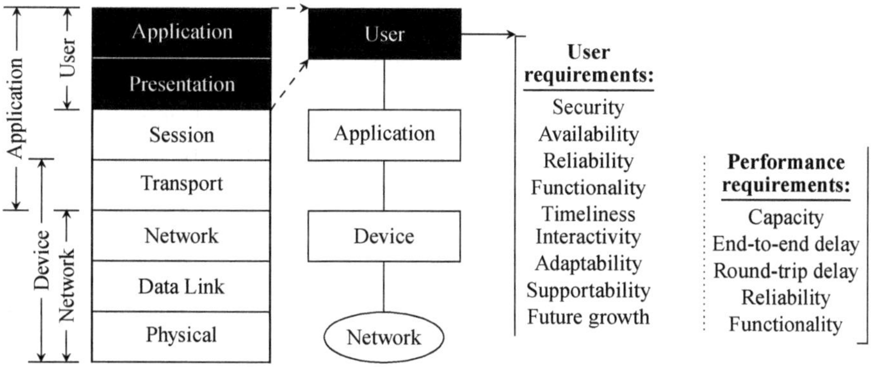

Fig. 3.2 User requirements

At the top layer of the generic network analysis model, the *user* component addresses the requirements from *end users* including network administrators and managers. It is associated with functions across layers 7 and 6 of the OSI seven-layer architectural model. A fundamental question to consider is: what do users need from the network to perform their tasks? User requirements can be developed from the end-user perspective in order for the users to perform their tasks successfully.

User requirements can be approached from various aspects. A list of general user requirements is presented in Fig. 3.2. Undoubtedly, it is not an exhaustive list. Also, each of the listed requirements is qualitative, and thus needs to be further quantified through the development of detailed technical requirements. A good reference for the general user requirements listed in Fig. 3.2 is the book by McCabe [2, pp. 64–66]. Let us briefly discuss these user requirements in the following based on our understanding and practice.

Security is listed as the first user requirement because it is one of the main challenges in networks and should therefore be given top priority. From the user perspective, it refers to the Confidentiality, Integrity, and Availability (CIA) of users' information and network resources. This entails protecting the information and resources from unauthorized access and disclosure. User security requirements can also be characterized from the reliability and availability perspectives. They will affect delay performance and capacity planning due to the additional overhead introduced by security enhancements.

Availability and **Reliability** have been discusses previously in Sect. 2.5.4 of Chap. 2. Both concepts are fundamentally different. For example, a highly reliable network service may not be highly available to an end user. However, from the user perspective, reliability more or less means availability. To the user, network services should not only be highly available but also have a consistent level of QoS. Meeting user availability and reliability requirements may need additional network resources, such as redundant links or servers. As a result, they can have a potentially important impact on delay performance, capacity planning, and network management.

Functionality refers to any user requirement that network services or applications accessed by users must be functional, in additional to being reliable and available. Functionality is closely tied to services and applications, and thus will be linked to application requirements later on. It is a good practice to analyze what functions are required by which user or user group. It does not make any sense to plan the deployment of an application that nobody will actually use.

Timeliness indicates how tolerant a user is to the variation of an expected time frame for accessing, transferring, and modifying information. It emphasizes the need for tasks to be performed with a specific time frame. There is a common misconception that timeliness is synonymous with a fast response. While some users do prioritize a quick response, others may prioritize accurate timing over response time. Accurate timing means performing tasks at the right times regardless of the speed of the response. Therefore, depending on specific scenarios, timeliness may imply either or both of a fast response and accurate timing. Timeliness can be quantified using end-to-end delay, Round Trip Time (RTT), and other delay-related performance metrics.

Interactivity measures the response times of the system and network when direct interactions with users are required. It has an impact on network architecture when

- There is a high degree of interactions between the users and the system/network, or
- The response times of the system and network are comparable to the response times of users.

RTT is a suitable metric to characterize response times.

Adaptability refers to the capability of the system and network to adapt to the changing needs of users. For example, during the COVID-19 pandemic, organizations faced a surge in demand for high-speed VPN connections to their enterprise networks from off-premises locations over the Internet. As another example, potable and mobile devices are being increasingly used for accessing enterprise networks, leading to evolving requirements for user mobility and wireless access. The system needs to be flexible and adaptable to meet these changing user requirements.

Supportability characterizes the requirements of users regarding the support they expect from the system, network, and IT support team. This includes technical support, system operation, system maintenance, and troubleshooting. The analysis of user supportability will later be linked to network requirements and network management architecture.

Future growth clarifies whether and when users have plans to deploy and use new services, applications, and/or devices on the network. Network architecture should be designed to accommodate future growth. For example, IP address allocation to each network segment should have some addresses reserved for future use.

It is worth noting that some user requirements discussed above are common to many users, others may be specific to certain users or user groups. For example, a user may have a stricter latency requirement compared to others. While security is important to all users, it may hold significantly greater importance for a specific user who is running a safety-critical application. Differentiating user requirements among various user groups provides valuable insights into network architecture.

3.4 Application Requirements

The application requirements are a set of requirements necessary for applications to function effectively on the system. They are developed from the application perspective to meet the user requirements at a higher layer in the four-layer generic network analysis model. As applications connect users and devices to the network, the primary application requirements are related to application performance, especially in terms of connectivity and data transfer through the network. Moreover, as applications are associated with specific services or tasks that are directly accessible to users in general, most network requirements are defined based on application categories, groups, and locations. Figure 3.3 illustrates a few examples of application requirements in the four-layer generic network analysis model.

3.4.1 Application Category

Applications can be classified in various ways. Depending on the levels of performance requirements, applications can be categorized as mission-critical, rate-critical, and real-time and interactive applications. Classifying an application into one of these categories helps capture the main features and requirements of the application:

- A mission-critical application has predictable, guaranteed, and typically high-performance requirements in terms of Reliability, Maintainability, and Availability (RMA), which are statistical performance metrics discussed previously. Exam-

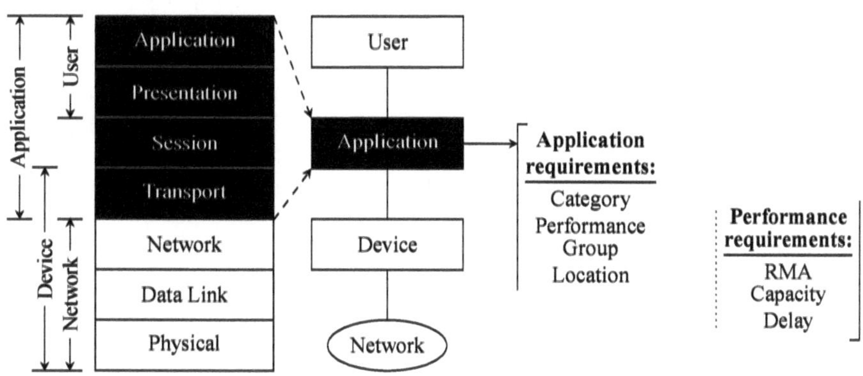

Fig. 3.3 Application requirements

ples of mission-critical applications include airline reservation systems, university teaching systems, and e-commerce Point of Sale (POS) systems.
- A rate-critical application has predictable, guaranteed, and/or high-performance capacity requirements. A typical example of a rate-critical application is a telemedicine system.
- A real-time or interactive application has predictable, guaranteed, and/or high-performance delay requirements. Applications used in industrial control systems, such as air traffic control, fall into this category.

Other applications that do not fit into any of these categories can be considered normal applications and are typically provided with best-effort services by default.

3.4.2 Performance Characteristics of Applications

The performance of applications is measured based on metrics such as RMA, capacity (bandwidth), latency (delay), and others, as discussed in detail previously. For the applications under consideration, the required level of the performance has a significant impact on the development of application requirements. Appropriate levels of RMA, capacity, and delay performance should be guaranteed for mission-critical, rate-critical, and time-critical applications.

That being said, RMA and capacity requirements tend to be more subjective and less technical. One may always argue for a high level of RMA or capacity, even if it is not really needed. However, it is worth noting that a higher level of RMA or capacity generally comes with increased costs and potentially more complex configurations. Therefore, determining a reasonable level of RMA or capacity for an application is beneficial in finding the right trade-offs among performance, costs, and network complexity.

In comparison with the RMA and capacity requirements, the timeliness requirements of real-time and interactive applications are more technical and less subjective. Quite often, they are inaccurately interpreted as fast responses only. While many real-time applications do require fast responses, there are many other real-time applications that require accurate timing control regardless of response speed. For example, in a scenario where an industrial process is controlled periodically every 10 s, precise timing for the period of 10 s is an essential requirement. This highlights the importance of performing right actions at the right times, neither too early nor too late. Real-time applications typically demand either or both of fast response and precise timing requirements [3, pp. v–vi].

Real-time requirements can be soft or hard. In hard real-time applications, the deadlines must be met. Missing a deadline would violate the system requirements and potentially lead to functional failures of the system or even catastrophic consequences. In comparison, soft real-time applications can tolerate occasional deadline misses without impacting the core functions of the applications though the overall system performance may deteriorate.

3.4.3 Application Group

In order to specify application requirements, it is helpful to group the applications under consideration based on their performance characteristics. Applications that share similar performance characteristics can be grouped together. Table 3.1 provides an example of application grouping.

3.4.4 Top N Applications

With the understanding of the application requirements discussed above, it becomes meaningful to identify top *N* key applications that serve the core business of the organization and drive the network planning. Depending on the specific network scenarios under consideration, *N* could be a number that suits the needs of the network planning project, such as 5, 10, or 15. From the list of top *N* applications, further identify to what extent an application is critical:

- A few applications may be extremely critical, which require a specific design for guaranteed QoS. Examples include mission-critical services (e.g., teaching

Table 3.1 Examples of application groups [2, pp. 73–75]

Group	Description	Examples	Scenario/s
Web development, access, and use	Accessing remote devices and downloading and/or uploading information with the aid of graphic interfaces	Same as traditional remote device and information access utilities telnet and FTP	Interactive, a mix of interactive burst and interactive bulk
Telemetry/ command and control	Transmission of data and command information between remote devices and control stations	Remotely piloted vehicles, commercial aircraft	Real-time and/or interactive delay
Operations, administration, maintenance, and provisioning	Those applications required for proper functioning and operation of the network	DNS services, network security	Mission-critical and interactive
Client–Server	Those applications whose traffic flows behave in a client-server fashion	Customer relationship management, enterprise resource planning	Mission-critical and interactive
Bulk data transport	Large amounts of data with less interactions	FTP, MFTP, ARCP	No high-performance requirements

and learning in a university, and financial transactions in a financial institution), safety-critical services like air traffic control, and time-critical services like military systems.
- Some applications may be somewhat critical, which require soft real-time or predictable QoS management. For example, an e-commerce POS works well with soft real-time QoS. It can tolerate occasional long delays but should not experience sustained long delays.

It is worth noting that an application that is critical or extremely critical to one organization may not be critical at all to another organization. For example, video streaming is particularly critical for remote surgery and should be managed with guaranteed QoS. However, it is typically configured as a best-effort service by default in the majority of other networks. Therefore, it is important to investigate the specific network under planning for its requirements.

3.4.5 Application Map

There are many applications that apply to everywhere and used by everyone. Examples include email, web browsing, word processing, and other general office applications. Some applications are transparent to, but used by, all users, such as DNS and DCHP services. Some of these applications (e.g., email) are configured as best-effort services. However, some others may be critical for a specific organization or network and thus should be identified and configured with predictable and guaranteed services.

There are also many applications that are specific to some segments of a network and used by particular groups of users. For these applications, it is useful to clarify their location dependencies physically and logically, resulting in an application map as illustrated in Fig. 3.4. The application map helps in determining the flow characteristics of the applications and mapping the traffic flows during traffic flow analysis. From the application map, a more general requirements map will be developed later as one of the two main outcomes from the overall requirements analysis.

For the development of the application map, the following questions need to be answered:

- Where an application will be applied, in users' environment or within the environment of the overall network system?
- On which devices will the application be used, on general end-user devices or specific devices?

Answering these questions will assist in clarifying the location information of the application.

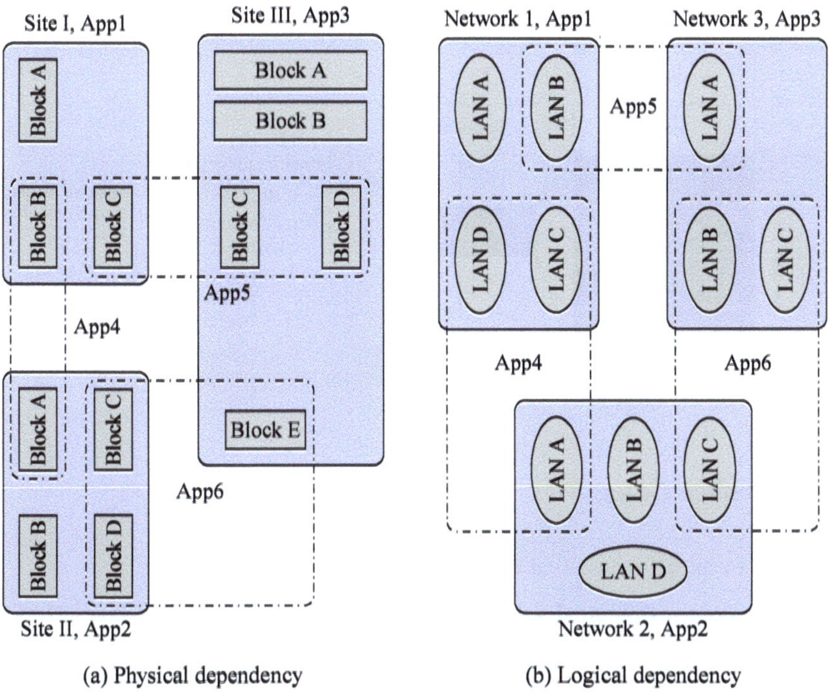

Fig. 3.4 Application map

3.5 Device Requirements

As discussed previously, traditional networking primarily focuses on network bandwidth capacity for the connectivity of network devices. In comparison, modern networking additionally considers the requirements of users and applications. This is due to the fact that the users and applications provide a complete vision of the overall system requirements. Therefore, the development of device requirements builds upon the user and application requirements discussed above.

Overall, the device requirements of a network encompass the set of requirements that network devices need to fulfill to enable the network to function in a manner that meets the user and application requirements. These requirements are illustrated in the logical diagram shown in Fig. 3.5 within the framework of the four-layer generic network analysis model.

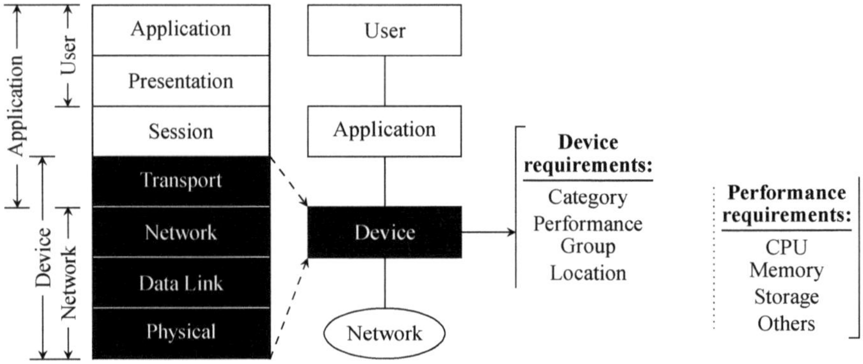

Fig. 3.5 Device requirements

3.5.1 Device Categories

Devices are typically classified into the following categories:

- **Generic computing devices** are those devices that most users have. They are typically end devices that serve individual users and provide access points to the network. Examples of generic computing devices include laptops and office personal computers.
- **Servers** are devices that serve one or more clients, such as storage and compute servers. They usually have a large amount of dedicated resources for specific purposes, such as memory, processing power, networking capabilities, and peripherals, which are superior to those found on generic computing devices. For example, consider a scenario where a cluster of powerful workstations used as servers are interconnected via dual Infiniband and Ethernet networking. The Infiniband networking provides high-speed data exchange among workstations, while the Ethernet networking provides general network connectivity to the enterprise network.
- **Specialized devices** are devices that collect, develop, and operate information for their users, such as supercomputers and networked cameras. These devices generally do not support direct physical access from users but can be accessed remotely through protocols like SSH or other interfaces. Medical facilities and smart traffic lights are examples of specialized devices. An additional example of a specialized device is a network device designed specifically for satellite communications. The device receives information from satellites, pre-processes the received data, and streams the pre-processed data to other devices for further processing.

Requirements for different categories of network devices vary in general. A good understanding of the requirements helps in developing strategies to address the so-called *last foot* problem. The last foot problem refers to delivering services with required performance to users and applications from network devices or their interfaces.

3.5.2 *Performance Characteristics of Devices*

The performance of network devices is not easy to determine because of two reasons:

- It is closely tied to hardware, firmware, and software that join users, applications, and other components of the system; and
- The components within a device are often proprietary, implying that detailed information about their performance may be limited or unavailable.

Consequently, there may be a lack of device performance details.

Recall that in computer networking, network performance, or more generally network QoS, is managed and measured end-to-end. From the end-to-end perspective, the performance characteristics of a device can be determined by considering the device's overall performance. They can be described based on various components of the device, such as processors, memory, storage, device drivers, and read/write speed, all of which impact the overall performance of the device.

By clarifying the performance characteristics of a device, potential performance problems or limitations within the devices can be identified. This enables the development of strategies to overcome these limitations. For instance, by identifying bottlenecks in the network interfaces of a device, it becomes possible to find out how to upgrade the device in order to achieve the required level of performance.

3.5.3 *Location Dependency of Devices*

General computing devices, such as personal computers, laptops, and mobile devices, are typically location-independent. They can be used as plug-and-play devices anywhere on the network or off-premise for remote access to the network via the Internet. Nevertheless, understanding where and how many generic computing devices are accessing the network is helpful in determining how network resources should be allocated and what the overall performance of the applications provided through the devices would look like for their users.

Different from general computing devices, servers are more location-dependent. For example,

- A computing laboratory is set up on a specific floor of a building to house a cluster of high-performance computers. This specific location is chosen to provide a controlled environment, optimized power supply, and efficient networking for the cluster of computers.
- A storage server is placed in a specific location that is in close proximity to another server running a database service. This close placement ensures reduced latency and improved data transfer between the storage server and the database server.

In these examples, the placement of servers or workstations is purposeful and strategic. Factors such as proximity, resource sharing, and specialized services are taken into account during the decision-making process.

Specialized network devices are typically location-dependent. This is understandable considering their specific purposes, such as smart traffic lights and medical facilities, as mentioned earlier. While routers and firewalls are general devices in networking, they are specialized devices to general users. They are strategically installed at specific locations to interconnect multiple networks and provide security protection.

For specialized devices, the information about their locations is also valuable in determining the relationships among users, applications, and networks, as well as the relationships between the components of the overall system. This location information plays an important role in developing flow characteristics and performance requirements specific to these devices. For example,

- Consider a specialized device that generates or receives a significant volume of network traffic. The location of this device has an impact on the bandwidth and latency requirements for its surrounding network environment.
- Similarly, a specialized device with a high level of security requirements may necessitate a design that includes dedicated security protection measures. Knowing the location of such a device enables the implementation of appropriate security mechanisms and protocols to safeguard sensitive information and protect against potential threats.

The location information of network devices, particularly servers and specialized devices, is of utmost significance when it comes to outsourcing system components or functions for the purpose of consolidating and relocating network services and applications within an organization. In such scenarios, the outsourcing agent can choose to operate, administer, maintain, and provision the required resources either on-site or remotely. For remote resource provisioning, the on-site resources will be removed, and the remote resources are provisioned in a cloud infrastructure owned by either the outsourcing agent or a third-party cloud service provider. As a result, the resource provisioning shifts from on-site Gigabit Ethernet LANs to cloud WANs. Consequently, some LAN-based applications now become WAN-based, resulting in increased round-trip delays due to the change in devices and service locations. The impact of the resource relocation needs to be carefully evaluated to ensure that the performance requirements expected from users and applications are met.

3.6 Network Requirements

Network requirements build upon, and are more technical in nature than, user, application, and device requirements, which are subjective to some extent as we have already understood. All user, application, and device requirements will be eventually reflected in, and translated into, network requirements. Therefore, for the development of network requirements, it is essential to conduct a detailed analysis

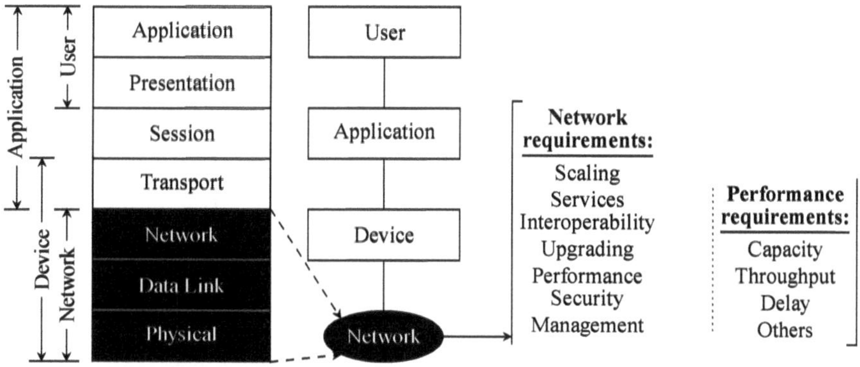

Fig. 3.6 Network requirements

from various technical aspects of networks. The derived network requirements will drive the subsequent development of detailed technical specifications for network planning.

In most cases, networks are not built from scratch. Rather, network planning projects typically consider extending or upgrading existing networks. Therefore, in the analysis of network requirements, it is critical to characterize existing networks, integrate new components and technologies into the existing infrastructure, and establish a clear pathway for migrating or upgrading the existing networks to the new ones being planned.

In the following, let us discuss how to incorporate existing networks into network planning. This will be followed by an analysis of network requirements in terms of management, performance, and security. The topic of characterizing existing networks will be addressed later in a separate section. Figure 3.6 provides an overview of the overall network requirements within the framework of the four-layer generic network analysis model.

3.6.1 Extending Existing Networks

In the analysis of network requirements, several key aspects need to be considered for extending existing networks. These include network scaling, performance constraints, support services, interoperability, and location dependencies.

Network scaling investigates how the planned network will scale well when new network components or technologies are added to the existing network. When the existing network is upgraded, it may continue to function well or face unexpected issues. If it fails to provide network services at the desired level of QoS, what strategies should be implemented to fix the problem?

Performance constraints exist in the existing network. Examples include the maximum number of LANs that can be interconnected to a router for the required capacity and performance, the lowest possible delay that can be achieved in the network, and the best possible security protection that the current firewalls can provide. Beyond these constraints, improved or additional mechanisms will need to be designed and provided for the desired level of network performance.

Support services are the services that support the functionality of the network and networked systems. Typical examples include strategies and mechanisms for addressing, routing, security, performance, and management. When an existing network is upgraded to a new one, it is essential to understand the network requirements for each of these support services.

Interoperability ensures the smooth transition from the existing network to the planned new network. If the planned network follows the same addressing and routing strategies as those in the existing network, no translation will be required at the boundary between the existing and new networks. Therefore, it is part of the network requirements analysis to clarify the technologies and media used in the existing network, as well as any performance or functional requirements for the upgrading of the existing network to the planned new one.

Location dependency of the existing network may change when it is upgraded to the planned new one. For example, a LAN-based service in the existing network may become a WAN-based cloud service in the planned new network. This change can impact the QoS of the service, and thus needs to be considered in the development of network requirements.

3.6.2 Addressing and Routing Requirements

Network addressing determines how to identify networks and network devices. It also largely determines the format of data packets for network communications. It further determines how to route traffic from one end the another. For TCP/IP networks, network addressing implies IP addressing, which can be either or both of IPv4 and IPv6.

Currently, the majority of existing networks still use IPv4. However, more and more networks adopt IPv6. They are typically configured with a dual stack that supports both IPv4 and IPv6 although there are networks that operate solely on IPv6. When a new network is designed or an existing network is upgraded, the choice between IPv4 and IPv6, or the utilization of both, becomes part of the addressing requirements.

Next, how to allocate IP addresses and how to organize networks in segments should be clarified. It is generally recommended that networks be organized hierarchically. Accordingly, IP addresses should be allocated in a hierarchical manner as well. This helps simplify not only network management but also traffic routing.

As a network-layer function in the 7-layer OSI network architecture, routing makes decisions to choose a communication path, typically from multiple ones, to route traffic from its source to its intended destination. While there are some pre-configured static routes, routing decisions are typically made dynamically by a routing protocol. Different routing protocols have different processes for making routing decisions. When developing routing requirements, it is important to take into account factors such as the network environment and the applications running on the network. This helps guide the selection of an appropriate routing protocol that aligns with the specific needs and characteristics of the network.

In some cases, there may be a requirement for multiple routing protocols to be used within the same network. This could be due to various reasons, such as a large-scale enterprise network that is integrated from multiple networks running different routing protocols. For such scenarios, what are the requirements for multiple routing protocols to work together? Protocol interoperability and route redistribution will need to be considered.

3.6.3 Performance and Management Requirements

Network performance is assessed from various aspects such as network capacity, latency, jitter, RMA, and many others. It is characterized by a set of metrics and their quantitative values, bounds, or thresholds. The primary focus of performance requirements is on the fulfillment of the required or desired QoS for network services and applications. More specifically, the question at hand is: What types and levels of QoS support are needed for specific services and applications? As network services are provisioned as best-effort services by default as we have already understood, network performance requirements mainly deal with non-best-effort services.

Some applications may require hard-real-time QoS management to meet hard deadlines in response to network events. Missing a deadline could lead to system failure at the best or cause catastrophic events at the worst. For such applications, it is essential to implement end-to-end hard-real-time performance management. This may indicate the adoption of Integrated Service (IntServ) as a potential QoS management mechanism.

Some other applications may be better served with soft-real-time QoS management. Occasional misses of deadlines can be tolerated as long as there are no frequent deadline misses. For example, in video streaming, missing a few frames of the video within a second has a minimal impact on video quality. Soft-real-time QoS management would be sufficient for such applications. Differentiated Service (DiffServ) can be considered as a potential QoS management mechanism.

Network performance can be managed at either or both of layer 2 and layer 3 in the 7-layer OSI network architecture. Examples include layer-2 tagging and layer-3 Differentiated Services field CodePoint (DSCP) for traffic prioritization. These techniques will be discussed later in a separate chapter. Any specific requirements for performance management at particular layers can be identified during the analysis of network requirements.

To effectively implement, improve, and enhance QoS, network resources need to be planned, monitored, and controlled through network management. Various management protocols and mechanisms are available for monitoring and evaluating network events, information flows, integrity and security, and other network performance metrics. In network analysis, develop network management requirements in relation to the following aspects, which are not listed exhaustively:

- What needs to be monitored and managed?
- Is monitoring for event notification or trend?
- What instrumentation methods, such as Simple Network Management Protocol (SNMP), should be used?
- To what level of detail should events or trend be monitored?
- Should management be performed in-band or out-of-band?
- Is centralized or distributed monitoring more suitable?
- What is the impact of network management on network QoS?
- How should the management itself and management data be managed?

By addressing these aspects, comprehensive network management requirements can be developed to ensure effective QoS implementation, management, and control.

3.6.4 Security Requirements

Security is one of the fundamental requirements in computer networking. Addressing the Confidentiality, Integrity, and Availability (CIA) of networks, it is a set of rules and configurations with the aim to safeguard computer networks and data from unauthorized access. Technically, security does not directly cover the concept of privacy, which emphasizes the protection of private information from unauthorized access and disclosure. Security can be achieved without privacy but privacy cannot be achieved without security. Practically, it is common for people to refer to both security and privacy when discussing security.

In the analysis and development of network security requirements, it is necessary to clarify what needs to be protected and what potential risks are. More importantly, security is managed from not only the technical perspective, but also the safety and social perspectives. For example, theft and physical damages are more relevant to safety. Security awareness is more about social activities. Nevertheless, all these aspects are discussed in the broad area of network security.

After security risks are analyzed, requirements should be clarified and developed for a security plan, security policies, and security procedures. A security plan is a high-level framework of broad guidelines that guide the design of security policies and procedures. Requirements also need to be developed for the separation of security services between external and internal provisioning, as well as among dedicated hosts or different network components.

From the technical perspective, network security requirements must also be developed for choosing security mechanisms and protocols, securing specific network segments or groups, and ensuring the security of security measures themselves. Examples include, but are not limited to, the following aspects:

- What servers need to be protected?
- What requirements are for the security of different types of servers, such as DNS, mail, and web servers?
- What are the requirements for firewalls, DeMilitarized Zone (DMZ), and packet inspection?
- What are the requirements for corporate edge security, such as Virtual Private Network (VPN)?

Clarifying these and other security requirements will facilitate the development of a robust security architecture for the network under consideration.

3.7 Characterizing Existing Networks

Characterizing existing networks will help develop realistic goals for network planning. Any bottlenecks, performance problems, and network devices or components that require replacement or improvement in the existing network can be identified. This will give some hints about potential solutions for the network planning project.

The main tasks involved in characterizing existing networks include the understanding of the physical and logical network architecture, addressing and routing architecture, performance baselines, management architecture, and security architecture. Analyzing protocols used in existing networks is also an important task.

3.7.1 Characterizing Network Architecture

The physical network architecture provides information about the geographical locations of network components, devices, and services. By using the top-down method, the top-level view of the physical network architecture can be developed. It shows physical network sites, and their geographical locations and connectivity such as WAN.

For each network site, one or multiple network maps can be developed to illustrate more detailed physical information such as:

- Buildings, floors, and rooms or areas.
- The physical location of main servers or server farms, such as web servers, DNS servers, mail servers, database servers, and storage servers.
- The physical locations of routers and switches, such as border routers and other routers.
- The physical locations of high-performance computing clusters, computing laboratories, and other specific computing facilities.
- The physical locations of network management components, such as enterprise edge management components (e.g., VPN servers).

- The locations of wireless access points and other network hotspots.
- The locations of important Virtual Local Area Networks (VLANs).

In addition to the physical network architecture, the logical topology of the existing network can be further clarified. This is an important step because network addressing, routing, and other strategies heavily rely on the logical topology of the network. By understanding the logical topology, potential constraints for upgrading the network can be identified. For example, a network with many devices interconnected in a flat topology does not scale well. Likewise, a hierarchically interconnected network without a core layer of high-end routers may face scalability issues. A LAN that connects an excessive number of hosts may suffer from broadcast storm. Key servers without an appropriate redundancy design will become a single point of failure. For instance, redundant servers are single-homed, i.e., connected to a single switch, or there are no redundant servers at all. Identifying all these issues helps inform the design of the planned new network, allowing for appropriate solutions and improvement.

3.7.2 Characterizing Addressing and Routing

For addressing, it is necessary to clarify whether IPv4 or IPv6 is in use. If the dual-stack configuration of both IPv4 and IPv6 is already implemented, continue to use it in the planned new network. If only IPv4 is being used, it is the right time to consider introducing IPv6 to the new network.

Investigate if IP addresses are allocated hierarchically with good scalability to accommodate future growth. Identify areas where IP address allocation can be improved through better subnetting. For example, discontiguous subnets should be avoided in IP address allocation.

IP addressing is tightly coupled with routing. By analyzing subnetting strategies, it is relatively easy to characterize how route summarization, i.e., router aggregation or supernetting, has been implemented in the existing network. Assess whether route aggregation can be further improved in the planned new network.

Evaluate if routers are appropriately placed with the desired security protection in the existing network. Analyze what routing protocol is being used in the existing network. Does the routing protocol continue to function effectively with any changes made to the network architecture and IP addressing in the planned new network? If more than one routing protocol is employed, how do they work together in the same network?

3.7.3 Characterizing Performance and Management

Characterizing the performance of the existing network will help establish performance baselines and determine the requirement improvement for network upgrading. The first set of performance metrics to consider is RMA (Reliability, Maintainability, and Availability), which has been discussed in detail previously. It may be expected that the current levels of RMA performance are maintained, or improved RMA performance is desired in the planned new network.

Other performance measures that could be characterized include latency, throughout, network utilization, network efficiency, and network accuracy, which have been briefly discussed in the previous chapter:

- Latency can be measured in different ways depending on application scenarios, such as one-way delay, Round Trip Time (RTT), and response time. In some cases, the boot time of machines, whether they are Physical Machines (PMs) or Virtual Machines (VMs), should also be considered. For example, if migrating a VM to a new PM that is currently off, it will take time to boot the new PM first and then boot the VM hosted on the new PM.
- Throughput reflects the capacity of the network for data communications. Its value in the existing network can be used as a performance baseline for network upgrading.
- Network utilization measures, typically in the percentage of capacity, how much bandwidth is in use during a specific period of time, such as 10 min or an hour. It is high during peak hours and low during off-peak hours. The pattern of network utilization versus time helps understand the normal traffic behaviors. The planned new network should be able to handled the expected traffic pattern.
- Network efficiency is commonly understood as the successfully transferred data expressed as a percentage of the total transferred data. This understanding has considered protocol overhead as part of the useful data, and thus is not accurate. More accurately, network efficiency measures how much payload, i.e., user data, is successfully transferred in comparison with the total transferred data including overhead no matter whether the overhead is caused by collisions, frame headers, acknowledgments, and re-transmission. For example, if 10 packets have been successfully transmitted without re-transmission and the overhead is 20% (e.g., from frame headers), then the network efficiency is 80%. Therefore, a larger packet size is beneficial in general. But any transmission errors will lead to re-transmission of large packets in this scenario.
- Network accuracy captures how correctly data packets can be transferred over the network. For WAN links, it is measured by Bit Error Rate (BER). which is typically around 1 in 10^{11} for fibre-optic links. For LAN links, the focus is on data frames rather than individual data bits. A typical network accuracy threshold for LANs is a bad frame per 10^6 data frames. If WAN links are provided by a third-party service provider, the desired network accuracy can be specified in a Service Level Agreement (SLA). Check if there is such an SLA for the existing network.

For network management, investigate what QoS management architecture is and what QoS mechanisms and strategies are in place in the existing network. Also, find out how the network management itself is currently managed. For critical network services, such as mission-critical, safety-critical, and time-critical services, identify what and where specific QoS designs have been implemented to provide QoS guarantee. Whether or not these designs are effective? If not, where and how they can be improved in the planned new network?

3.7.4 *Characterizing Security*

The security of the existing network can be characterized from various aspects. For example,

- What assets are being protected?
- What are the security risks to the assets?
- What physical security measures are in place?
- What methods are being used for security awareness?
- What is the security plan currently in use?
- What security policies have been implemented?
- What security procedures are designed?
- How is DMZ designed for security protection, and which servers are placed in the DMZ?
- How is the overall security managed?

Through such a process of characterizing the security of the existing network, it will become clear which security mechanisms and strategies could be inherited in the planned new network. Also, potential security issues that require attention or enhancement can be identified. Meanwhile, potential solutions to these issues may be proposed. For instance, a current firewall should be replaced with a new one that provides enhanced security protection.

3.7.5 *Analyzing Protocols in Use*

Developing a basic understanding of the protocols in use helps upgrade existing networks for better scalability and expected performance behavior. To achieve this, it is necessary to clarify the following aspects:

- What protocols are currently in use?
- How many users are using each protocol?
- How many devices and servers are using each protocol?
- Which applications does each protocol support?

- Are there any other perceived issues, such as the scalability of using a specific protocol?

It may not be realistic or necessary to list all protocols used in the existing network. For example, IPv4 or IPv6 is always present regardless of the applications being executed. Therefore, it is a general practice to focus on the most important protocols or the protocols with specific use cases. For instance, IntServ is used to provide end-to-end QoS guarantee for a specific application, highlighting the importance of that application in the existing network. Similarly, DiffServ is being used by a group of video streaming applications, showing the soft-real-time QoS nature of these applications.

3.8 Requirements Specifications and Trade-Offs

Through a comprehensive analysis of various requirements, the main objectives and requirements of a network project are identified, developed, and prioritized. This leads to a set of requirements specifications as one of the two main outputs of requirements analysis. The other output is the requirements map, which is extended from the application map discussed previously (Fig. 3.4). The requirements map describes the location dependencies of application requirements as well as other requirements.

3.8.1 Requirements Specifications

An example of requirements specifications is illustrated in Table 3.2. Additional examples can be found in [2, pp. 90–94]. Figure 3.7 shows an simplified requirements map that depicts the location dependencies of the requirements and applications. As indicated in the caption of the figure, a complete requirements map would be more complex with detailed information about the location dependencies of comprehensive requirements and a large number of important applications.

3.8.2 Technical Goals and Trade-Offs

From a high-level view, the requirements developed in the requirements specifications focus on the following aspects:

- Connectivity,
- Scalability,
- RMA,
- Performance,

Table 3.2 An example of requirements specifications

ID	Date	Type	Description	Source	Where	Priority
R1.0	03 Jan 23	User	Users 15000 (Seniors 100, academics 2000, admin 1500, students 12000, others 400)	Mgmt	Site Map	Normal
R1.1	03 Jan 23	User	Seniors 100 (20 Bldg A, 30 Bldg B, 50 Bldg C)	Mgmt	Site Map	Normal
R1.2	03 Jan 23	User	Academics 2000 (200 Bldg B, 300 Bldg C, 500 Bldg D, 500 Bldg E, 300 Bldg F, ...)	Mgmt	Site Map	Normal
...	...	...	...	...	...	...
R2	03 Jan 23	Netw	Gigabit Ethernet connections to backbone	IT Team	All Bldgs	Normal
R3	03 Jan 23	Netw	Redundant access to Internet	IT Team	All Bldgs	Normal
R4.0	04 Jan 23	App	Teaching system Canvas mission-critical	T&L	All bldgs	High
R4.1	04 Jan 23	App	Canvas response time $<10\,s$	T&L	All bldgs	High
R4.2	04 Jan 23	App	Student record databases, finance databases, & payroll mission-critical	Mgmt	Data center	High
R5	05 Jan 23	Device	A cluster of 10 HPCs interconnected with Infiniband and Ethernet	CS Dept	Bldg D	Normal
R6	05 Jan 23	Device	Connection of a GNSS receiver and data processing system	Eng Dept	Bldg E	Normal
R7	06 Jan 23	Netw	Critical databases incl. Canvas, student record, and finance in a separate DMZ	IT Team	Data center	High++
R8	09 Jan 23	Netw	Connection of a new network in a new building to the existing network	Mgmt	Bldg H	Normal
...	...	...	...	...	...	...

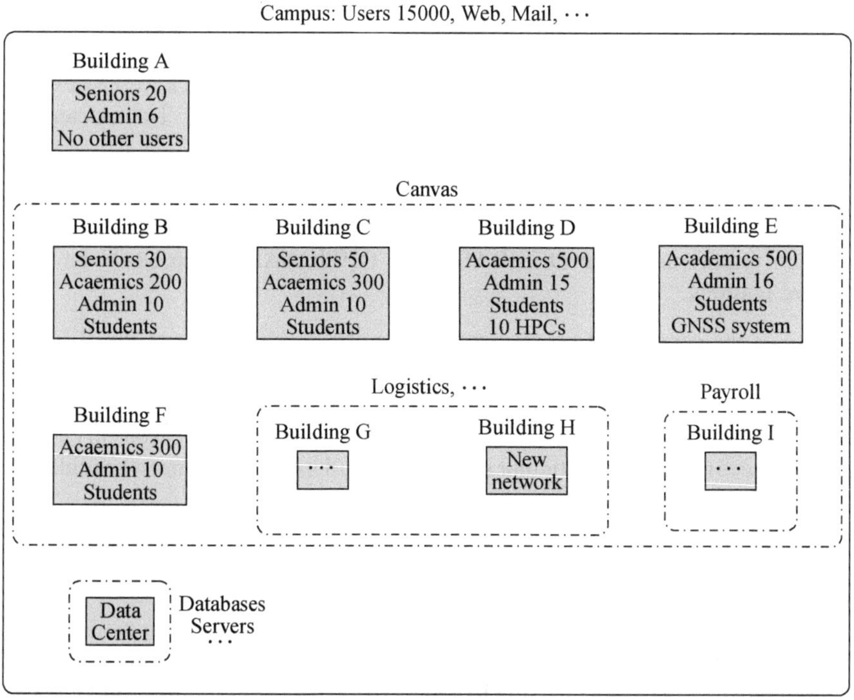

Fig. 3.7 A simplified requirements map. A complete requirements map has detailed information about the location dependencies of comprehensive requirements and a large number of important applications

- Manageability,
- Security, and
- Affordability.

All of these aspects are important. However, it should be understood that optimizing all of them simultaneously is practically impossible in a real-world network. Improving one aspect may lead to the sacrifice of one or more other aspects. For example, enhancing security through deep packet inspection will inevitably introduce additional time delay. Similarly, enhancing the reliability of a network may require the implementation of additional redundancy mechanisms and strategies. Consequently, this will result in increased capital and operational costs for the network, as well as more complicated network management. Therefore, in order to achieve a satisfactory solution, trade-offs need to be made for the identified requirements and technical goals. This can be achieved by carefully evaluating and prioritizing various aspects, taking into account the specific needs and objectives of the network project.

Categorizing Requirements

How to develop trade-offs? An effective approach is to refer back to the developed requirements specifications, such as those presented in Table 3.2. Then, determine

- Which requirements are core and fundamental to the network, and thus are REQUIRED and MUST be met.
- Which requirements are RECOMMENDED and therefore SHOULD be met.
- Which requirements are DESIRABLE or OPTIONAL which MAY be met as future features.

The use of the capitalized words here follows the specifications of the IEFT RFC 2119 [4] and RFC 8174 [5] regarding the keywords that indicate requirement levels. It is worth mentioning that, according to RFC 2119 [4], the word 'SHOULD' or the adjective 'RECOMMENDED' means "that there may exist valid reasons in particular circumstances to ignore a particular item, but the full implications must be understood and carefully weighed before choosing a different course".

Prioritizing Requirements in Each Category

With these three categories of requirements, assign top priority to the REQUIRED requirements, normal priority to the RECOMMENDED requirements, and low priority to the DESIRABLE/OPTIONAL requirements, respectively. Within the RECOMMENDED and DESIRABLE/OPTIONAL categories, prioritize the requirements by determining the order in which they will be implemented. In this way, high-priority requirements can always be met and low-priority requirements will be met whenever possible.

Resolving Conflicts and Refining Requirements

When conflicts arise in implementing the requirements, several options exist. Here are some examples:

- Meet the high-priority requirements and relax the low-priority requirements. For example, the time delay of 10 s specified in Requirement R4.1 of Table 3.2 might be slightly relaxed, e.g., 10.5 or 11 s, with a minimal impact on the QoS of the Canvas application for teaching and learning in a university.
- Investigate whether the conflicting requirements are appropriately specified. If not, consider re-defining them. For example, find out what causes the delay in accessing Canvas. Is it because of excessive simultaneous sessions or users, or is it due to an inefficient authentication process? Can on-campus access to Canvas be streamlined for faster access? A detailed investigation can help refine the requirements and potentially eliminate the conflicts.

- Use cost-effective solutions to resolve cost-related conflicts. For example, consider the redundant access to Internet specified in Requirement R3 of Table 3.2. Using two Internet Service Providers (ISPs) for redundancy might be more expensive. Instead, an alternative and cost-effective design can be considered: using two different physical paths and routers to the same ISP. It provides Internet redundancy without incurring significant cost implications.

It is worth mentioning that different groups of users may have different expectations. For example, the reliability of the Canvas system for teaching is more important for a university compared to its students. If the system is down on a weekday, the students may not be greatly affected. However, to the university, this will have a significant impact on not only the teaching itself but also many related systems, applications, and resources, such as timetabling, classrooms, and academic calendar. In this scenario, meeting the university's overall requirements for Canvas will automatically satisfies the students' expectations without the need of additional effort. Therefore, it is essential to develop appropriate requirements that accurately capture the desired features and behaviors of the network.

Through all these efforts, a comprehensive set of requirements specifications can be developed. These requirements are prioritized by considering technical goals and trade-offs. Together with the requirements map and traffic flow specifications, which will be discussed in the next chapter, they will be used to drive network planning.

3.9 Summary

Requirements analysis plays a significant role in network planning. It helps clarify and define the network problems that need to be addressed, especially with regard to network architecture and design. These defined problems are presented in a set of requirements specifications along with a requirements map, which shows the location dependencies of the requirements within the network. Since it is unrealistic to solve a problem that has not been properly defined, requirements analysis must be conducted as the initial phase of systematic network planning.

Network planning must align with business goals while considering various constraints. Therefore, capturing business goals and constraints is an important part of requirements analysis, providing a comprehensive understanding of several key aspects including the core business, organizational structure, location dependencies of key applications, potential future changes, and project scope.

Furthermore, since it is uncommon to build a network from scratch, it is essential to consider the extension and upgrade of the existing network when planning a new network. Therefore, it is critical to understand the existing network and integrate it with any new components to be introduced in the planned network. Detailed investigations and evaluations should be conducted with the focus on scalability, performance constraints, support services, interoperability, and location dependency.

Given the complexity of networks and network services, network requirements can be analyzed in separate but interconnected components. In this chapter, requirements analysis has been conducted within the framework of a four-layer generic network analysis model. Accordingly, requirements are grouped into user, application, device, and network requirements. As we ascend the layered hierarchy towards the top user layer, the requirements become more subjective. By contrast, as we move down the layers towards the bottom network layer, the requirements becomes more technical and objective.

The requirements identified from the user, application, device, and network components are integrated to form a complete set of requirements. The resulting requirements are further analyzed for their prioritization and refinement, which include resolving conflicts and making trade-offs. Ultimately, detailed requirements specifications, along with a requirements map, are developed with technical trade-offs and constraints. They will serve network planning, specifically in relation to network architecture and design.

References

1. Oppenheimei, P.: Top-Down Network Design, 3rd edn. Cisco Press, Indianapolis, IN 46240, USA (2011). ISBN 978-1-58720-283-4
2. McCabe, J.D.: Network Analysis, Architecture, and Design, 3rd edn. Morgan Kaufmann Publishers, Burlington, MA 01803, USA (2007). ISBN 978-0-12-370480-1
3. Tian, Y.C., Levy, D.C.: Handbook of Real-Time Computing. Springer, Singapore 189721 (2022)
4. Bradner, S.: Key words for use in RFCs to indicate requirement levels. RFC 2119, RFC Editor (1997). https://doi.org/10.17487/RFC1112
5. Leiba, B.: Ambiguity of uppercase vs lowercase in rfc 2119 key words. RFC 8174, RFC Editor (2017). BCP 14. https://doi.org/10.17487/RFC1112

Chapter 4
Traffic Flow Analysis

Many of the requirements developed from the requirements analysis in the previous chapter are directly or indirectly related to performance for users, applications, devices, and the network in the four-layer generic network analysis model. Along with their location dependencies, they are affected by the patterns and behaviors of traffic flows. More importantly, the implementation of various Quality of Service (QoS) management mechanisms and strategies relies on traffic flow management. Therefore, traffic flow analysis is an important step in the development of flow requirements specifications. It is an integral part of requirements analysis in network planning projects.

Traffic flow analysis characterizes traffic flows within a network to understand the following aspects:

- *Traffic flows*: identifying where the flows will likely occur,
- *Traffic QoS*: determining what levels of QoS the flows will require,
- *Traffic models*: comprehending what types of traffic flows have been well understood,
- *Traffic measurement*: clarifying how traffic flows are measured and quantified,
- *Traffic load*: assessing how much load the flows carry,
- *Traffic behavior*: examining how the flows behave, and
- *Traffic management*: defining how the flows should be managed.

Through the process of flow analysis, a comprehensive set of flow requirements specifications will be identified. These requirements indicate where traffic flows are likely to occur and how flow requirements will combine and interact. They also offer valuable insights into network hierarchy and redundancy, and may even suggest interconnection strategies. More specifically, the developed flow specifications will be used later in network architecture planning, particularly for the management of network performance and QoS.

As we have already understood, the majority of network services are typically provisioned as best-effort services by default, implying that they do not offer any explicit

Y.-C. Tian and J. Gao, *Network Analysis and Architecture*, Signals and Communication Technology, https://doi.org/10.1007/978-981-99-5648-7_4

guarantees for performance or QoS. Therefore, there is no need to conduct a detailed analysis or characterization of every single traffic flow within a network. Instead, the focus should be on the flows that have the most significant impact on the network architecture and QoS management. These flows are mostly related to predictable and/or guaranteed services, which necessitate specific levels of QoS assurance.

With this understanding in mind, let us now embark on our discussion, starting from the fundamental concepts of traffic flows.

4.1 Traffic Flows

Traffic flows have been described from different perspectives. Let us examine in the following how traffic flows have been defined.

4.1.1 Concepts of Traffic Flows

In the IETF RFC 2722 [1, p. 5], a traffic flow is defined as "an artificial logical equivalent to a call or connection". It is "a portion of traffic, delimited by a start and stop time", that belongs to "a user, a host system, a network, a group of networks, a particular transport address (e.g. an IP port number), or any combination of the above". "Attribute values (source/destination addresses, packet counts, byte counts, etc.) associated with a flow are aggregate quantities reflecting events which take place in the DURATION between the start and stop times. The start time of a flow is fixed for a given flow; the stop time may increase with the age of the flow." [1, p. 5].

In IPv6 networks, the IETF RFC 3697 [2, p. 1] defines a flow as "a sequence of packets sent from a particular source to a particular unicast, anycast, or multicast destination that the source desires to label as a flow. A flow could consist of all packets in a specific transport connection or a media stream. However, a flow is not necessarily 1:1 mapped to a transport connection."

In the IETF RFC 3917 [3], which specifies the IP Flow Information eXport (IPFIX) [3, pp. 3–4], a flow is defined as "a set of IP packets passing an observation point in a network during a certain time interval. All packets belonging to a particular flow have a set of common properties." The flow *properties* mentioned here refer to flow *attributes* that are described in the IETF RFC 2722 [1]. Each property, or attribute, results from

- The packet header field (e.g., destination IP address), transport header field (e.g., destination port number), or application header field,
- The characteristics of the packet itself, such as QoS levels, and/or
- The fields derived from packet treatment, e.g., the next hop IP address.

A packet is considered to belong to a flow if it fully satisfies all the defined properties of that flow. For example, traffic originating from the same application, e.g., video streaming or VoIP, can be effectively managed within a single flow. Similarly, traffic with identical QoS requirements can be grouped together within the same flow, facilitating streamlined QoS management.

From the above discussions, it is seen that the concept of traffic flows is applied to "any protocol, using address attributes in any combination at the *adjacent*, network, and transport layers of the network stack" [1, p. 3]. The term *Adjacent* refers to "the next layer down in a particular instantiation of protocol layering" though it usually means the link layer [1, p. 3]. More specifically:

- A traffic flow is a sequence of packets with *common attributes* or properties, which "are defined in such a way that they are valid for multiple network protocol stacks" [1, p. 3],
- It is measured during *a time interval* and in *a single session* of an application,
- It is characterized at *an observation point*, and
- It is *end-to-end* between source and destination applications/devices/users.

To broaden the scope of Real-Time Flow Measurement (RTFM) beyond simple traffic volume measurements defined in the IETF RFC 2722 [1], new flow attributes are introduced in the IETF RFC 2724 [4]. They include performance attributes, such as throughput, packet loss, delays, jitter, and congestion measures. These performance attributes are calculated as extensions to the RTFM flow attributes according to the following three general classes [4, p. 5]:

- *Trace*: Attributes of individual packets within a flow or a segment of a flow, e.g., last packet size,
- *Aggregate*: Attributes derived from the flow considered as a whole, e.g., mean rate, and
- *Group*: Attributes calculated from groups of packet values within the flow, e.g., inter-arrival times.

Figure 4.1 illustrates a traffic flow over a network. Since traffic flows are end-to-end, they can also be examined and evaluated on a link-to-link or network-to-network basis. This will help integrate flow requirements at the link or network level.

Most flows are bidirectional, with the same set of attributes for both directions or a different set of attributes for each direction. However, there are instances where flows are unidirectional with a single set of attributes. The flow shown in Fig. 4.1 represents an example of such a unidirectional flow.

4.1.2 Individual and Composite Flows

The traffic flow shown in Fig. 4.1 captures the features of an individual flow. *An individual flow* is defined as a flow of protocol and application information transmitted over the network during a single session. It is the basic unit of traffic flows, and can be aggregated with other individual flows to form aggregated traffic flows, such as composite flows, which will be discussed later.

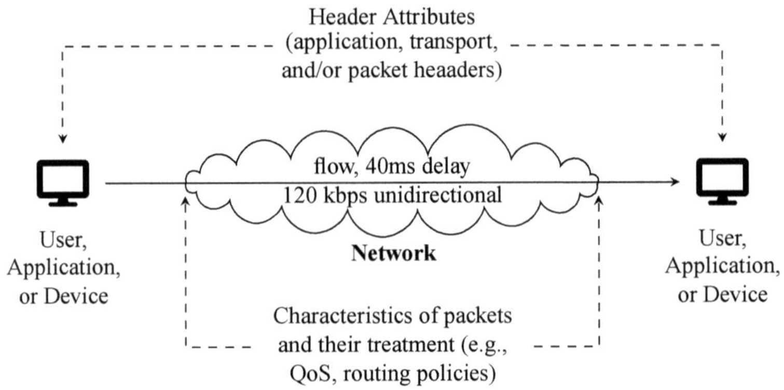

Fig. 4.1 A traffic flow with attributes applied end-to-end and over network

An individual flow is considered when there are separate requirements unique to that flow. For example, in the case of the flow depicted in Fig. 4.1, where a guaranteed 40 ms delay and 120 kbps bandwidth are required, it is necessary to treat this individual flow separately from others. Mechanisms and strategies must be designed to meet the requirements of this flow.

The attributes of an individual flow can be characterized by factors such as flow direction, routing features, address information, the numbers of packets, and the number of bytes or bits. As mentioned earlier, these flow attributes are described in detail in the IETF RFC 3917 [3, pp. 3–4]. Among these attributes used to characterize traffic flows, the simplest one is to count the number of bits transmitted per second between the communicating entities, e.g., 120 kbps. This information is employed to quantify the bandwidth requirement for a traffic flow or a network link when all traffic flows passing this link are considered as a whole.

When the requirements of multiple individual flows that share a same link, path, or network are considered together, these flows are aggregated to form *a composite flow*. Figure 4.2 shows an example of a composite flow, which is formed by aggregating three individual flows:

1. Individual flow 1: 120kbps upstream, 500 kbps downstream,
2. Individual flow 2: unidirectional, 1 Mbps, 20 ms one-way delay, and
3. Individual flow 3: 10 Mbps bidirectional, 100 ms round-trip delay.

These three individual flows may originate from the same source or different sources. They can be directed towards the same destination or different destinations. However, they are merged at the same observation point in the network, i.e., at router R1 in Fig. 4.2, to form a composite flow.

The concepts of *upstream* and *downstream* traffic are essential to the understanding of traffic flows in a network. Traffic flowing *upstream* indicates the direction towards the core of the network in general. By contrast, traffic flowing *downstream* refers to the direction towards the edge of the network. For example, when downloading files from the Internet, the data traffic flows downstream towards the user. In

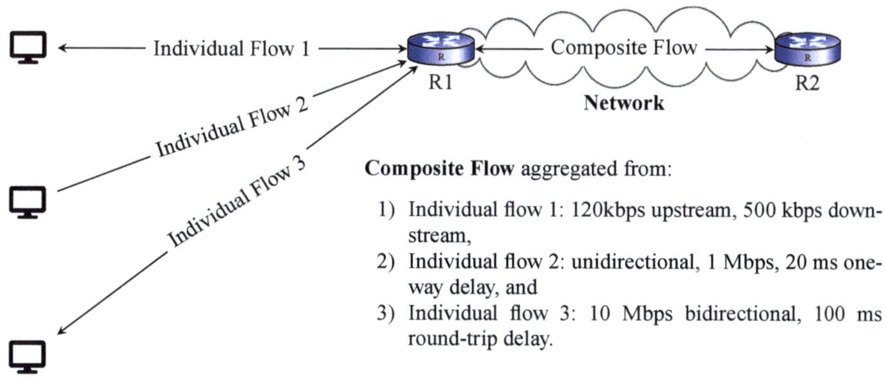

Fig. 4.2 A composite flow aggregated from individual flows

comparison, when uploading a file to a File Transfer Protocol (FTP) server over the Internet, the traffic flows upstream from the user. Internet access speeds are typically asymmetrical with the upstream rate being considerably slower than the downstream rate.

In the process of flow analysis, not all traffic flows need to be considered. Therefore, it is important to clarify questions such as:

- Which individual flows should be considered?
- Where and what requirements should be applied to these flows,
- When do these flows contribute to composite flows, and
- How are the requirements of individual flows that contribute to the composite flows aggregated to form the requirements for the composite flows?

By addressing these questions, flow analysis can focus on the flows that have the most significant impact on the network architecture and QoS management.

4.1.3 *Critical Flows*

Critical flows are traffic flows that are considered more important than others and, as a result, have higher levels of QoS requirements. Traffic flows associated with mission-critical, safety-critical, and time-critical services and applications are typical examples of critical flows. Traffic flows that serve more important users, applications, and and devices can also be treated as critical flows, even if they are not as critical as mission-critical, safety-critical, or time-critical flows. Traffic flows that have high, predictable, and guaranteed performance requirements drive the architectural design of a network, especially in service-based networking. They dictate resource allocation, prioritization schemes, and overall network QoS management to ensure that the required level of network performance is maintained.

4.1.4 Flow Sources and Sinks

Identifying the sources and sinks of traffic flows will help characterize the flows and determine their directions. A *flow source* or *data source* is a network entity where a traffic flow originates. A *flow sink* or *data sink* is a network entity where a traffic flow terminates. Graphically, a flow source can be represented by a circled dot, whereas a flow sink can be denoted by a circled asterisk, as shown in Fig. 4.3. Figure 4.4 provides some examples of flow sources and sinks.

In a computer network, some network devices only generate data and traffic flows, making them pure flow sources. A typical example of such devices is video cameras used in a networked surveillance system.

There are also network devices that only consume data and terminate traffic flows, making them pure flow sinks. Video monitors in a network surveillance system are examples of pure data sinks. Global Positioning System (GPS) devices used in general vehicles receive satellite signals for positioning and navigation. They are pure data sinks. However, high-end GPS devices may also communicate with satellite base stations and other network devices, and thus are not pure flow sinks.

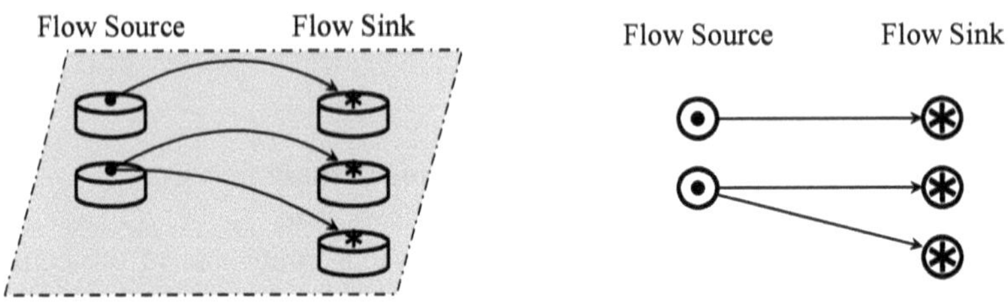

(a) Three-dimensional representations (b) Two-dimensional representations

Fig. 4.3 Representations for flow sources and sinks

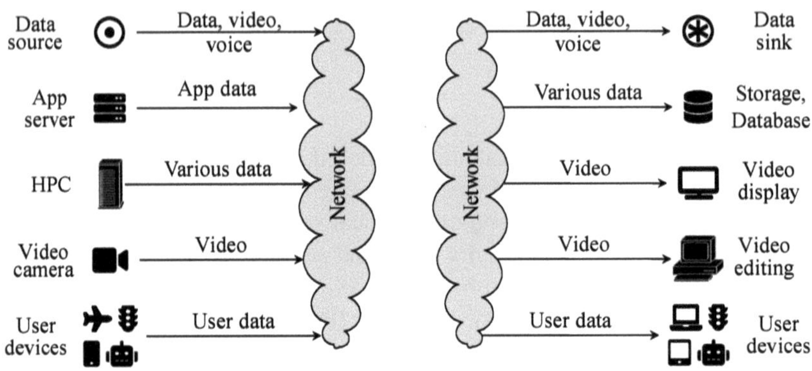

(a) Flow sources (b) Flow sinks

Fig. 4.4 Examples of flow sources and sinks

Almost all network devices in a network generate and consume data, and thus play dual roles as both flow sources and sinks. For example, consider a web server that receives requests from visitors and responds by providing the requested information. When receiving requests, the web server functions as a flow sink. However, when sending out information in response to the requests, it acts as a flow source. As the traffic of web server responses is generally much larger than that of requests, the web server primarily acts a flow source. Similarly, a storage server primarily functions as a flow sink because the incoming traffic directed towards the server is generally much higher than the outgoing traffic originating from the server.

4.1.5 Flow Boundary

To accurately characterize traffic flows, it is essential to identify and define *flow boundaries*. The concept of flow boundaries offer several benefits including:

- Assisting in traffic flow consolidation, which is particularly useful when managing the QoS of traffic flows through traffic flow aggregation, e.g., in a Per-Hop Behavior (PHB) manner in Differentiated Service (DiffServ) [5].
- Geographical separation of the network for network performance and QoS management. The separation can be:
  - LAN/MAN/WAN,
  - Multiple physical sites, buildings, or floors,
  - Internet or cloud service providers, backbones, and access points,
  - Specialized areas or regions, e.g., an HPC cluster for scientific computing, a computer lab for the reception and processing of satellite signals, and a DeMilitarized Zone (DMZ) for security management.
- Logical separation of the network to understand where flows can be consolidated or what specific requirements are needed for certain flows. The logical separation can include:
  - Multiple flows transiting a region or area,
  - The convergence of traffic flows, e.g., for Internet access points,
  - WANs where Internet and cloud service providers may be involved,
  - Specific services that may be required,

An example of using the concept of flow boundaries to identify traffic flows is depicted in Fig. 4.5, which is self-explained.

When dealing with traffic flows from LAN to MAN, how do we know how much traffic will remain within the LAN and how much will be routed to WAN? In case where no additional information is available about flow distribution between the LAN and WAN, the traditional 80/20 rule may be employed, i.e., 80% of the traffic will stay within the LAN and the remaining 20% will go to WAN. However, with the increasing deployment of cloud-based computing, it may be necessary to modify

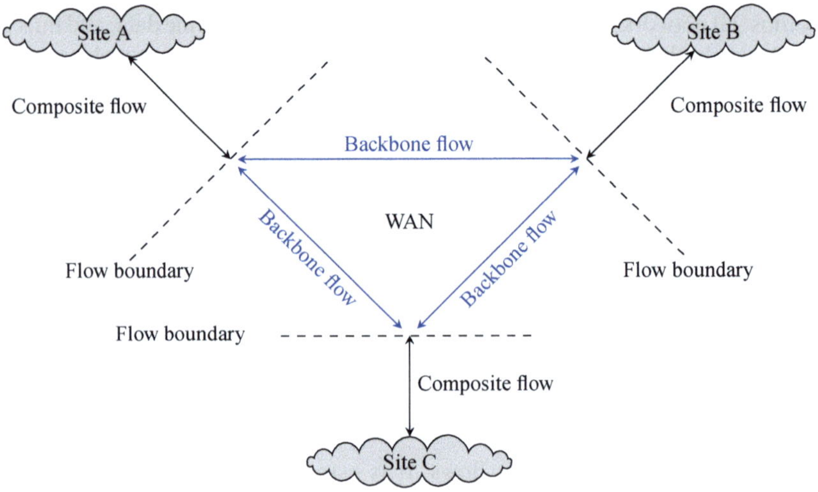

Fig. 4.5 Flow boundaries in a network with three sites

this 80/20 rule. This is because, from the user's perspective, it is often transparent where and how much resources come from the WAN, or users do not care where the resources come from as long as they are readily available. Consequently, the growing demand for WAN resources may require a revised rule, such as a 50/50 split or even a 20/80 distribution. The decision to use local network resources within the LAN or remote resources through the WAN should consider various factors such as cost, performance, and other specific requirements.

4.1.6 Identifying Traffic Flows

In general, traffic flows can be identified based on the requirements specifications developed in the previous chapter, i.e., the requirements for users, applications, devices, and the network, as well as location dependencies and performance requirements. There are various approaches to identifying traffic flows. These approaches can be used separately and in combination. Here are some examples of different ways to identify traffic flows:

- Focus on specific devices or functions such as flow sources and sinks (e.g., video cameras that send videos to a central server).
- Identify traffic flows associated with a particular application or multiple applications used by a single user or a group of users, e.g., web browsing, mail services, and file transfer for all staff in a company. For sets of applications that ae common to a group of users, or sets of traffic flows with similar requirements, a require-

Table 4.1 Flow characteristics of applications [7, p. 96]

Name of application	Type of flow	Protocols used by application	User or user group	Data stores (servers, hosts, etc)	Bandwidth requirements	QoS requirements
App1						
...						
...						

ments profile, also known as a template, can be developed to simplify requirements descriptions [6, pp. 172–173].

- Identify top *N* applications that are critical with higher performance requirements compared to others and thus will drive network planning. For example, Canvas and other related applications for teaching and learning are among mission-critical applications for a university.

It is a common recommendation from both references [6, p. 168] and [7, p. 95] that traffic flows be identified and documented from the application perspective. It is important to understand that flow identification is a high-level network analysis for network architectural design. Therefore, it should be independent of specific network technologies and their physical implementations.

A requirements template used in [7] to document traffic flows for applications is presented in Table 4.1. When filling out this template, select a flow mode from the well-established flow models, namely peer-to-peer, client-server, hierarchical client-server, and distributed computing, as the type of flow. These flow models will be discussed later in this chapter. The QoS requirements listed in the last column of Table 4.1 will be described in detail in the next section.

4.2 QoS Requirements

In computer networking, QoS is a fundamental concept for network performance and management. It drives service-based networking. This section aims to provide a clear understanding of QoS in computer networking by addressing the following aspects: the concept, importance, and objective of QoS, as well as the characterization and prioritization of traffic flows from the QoS perspective.

4.2.1 The Concept of QoS

The concept of QoS can vary in its interpretation across different application domains or use cases. In telephony systems, QoS refers to the overall performance of the services that are provided by the system. It is quantified using various performance metrics. These metrics include packet loss rate, bit rate, throughput, latency, jitter, reliability, and availability. Quite often, this interpretation of QoS is also applied in computer networking to indicate the quality of various services provided by a network.

In computer networking, QoS specifically refers to a set of mechanisms and technologies to control network traffic and related resources such that the performance of high-priority and critical applications is ensured within the resource capacity of the network. This interpretation of QoS is adopted in this chapter for discussing performance architecture of computer networks.

QoS is a feature of routers and switches. Therefore, QoS mechanisms are implemented in routers and switches to prioritize traffic and control resources so that more important traffic can pass first over less important traffic. Traffic with QoS requirements should be marked by te applications that generate the traffic. When routers and switches receive the marked traffic, they will be able to categorize the traffic into different groups for QoS management.

Understandably, it is necessary to characterize, quantify, measure, and manage the achieved performance quality of a computer network with QoS control, as well as the quality of the applications that the network serves. This should be addressed in conjunction with performance management as part of overall network management architecture, which will be discussed later in a separate chapter. Various performance metrics are used for this purpose, such as packet loss rate, bit rate, throughput, latency, and jitter, as mentioned above.

4.2.2 The Importance of QoS

In computer networking, network service requests from end users, applications, and devices are supported through network service offerings by the network. By default, network service offerings are provisioned as best-effort service responses. This means that the network does not differentiate traffic flows from different users, applications, and devices. As a result, the network simply offers whatever network resources that are available at any given time without providing any performance guarantee to network services. As the available network resources fluctuate over time, the best-effort allocation of network resources to each user, application, and device also changes from time to time. Therefore, it becomes the responsibility of the system or underlying applications to adapt their traffic flows to the available services. The flow control function of TCP is an example of self-adaptation to a dynamic network environment.

While the majority of network services are best-effort services, there are also network services that are time-critical, safety-critical, and mission-critical services. VoIP and video streaming services integrated with data networks are typical examples, which are sensitive to delay and jitter. Due to their real-time requirements, both VoIP and video streaming use UDP for the transport of voice and video datagrams. In the best-effort service environment, the delay and jitter may exceed their maximum tolerable thresholds, resulting in poor or even unacceptable quality of voice and video. More severely, if the loss rate of datagrams becomes high, as there is no mechanism for re-transmitting the lost datagrams in UDP, the voice and video services may become completely unusable. Even if a re-transmission strategy is designed at the application layer, the re-transmitted datagrams may arrive too late to be useful at the receiver.

Traditionally, over-provisioning network bandwidth has been used to partially address such problems in network services with specific performance requirements. When network utilization, which is the used bandwidth relative to the total available bandwidth, is relatively low, burst traffic in the network can be handled without a major impact on the performance of voice, video, and other services with performance requirements. Thus, throwing more bandwidth is helpful to mitigate some network service problems.

However, due to the limited network resources, meeting the performance requirements for some network services remains a challenge. This challenge become more severe when more critical network services are integrated into the data network. Therefore, there is a demand for systematic mechanisms to control and manage the performance of network services. This is where QoS comes into play.

The most common use cases for QoS in computer networking are voice and video streams, as discussed earlier. In addition to voice and video services, many other network services also require QoS, particularly for real-time applications as well as safety-critical and mission-critical systems. For example, in large-scale manufacturing and agricultural applications, numerous real-time monitoring and control tasks operate with the support of network communications integrated with industrial Internet and IoT networking. Some tasks have higher priority than others, such as sending out emergency commands versus receiving regular sensing measurement data. With QoS management, tasks with higher priority can be executed earlier than other tasks.

As the demand for network connectivity continues to increase, more and more network services are being provisioned over networks. This trend motivates and necessitates the increasing deployment of QoS control in computer networks to provide differentiated services to various end users, applications, and devices. Therefore, QoS will play an increasingly important role in future network systems, ensuring that certain data streams are handled with higher priority over others within the given network resource capacity.

4.2.3 *The Objective of QoS*

In general, the intention of QoS in computer networking can be categorized into two types of responses: application-centric responses and network-centric responses. These two types of responses are described in the IETF RFC 2990 [8, pp. 18–19]. They are briefly discussed below.

The application-centric responses summarized in RFC 2990 include:

- To control the network service response such that
  - The response to a specific service element is consistent and predictable; and/or
  - A service element is provided with a level of response equal to, or better than, a guaranteed minimum;
- To allow a service element to establish in advance the service response that can be obtained from the network.

The network-centric responses described in RFC 2990 are:

- To control the contention for network resources such that
  - A service element is provided with a superior level of network resource; and/or
  - A service element has a fair share of resources;
- To allow for efficient total utilization of network resources while servicing a variety of directed network service outcomes.

It is worth pointing out that none of the above-described responses can be addressed alone without considering other responses in any effective QoS architecture. A few or all of these responses need to be considered in computer networking to fulfill the overall QoS control requirements.

4.2.4 QoS Requirements of Flows

For identified traffic flows, it is not sufficient in network planning to know only simple QoS characteristics such as load (bandwidth) and behavior, which will be discussed later in this chapter. As network services and QoS requirements drive service-based networking, it is important to comprehensively understand QoS requirements, particularly for critical flows and applications.

The majority of network services and applications in a network are served with best effort by default. No specific QoS requirements will be applied to these best-effort services. If sufficient bandwidth is available, they may perform well. But if not, they may function poorly. This is the behavior that we would expect for these services and applications. For example, web browsing and mail services are generally deployed as best-effort services.

Some flows or applications can be served with service differentiation such as the DiffServ mechanism specified in the IETF RFC 2475 [5]. DiffServ is a layer-3 QoS control mechanism. It does not support end-to-end QoS management for individual flows. Rather, it aggregates traffic flows between two hops and manages the aggregated flows in a PHB manner. This is particularly suitable for soft-real-time services and applications.

There are flows that require end-to-end QoS support. These flows can be served with Integrated Service (IntServ), which is originally defined in the IETF RFC

1633 [9]. Like DiffServ, IntServe is also a layer-3 QoS control mechanism. But different from DiffServ that manages flows in a PHB manner, IntServ manages each individual flow from end to end. Therefore, it requires all routers along the path of the flow to reach an agreement on reserving resources, e.g., bandwidth, for the flow. For this purpose, a signaling system is essential for communicating the flow requirements with all participating routers. Obviously, it is also essential that all participating routers must be able to reserve the required resources. Due to its end-to-end flow support, IntServ is well suitable for hard-real-time services and applications such as those with QoS requirements for guaranteed flows.

In IntServ, two major classes of QoS services have been defined: *guaranteed service* and *controlled-load service*:

- *Guaranteed service* is introduced with IntServ. It is defined in the IETF RFC 2212 [10]. Its objective is to provide "firm (mathematically provable) bounds" on the queuing delays that a packet will experience in a router. As a result, guaranteed service guarantees both delay and bandwidth for a flow.
- *Controlled-load service* is defined in the IETF RFC 2211 [11]. It aims to provide the client traffic flow with a QoS "closely approximating the QoS that same flow would receive from an unloaded network element". It operates effectively for the served flow regardless of the traffic load of the router through which the flow is passing. Thus, admission control is used to ensure that the controlled-load service performs well even if the router is heavily loaded or overloaded. Controller-load service does not specify any specific performance guarantees, making it suitable for real-time multimedia applications such as video streaming.

More details of the controlled-load service and guaranteed service developed for IntServ will be discussed later in the performance component architecture of this book.

4.2.5 Flow Prioritization

From the QoS requirements, it is seen that some flows are more important than others. Therefore, the technique of flow prioritization is used to formally determine which flows should receive the most resources or which flows should be allocated resources first. While network engineers can always request more funding for additional network resources, it is important to acknowledge that network resources are not infinite in the real world, as discussed in the IETF RFC 1633 [9, p. 4]. Consequently, flow prioritization should be conducted under the constraints of network resources. The following discussions on flow prioritization does not consider any budget constraints in relation to the acquisition of additional network resources.

Critical flows, such as mission-critical, safety-critical, and time-critical flows, should ne assigned higher levels of priority than other flows. They may not necessarily require a significant amount of resources, but will require service guarantees such as an upper delay bound or a lower bandwidth bound. This necessitates the use of guaranteed service provided by IntServ. Moreover, among the critical flows, certain flows may be more important than others.

For traffic flows with soft QoS requirements, lower levels of priority can be assigned than those for critical flows. Typical examples include non-critical video streaming and other multimedia flows. Depending on the application scenarios, they can be managed through controlled-load service within the DiffServ framework if end-to-end support is not necessary, or through guaranteed service within the IntServ framework if end-to-end support is essential.

Traffic flows without specific QoS requirements can be treated as best-effort flows, as we have already understood. They can be managed through best-effort services. When the path of a flow experiences heavy traffic load, the service provided to the flow may be much slower than it would be under normal traffic conditions.

There are layer-2 and layer-3 mechanisms for flow prioritization. The basic idea behind flow prioritization mechanisms is to mark the outgoing traffic flows before they are transmitted. At layer 2, a 3-bit Priority Code Point (PCP) in the frame header represents eight levels of priority for data frames. In comparison, at layer 3, a 6-bit Differentiated Services field CodePoint (DSCP) is embedded into the IP header to differentiate 64 levels of priority for data packets. What flow prioritization mechanisms should be chosen and how to control the marked traffic flows with QoS requirements will be the topics of performance component architecture, which will be covered later in a separate chapter of this book.

4.3 Traffic Flow Models

There are a few well-established traffic flow models, each showing specific and consistent flow behaviors. An effective approach to flow analysis is to map network flows to one of these flow models. This section introduces the peer-to-peer, client-server, hierarchical client-server, and distributed-computing flow models. These flow models have been characterized with examples in the book by McCabe [6, pp. 180–191].

4.3.1 Peer-to-Peer Flow Model

The *peer-to-peer* flow model describes fairly consistent traffic flows in a physically or logically flat network topology. Shown in Fig. 4.6, it exhibits the following two key features:

- No peers are more important than others.
- There is no manager among the peers.

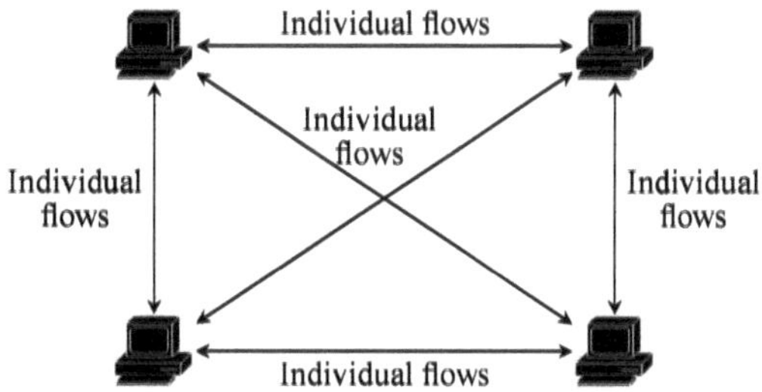

Fig. 4.6 Peer-to-peer flow model in which all flows are individual flows

As a result, all flows among the peers are considered to be equal in terms of their importance. Either all or none of the flows are critical. A single set of flow requirements, known as a profile, applies to all flows.

Typical examples of peer-to-peer flow behavior include:

- Web browsing across the Internet: A huge number of users browse numerous websites over the Internet. No users or websites are considered to be more important than others, and there are no managers to coordinate web browsing activities over the Internet.
- FTP services over the Internet: This is similar to web browsing over the Internet. A large number of FTP servers serve a vast user base for file transfer across the Internet. No FTP servers or users are more important than others.
- Email services over the Internet: This is also similar to web browsing over the Internet. A vast number of email users use a large number of email servers over the Internet for email services. Nobody on the Internet is able to coordinate such email services.
- Social networking among peers: Examples include Twitter, Facebook, TikTok, and WeChat. While these social networks have servers as back-end infrastructure support, users over the Internet are considered to be equally important. They engage in peer-to-peer conversations where no one is more important than others. Traffic flows to or from each user are treated without differentiation.
- A combination of web browsing, FTP, email services, and social networking over the Internet.

4.3.2 Client-Server Flow Model

The *client-server* flow model is graphically depicted in Fig. 4.7. It is the most generally applicable traffic flow model in computer networking. The client-server model consists of a centralized server and one or more distributed clients. It has a request/response communication feature with bidirectional traffic flows:

- Clients send requests to the server for data and services, and the server responds to the requests by providing data and services to the clients.

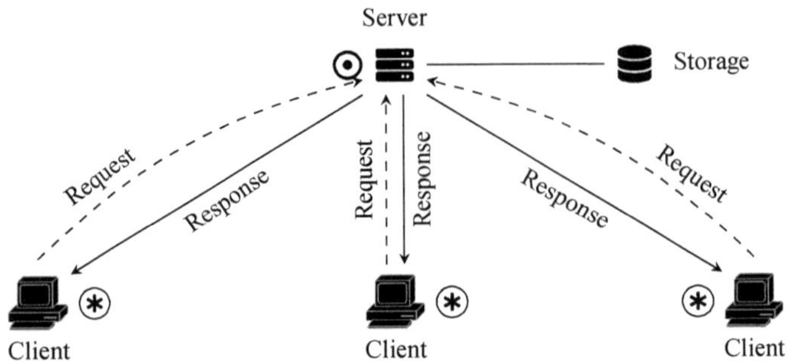

Fig. 4.7 Client-server flow model with asymmetric traffic flows [6, p. 184]. The traffic of responses is much bigger than that of requests. Thus, the server is more likely a flow source whereas the clients are more likely flow sinks

- The downstream traffic of the responses from the server to the clients is typically much bigger than the upstream traffic of the requests from the clients. Therefore, the bidirectional upstream ad downstream traffic flows are asymmetric. As a result, the server is considered more likely as a flow source whereas the clients act more likely as flow sinks.
- In comparison with the requests from the clients, the responses from the server are more important because they are being expected by the clients.

Many network services and applications fit well into the client-server model. A typical example is user access to the centralized servers like web server, FTP server, mail server, and other servers within an enterprise network. In this model, users send requests to the server for data and services, and the server responds to the user by providing the requested data and services. This necessitates highly reliable servers, secure communication for requests and responses, and the ability to handle large downstream traffic. Therefore, a network must possess the capability to support client-server services and applications. It is worth noting that certain services and applications, such as web browsing, FTP downloading, and mail services, can fit into different flow models depending on specific scenarios. While they operate in a client-server manner within an enterprise network, they exhibit peer-to-peer traffic flows across the Internet, as discussed previously in relation to the peer-to-peer flow model.

Additional examples of the client-server flow model include various web-based or similar applications, such as Overleaf as an online LaTeX editor, SAP as an Enterprise Resource Planning (ERP) application, cloud services from cloud service providers, GitHub as a code hosting platform, arXiv as an open-access repository of electronic preprints and postprints, and ChatGPT as an online chatbot (launched on the 30th of November in 2022). These applications have centralized servers to offer data and services to distributed clients worldwide. Communication between the server and clients occurs through requests and responses, with traffic flows being bidirectional and typically asymmetric.

Terminal-host traffic flows were popular many years ago. They can be considered as a special type of client-server traffic flows in modern computer networks. They appear in the communications between a mainframe computer and its remote terminals, as well as in other Telnet applications. Typically, a terminal sends a single or a few characters to the host, and the host responds with many characters. Thus, terminal-host traffic flows are typically asymmetric. However, there are instances where a terminal sends a character to the host and receives a character in return, such as in the *vi* editor. There are also scenarios where a complete screen is updated at a time, for instance, in some mainframes like IBM-3270. Therefore, the efficiency performance of terminal-host traffic flows may vary depending on the specific application. Further discussions on terminal-host traffic flows can be found in the book by Oppenheimer [7, p. 91].

4.3.3 Hierarchical Client-Server Flow Model

When more tiers of hierarchy are added to the client-server model, the characteristics of network communication traffic can be better described using a *hierarchical client-server* flow model, which is also known as a *cooperative computing* flow model. This is illustrated in the logical diagram of Fig. 4.8. In the upper tiers of the hierarchical client-server flow model, there exist multiple hierarchical tiers of servers. These servers engage in communication with one another, and function as both flow sources and sinks. The top-tier server performs as the global manager. At the lowest tier of servers, each server serves one or more clients within a client-server flow model, forming multiple client-server systems.

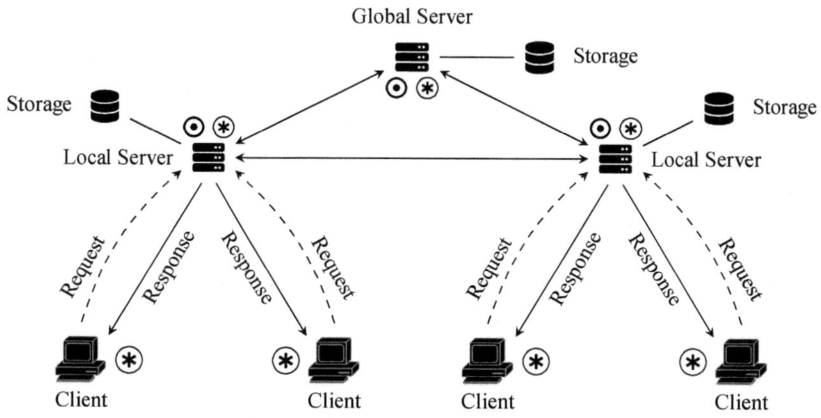

Fig. 4.8 Hierarchical client-server flow model [6, p. 186], also known as cooperative computing flow model, with two or more hierarchical tiers of servers acting as both flow sources and sinks. The distributed clients act more likely as flow sinks

As in the simple client-server flow model, the server-to-client traffic flows are considered more important than the client-to-server flows in the hierarchical client-server flow model. In addition, without more detailed information, the server-to-server traffic flows are also considered more important than the client-to-server flows.

With the increasing deployment of web applications over the Internet, the demands for server reliability and performance are growing. As a result, many web servers are replicated and then distributed across various physical locations. These servers are often managed and coordinated by a global server. As a result, numerous applications that were originally served within the client-server model are now being served within a hierarchical client-server model. In this model, the traffic flows between servers, and servers and managers, become more important than before in ensuring the full functionality, reliability, and security of web applications.

In the hierarchical client-server model, servers at multiple tiers may offer similar functions and services as discussed earlier. It is also possible for them to provide different functions and services. For example, one server may primarily be used for scientific computing, while another may be predominately used for e-commerce. Nevertheless, the servers communicate with each other for resource sharing, data replication, task migration, and other purposes.

4.3.4 Distributed-Computing Flow Model

The distributed-computing flow model is the most specialized flow model. Briefly speaking, depending on the application scenarios, it may exhibit traffic flow characteristics that resemble a combination of both peer-to-peer and client-server flow models, or demonstrate the opposite characteristics of the traffic flows observed in the client-server flow model. A logical diagram of the distributed-computing flow model is presented in Fig. 4.9. This model comprises a central task manager and multiple distributed computing nodes. The task manager assumes responsibility for managing the overall computing task, dispatching subtasks to the distributed computing nodes, and collecting computing results from them. The distributed computing nodes conduct the computations assigned by the task manager.

Let us have a detailed discussion on the features of the distributed-computing flow model. First of all, distributed computing is a specialized computing method that deals with the computation of computing tasks across multiple distributed computing nodes. It decomposes a computing task into multiple smaller subtasks and then distributes these subtasks to different computing nodes for computation. Then, it collects the computing results from the distributed computing nodes and derives the final computing result for the overall computing task.

There are several scenarios that necessitate distributed computing. Here are a few examples:

- When the data required for a computing task are distributed in multiple nodes and cannot be consolidated onto a single node due to factors such as data ownership

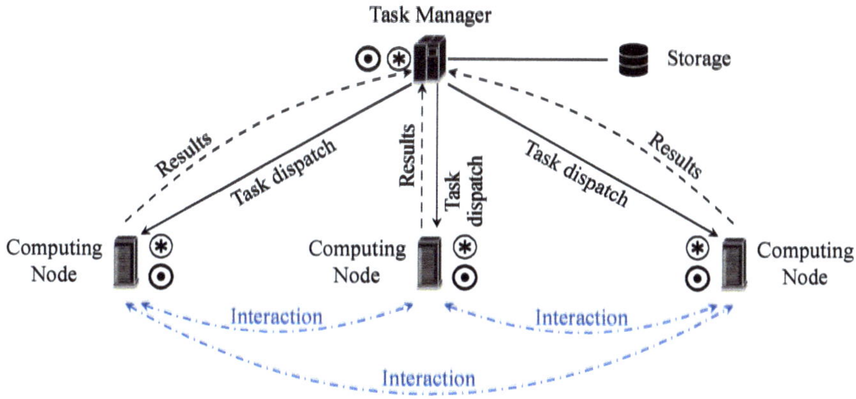

Fig. 4.9 Distributed-computing flow model with a task manager and multiple computing nodes [6, p. 189]. Depending on the application scenarios, the task manager can be either or both of flow sources and sinks, and so does each of the computing nodes. Direct interactions among the computing nodes may or may not exist for information exchange

or other constraints, distributed computing becomes the only viable approach to processing the distributed data and achieving the desired computing objective.

- In case where the datasets for a computing task are huge in volume, it may become physically or technically infeasible to transfer all the required data to a central storage server for management and further processing. Distributed computing is required to handle the computing task effectively and efficiently.
- Some computing tasks are too big to be executed on a single computing node. An example is the all-to-all comparison problem [12, 13], which involves pairwise comparisons of all data items, e.g., for the computation of a similarity matrix. When the number of data items is huge, the computation requirements become resource-intensive, thereby motivating the use of distributed computing.

A distributed computing task can be data-intensive, computing-intensive, or both.

- For a computing-intensive task, the task manager does not have a significant amount of data to distribute to the computing nodes. Rather, it may need to collect a substantial amount of data from the computing nodes. In this scenario, the task manager primarily acts as a flow sink, while the computing nodes mainly function as flow sources.
- In the case of data-intensive distributed computing, e.g., for the all-to-all comparison problem [12, 13], a large volume of data needs to be distributed to the distributed computing nodes. As a result, the task manager assumes the role of both flow source and sink, either simultaneously or at different stages.

Distributed computing systems typically involve interactions between any pairs of the distributed computing nodes for information exchange, as shown in the lower part of Fig. 4.9. When the subtasks executed on the computing nodes are tightly coupled, these interactions become necessary to simplify the distributed computing algorithms and speed up the overall execution of the computing. If the subtasks are additionally fine granulated, the distributed computing system behaves akin to a parallel computing system. In such cases, the task manager may dynamically dispatch subtasks to the computing nodes.

When the subtasks dispatched to the distributed computing nodes are loosely coupled, interactions among the computing nodes may not be necessary. Even if some interactions are needed, they can be achieved indirectly through the task manager. If the subtasks additionally have a coarse granularity, it is feasible to statically allocate the subtasks to the computing nodes at the beginning of the overall computing task based on the computing capacity and resources of the nodes. This distributed computing scenario resembles cluster computing.

Regarding the traffic flows in the distributed-computing flow model, the task manager actively dispatches subtasks to the distributed computing nodes and passively receives results from them. This is different from the client-server flow model, in which clients send requests to the server and receives responses from it. However, in terms of one-to-many communication with asymmetric traffic flows, both the distributed-computing and client-server flow models behave similarly. As for the interactions appearing in the distributed-computing flow model, their traffic flows are similar to those in the peer-to-peer flow model. No single interaction is more important than others, and none of the computing nodes manage all these interactions.

4.4 Traffic Flow Measurement

The RTFM Working Group (WG) has developed a system for measuring and reporting information about traffic flows over the Internet. The system is specified in the IETF RFC 2722 [1]. Measuring traffic flows serves several purposes. Here are some use cases:

- To characterize and understand the behavior of existing networks,
- To plan for network development and expansion,
- To quantify network and application performance,
- To verify the quality of network service, and
- To attribute network usage to users and applications.

This section discusses how traffic flows are measured.

4.4.1 *Flow Measurement Architecture*

The RTFM architecture, as specified in the IETF RFC 2722 [1], is illustrated in Fig. 4.10. It consists of four main components: meter, meter reader, manager, and analysis application. Each component has dedicated functions and responsibilities, which are described in the following.

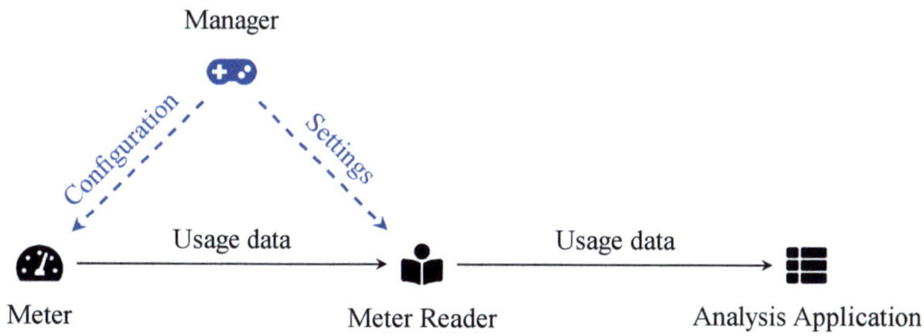

Fig. 4.10 The real-time flow measurement architecture [1, p. 6]

Meters are placed at flow measurement points in order to (1) observe packets as they pass by the points on their way through the network and (2) classify them into certain groups. A group may correspond to a user, a host, a network, a group of networks, a transport address, or any combination of the above. For each of such groups, a meter will accumulate relevant attributes for the group. Each meter selectively records network activity as directed by its configuration that is set by the manager. It can also process the recorded activity, such as aggregation and transformation before the data is stored.

A *meter reader* is responsible for reading and transporting a full copy, or a subset of, usage data from one or multiple meters so that the data is available to analysis applications. What, where, when, and how to read will be directed by the configuration set by the manager.

A flow measurement *manager* is an application that configures meter entities and meter reader entities. It sends configuration commands to the meters, and supervises each meter and meter reader for their proper operation. For convenience, the functions of meter reader and manager may be combined into a single network entity. It is worth mentioning that the manager of a meter is the master of the meter. Therefore, the parameters of the meter can only be set by the manager.

An *analysis application* analyzes and processes the collected usage data, and reports useful information for network management and engineering. For example, the following information may be reported: traffic flow matrices (e.g., total flow rates of many paths), flow rate frequency distribution (i.e., flow rates over a duration of time), and usage data showing the total amount of traffic to and from specific hosts or users. These reports assist in network monitoring, planning, and optimization.

4.4.2 Interactions Between Measurement Components

To better understand the operation of the flow measurement system, let us consider the interactions between the flow measurement components as depicted in Fig. 4.10. The interactions between a meter and a meter reader involve the transfer of usage date captured by the meter. This data is organized in a Flow Table. The meter reader can read a full copy or a subset of the usage data by using a file transfer protocol. The subset of the usage data can be a reduced number of records with all attributes, or all records with a reduced number of attributes. A meter reader may collect usage data from one or multiple meters.

The flow measurement manager is responsible for configuring and controlling flow meters and flow meter readers. It sends configuration and setting information to the meters and meter readers. The configuration for each meter includes the following aspects [1, pp. 7-8]:

- Flow specifications indicating which flows are to be measured, how they are aggregated, and what data the meter is required to compute for each flow being measured;
- Meter control parameters such as the inactivity time for flows; and
- Sampling behavior, which determines whether all packets passing through the measurement point are observed or only a subset of them.

It is worth mentioning that a meter can execute several rule sets concurrently on behalf of one or multiple managers.

The configuration for each meter reader includes specific information about the meter from which usage data is to be collected. This information is defined in the IETF RFC 2722 [1, p. 8]:

- The unique identity of the meter, i.e., the meter's network name and address,
- How frequently usage data is to be collected from the meter,
- Which flow records are to be collected, and
- What attributes are to be collected for the above flow records.

4.4.3 Multiple Managers, Meters, and Meter Readers

It is feasible to have multiple managers, meters, and meter readers in traffic flow measurement for the same network. This allows for more flexibility and scalability in managing and collecting usage data from different parts of the network. In the example depicted in Fig. 4.11, Meter 1, Meters 2 and 3, and Meter 3 are placed in three separate network segments, respectively. Manager A configures and controls Meters 1, 2, and 4, as well as Meter Reader II. Manager B manages Meters 3 and 4, and Meter Reader I. Meter Reader I collects usage data from Meters 1, 2 and 3, whereas Meter Reader II collects usage data from Meters 2, 3, and 4.

We have the following observations from this example:

- A manager can manage several separate meters. For example, Meters 1, 2, and 4 are managed by Manager A as shown in Fig. 4.11.
- A meter can have several rule sets from multiple managers. For example, Meter 4 is managed by both Meter Readers I and II in Fig. 4.11.
- Multiple meters can report to one or more meter readers. For example, Meters 2 and 3 report to both Meter Readers I and II as illustrated in Fig. 4.11, providing

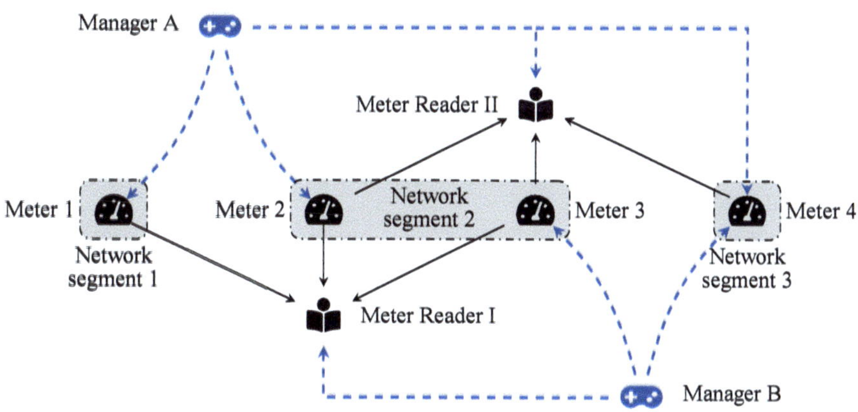

Fig. 4.11 Multiple managers, meters, and meter readers

redundancy to meter readers. If a meter reader fails, the other can still collect usage data from both Meters 2 and 3.

- Placing both Meters 2 and 3 within the same network segment also adds redundancy to the traffic metering of the segment. If one meter fails, the other can still report the usage data for the network segment.
- In this example, no synchronization is required between the two Meter Readers, indicating that they can operate independently.

In a flow measurement configuration with multiple Meter Managers, it is necessary to have communication between the managers. However, the interactions between Meter Managers cannot be fully addressed solely from the flow measurement perspective. They should be explored in the broader context of network management, which will be covered in later chapters of this book.

4.4.4 *Granularity of Flow Measurements*

There are typically large volumes of traffic flows in a network. Capturing all of them for flow measurements can be resource-demanding. In many cases, it may not be feasible or necessary to capture every single flow. The flow *granularity* of flow measurements controls the trade-off between the 'overhead' associated with the measurements and the 'level of details' provided by the usage data. A higher level of details implies higher overhead, and vice versa.

How to control the flow granularity? It is controlled by adjusting the level of details for various factors, such as those listed below [1, p. 13]:

- The metered traffic group, which can be based on address attributes,
- The category of packets, such as the attributes other than addresses, and

- The lifetime or duration of flows, i.e., the reporting interval that may need to be sufficiently short for accurately measuring the flows.

The rule set that determines the traffic group of each packet is known as the *current rule set* for the meter. It is an essential part of the reported information. This means that the reported usage data information cannot be properly interpreted without the current rule set.

4.4.5 Meter Structure

The structure of flow measurement meters is outlined in the logical diagram of Fig. 4.12 [1, p. 18]. The headers of incoming packets to the meter are fed into the packet processor. They are forwarded by the packet processor to the packet matching engine for classification as directed by the current rule set. The classification result returned to the packet processor can be either a 'flow key' or an indication for the packets 'to be ignored':

- If a packet is classified as 'to be ignored', the packet processor will discard the packet;
- Otherwise, a 'flow key' is returned, which provides the information about the flow to which the packet belongs.

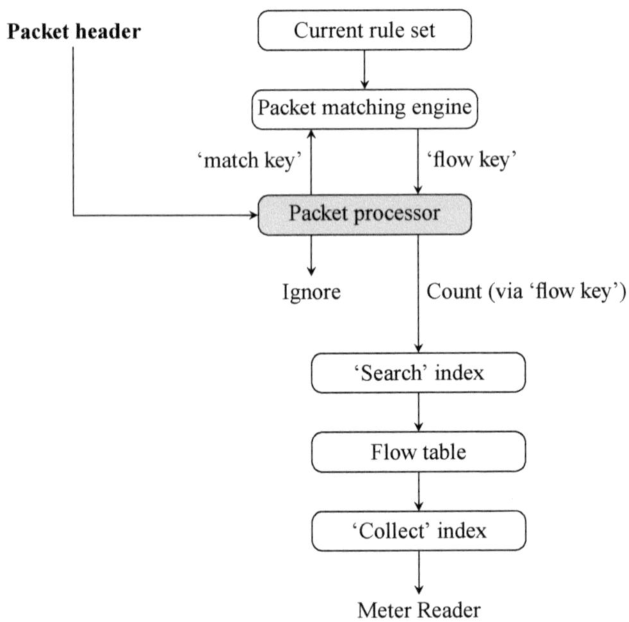

Fig. 4.12 Meter structure [1, p. 18]

The ‘flow key’ is used to locate the entry of the flow in the flow table. If no such an entry is found in the flow table, create one and add it to the flow table. Then, update the data fields of the entry, e.g., packet and byte counters. The information shown in the flow table can be collected at any time by a meter reader. To locate specific flows to be collected within the flow table, the ‘collect’ index can be used by the meter reader.

4.5 Traffic Load and Behavior

After identifying traffic flows, the next step is to analyze and quantify these flows. This analysis enables a better understanding of the behavior of the protocols and applications that generate the flows, providing valuable insights into network traffic patterns, usage trends, and performance characteristics. It also assists in designing appropriate network architecture and selecting suitable network technologies for a network planning project, especially for capacity planning.

4.5.1 Characterizing Traffic Load

Before we delve into detailed discussions on traffic load, it is helpful to acknowledge that traffic load estimates are approximate due to various factors. For example, when considering a specific application, assumptions need to be made in order to estimate its traffic load, such as the number of simultaneous sessions, the average duration of a session, and the amount of data that needs to be transferred upstream and downstream in a session. Since these assumptions are not precise, it is unrealistic to expect an accurate traffic load estimate. In fact, it is not necessary to have a precise estimate of the traffic load because the resulting estimate will need to be scaled up in capacity planning to provide a performance margin and particularly account for future network growth.

The primary purpose of traffic load analysis is not to obtain precise estimates, but rather to avoid a network design with critical bottlenecks. Traditionally, network problems are addressed by throwing more bandwidth, i.e., *over-provisioning* bandwidth. But this approach is not an appropriate solution to service-based networking, as discussed in the IETF RFC 1633 [9, p. 4]. It is worth noting that bandwidth has become increasingly affordable, especially for LANs where 10 Gigabit Ethernet has become a standard deployment in modern networking. As a result, capacity planing for LANs is relatively straightforward: simply choose 10 Gigabit Ethernet. While WAN bandwidth is still relatively expensive, it is also becoming more affordable. Therefore, bandwidth is no longer a major constraint in network planning.

Nevertheless, estimating traffic load remains important for critical flows with strict QoS requirements. These flows may require end-to-end bandwidth reservation, such as in IntServ QoS management. Over-provisioning bandwidth does not guarantee end-to-end IntServ QoS.

Calculating the traffic load of flows generated by an application or group of applications requires only a few parameters. To illustrate this, let us consider a simple example with the following assumptions:

- There are 30 simultaneous interactive sessions in the application,
- Each session has an average duration of 10 minutes,
- Each session transfers a total of 1 kB data passing the point at which traffic flows are measured.

With these assumptions, we can perform the following calculations:

- The total amount of data transferred in 30 simultaneous interactive sessions is:

$$1 \text{ kB per session } \times 30 \text{ sessions } = 30 \text{ kB}$$

- The amount of data transferred per minute is:

$$\frac{30 \text{ kB}}{10 \text{ min}} = 3 \text{ kB/min}$$

- The traffic load in bits per second (bps) is:

$$3 \text{ kB/min} = \frac{3 \text{ kB } \times 8 \text{ bits per byte}}{1 \text{ min } \times 60 \text{ s/min}} = 0.4 \text{ kbps} = 400 \text{ bps}$$

It is worth mentioning that the unit 'kB' (kilo-byte) can have slightly different meanings in different contexts. In base 10, which aligns with the *International System of Unit* (SI), 1 kB = $1,000$ bytes, i.e., 10^3. However, 1 kB = $1,024$ bytes in base 2, i.e., 2^{10}. The base 2 representation is particularly used to measure the size of data files and the space capacity of hard drives and memory. Regardless, in our calculations, we use the base 10 representation, i.e., 1 kB = $1,000$ bytes, for several reasons:

- It is more convenient for calculations;
- The difference in the calculation results is small enough; and
- The small difference can be well accommodated because
  - The calculation result is only an approximate estimate under simplified assumptions, and
  - The result will need to be scaled up in general for capacity planning.

When the protocols that the application uses for data transmission are considered, the estimate of the traffic load can be refined by taking into account the protocol overhead. Table 4.2 tabulates the overhead of some commonly used protocols [7, p. 100].

Moreover, it is necessary to consider any additional traffic load that may be generated by running an application. Depending on the application scenario, this additional traffic load may or may not have an impact on the performance of the application. Some sources of additional traffic load include:

Table 4.2 Protocol overhead measured in the number of bytes

Protocol	Overhead (#bytes)	Details
Ethernet version 2	38	Preamble 8, header 14, CRC 4, IFG 12
IEEE 802.3 with 802.2	46	Preamble 8, header 14, LLC 3 or 4, SNAP if present 5, CRC 4, IFG 12
IP	20	Header with no options
TCP	20	Header with no options
UDP	8	Header

CRC: Cyclic Redundancy Check; IFG: InterFrame Gap;
LLC: Logical Link Control; SNAP: SubNetwork Access Protocol

- Address Resolution Protocol (ARP),
- Dynamic Host Configuration Protocol (DHCP),
- Domain Name System (DNS),
- Internet Control Message Protocol (ICMP),
- Network Time Protocol (NTP),
- Simple Network Management Protocol (SNMP), and
- Encryption protocols.

It is worth noting that routing protocols also contribute to additional traffic load as they require a significant amount of bandwidth for establishing and updating routing information in general. For example, the Routing Information Protocol (RIP) updates the routing information every 30 s by sending out one or more routing packets. Each routing packet consists of a packet header and 25 routes of 20 bytes each, implying a packet size of 532 bytes. More details about various routing protocols will be discussed later in the chapter on network routing architecture.

4.5.2 *Examples of Traffic Load Analysis*

To demonstrate traffic load analysis, let us consider a few examples of local and remote database access.

Local Access to a Database

Consider a simple scenario where a user accesses to a local database, as shown in Fig. 4.13. This scenario can be well fitted into a client-server flow model. It is assumed that

- There is an interactive user interface with
  - 100 simultaneous interactive sessions,
  - 30 queries per session, and
  - An average session duration of 10 min.

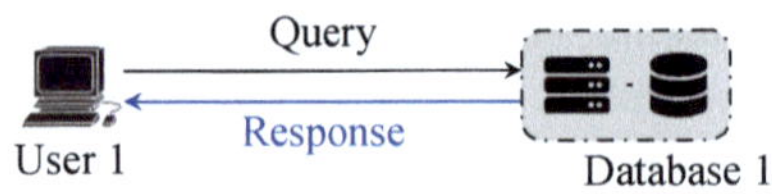

Fig. 4.13 Access to a local database

- A query request is 1 kB in size, and
- A query response is 10 kB in size.

Then, we aim to calculate the traffic load for queries and responses in the client-server model.

We have the following calculation steps:

(1) The total number of simultaneous queries to the database:

$$100 \text{ sessions } \times 30 \text{ queries/session } = 3{,}000 \text{ queries}$$

(2) The number of queries per minute:

$$\frac{3000 \text{ queries}}{10 \text{ min duration}} = 300 \text{ queries/min}$$

(3) The total size of the queries in bytes per minute:

$$300 \text{ queries/min } \times 1 \text{ kB/query} = 300 \text{ kB/min}$$

(4) The traffic load for the queries expressed in bps:

$$\text{Queries: } \frac{300 \text{ kB/min } \times 8 \text{ bits/byte}}{60 \text{ s/min}} = 40 \text{ kbps}$$

(5) The traffic load for local database responses is 10 times as big as the query traffic load. Thus, we have:

$$\text{Resposes: } 40 \text{ kbps } \times 10 = 400 \text{ kbps}$$

In the above calculations, we have used base 10 to convert 1 kB to $1,000$ bytes for the reasons discussed earlier.

Remote Access to Multiple Databases

Now, we extend the single database scenario discussed above to three database sites as depicted in Fig. 4.14. The user and application requirements for each site are the same as those for the single site scenario discussed above. Additional assumptions for each of the three sites are given below:

- 80% of the queries can be answered locally from the local database;
- 20% of the queries have to be answered remotely from another database
  - The transfers of the queries and responses are server-to-server, and

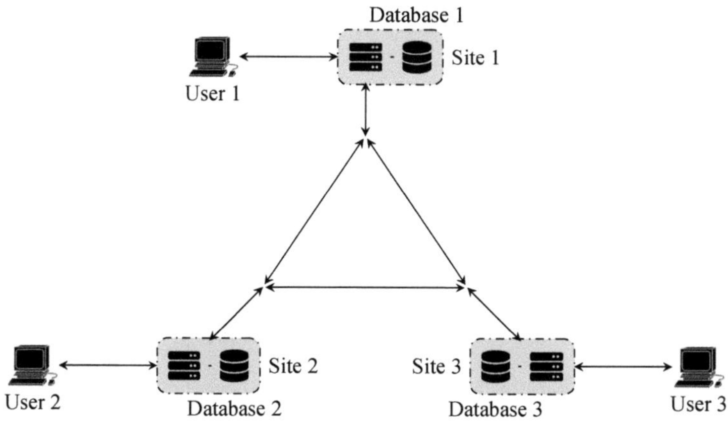

Fig. 4.14 An example of traffic flow calculation

– The sizes of the queries and responses are respectively the same as those for local database access.

From these assumptions and the logical diagram in Fig. 4.14, each of the three sites for access to the local database fits well into the client-server flow model. However, the communication among the three database servers is peer-to-peer because no server is a manager and all flows between any pairs of servers are considered equally important. Therefore, Fig. 4.14 shows a combination of the client-server flow model and peer-to-peer flow model.

For each of the three sites, the local database access generates the following traffic load:

$$\begin{aligned} &\text{Queries from a user: } 40 \text{ kbps} \\ &\text{Responses to a user: } 400 \text{ kbps} \end{aligned}$$

For each site, the queries to, and responses from, remote servers, generate the following traffic flows:

$$\begin{aligned} &\text{Queries to remote sites: } 40 \text{ kbps} \times 20\% = 8 \text{ kbps} \\ &\text{Responses from remote sites: } 8 \text{ kbps} \times 10 = 80 \text{ kbps} \end{aligned}$$

For the calculation of backbone flows between remote servers, how are the traffic flows from and to a server distributed to the paths connecting the other remote servers? In the absence of additional information about the specific distribution, we consider the worst case scenario, which is an even distribution of traffic flows to all connecting paths, as shown in Fig. 4.15. In the given example, the 8 kbps flows for queries to remote servers are evenly distributed to the two paths connecting the other two sites, resulting in 4 kbps for each of the two paths. Similarly, the 80 kbps flows

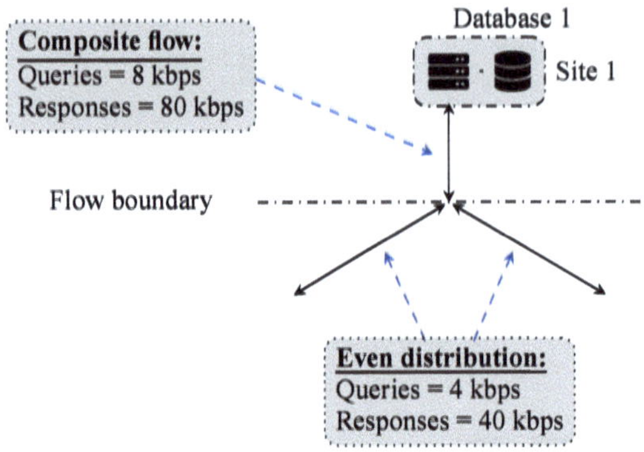

Fig. 4.15 Flow distribution for access to remote databases

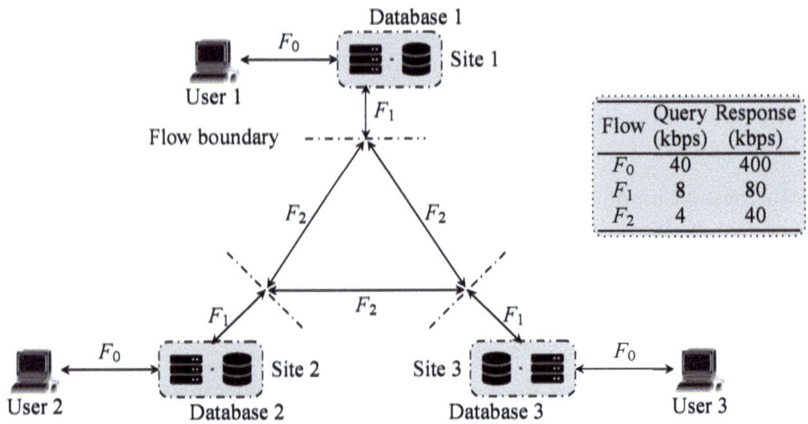

Fig. 4.16 Calculation results of traffic flows from a single pair of query and response for the example in Fig. 4.14

for responses from remote servers are evenly distributed to the two paths, resulting in 40 kbps for each path.

The calculation results of the traffic load for the backbone flows generated from a single pair of query and response are summarized in Fig. 4.16. Traffic flow F_0 represents the flow for local database access. Traffic flow F_1 is a composite flow, which carries query information to, and responses from, remote database servers. Traffic flow F_2 is the traffic flow to, and from, a remote database server.

Additional Database Synchronization

Let us additionally consider database synchronization. It is assumed that the databases at the three remote sites are synchronized with each other once every 30 minutes. In the synchronization process, the amount of data that needs to be transferred is 9 MB. For constant-rate synchronization, the traffic load is calculated as follows:

Table 4.3 Traffic load with database access and synchronization for the example shown in Fig. 4.14

Flow	Database access only (kbps)	Database synchronization only (kbps)	Database access and synchronization (kbps)
F_0(query)	40	–	40
F_0(response)	400	–	400
F_1(outbound)	8+40+40=88	80	88+80=168
F_1(inbound)	80+4+4=88	80	88+80=168
F_2(one direction)	4+40=44	40	44+40=84
F_2(the other direction)	4+40=44	40	44+40=84

$$\frac{9 \text{ MB}}{30 \text{ min} \times 60 \text{ s/min}} = 0.05 \text{ MB/s}.$$

Converting to bits per second yields:

$$0.05 \text{ MB/s} \times 8 \text{ bits/byte} \times 10^6 \text{ bytes/MB} = 40 \text{ kbps}.$$

Therefore, the traffic load for constant-rate synchronization is 40 kbps.

When additional database synchronization is considered in the example depicted Fig. 4.14, the flow F_0 shown in Fig. 4.16 for each site remains the same. But the flows F_1 and F_2 shown in Fig. 4.16 will change due to the additional database synchronization. They are calculated as follows:

$$\begin{aligned}
&F_1 \text{ (outbound):} && 88 + 40 \times 2 \text{ kbps} && = 168 \text{ kbps}\\
&F_1 \text{ (inbound):} && 88 + 40 \times 2 \text{ kbps} && = 168 \text{ kbps}\\
&F_2 \text{ (one direction):} && 44 + 40 \text{ kbps} && = 84 \text{ kbps}\\
&F_2 \text{ (the other direction):} && 44 + 40 \text{ kbps} && = 84 \text{ kbps}
\end{aligned}$$

The calculated traffic load for the overall system is tabulated in Table 4.3. The results in the table indicate the minimum bandwidth requirements for the network. In network planning, these requirements will need to be scaled up for future growth.

4.5.3 *Characterizing Traffic Behavior*

A solution to a network planning project depends on not only traffic flows and traffic load, but also traffic behavior. Traffic behavior is largely influenced by protocol behavior, application behavior, and bandwidth usage patterns. Particularly, to a large extent, broadcast traffic may dictate network architecture, such as LAN topology and network segmentation. Moreover, application performance can be affected by various factors like frame size, protocol interactions, and traffic patterns.

Broadcast Behavior

A binary IP address with all 1s in its host portion is a broadcast address. Here are a few examples of broadcast IP addresses: 132.100.255.255/16, 132.100.63.255/18, 200.100.100.255/24, and 200.100.100.31/27. When a frame is sent to a broadcast address, it goes to all network devices on the LAN to which the broadcast address belongs. Therefore, broadcast can cause increased bandwidth consumption and consequently network congestion.

In comparison, a multicast frame is sent to a subset of network devices on a LAN. However, it is actually forwarded out all ports by a layer-2 internetworking device such as as a switch. Therefore, this behavior of multicast is similar that of broadcast. It can also cause network congestion, known as multicast flooding. The issue of multicast flooding can be fixed by using the mechanisms specifically developed in the Internet Group Management Protocol (IGMP) [14] and Multicast Listener Discovery (MLD) [15]. These protocols have been updated with Source-specific Multicast (SSM) awareness in the IETF RFC 4604 [16]. Further details on the mechanisms specified in these protocols will be discussed in a later chapter on network addressing architecture.

Returning to the topic of broadcast, let us consider what devices do not forward broadcast frames. The answer is clear: layer-3 routers do not forward broadcast frames. Devices on one side of a router belong to a *broadcast domain*. Understanding the behavior of broadcast within individual broadcast domains is valuable for network segmentation, which also aids in network management and security. Excessive broadcast traffic has a negative impact on network and application performance. An effective approach to reducing broadcast traffic is to reduce the size of individual broadcast domains through appropriate network segmentation.

An alternative approach to reducing broadcast traffic is to use the Virtual Local Area Network (VLAN) technology. A VLAN does not forward broadcast frames to other VLANs. However, VLANs are typically implemented for other purposes, such as grouping users across different floors of a building. Therefore, VLANs are not primarily considered as a solution to the broadcast congestion problem, even though they do help reduce broadcast traffic.

It is worth indicating that broadcast is necessary and unavoidable although it may cause some issues. Quite a few protocols rely on broadcast for them to function, such as ARP, DHCP, and Open Shortest Path First (OSPF). Therefore, broadcast is essential for the whole network to be fully functional.

Network Efficiency

Network efficiency characterizes how efficiently network applications and protocols use network bandwidth resources. It affects network behavior and network resource utilization. Network efficiency depends on multiple factors, such as frame size, flow control, protocol interactions, error handling mechanisms, and reliability enhancements.

If no transmission errors are present, the larger the frame size is, the higher the network efficiency will be because a higher portion of the frame can be dedicated to the payload. However, if an error occurs in the transmission of the frame, the frame may need to be re-transmitted, leading to a waste of bandwidth. In this case, a bigger frame size means a higher bandwidth waste. To optimize the use of bandwidth resources, the concept of Maximum Transmission Unit (MTU) is adopted in computer networking. Some applications allow for MTU configuration. If the frame size exceeds the MTU, fragmentation occurs in an IP environment. Fragmentation splits the large frame into multiple smaller frames, each of which is equal to, or shorter than, the MTU. While fragmentation ensures data transmission in the network, it can also slow down data communications.

IPv6 supports MTU discovery, which discovers the largest frame size that can be used without the need of fragmentation. This improvement over IPv4 networking eliminates the need for MTU configuration.

Flow control in TCP/IP communications plays a critical role in controlling network efficiency within the constraints of communication paths. A TCP sender can fill a *send window*, or buffer, with data for transmission without waiting for an ACK from the receiver. The receiver will place the received data into the *receive window*, or buffer, with a maximum size of 65, 535 bytes for processing. The bigger the receive window, the higher the network efficiency will be. But this requires more memory and CPU resources on the receiver side.

Unlike TCP, UDP-based communications do not have built-in flow control at the transport layer (layer 4). However, user-defined flow control can be implemented in application programs at higher layers, such as session layer (layer 5) or application layer (layer 7). This allows for customized flow control mechanisms tailored to specific requirements. Table 4.4 provides a list of protocols or services that use either TCP or UDP as the underlying transport protocol.

To evaluate the impact of protocol interactions on network efficiency, it is essential to understand what protocols are used by an application and what relationships and dependencies they have. For example, an email client that uses IMAP to retrieve email messages from the mail server relies on TCP as the underlying transport protocol. This requires a three-way handshaking process for TCP connection, and a four-way handshaking process for TCP disconnection. IMAP can be configured without encryption on port 143 or with encryption using Transport Layer Security (TLS) on port 993 (Table 4.4). This implies that the email client will interact with TLS in addition to IAMP when encryption is enabled. It is seen that multiple protocols work together to enable the functionality of a specific application. The interactions among these protocols affect traffic load and behavior, and ultimately network efficiency. Hence, it is worthwhile to check the feature configurations of the application to ensure their appropriateness without significantly impacting network efficiency.

Effective error handling is important in network communications to ensure the successful data transmission from one end to the other. Errors in data communications include packet corruption and dropout. While it is possible to encode a packet to achieve full error recovery solely from the received data, this will require excessive resources and thus is impractical in most cases. The most common method of error

Table 4.4 Some protocols that use TCP or UDP

Protocol			Port
TCP-based	FTP	(File Transfer Protocol)	20 (data), 21 (control)
	HTTP	(HyperText Transfer Protocol)	80
	IMAP	(Internet Message Access Protocol)	143, 993 (encryption)
	POP3	(Post Office Protocol version 3)	110, 995 (encryption)
	SMTP	(Simple Mail Transfer Protocol)	25, etc
	SSH	(Secure Shell)	22
	Telnet		23
UDP-based	DHCP	(Dynamic Host Configuration Protocol)	67 (server), 68 (client)
	DNS	(Domain Name System)	53
	NTP	(Network Time Protocol)	123
	SNMP	(Simple Network Management Protocol)	161 (agent), 162 (trap)
	TFTP	(Trivial File Transfer Protocol)	69

handling is the use of ACK and re-transmission mechanisms. These mechanisms can be implemented at different layers, such as layer 2 (link layer), layer 4 (transport layer), and layer 7 (application layer). A typical example is TCP at layer 4, which supports reliable communication due to its connection-oriented nature. By contrast, most layer-2 and layer-3 protocols, as well as layer-4 UDP, are connectionless and do not have built-in error recovery mechanisms. When UDP is used as the underlying transport protocol in an application, reliable communication can still be achieved by implementing ACK and re-transmission at a higher layer, such as layer 7 (application layer).

4.5.4 Bandwidth Usage Pattern

Bandwidth usage in an enterprise network exhibits a clear pattern at different time scales. Let us consider the weekly bandwidth usage pattern demonstrated in Fig. 4.17.

Day-Night Pattern

It is seen from Fig. 4.17 that the network bandwidth usage follows a specific pattern, with peak and off-peak hours occurring around midday and midnight, respectively. Bandwidth usage tends to increase in the morning, reaching its maximum around midday. Then, it gradually decreases in the afternoon, reaching its minimum after midnight. This day-night pattern can be attributed to employees starting to work in the early morning and returning home in the late afternoon, resulting in significantly

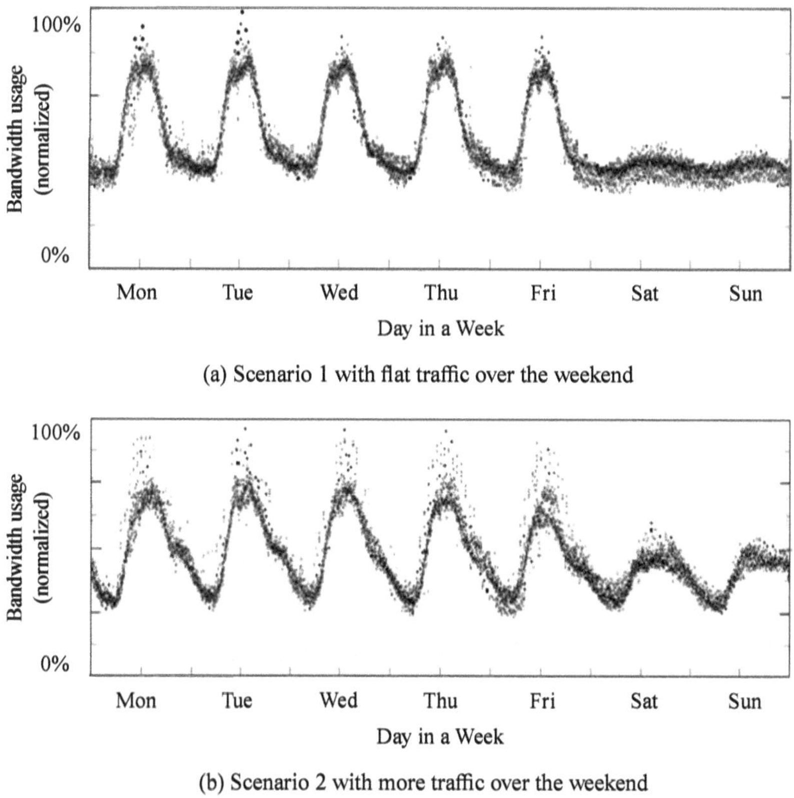

Fig. 4.17 Weekly bandwidth usage pattern normalized as a percentage of the maximum bandwidth usage over a week

different network traffic volumes during peak and off-peak hours. This bandwidth usage pattern should be taken into account in network architectural design.

Figure 4.17 also shows that the bandwidth usage on weekends is almost flat without exhibiting the same day-night pattern observed on weekdays. This is because fewer users use the network during weekends compared to weekdays.

A network usage analysis conducted for the student residence of a university shows interesting findings [17]. The study reveals that peak network usage occurs around midnight, while off-peak hours are in the early morning typically around 6–7 am. Notably, student residents exhibit similar levels of activities on weekends compared to weekdays, indicating a consistent network usage pattern throughout the week. In terms of overall traffic, incoming traffic dominates, accounting for approximately 80%, while outgoing traffic constitutes around 20%.

From the application perspective, the study mentioned above [17] reports that HTTP-related applications contribute to one third of the overall traffic. Video services including YouTube account for roughly one fifth of the traffic. Social network

Table 4.5 Application usage from 27 Mar to 3 Apr 2011 [17]

Application	Percentage
HTTP-related	>35%
Video services, e.g., YouTube	≈20%
Social networking, e.g., Skype	≈10%
All others	≈35%

applications including Skype represent around one tenth. These statistics are summarized in Table 4.5. As these statistics were measured many years ago, they do not reflect the current popularity of applications such as TikTok and Zoom. Nevertheless, these statistics provide some insights into historical application usage patterns.

4.5.5 *Weekday Pattern*

From the day-to-day perspective, it is seen from Fig. 4.17 that the bandwidth usage on weekdays from Monday to Friday exhibits a high degree of similarity and consistency. However, there is a notable distinction between the bandwidth usage on weekdays and weekends, primarily because a significant portion of network users do not use the network during weekends.

The traffic analysis for the student residence discussed above [17] shows a consistent day-night pattern across the whole week from Monday through Sunday. This consistency is attributed to the active network usage by students, even on weekends.

4.6 Flow Specification

The results from the process of identifying, characterizing, and developing traffic flows are integrated to form a full traffic flow specification known as *flowspec*. Let us describe flowspecs from three perspectives:

- Flow analysis as part of requirements analysis,
- QoS management, and
- Traffic routing for efficient traffic forwarding.

This section primarily focuses on the flowspec from the flow analysis perspective for the purpose of top-level logical network planning. Flowspecs for QoS management and routing serve specific component-based network architecture.

4.6.1 *Flowspec from Flow Analysis*

From the flow analysis perspective, a flowspec describes traffic flows of a network, and the performance requirements and prioritization of the flows.

The performance requirements for a flow can be simply classified into three types: best-effort, predictable, and guaranteed requirements. They describe how critical the flow is (e.g., mission-critical, safety-critical, or time-critical). A specific performance threshold or range can also be given. When specifying traffic flows, the requirements for a composite flow aggregated from multiple individual flows can be integrated from the requirements for individual flows. The combined requirements of all flows can later be used for performance management, e.g., through the implementation of QoS mechanisms like DiffServ.

To characterize different levels of performance requirements for traffic flows, a flowspec can be presented in a one-part, two-part, or multi-part form. A one-part flowspec, also known as a *unitary* flowspec, provides a basic level of detail and performance requirements. As we move to two-part and multi-part flowspecs, the level of detail and performance requirements increases. The three types of flowspecs and their characteristics are summarized in Table 4.6.

One-Part Flowspec

A *one-part flowspec* lists traffic flows with best-effort performance requirements. Best-effort performance implies no specific requirements are specified because all network services and flows are provisioned with best effort by default in computer networking. Therefore, a one-part flowspec suits general traffic flows that do not have specific performance requirements. Typical examples that can be characterized by using a one-part flowspec include general web browsing, mail services, FTP downloading, and network printing. These types of applications typically do not require specific performance guarantees. They can operate effectively with best-effort services as they can dynamically adapt to available network resources such as bandwidth.

Table 4.6 Three types of flow specifications

Flowspec Type	Flows	Characteristics
One-part flowspec	Best-effort individual and composite flows	Bandwidth capacity
Two-part flowspec	Best-effort and stochastic flows, which are individual or composite flows	Capacity, reliability, delay, jitter, etc
Multi-part flowspec	Best-effort, stochastic, and guaranteed flows, which are individual or composite flows	Capacity, reliability, delay, jitter, etc

Two-Part Flowspec

A *two-part flowspec* describes traffic flows with two parts of requirements:

- One part for predictable performance requirements, and
- The other part for best-effort performance requirements.

Therefore, a two-part flowspec can be seen as a natural extension of the one-part flowspec by adding traffic flows with predictable performance requirements.

Predictable performance requirements, such as delay and jitter, are necessary for certain soft real-time applications like video streaming and voice services. These applications are sensitive to delay and jitter because they require timely delivery of data packets to maintain their QoS. However, video and voice services can tolerate a certain range of delay and jitter. Occasional losses of a few video frames may not significantly impact the user experience in general video applications, such as YouTube. Therefore, it is not necessary to provide guaranteed performance for delay and jitter in such cases.

Multi-Part Flowspec

In addition to best-effort and predictable requirements, there are cases where guaranteed performance requirements need to be considered. The addition of guaranteed performance requirements to the two-part flowspec results in a *multi-part flowspec*. Guaranteed requirements are specifically developed for critical applications or services that demand a certain level of performance assurance. These requirements define specific thresholds or ranges that must be met to ensure the desired performance. For example, a critical application may require a maximum delay of 20 ms and a minimum bandwidth of 100 kbps at all times.

Flowspec Algorithm

When multiple flowspecs have been developed, how do we combine the requirements from all these flowspecs? The *flowspec algorithm* provides a set of rules to combine the requirements. These rules address the best-effort, predictable, and guaranteed performance requirements specified in the flowspecs:

R1 Add up the capacity requirements of all flows to form the overall capacity requirement for the network.
R2 For performance characteristics other than bandwidth, such as delay, jitter, and reliability, the best performance requirements among all predictable flows are selected and applied to these flows. This rule ensures that the flowspecs with predictable performance requirements are met with optimized performance.
R3 Guaranteed flows must be handled individually by considering their specific requirements separately.

By applying these rules to all flowspecs, the flowspec algorithm determines the overall capacity requirements and performance characteristics needed for QoS management. Rule R1 applies to all types of flowspecs including one-part, two-part, and multi-part flowspecs. Rule R2 is used for predictable flows found in two-part

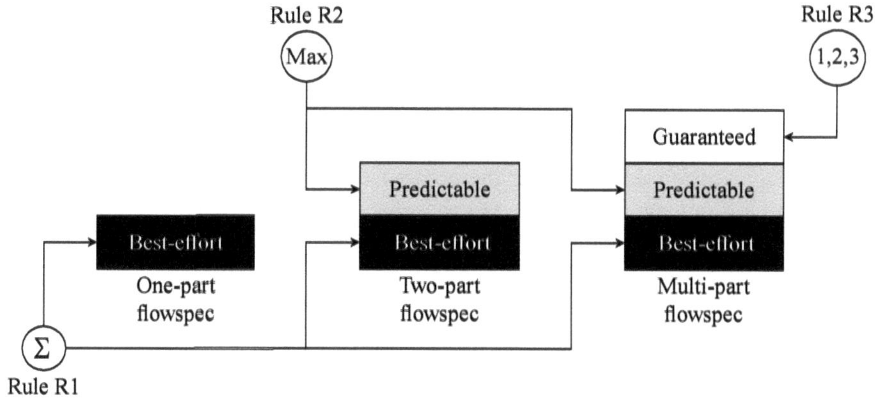

Fig. 4.18 Applications of flowspec rules to flowspecs

and multi-part flowspecs. Rule R3 addresses guaranteed flows listed in multi-part flowspecs. The applications of the three flowspec algorithm rules to the three types of flowspecs are illustrated in the logical diagram of Fig. 4.18.

The use of Rules R1 and R3 are straightforward. To illustrate how to use Rule R2, lect us consider an example with four flows:

- Flow I: bandwidth 200 kbps (best-effort)
- Flow II: bandwidth 100 kbps, delay 20 ms, reliability 99%,
- Flow III: bandwidth 200 kbps, delay 50 ms, reliability 99.2%, availability 90%,
- Flow IV: bandwidth 1 Mbps, delay 80 ms, reliability 99.5%

To combine all requirements from the four flows, add up the capacity requirements first, leading to the baseline of the overall capacity requirement of 1.5 Mbps. Then, for the predictable Flows II, III, and IV, consider the performance characteristics like delay, reliability, and availability. Choose the best performance among these three flows, yielding a delay of 20 ms, reliability of 99.5%, and availability of 90%. These optimized performance requirements are then applied to these three flows.

4.6.2 Flowspec for QoS Management

In IntServ QoS management, flowspecs help each router that participates in an IntServ session to determine whether it has sufficient resources to meet the IntServ QoS requirements. For an identified flow that requires IntServ QoS control, the corresponding flowspec must describe

- What the required QoS is, and
- What the characteristics of the QoS are.

These two aspects are described by the Traffic SPECification (TSPEC) and Request SPECification (RSPEC), respectively. Both TSPEC and RSPEC are essential components of flowspecs in IntServ. Detailed technical specifications about TSPEC and RSPEC can be found in the IETF RFC 2210 [18] and RFC 2215 [19]. These RFCs provide comprehensive explanations of the TSPEC and RSPEC formats, their parameters, and their usage within the context of IntServ. The use of TSPEC and RSPEC in QoS management will be further discussed in a separate chapter on network performance architecture.

4.6.3 Flowspec for Traffic Routing

Routers are responsible for making routing decisions and forwarding traffic over networks. In modern IP networks, routers have evolved to be more powerful with additional capabilities such as traffic management, security policy enforcement, and other functions beyond basic routing. These capabilities allow network operators to apply various rules and actions to packets based on specified criteria defined by network policies. Such rules are known as match rules because their use is through matching multiple fields of the packet header. Traffic classification and shaping, as well as other traffic management actions can be associated with these matching rules.

To make traffic routing more efficient with required actions such as those mentioned above, a flowspec that a router receives is expected to have well-defined match criteria. For this purpose, the IETF RFC 8955 [20] has defined a flowspec as "an n-tuple consisting of several matching criteria that can be applied to IP traffic". A packet that matches all the specified criteria is said to match the defined flowspec. In the IETF RFC 8955 [20], this n-turple flowspec is encoded into the Network Layer Reachability Information (NLRI) of the Border Gateway Protocol (BGP). The encoding format can be used "to distribute (intra-domain and inter-domain) traffic Flow Specifications for IPv4 unicast and IPv4 BGP/MPLS VPN services".

4.7 Summary

As part of the requirements analysis process, flow analysis identifies and characterizes traffic flows and further develops a set of flow specifications that highlight the QoS performance requirements of these flows. The developed flow specifications provide insights into network architecture and QoS management mechanisms. In addition to serving network capacity planning, they are particularly important for network applications and services that have predictable and guaranteed QoS performance requirements.

From the descriptions provided in various RFCs, this chapter has established that a traffic flow is a sequence of packets with common attributes, measured during a period of time in a single session of an application, characterized at a measurement point, and spans from a source to a destination in an end-to-end manner. In the presence of network hierarchy, multiple individual flows may be aggregated to form a composite flow.

If a flow does not have specific performance requirements, it is treated as a best-effort flow and thus can be managed with best-effort service. This is the default configuration for traffic flows and network services in computer networking when no performance requirements are specified. QoS performance requirements for a flow may include capacity, delay, reliability, manageability, availability, and other factors. In traditional networking, these requirements are often addressed through over-provisioning of bandwidth. However, in modern computer networking where the number of network services is increasing, simply throwing more bandwidth is insufficient to provide predictable and guaranteed services, especially for flows that required end-to-end IntServ QoS support. Therefore, developing QoS performance requirements is essential for flow prioritization and overall network QoS management.

An effective approach to flow analysis is to map network applications and services into a well-established traffic flow model. This chapter has discussed four popular flow models: peer-to-peer, client-server, hierarchical client-server, and distributed computing flow models. Conventional terminal-host traffic flows are no longer popular in modern networking although they still appear in some networks. They are treated as a special type of client-server flows in this chapter.

Traffic load and behavior can affect the performance requirements of traffic flows. When estimating traffic load, considerations should include protocol overhead as well as the interactions between protocols used by an application. Also, various patterns of traffic, users, and applications should be taken into account, such as peak and off-peak traffic requirements.

The flowspec developed from flow analysis describes the best-effort, predictable, and guaranteed performance requirements. The flowspec algorithm combine these requirements through using flowspec rules, forming the overall QoS performance requirements for the network. These results will be used for network architecture planning.

References

1. Brownlee, N., Mills, C., Ruth, G.: Traffic flow measurement: Architecture. RFC 2722, RFC Editor (1999). https://doi.org/10.17487/RFC2722
2. Rajahalme, J., Conta, A., Carpenter, B., Deering, S.: Ipv6 flow label specification. RFC 3697, RFC Editor (2004). https://doi.org/10.17487/RFC3697
3. Quittek, J., Zseby, T., Claise, B., Zander, S.: Requirements for IP flow information export (IPFIX). RFC 3917, RFC Editor (2004). https://doi.org/10.17487/RFC3917
4. Handelman, S., Stibler, S., Brownlee, N., Ruth, G.: RTFM: New attributes for traffic flow measurement. RFC 2724, RFC Editor (1999). https://doi.org/10.17487/RFC2724
5. Blake, S., Black, D., Carlson, M., Davies, E., Wang, Z., Weiss, W.: An architecture for differentiated services. RFC 2475, RFC Editor (1998). https://doi.org/10.17487/RFC2475
6. McCabe, J.D.: Network Analysis, Architecture, and Design (3rd ed.). Morgan Kaufmann Publishers, Burlington, MA 01803, USA (2007). ISBN 978-0-12-370480-1
7. Oppenheimei, P.: Top-Down Network Design (3rd ed.). Cisco Press, Indianapolis, IN 46240, USA (2011). ISBN 978-1-58720-283-4
8. Huston, G.: Next steps for qos architecture. RFC 2990, RFC Editor (2000). https://doi.org/10.17487/RFC2990
9. Braden, R., Clark, D., Shenker, S.: Integrated services architecture. RFC 1633, RFC Editor (1994). https://doi.org/10.17487/RFC1633
10. Shenker, S., Partridge, C., Guerin, R.: Specification of guaranteed quality of service. RFC 2212, RFC Editor (1997). https://doi.org/10.17487/RFC2212
11. Wroclawski, J.: Specification of the controlled-load network element service. RFC 2211, RFC Editor (1997). https://doi.org/10.17487/RFC2211
12. Zhang, Y.F., Tian, Y.C., Fidge, C., Kelly, W.: Data-aware task scheduling for all-to-all comparison problems in heterogeneous distributed systems. J. Parallel Distrib. Comput. **93–94**, 87–101 (2016)
13. Zhang, Y.F., Tian, Y.C., Kelly, W., Fidge, C.: Scalable and efficient data distribution for distributed computing of all-to-all comparison problems. Futur. Gener. Comput. Syst. **67**, 152–162 (2017)
14. Cain, B., Deering, S., Kouvelas, I., Fenner, B., Thyagarajan, A.: Internet group management protocol, version 3. RFC 3376, RFC Editor (2002). https://doi.org/10.17487/RFC3376
15. Vida, R., Costa, L.: Multicast listener discovery version 2 (MLDv2) for IPv6. RFC 3810, RFC Editor (2004). https://doi.org/10.17487/RFC3810
16. Holbrook, H., Cain, B., Haberman, B.: Using Internet group management protocol version 3 (IGMPv3) and multicast listener discovery protocol version 2 (MLDv2) for source-specific multicast. RFC 4604, RFC Editor (2006). https://doi.org/10.17487/RFC4604
17. Lam, A.: Network usage analysis at student residence. Online report (2011). https://www.cityu.edu.hk/its/news/2011/06/27/network-usage-analysis-student-residence. Accessed 26 Jan 2023
18. Wroclawski, J.: The use of RSVP with IETF integrated services. RFC 2210, RFC Editor (1997). https://doi.org/10.17487/RFC2210
19. Shenker, S., Wroclawski, J.: General characterization parameters for integrated service network elements. RFC 2215, RFC Editor (1997). https://doi.org/10.17487/RFC2215
20. Loibl, C., Hares, S., Raszuk, R., McPherson, D., Bacher, M.: Dissemination of flow specification rules. RFC 8955, RFC Editor (2020). https://doi.org/10.17487/RFC8955

Part II
Network Architecture

This part is composed of six chapters:

- Chapter 5: Network Architectural Models.
- Chapter 6: Network Addressing Architecture.
- Chapter 7: Network Routing Architecture.
- Chapter 8: Network Performance Architecture.
- Chapter 9: Network Management Architecture.
- Chapter 10: Network Security and Privacy Architecture.

Following network analysis in the previous part, this part conducts architectural planning for large-scale computer networks. It will begin with an overall network architecture design. Then, component-based network architecture will be investigated by following international standards and best industrial practices. The investigations will cover important network components including addressing, routing, performance, management, and security.

Chapter 5
Network Architectural Models

Network architecture is a high-level view of network structure from various perspectives. It describes the logical relationships of the topological and functional components of the network for network services and applications. Therefore, the development of network architecture mainly focuses on a top-level view of the logical integration of topological and functional components into the network. This implies that physical locations and detailed technical implementations of the network are not the primary purpose of network architecture, although an abstraction of some physical and technical information may appear in the architectural structure.

Network architecture should not be confused with network design. While both network architecture and network design aim to solve the complex problem of network development, they address different aspects of the network. Architecture is more concerned with providing a high-level map of the logical relationships of the topological and functional components of the network. Network design, on the other hand, deals with detailed technical implementation, vendor or equipment selection, physical design, structured cabling, and floor plan. For a specific segment of the network, network design determines where to install routers, switches, servers, computing nodes or hosts, wireless access points, and other network devices. It also determines what types of equipment, operating systems, and protocols should be used in that network segment. These aspects are not part of network architecture planning.

Aiming to optimize everything for a network will not be effective. Instead, network architecture planning should balance the requirements from users, devices, networks, and applications by considering the network technologies and resources currently available, as well as technical and non-technical constraints. Technical constraints include the capacity of existing technologies, while non-technical constraints may include budget limitations, the availability of skilled IT support, and local legislation and policies.

Network architecture needs to be approached in a systematic and reproducible manner. This is particularly necessary for large-scale networks, where the logical

Y.-C. Tian and J. Gao, *Network Analysis and Architecture*, Signals and Communication Technology, https://doi.org/10.1007/978-981-99-5648-7_5

relationships among various physical and functional components are complex. This chapter discusses network architecture using the top-down methodology, systems approach, and hierarchical network architecture models. It also includes geographical models, functional models, flow-based models, and component-based models of networks. Enterprise edge topology and redundant network models will also be discussed in this chapter.

5.1 Hierarchical Network Architecture

An organization has a hierarchical structure. To better understand the organization, gaining a top-level view of the organization and analyzing its structural hierarchy can provide valuable insights. This naturally leads to the adoption of the top-down methodology and hierarchical models in the development network architecture to meet the business goals and network service requirements of the organization. The top-down methodology and systems approach facilitate a *divide and conquer* method to network architecture development. The hierarchical models allow for the development of network architecture in layers, ensuring connectivity, scalability, manageability, performance, and accomodation of future growth for the planned network.

5.1.1 The Core/Distribution/Access Architecture

A typical hierarchical network topology is Cisco's three-layer architectural model, i.e., the core/distribution/access model. It is depicted in Fig. 5.1 with a logical view and an implementation example. Each of the three layers has its own functions, which are discussed in the following.

The Core Layer

The core layer of the network is the Wide Area Network (WAN) backbone of the network. Consisting of high-end routers and switches, it is optimized for bulk transport of traffic between sites, and thus individual flows are not visible in the core layer unless specifically planned for critical or guaranteed services. The core layer aims to provide high reliability and availability, good redundancy and fault tolerance, easy management, and low latency in the transport of traffic. Therefore, it has a limited and consistent diameter and requires some degree of quick adaptivity to network changes. Slow packet manipulation should be avoided such as packet filtering.

The Distribution Layer

The distribution layer forms the backbone of a campus network. Consisting of routers and switches, it can be designed to perform a number of functions, e.g.,

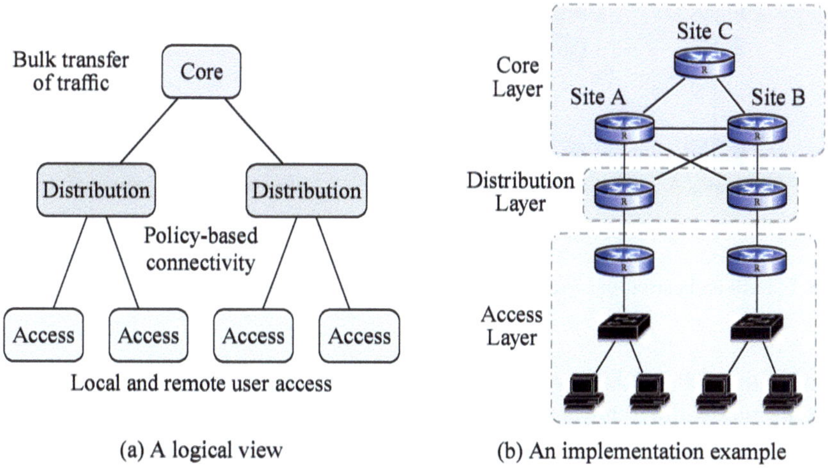

Fig. 5.1 Cisco's core/distribution/access architecture

- Traffic forwarding policies, for example, to forward traffic from a specific network out of one interface of a router whilst forwarding all other traffic out of another interface of the router;
- Address or area aggregation though supernetting, thus improving routing efficiency;
- The definition of broadcast and multicast domains through appropriate segmentation of the network;
- Virtual Local Area Network (VLAN) traffic routing;
- Media transition, e.g., from Ethernet to Asynchronous Transfer Mode (ATM).
- Security mechanisms;
- Route redistribution between routing domains if different routing protocols are used; and
- Demarcation between static and dynamic routing.

The distribution layer is the demarcation point between the core and access layers. It does not have direct connection to end users in general unless purposely designed for a specific reason. Therefore, it usually consolidates traffic flows from the access layer. Performance management for individual and consolidated flows can be implemented here.

The Access Layer

The access layer in computer networks connects users and their applications to the network via low-end switches and wireless access points. Most traffic flows originate and sink at the access layer. They can be treated easily on an individual basis. Therefore, access lists and packet filters can be implemented here. In a campus network environment, the access layer can be designed to manage segmentation, micro-

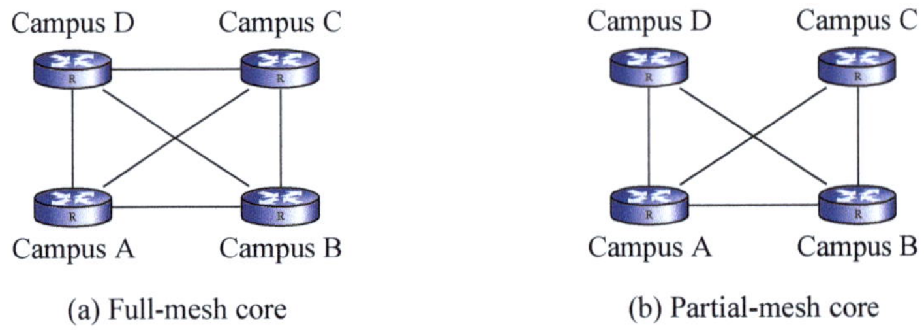

Fig. 5.2 Full- and partial-mesh architectural models for core layer

segmentation, shared bandwidth, switched bandwidth, and MAC-layer filtering to meet the bandwidth requirements for specific users, user groups, and applications. In small-scale network environments such as a small branch office, the access layer can be designed to allow remote access to the corporate network via WAN technologies such as Integrated Services Digital Network (ISDN) and leased lines.

5.1.2 Mesh Topology for Core Layer

As the core layer provides high reliability and availability for high-speed bulk transport of traffic, it must be designed with redundancy. Two options of the core layer topology are full-mesh and partial-mesh topology, as shown in Fig. 5.2. Full-mesh topology offers the highest reliability due to multiple paths between any pair of routers. But it is relatively expensive and also complex to manage. Also, full-mesh topology lacks scalability. Therefore, it is not suitable for a core with many sites that are interconnected through WANs.

A simpler and cost-effective alternative is the partial-mesh topology. In partial mesh, routers have fewer links to other routers compared to full mesh. The scalability of partial mesh is also improved over full mesh. Nevertheless, how to develop a partial mesh still needs to be carefully considered. For example, the partial mesh shown in Fig. 5.2b is preferable over a loop topology where the four routers form a circular connection. The loop topology with many sites is not recommended in general. This is due to the fact that if a link is down there will be many hops between routers on opposite sides of the broken loop, negatively impacting overall network performance.

5.1.3 Hierarchical Redundancy for Distribution Layer

The distribution layer handles many servers and specialized devices, and also implements performance mechanisms. Therefore, it also requires redundant connections to both the core and access layers. Mesh or even partial-mesh is not common between routers at the distribution layer. Rather, a distribution-layer router is connected to

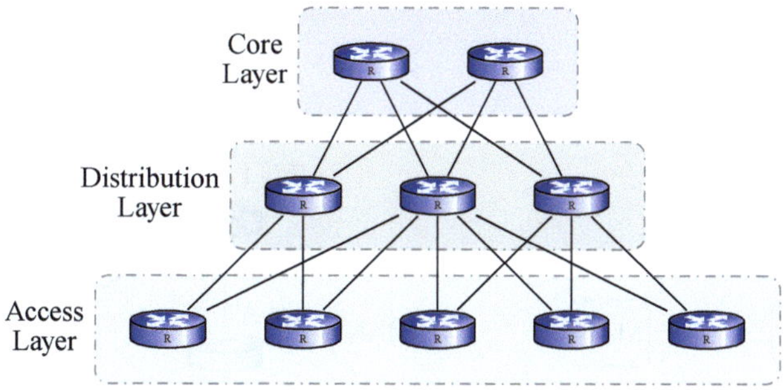

Fig. 5.3 Hierarchical redundant architecture for distribution layer

more than one core-layer router, and an access-layer router is connected to multiple distribution-layer routers. This forms a hierarchical redundant topology for the distribution layer, as illustrated in Fig. 5.3. In this figure, if a link failure occurs between the distribution layer and either the access layer or core layer, there is always an alternative link available for transporting traffic. This significantly improves the overall reliability of the network.

5.1.4 The LAN/MAN/WAN Architecture

A large-scale network can be viewed from the functional perspective as shown previously with the core/distribution/access model. It can also be approached from the geographical perspective with natural physical separation of Local Area Network (LAN), Metropolitam Area Network (MAN), and WAN. This motivates the popular use of another high-level topological model: the LAN/MAN/WAN architecture, as shown in Fig. 5.4. Some comparisons of LAN, MAN, and WAN are tabulated in Table 5.1.

Similar to the functional core/distribution/access architecture introduced from Cisco, the LAN/MAN/WAN architecture also has a hierarchical topology. At the bottom layer of this topology, LANs implement local user access to the network. MANs, on the other hand, interconnect multiple sites or campus networks within the same metropolitan area through the MAN infrastructure such as twisted-pair and fibre-optic cables, some of which may be owned by third parties. The WAN interconnects geographically remote sites through WAN infrastructure like fibre-optic cables, radio waves, and satellites. WAN infrastructure is typically owned by third parties.

However, different from the functional core/distribution/access architecture, the LAN/MAN/WAN architecture concentrates on physical locations of network components. Particularly, it focuses more on the boundaries between LANs, WANs, and the WAN, as well as the specific features and requirements associated with these boundaries. For example, border routers of a campus network are installed on the

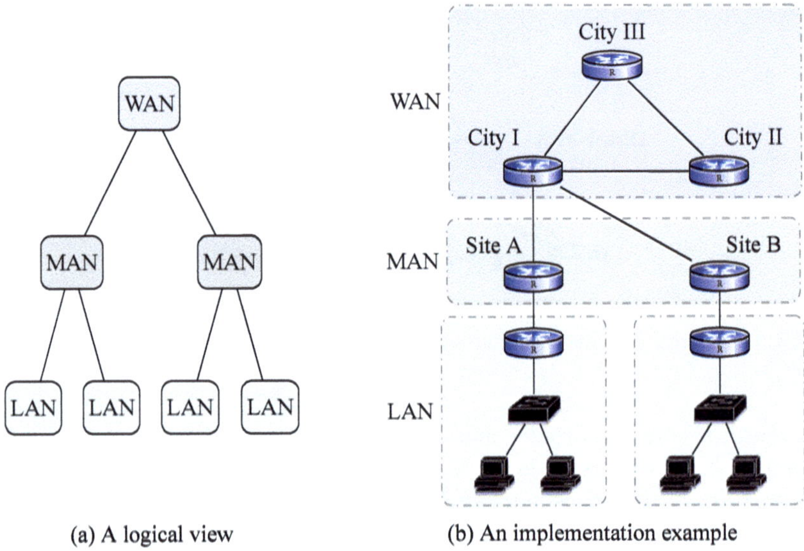

Fig. 5.4 LAN/MAN/WAN architecture

Table 5.1 Comparisons of LAN, MAN, and WAN

Criterion	LAN	MAN	WAN
Cost	Low	Medium	High
Size	Small	Medium	Largest
Speed	Fastest	Medium	Slowest
#computers	Smallest	Medium	Largest
Media	Twisted-pair	Twisted-pair, and Fibre-optic cables	Fibre-optic cables, radio wave, and satellite

boundary of the network. Network performance and service requirements specified in SLAs between an enterprise and its ISP service provider can be measured at the edge of the enterprise network.

Both the core/distribution/access model and LAN/MAN/WAN model can be used together or independently to describe network architecture from different perspectives. However, there is no one-to-one mapping between the layers of the core/distribution/access model and the LAN/MAN/WAN model. Generally, the core layer corresponds to the WAN and part of MANs. The distribution layer may cover MANs and part of LANs. The access layer has overlapping with LANs and part of MANs. The actual relationships between the layers of the core/distribution/access model and LAN/MAN/WAN model depend on the specific network architecture planning.

In the LAN/MAN/WAN architectural model, the LAN layer does not necessarily cover LANs only. It often extends to multiple buildings and floors. The resulting LANs can be interconnected with routers at one or more layers in a hierarchical manner.

It is common for the LAN/MAN/WAN architecture to be implemented without a separate MAN layer, leading to a LAN/WAN model. In this LAN/WAN model, multiple sites of an enterprise network within the same metropolitan area are interconnected through WAN connections. Other sites located in different metropolitan areas are also connected via WAN connections.

5.2 Enterprise Edge Architecture

The edge of an enterprise network is expected to ensure reliable and secure connectivity to external networks and the Internet. It should also provide authorized users with secure access to the enterprise network from outside through a public WAN connection or the Internet. Therefore, enterprise edge architecture may include redundant WAN segments, multihomed Internet connectivity, and secure Virtual Private Network (VPN) links.

5.2.1 *Redundant WAN Segments*

WAN links are critical for a reliable core layer in the core/distribution/access architecture. This necessitates redundant WAN connections at the edge of an enterprise network. As shown previously in Fig. 5.2, a full-mesh or partial-mesh topology should be designed to provide redundant WAN links.

When a full or partial mesh is provisioned for WAN connections, it is necessary to ensure that the backup links are indeed functional paths. This requires the implementation of *circuit diversity*, which refers to the use of different physical paths. Different ISPs may use the same WAN infrastructure from a third party. In such cases, if two such ISPs are chosen for WAN backup, the backup may not work as expected. If the WAN link from an ISP fails, the WAN link from the other ISP also fails.

Meanwhile, it is also necessary to ensure that the local cabling system of the WAN segments has different physical paths for backup purposes. If one physical path encounters a failure, the alternative physical path can serve as a functional backup. Make sure that the links to the ISPs are reliable with redundant physical cabling.

5.2.2 Multihomed Internet Connectivity

Multihoming is a network technology that provides more than one connection for a system to access and offer network services. In the context of network connectivity, the term *multihoming* refers to the use of multiple network connections, which specifically indicate more than one Internet entry here. Multihoming offers Internet redundancy and thus enhances the reliability and availability of network access. It is worth mentioning that the same term *multihoming* is also used to describe multihomed servers and multihomed Content Delivery Networks (CDNs). A multihomed server is connected to multiple networks or has multiple network interfaces, allowing it to communicate and provide services through different network paths. A multihomed CDN has more than one Point of Presence (PoP) or edge server, enabling it to deliver content and services to end users through the most optimal and geographically closer PoP.

There are different ways to achieve multihoming for Internet connectivity. The enterprise edge can be connected to a single ISP through a single edge router or multiple edge routers. Alternatively, it can be connected to two different ISPs. Therefore, there are four basic options:

(1) A single edge router connecting to two routers from the same ISP,
(2) Two edge routers connecting to two routers of the same ISP via different links,
(3) A single edge router connecting to two routers each from a different ISP, and
(4) Two edge routers connecting to two routers each from from a different ISP.

These options are depicted in Fig. 5.5.

Each of the four options presented above has its own advantages and disadvantages. The selection of the appropriate option depends on the specific requirements of the network. In general, working with a single ISP is simpler and easier compared to dealing with multiple ISPs, but it lacks ISP redundancy. Choosing an option with a single edge router is a cost-effective solution, but it also introduces a single point of failure. In comparison, using multiple edge routers to connect to multiple ISPs offers the highest reliability for Internet connectivity, but it is more expensive and complex.

5.2.3 Secure VPN Links

Nowadays, working off-premises has become increasingly important for organizations. There are many scenarios of working off-premises, such as inter-state or overseas business travels. In such cases, authorized users such as employees, customers, and business partners, need to access to their enterprise networks securely over third-party networks or the Internet.

VPNs enable organizations to establish secure, end-to-end, and private networks over a third-party network or the Internet. This is achieved through advanced encryption and tunneling protocol. The tunneling technique encapsulates data packets of one

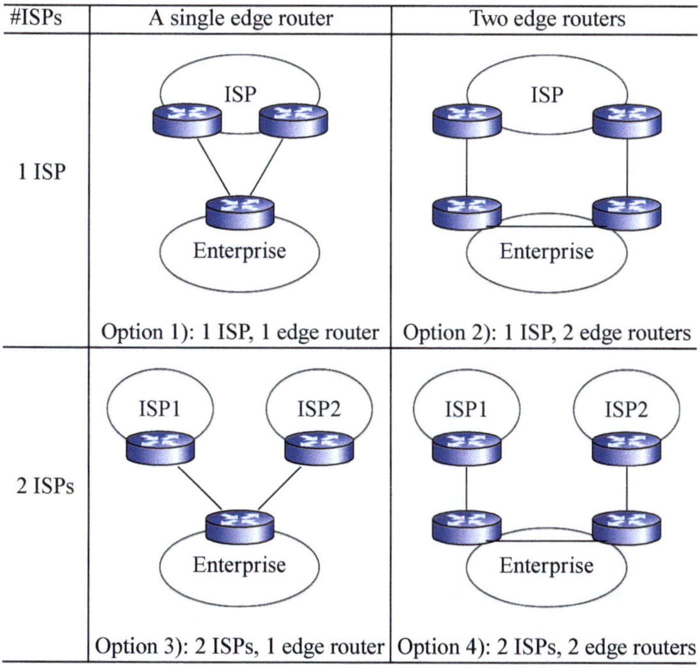

Fig. 5.5 Multihomed Internet connectivity

protocol within another. A typical use case of tunneling is to carry IPv6 packets over IPv4 networks. But for a secure VPN connection, the data packets are encapsulated and encrypted for their secure transport over third-party networks or the Internet.

VPNs are a cost-effective solution for users and telecommuters to connect to enterprise intranet or extranet. Because local Internet connectivity is widely available worldwide, connecting to an enterprise network through VPN over the Internet is simple. In general, a VPN server needs to be set up within the enterprise network for users to connect from outside. Also, a VPN client needs to be installed on the computer of the VPN user. The VPN server and client work together to establish and maintain a secure point-to-point VPN tunnel between the user and the enterprise network. A typical scenario of using VPN for secure connections to an enterprise network is shown in Fig. 5.6. Further details on the VPV technology will be discussed later in the chapter on security component architecture.

5.3 Flow-based Architectural Models

Network architecture can be analyzed from the traffic flow perspective. From the flow characteristics developed through traffic flow analysis in the previous chapter, this section focuses on architectural features in flow-based architectural models. Four

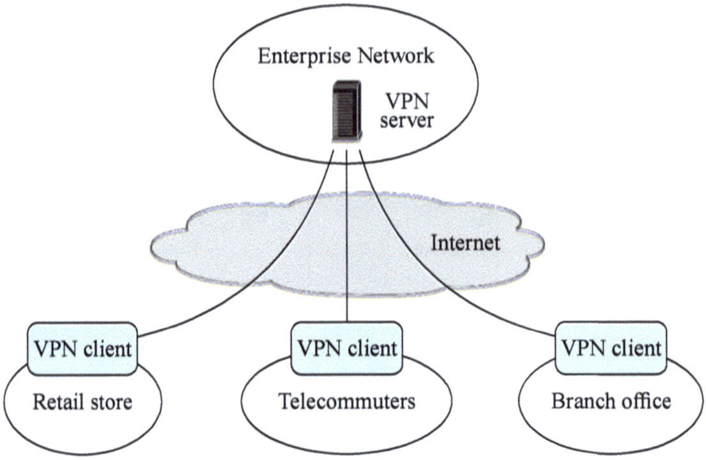

Fig. 5.6 Using VPN to connect to an enterprise network

popular types of flow-based architectural models used in computer networks will be discussed: peer-to-peer, client-server, hierarchical client-server, and distributed computing architectural models.

5.3.1 Peer-to-Peer Architectural Model

From the peer-to-peer traffic flow model discussed in the previous chapter, a *peer-to-peer architectural model* can be developed, in which no centralized control exists. Depending on the application scenarios, requirements, and/or constraints, the peer-to-peer architecture can be either fully meshed or partially meshed. This is shown in Fig. 5.7, in which a full-mesh peer-ro-peer core network and a partial-mesh peer-to-peer ad-hoc network are illustrated.

In the peer-to-peer core network architecture, the functions and features of the architecture are pushed to the edge of the enterprise network. Therefore, architectural planning should focus on the core layer and edge of the enterprise network.

In the ad-hoc network architecture, since there is a lack of support from a fixed infrastructure, the architectural features should concentrate on the connectivity of end nodes to the network to ensure effective network communication. In such cases, various factors will need to be considered such as bandwidth resources, throughput performance, latency, packet loss, and energy consumption.

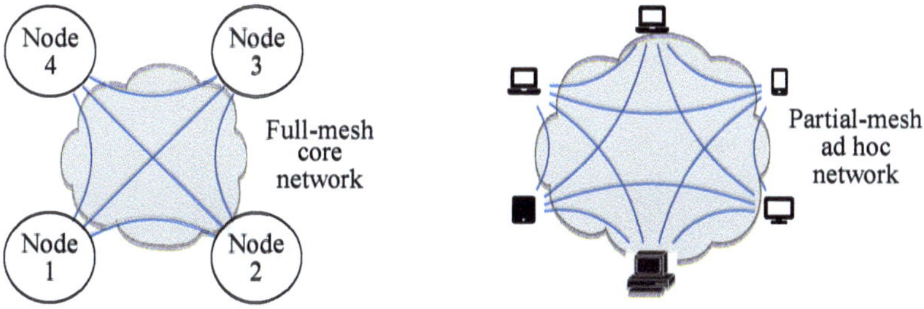

(a) Core network (b) Ad hoc network

Fig. 5.7 Peer-to-peer architectural model

5.3.2 Client-Server Architectural Model

Client-server architecture is a widely used architectural model in computer networking. Following the client-server flow model, the *client-server architectural model* is developed, in which a server responds to requests from clients through network communication. It is graphically shown in Fig. 5.8.

In this model, all traffic flows generated by the requests from clients are directed towards the location where the server is connected to the network. Therefore, the architectural features and functions are particularly applied to the server location in addition to the interfaces to client LANs and client-server flows. It is essential to ensure that the network has sufficient resources and capabilities to handle the traffic flows originating from the server and clients.

5.3.3 Hierarchical Client-Server Architectural Model

The *hierarchical client-server architectural model*, as illustrated in Fig. 5.9, is a variation of the client-server architectural model. It shares similarities with the client-server architectural model in the sense that architectural features, functions, and

Fig. 5.8 Client-server architectural model

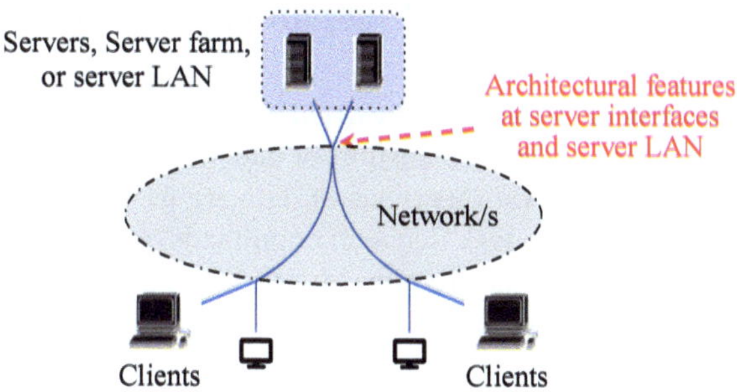

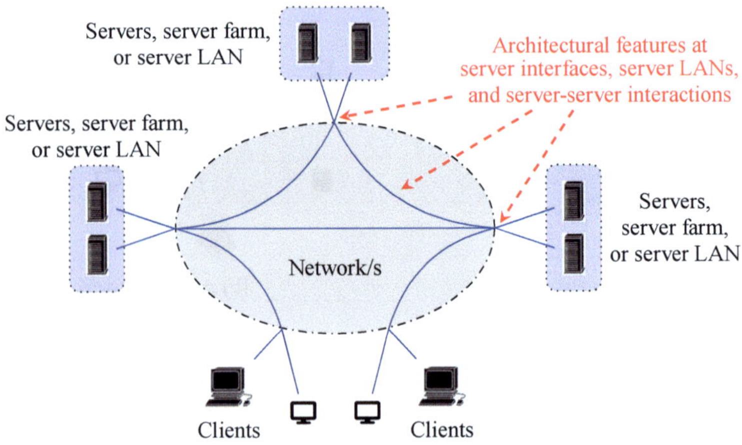

Fig. 5.9 Hierarchical client-server architectural model

characteristics are applied to server locations, interfaces to client LANs, and client-server flows.

However, in hierarchical client-server systems, multiple layers of servers work together to fulfill the requirements of the clients. This introduces traffic flows and interactions not only between clients and servers but also between the servers themselves. Therefore, when designing the hierarchical client-server architecture, it is necessary to consider architectural features, functions, and characteristics for both client-server and server-server interactions.

Efficient communication and coordination between servers at different layers are essential to ensure smooth operation and optimal performance of the hierarchical client-server system. This includes considerations such as load balancing, resource allocation, data synchronization, and inter-server communication protocols. By addressing the requirements of both client-server and server-server interactions, the hierarchical client-server architectural model provides a scalable and organized approach to handling complex network services and applications.

5.3.4 Distributed Computing Architectural Model

The *distributed computing architectural model*, as depicted in Fig. 5.10, is designed to facilitate efficient transport of data over a network to distributed computing tasks. In this model, various types of data are transferred over the network. They include job submission, task dispatch commands, monitoring information of the computing process, source data required for the computing, data generated from the computing, and final results. Therefore, architectural features are applied to the locations of data sources and sinks to ensure smooth and reliable data transfer.

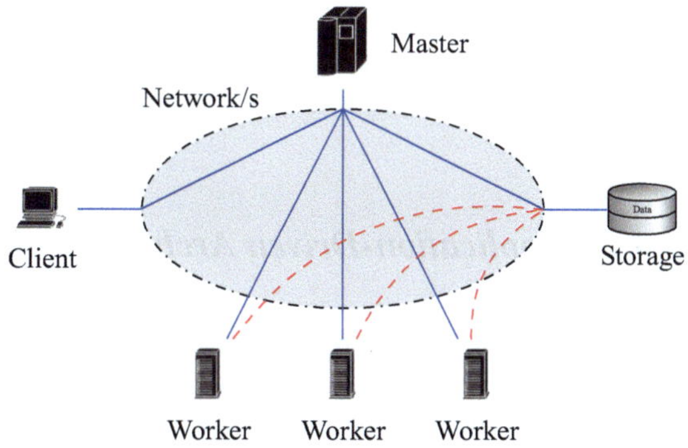

Fig. 5.10 Distributed-computing architectural model

The deployment of distributed computing can vary depending on the use cases and application scenarios. For example, within a university, distributed computing laboratories are often built as a cluster of nodes within a LAN. In such cases, Ethernet networking is typically used to interconnect low-end manager and worker nodes, such as desktop computers. High-end manager and worker nodes, such as workstations, HPCs, and storage servers, can be interconnected through Infiniband networking with fiber-optic Infiniband switching, in addition to Ethernet networking.

This dual-networking configuration allows for high-speed data transfer via Infiniband, while normal network communication is carried out through Ethernet. By combining these networking technologies, the distributed computing environment can achieve efficient data transfer and effective network communication, catering to the requirements of different types of nodes and distributed computing tasks.

5.4 Functional Architectural Models

The core/distribution/access mode, as mentioned before, provides a high-level functional architectural model with the focus on the functions and traffic flows within the network. However, there are additional functional architectural models that are designed to support specific functions and services from the application perspectives. These models provide frameworks for designing network architectures that meet the specific requirements and objectives of different applications, services, and network environments. By considering the unique features and needs of each model, network planners can develop architectures that effectively support the desired functionalities and services.

Some examples of functional architectural models are briefly discussed in [1, pp. 237–238]. This section will introduce a few functional architectural models commonly found in computer networks: application-driven model, end-to-end service model, intranet-extranet model, and service-provider model.

Recall that network services are provisioned end-to-end. Therefore, it is essential to understand the end-to-end features and requirements of important and critical network services in the development of functional architectural models.

5.4.1 Application-Driven Architectural Model

The *application-driven architectural model* focuses on providing predictable or guaranteed network performance for specific applications such as mission-critical, safety-critical, and time-critical applications. These applications rely on a network infrastructure that can provide predictable or guaranteed network performance, such as reliable connectivity, dedicated bandwidth, and bounded latency.

There are use cases that demand application-driven networks to connect distributed cloud application components with localized on-premises databases, as in health-care and hospital applications. These networks integrate various elements such as Internet of Things (IoT) devices, edge devices, and cloud services into an enterprise network, enabling distributed application components with local and remote datasets to collaborate effectively towards achieving the overall objective of the application (Fig. 5.11).

However, traditional networking, particularly WAN communication, is not inherently designed to support the connectivity requirements of distributed application components with local and remote data. This necessitates the development of an application-driven network architecture tailored to the specific needs of the application. Key considerations for such an architectural model include customized data paths based on the location of application components, application-based access policies, Quality of Service (QoS) provisions, and robust security measures.

From the implementation perspective, Software Defined Networking (SDN) offers a viable option for building application-driven networks. This can be accomplished by using APIs available in most SDN platforms. Leveraging SDN technologies makes the network programmable and flexible, thus helping ensure the alignment of the network infrastructure with the unique requirements of the application.

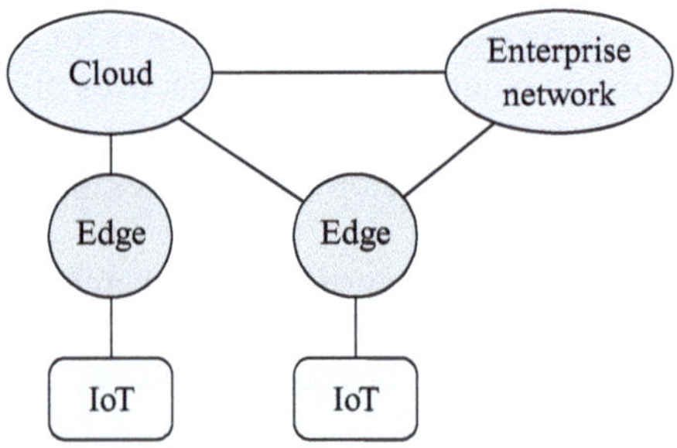

Fig. 5.11 An example of application-driven networking with distributed application components everywhere and localized/remote datasets

5.4.2 End-to-End Service Architectural Model

The *end-to-end service architectural model* focuses on the components that support end-to-end traffic flows within a network. It recognizes that network services are provisioned end-to-end. The performance of individual network services is also measured end-to-end. Some critical network services and applications require end-to-end performance guarantees.

To achieve end-to-end performance guarantees, end-to-end performance management such as Integrated Service (IntServ) can be implemented. IntServ allows for the measurement, control, and assurance of end-to-end performance. Identifying the specific services or applications that require end-to-end support is part of the system requirements analysis process. Designing an architectural model that supports the end-to-end traffic flows for those services or applications becomes a key task in network architecture design. Figure 5.12 shows an example of end-to-end architectural model that considers all components along the end-to-end path of the traffic flow.

5.4.3 Intranet-Extranet Architectural Model

In computer networking, it is a common practice to design and deploy intranet and extranet. Both intranet and extranet are private networks that serve specific purposes. An intranet creates connections within an organization for communication and information sharing among employees. An extranet extends a network beyond the boundaries of the organization, allowing external users to get limited access to authorized information. External users can be business partners, customers, collaborators, and other authorized individuals. Beyond the extranet lies the Internet, which

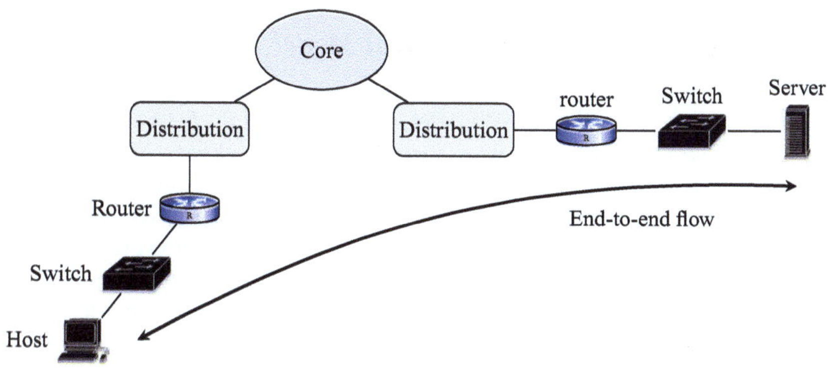

Fig. 5.12 An example of end-to-end architecture model that considers all components along the path of the traffic flow

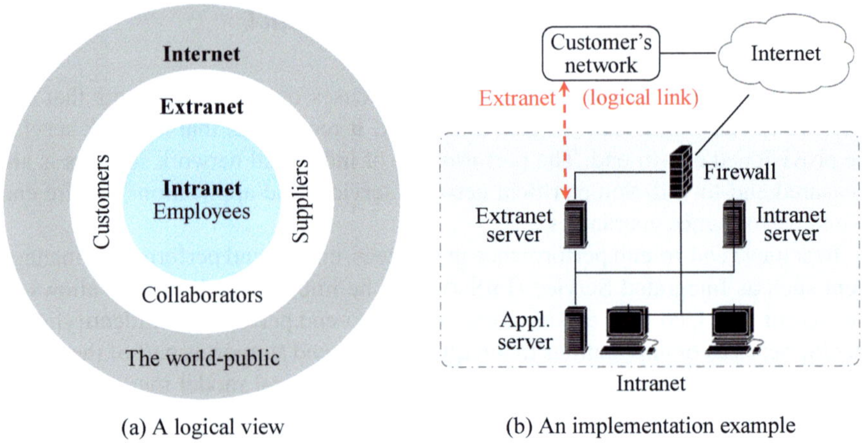

Fig. 5.13 Intranet/extranet architectural model

provides global network connectivity to a vast number of users and organizations worldwide. Figure 5.13 shows the *intranet-extranet architectural model.*

When setting up an extranet, particular attention must be given to security and privacy protection because external users can access the extranet through public networks. Separating users, applications, and devices into different groups will help manage security and privacy effectively. A firewall must be designed together with appropriate security policies for various application scenarios to safeguard against unauthorized access and potential threats. Depending on the specific requirements, multi-level security mechanisms may be necessary to enhance the overall security of the extranet environment.

5.4.4 Service-Provider Architectural Model

The *service-provider architectural model*, as shown in Fig. 5.14, focuses on the fulfillment of various requirements associated with the delivery of services from service providers to service customers. These requirements encompass aspects such as security and privacy, performance, and billing. Traditionally, service providers are seen as external entities that offer services to customers. However, in modern networks, there is a growing trend of internal service providers within organizations. Service customers can be individuals, work groups, or organizational departments within the organization. They subscribe to network services provided by service providers and receive services according to predefined SLAs.

It is worth mentioning that the service-provider architectural model discussed here should not be confused with the architecture for service provider networks such as those operated by ISPs. The architecture for service provider networks will not be

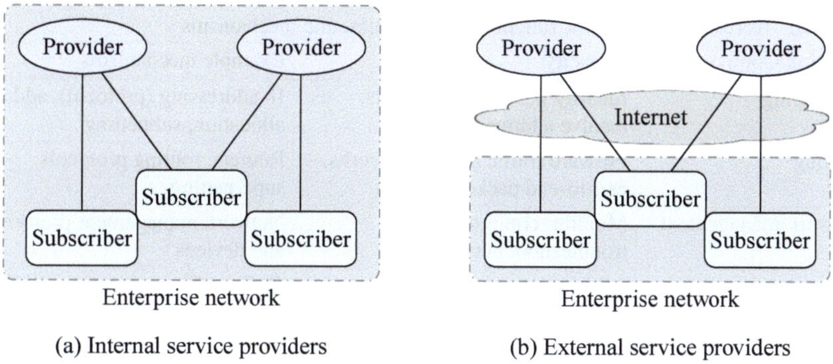

Fig. 5.14 Service-provider architectural model

discussed in this chapter. The focus of our discussions here is on the service-provider architectural model within enterprise and campus networks.

5.5 Component-Based Architectural Models

The functional architectural models described in the previous section address functions from the perspectives of applications, end-to-end services, intranet and extranet, and subscriber and provider. Therefore, these models do not concentrate on individual components of the network, but rather spread across multiple network components. Component-based architectural models can also be seen as functional architectural models in the sense that they describe how and where each network function is applied within the network. However, they have a specific focus on the functions of individual components of the network for network connectivity, scalability, manageability, performance, and security. These component-based architectural models provide a detailed understanding of the capabilities and requirements of each network component.

5.5.1 Component-Based Architecture Design

Each network component represents a major type of capacity within the network. It is supported by a set of mechanisms. These mechanisms encompass a range of hardware, software, protocols, policies, and techniques designed and deployed in the network to achieve its capacity under various constraints.

In this book, we will examine the following network components for network architecture:

Table 5.2 Network components, functions, capabilities, and mechanisms

Function/Component	Capacity	Example mechanisms
Addressing	Identify networks and hosts, resolve addresses	IP addressing (protocol), address allocation, subnetting
Routing	Forward traffic across networks, end-to-end packet delivery	Routers, routing protocols, supernetting
Network management	Monitor, configure, and troubleshoot network	Network management protocols and devices
Performance	Allocate resources in support of performance requirements	QoS, traffic engineering, DiffServ, IntServ, SLAs, policies, single- or multi-tier performance
Security	Manage network access, protect privacy	Firewalls, ACL, filters, IPsec, security policies
Enterprise Edge	Connect to external networks	WAN, Multihomed Internet, VPN
Redundancy	Provide reliable services and failover support	Redundant topology, protocols for redundancy, DNS settings

(1) addressing component,
(2) routing component,
(3) network management component,
(4) network performance component,
(5) network security component,
(6) enterprise edge component, and
(7) network redundancy component.

The functions of these network components are summarized in Table 5.2. The enterprise edge component has already been discussed previously, and the redundancy component will be discussed later in the next section. Below, we will provide a brief overview of the remaining components from the above list. However, a more comprehensive investigation into these components will be carried out in the upcoming chapters.

Developing a network component architecture requires to choose and determine the underlying mechanisms that support the component functions. Since multiple mechanisms work together within a component, it is important to understand and determine where each mechanism is deployed, how they interact, and what potential conflicts there might be. For example, if IntServ QoS is applied to a few specific services and Differentiated Service (DiffServ) is deployed for some other services in the same network, it becomes important to ensure that both IntServ and DiffServ coexist smoothly without compromising the QoS performance of each service.

Similarly, interactions between network components should also be taken into account. For instance, implementing security policies will introduce additional latency. Are the security solution and policies well designed without violating the latency requirements of critical applications? Will the management component affect the SLAs and other performance requirements? Will the addressing and routing components help enhance network performance and management? From the systems approach perspective, not only the decomposed individual components, but also their interactions and dependencies, determine the behavior of the overall network performance.

In general, trade-offs or balances are necessary among the functional components and the multiple mechanisms within each component. They are a fundamental requirement for network architecture design and overall network planning. By developing trade-offs or balances, more important services and their underlying mechanisms can be prioritized with sufficient resource support. Meanwhile, all other services and their underlying mechanisms are still functional under various network constraints.

5.5.2 Addressing Component Architecture

The addressing component architecture focuses on the selection of an addressing scheme to identify networks and network devices, the assignment of addresses to networks and devices, and the use of built-in features within the selected addressing scheme. Here are some key points regarding the addressing component architecture:

- It is widely recognized that IP addressing should be used in general networks. The choice between IPv4 and IPv6, will largely determine the format of data packets transmitted over networks or the Internet.
- Addresses are assigned hierarchically in blocks of a power of two. In residential, office, and enterprise networks, IPv4 private addresses can be utilized with the support of NAT.
- Both IPv4 and IPv6 have built-in features, which should be taken into account when developing an addressing component architecture. For example, the Differentiated Services field CodePoint (DSCP) embedded in the IP header can be used in conjunction with DiffServ for QoS performance management. IPv6 offers built-in security and true QoS, which can be leveraged to enhance network security and QoS management, respectively.

5.5.3 Routing Component Architecture

Working with a layer-3 addressing scheme, routing largely determines the efficiency of end-to-end delivery of data packets. *The routing component architecture* concentrates on the planning of traffic routing and forwarding across networks and the Internet from the traffic source to the intended destination. This requires to choose appropriate routing protocols, design route distribution if multiple routing protocols are used, and consider potential separation of routing decision and data forwarding.

There are different categories of routing protocols each with its own advantages and limitations. To choose a routing protocol, it is necessary to examine the suitability of candidate routing protocols based on the performance requirements of the network. The size or scale of the network will determine the requirements of scalability and convergence speed when selecting a routing protocol. Some routing protocols, such as OSPF, are simpler, while others, like BGP, are more complex. Some routing protocols are more suitable for larger networks with hierarchical structures.

A single routing protocol is generally recommended for an enterprise or campus network. However, there are situations where more than one routing protocol must be used, such as when integrating networks from two different companies each using a different routing protocol. In such cases, it becomes necessary to design route redistribution between the two different routing protocols.

Software Defined Networking (SDN) has been developed for programmable networking. It separates routing decision from data forwarding. For a specific network, it is worth investigating whether SDN is a better option than traditional routing. If SDN is considered to be a suitable choice, find out whether or not the network is SDN-ready.

5.5.4 Network Management Component Architecture

The network management component architecture aims to provide functions for implementing network monitoring, configuration, and troubleshooting. Therefore, it identifies the necessary functions, describes where and how they are applied, and chooses or designs appropriate mechanisms to support them.

As operating systems offer utilities, such as `ping`, to check network status and statistics, network management can be easily thought as an operational issue. However, before deploying and operating a network, several questions need to be answered. These include:

- What needs to be managed?
- What is the impact of the management on network performance and resource allocation?
- How are management data exchanged and stored?
- Which part of the management can be automated?

Moreover, not only networks need to be managed, but also the process of the network management itself needs to be managed, i.e., the management of the network management. Answering all these questions goes beyond operational issues. The network management component architecture will help clarify all these questions.

Topologically, network management operates within a layered hierarchy, which is consistent with the requirement of hierarchical networks. There are two main frameworks for network management: the IETF RFC 6632 [2], and the ISO CCITT

X.700 [3]/ISO/IEC 7498-4 [4]. Each framework is supported by various mechanisms, techniques, and protocols. In specific network scenarios, it is also possible to deploy protocols from both frameworks in a single network.

5.5.5 Performance Component Architecture

The performance component architecture focuses on meeting performance requirements through resource allocation under various constraints. As a set of levels of network capacity, network performance exhibits different requirements for different network services and applications. Since network services are best-effort by default, any additional performance requirements will be built upon this default best effort. Therefore, for a specific network, it is necessary to determine whether additional performance requirements can be fulfilled in a single-tier or multi-tier performance architecture.

To effectively address performance requirements, it is important to investigate and clarify what types of QoS requirements need to be fulfilled. To support the performance component architecture, it is advisable to use existing mechanisms if they are suitable for the requirements. However, if the existing mechanisms are not sufficient or do not meet the desired performance, it may be necessary to design new mechanisms specifically tailored to the network performance requirements.

Tiered-performance

Network performance can be managed in one or multiple tiers, as depicted in Fig. 5.15. If a group of services is required to meet a specific performance threshold, and if this is the only performance requirement, a single-tier performance architectural model can be designed to meet the performance requirement for the group of services. It will treat each individual service of the group equally with the same level of priority.

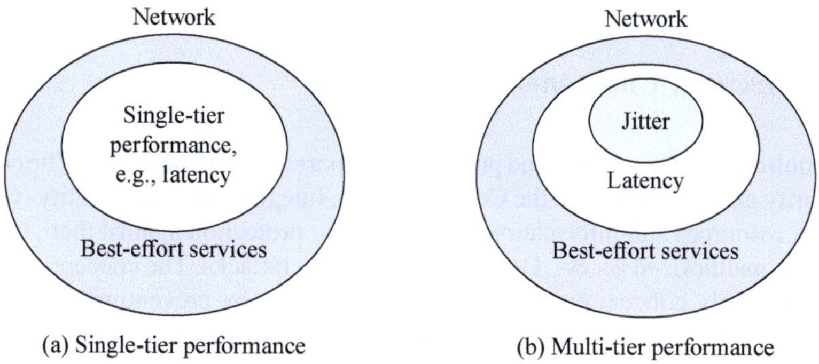

Fig. 5.15 Tiered-performance architectural models

If some services in the group of services have higher performance requirements than others, e.g., jitter in addition to latency, it is still possible to design a single-tier performance architectural model to meet the original (latency) and this additional (jitter) requirements. But this will lead to a waste of network resources because the services without the additional performance requirement will be over-provisioned with network resources. In such cases, a multi-tier performance architectural model would be a better design, which is able to provide network resources with differentiated levels of priority. It is implemented with performance mechanisms to fulfill the performance requirements.

Types of QoS requirements

There are different types of QoS requirements, each with its own characteristics. For example, SLAs document QoS requirements of different types between service providers and service customers, outlining the agreed-upon performance levels. Some services have requirements that can be defined by statistical or average performance metrics, such as average latency and bandwidth. These services can be managed based on Per-Hop Behavior (PHB) and traffic aggregation techniques. Other services may require the fulfillment of absolute performance thresholds, such as maximum end-to-end latency for individual flows. These services must be managed end-to-end to ensure that the desired performance levels are met. In the design of the performance component architecture, it is necessary to capture all these different types of performance requirements.

Performance mechanisms

The identified performance requirements can be supported by applying appropriate mechanisms, such as capacity planning, traffic engineering, and many others. For example, DiffServ can support PHB-based QoS performance management. In comparison, IntServ implements end-to-end performance management for individual flows. To support DiffServ, InteServ, and other QoS performance management mechanisms, appropriate resource reservation techniques and signaling protocols will need to be chosen in the development of performance component architecture.

5.5.6 Security Component Architecture

The requirements for security and privacy are important for all networks. The concept of security generally refers to the Confidentiality, Integrity, and Availability (CIA) of network resources and information, encompassing protection against theft, physical damage, unauthorized access, DoS, and various cyberattacks. The concept of privacy aspect primarily concentrates on safeguarding privacy by preventing unauthorized access and disclosure of sensitive information. *The security component architecture* addresses the requirements for both security and privacy aspects. Therefore, it is essential to investigate what security threats and risks there might be, what security and privacy mechanisms are available to mitigate these threats, and how these mechanisms can be integrated into the network to provide security and privacy protection.

Security risks in networks can originate from both internal and external sources. Information leakage and inappropriate uses of network resources by internal users within the network cause significant risks. These risks should be carefully addressed during various stages of network planning, design, management, and operation. While network security technologies continue to evolve, networks are becoming susceptible to external cyberattacks. Although some attacks can be detected, many others are not unknown to existing Intrusion Detection Systems (IDSs). To build an effective security and privacy protection system, it is necessary to develop a deep understanding of various cyberattacks.

Mechanisms currently available for network security include firewalls, ACLs, filters, IPsec and other security protocols, cryptography, and security policies. Some people consider NAT as a security mechanism because of the fact that private IP addresses behind NAT are not visible to external networks. This is a misconception about NAT. It is worth mentioning that NAT is not designed for security, and should not be relied on as a security solution. NAT is primarily adopted in IPv4 networks to conserve public IP address resources by using private IP addresses.

Integrating security mechanisms into the security component architecture is a complex task due to the diverse range of security risks that need to be addressed. There is no one-size-fits-all solution for security and privacy protection. Each security mechanism may be applied to specific security scenarios or targeted areas of the network. Multiple mechanisms can work together to provide comprehensive security and privacy protection for the entire network.

5.6 Redundancy Architectural Models

Network redundancy refers to the practice of incorporating additional resources into a network to enhance its reliability and availability, ultimately ensuring the protection of business functions. Because of this reason, it is specifically implemented for critical systems, services, devices, and paths. By introducing redundant components, the overall network system can continue to work even if one or more components fail. Therefore, network redundancy is a vital aspect of modern networking, enabling organisations to provide highly available services and maintain business continuity in the event of unexpected failures. In addition, network redundancy can also be used for load balancing, helping prevent traffic congestion and network bottlenecks. This section will discuss four types of network redundancy: router redundancy, workstation-to-router redundancy, server redundancy, and route and media redundancy.

5.6.1 Router Redundancy

In computer networking, a host relies on its gateway, which is the first-hop router, to communicate with other hosts outside its LAN. If the gateway fails, the host loses its network connectivity to all networks external to its LAN. Therefore, implementing redundant gateways are necessary to ensure uninterrupted network connections.

In an enterprise network, there are typically multiple routers that are strategically placed in different locations to serve different purposes. For example, an edge router installed at the network boundary establishes a connection between internal and external networks. If the edge router malfunctions, the entire enterprise network may lose its Internet connectivity. To mitigate this risk, having a backup edge router becomes essential, particularly when Internet connectivity is critical to the operations of the organization.

To achieve router redundancy, physical backup routers must be deployed at different network locations. They must also be reachable by workstations, switches, and other routers. Furthermore, mechanisms should be designed to effectively manage multiple routers and provide router redundancy. For example, how to find an alternative router to forward traffic in the event of a router failure.

In the following, we will discuss mechanisms related to first-hop redundancy, i.e., workstation-to-router redundancy. These mechanisms manage multiple routers and enable hosts to discover an alternative gateway should the primary router they are currently using fail.

5.6.2 Workstation-to-Router Redundancy

To ensure uninterrupted connectivity and access to external networks and the Internet, it is critical to establish reliable workstation-to-router connectivity in network architecture design. This highlights the importance of workstation-to-router redundancy in network planning.

For workstations to communicate with other networks and the Internet, they need to discover a suitable router. In scenarios where the currently used router becomes unreachable for any reason, it becomes necessary for the workstation to identify an alternative router. Various methods have been developed to facilitate workstation router discovery within a network. In some implementations, workstations are configured with an explicit and static setting of default (or primary) gateway and backup (or secondary) gateway, allowing workstations to automatically switch to the backup gateway if the primary gateway becomes unavailable. Other implementations let workstations discover a router automatically by using some protocols. Example protocols include:

- Address Resolution Protocol (ARP), which is a protocol implemented in operating systems, allowing workstations to discover and associate IP addresses with MAC addresses within the same local network.

- ICMP Router Discovery Protocol (IRDP): which is an IETF standard protocol that extends ICMP to enable hosts on multicast or broadcast networks to discover neighboring routers [5].
- Hot Standby Router Protocol (HSRP), which is a Cisco proprietary redundancy protocol that provides a fault-tolerant default gateway [6].
- Virtual Router Redundancy Protocol (VRRP), which provides first-hop redundancy for equipment from multiple vendors [7].
- Gateway Load Balancing Protocol (GLBP), which offers load balancing in addition to workstation-to-router redundancy.

These protocols and mechanisms ensure that workstations can automatically discover and select an appropriate router, thus maintaining connectivity and achieving redundancy in the network.

The Cisco HSRP

HSRP is a Cisco proprietary protocol, which manages multiple routers on the same subnet to provide workstation-to-router redundancy. Two versions of HSRP have been released so far, version 1 and version 2.

- HSRP version 1 is specified in the IETF RFC 2281 (March 1998) [6]. It will be discussed below in more detail.
- HSRP version 2 enhances version 1 by providing IPv6 support and increasing the number of HSRP groups available from 256 to 4, 096. Interestingly, no RFC has been found for HSRP version 2, indicating that it has not been formally adopted by the IETF.

It is worth mentioning that HSRP versions 1 and 2 are mutually exclusive. This means that HSRP version 2 is not backward compatible with HSRP version 1.

HSRP Functions

According to the descriptions in the IETF RFC 2281, with multiple routers physically reachable by hosts, HSRP enables the hosts to appear as if they were using a single router, thus maintaining network connectivity even if the current first-hop router fails. Multiple participating routers in the protocol work in concert and remain transparent to the hosts, creating the illusion of a single virtual router configured with an IP address and a MAC address. End hosts send their packets to the virtual router. At any given time, one and only one of the physical routers forwards packets on behalf the virtual router. If the active router fails, a standby router is activated to take over packet forwarding. The switching from the active router to the standby router is seamless and transparent to the communicating hosts, thus enhancing network connectivity.

Physical and Logical Views of HSRP

The physical and logical views of HSRP are demonstrated in Fig. 5.16. In the physical view, three routers are connected to a /24 LAN, for instance, 192.198.10.0/24. As depicted in Fig. 5.16a, the three routers are configured as normal routers with router interfaces and IP addresses within the network 192.168.10.0/24. For example, they are assigned the following IP addresses: R1 with 192.168.10.128/24, R2

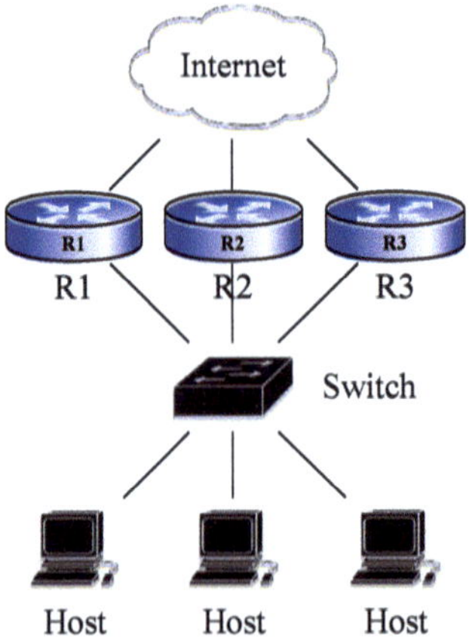

(a) Physical topology. IP addresses of the three physical routers are:
R1: 192.168.10.128/24;
R2: 192.168.10.160/24;
R3: 192.168.10.192/24.

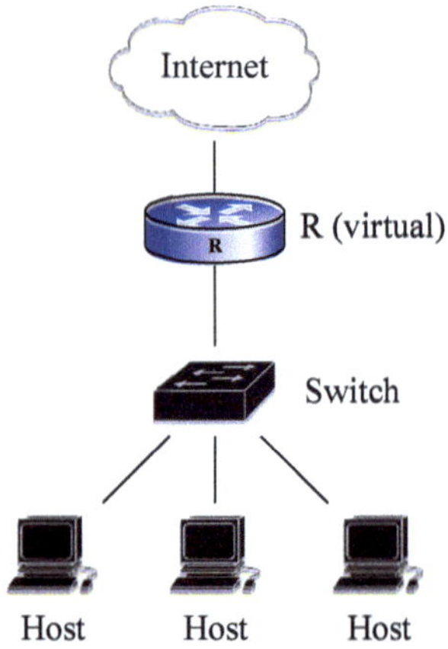

(b) Logical view to the hosts. The IP address 192.168.10.1/24 is assigned to the (virtual) router R, which logically forwards packets to the active router elected from the physical routers.

Fig. 5.16 The physical and logical views of HSRP in a local area network 192.168.10.0/24 with the default gateway 192.168.10.1/24. All hosts have IP addresses on this network 192.168.10.0/24

with 192.168.10.160/24, and R3 with 192.168.10.192/24, respectively. As physical devices, each physical router has its own unique MAC address.

If the three routers in Fig. 5.16a participate in HSRP, they form an HSRP group or a standby group. Within this group, the routers communicate with each other by using their own IP addresses. This communication takes place over UDP with port number 1985. From this HSRP group, one router is elected as the active router for packet forwarding, and the others become standby routers. If the active router fails or resigns from its role, another router in the HSRP group will become the active router and consequently take over the role of packet forwarding.

HSRP provides the advantage of transparency to the hosts on the LAN. The physical routers and their IP addresses within the HSRP group are hidden from the hosts. Instead, they work in concert to create the illusion of a single virtual router with its own IP address and MAC address. The MAC address of the virtual router, except for Token Ring, is 0000.0c07.ac**, where ** represents the HSRP group number. For Token Ring, three MAC addresses are available for HSRP: c000.0001.0000 for group 0, c000.0002.0000 for group 1, and c000.0004.0000 for group 2.

Corresponding to the physical view of HSRP in Fig. 5.16a, a logical view of HSRP is shown in Fig. 5.16b. In this figure, the virtual router is configured with an IP address 192.168.10.1/24, which is the default gateway of the LAN 192.168.10.0/24. This virtual router is the only router that is logically visible and accessible to the hosts on the network. Therefore, logically, regardless of the state (active or standby) of the physical routers in the HSRP group, network traffic from the hosts to the

```
 0                   1                   2                   3
 0 1 2 3 4 5 6 7 8 9 0 1 2 3 4 5 6 7 8 9 0 1 2 3 4 5 6 7 8 9 0 1 2
|-+-+-+-+-+-+-+-+-+-+-+-+-+-+-+-+-+-+-+-+-+-+-+-+-+-+-+-+-+-+-+-+-|
|   Version     |   Op Code     |     State     |   Hellotime   |
|-+-+-+-+-+-+-+-+-+-+-+-+-+-+-+-+-+-+-+-+-+-+-+-+-+-+-+-+-+-+-+-+-|
|   Holdtime    |   Priority    |     Group     |   Reserved    |
|-+-+-+-+-+-+-+-+-+-+-+-+-+-+-+-+-+-+-+-+-+-+-+-+-+-+-+-+-+-+-+-+-|
|                   Authentication  Data                        |
|-+-+-+-+-+-+-+-+-+-+-+-+-+-+-+-+-+-+-+-+-+-+-+-+-+-+-+-+-+-+-+-+-|
|                   Authentication  Data                        |
|-+-+-+-+-+-+-+-+-+-+-+-+-+-+-+-+-+-+-+-+-+-+-+-+-+-+-+-+-+-+-+-+-|
|                   Virtual IP Address                          |
|-+-+-+-+-+-+-+-+-+-+-+-+-+-+-+-+-+-+-+-+-+-+-+-+-+-+-+-+-+-+-+-+-|
```

Fig. 5.17 The format of HSRP datagram [6, p. 5]

Internet is always routed through this virtual router. But physically, the traffic is forwarded by the active router in the HSRP group. As mentioned previously, if a changeover of the active router occurs, HSRP ensures that network traffic to the Internet is routed through the new active router. This minimizes interruptions in the network connectivity of the hosts to the Internet.

HSRP Datagram

The HSRP datagram is 40-byte long. Its format is summarized in Fig. 5.17 [6].

Some of the octets in the datagram are described below:

- **Op Code**: The 1-octet Op Code specifies the type of message in the packet. The values 0, 1, and 2 represent Hello, Coup, and Resign messages, respectively. A Hello message from a router indicates its capability of becoming the active or standby router. A Coup message is sent when a router wishes to become the active router. A Resign message indicates that a router wishes to resign from its active router role.
- **State**: Each of the standby group routers maintains a state machine. The 1-octet State indicates the current state of a router at the time of sending the message. Possible values of the State are 0 for Initial, 1 for Learn, 2 for Listen, 4 for Speak, 8 for Standby, and 16 for Active, respectively.
- **Hellotime**: The 1-octet Hellotime is used only in Hello messages. It indicates the approximate period of time measured in seconds between Hello messages. If it is not configured or learned, it is recommended that a default value of 3 seconds be set.
- **Holdtime**: Also meaningful only in Hello messages, the Holdtime field is 1 octet. It indicates the amount of time measured in seconds that the current Hello message is considered valid. If the Holdtime is not configured or learned, it is recommended to set a default value of 10 seconds.
- **Priority**: The 1-octet Priority field is used to elect the active and standby routers. A router with a higher numerical value of priority supersedes other routers with lower numerical priority values. In the case of two routers with the same priority, the router with a higher IP address wins. This priority setting should not be confused

with the priority settings in real-time system scheduling, where a smaller integer value represents a higher task priority.

- **Authentication**: The Authentication field is 8-octet long, which contains a clear-text 8-character reused password. If no authentication data is configured, a default value is recommended, which is

  `0x63 0x69 0x73 0x63 0x6F 0x00 0x00 0x00`

  It is worth mentioning that while the authentication field helps prevent misconfiguration of HSRP, it does not provide security. HSRP can be easily subverted on the LAN, for example, through DoS attacks. However, "it is difficult to subvert HSRP from outside the LAN as most routers will not forward pakctes addressed to all-routers multicast address (224.0.0.2)" [6].
- **Virtual IP address**: The last field in the HSRP datagram is the 4-byte virtual IP address used by the group. During HSRP operation, at least one router in a standby group must know the virtual IP address. If a router in a standby group does not know the virtual IP address, it stays in the Learn state in the State field. In the Learn state without knowing the virtual IP address, a router is not allowed to Speak (in the State field), meaning that it cannot send periodic Hello messages. Otherwise, if a router knows the virtual IP address but is neither the active router nor the standby router, it remains in Listen state in the State field.

Other First-hop Redundancy Protocols

In addition to the Cisco HSRP, other fist-hop redundancy protocols are also available with different use case scenarios. Two such protocols are VRRP and GLBP, both of which are developed by Cisco.

VRRP is an open standard that can be used for first-hop redundancy in networks with equipment from multiple vendors. It is specified in IETF RFC 5798 (March 2010) [7]. Similar to HSRP, VRRP is also configured for a group of routers, i.e., gateways. But VRRP differs from HSRP in several aspects. Firstly, the master router in VRRP is manually configured by the network administrator. Also, unlike HSRP that uses a virtual IP address, VRRP uses the real IP address of the master's interface that connects to the subnet as the default gateway for clients. The backup members of the VRRP group keep communicating with the master router. If the master router is detected to be down, the backup members of the VRRP group will take over the role of the gateway to forward traffic. When the master router recovers from failure, it resumes its role as the gateway for forwarding traffic. It is interesting to note that when a backup gateway is functioning as the active gateway to forward traffic, the IP address it uses still belongs to the master gateway, which is the owner of the IP address.

GLBP is also a Cisco's proprietary protocol that can provide first-hop redundancy. While it shares similarities with HSRP and VRRP, it also has several distinct features. For example, GLBP uses virtual IP and MAC addresses as in HSRP. Also, similar to HSRP and VRRP, GLPBP maintains a group of routers. However, unlike HSRP and VRPP, all routers in a GLBP group are active and forward traffic. Therefore, this unique feature of GLBP allows for load balancing, as implied in the name of the protocol.

Within a GLBP group, each of the routers is referred to as an Active Virtual Forwarder (AVF). One of the routers will be elected as the Active Virtual Gateway (AVG), and the others serve as backup routers. One of the backup routers will assume the AVG role should the current AVG fail. The AVG has to main responsibilities:

- It assigns a *virtual* MAC address to each of the AVFs in the GLBP group, and
- It hands ARP requests by subnet devices, chooses an AVF to forward the traffic for each ARP request, and responds with the virtual MAC address of the chosen AVF to the ARP request.

When GLBP is configured, the default gateway is the AVG for the subnet. It has a *virtual* IP address, which remains the same across all devices within the subnet.

5.6.3 Server Redundancy

Depending on how critical a server is in a network, redundancy may need to be provided. In general, residential networks do not deal with critical network services and applications. Thus, redundancy is generally not required for servers. For enterprise networks, some network services and applications are mission-critical and thus require a certain degree of server redundancy.

What servers do we need to consider redundancy for? Potential servers are DNS, DHCP, file, web, mail, database, and other servers. If DNS servers fail, the IP addresses of Fully Qualified Domain Names (FQDNs) would not be resolved. Therefore, hosts within the subnet where the DNS servers are located may not be able to access remote sites and services. In an enterprise networks, database servers may be mission-critical and thus need to be protected through a redundancy design. Network requirements analysis will help identify critical servers that necessitate redundancy.

Strategies for Server Redundancy

There are various methods to achieve server redundancy. Some typical examples include:

- Fully-redundant servers with separate power supplies and separate networks,
- Mirrored disks or data copies, and
- Duplexed disks or data copies.

Full redundancy is typically recommended for mission-critical servers. Mirrored disks or data copies is a cost-effective solution that can be suitable for less critical applications. Duplexing offers an additional level of fault tolerance with redundant disk controllers.

Depending on the desired configuration and application requirements, server failover can operate in either an active-standby or active-active mode:

- In the active-standby mode, there is a primary active server, while one or more secondary and standby servers remain in a standby state. The standby servers are always ready to take over the role of the primary server, and a secondary server will assume the role should the primary server fail. Once the primary server is restored from failure, it resumes its role, and the secondary server returns to standby.

- In the active-active mode, two redundant servers are configured to be active. This configuration is typically related to load balancing, i.e., the two servers share the workload in normal operation. However, when a server fails, all traffic and workload will be shifted to the operational server.

DHCP Server Redundancy

In general, redundant DHCP servers are deployed in enterprise networks. This means that there are multiple DHCP servers in a LAN to ensure redundancy. Also, it is recommended that DHCP servers maintain mirrored copies of the DHCP database, which contains IP configuration information.

Where should DHCP servers be placed in a network? Here are some general guidelines:

- For small networks, place redundant DHCP servers at the distribution layer. This is based on the understanding that small networks typically do not experience excessive traffic when communicating with DHCP servers.
- For large networks, it is necessary to limit the traffic between the access layer and distribution layer. Therefore, place redundant servers at the access layer in large networks. This arrangement allows for localized DHCP service, reducing the need for DHCP requests to traverse the network core.
- For large campus networks, it is recommended to place DHCP servers on a different network segment from where the end systems are located. This often implies that the DHCP servers are located on the other side of a router. To enable DHCP functionality in such a scenario, the router must be configured to forward DHCP request broadcasts.

DNS Server Redundancy

DNS servers can be placed at either the access layer or distribution layer to resolve IP addresses of FQDNs. While they are important for network communications, they are less critical compared to DHCP servers. This is because we are able to communicate with a remote host through its IP address if the IP address is known. Many network servers are configured with static IP addresses, which are already known to us. For example, some ISPs have assigned static IP addresses to their mail servers and have publicly advertised these IP addresses. Nevertheless, it is necessary to plan for DNS server redundancy because it is not a reasonable assumption that the IP addresses of all remote sites are known.

It is worth mentioning that there are public DNS servers that can be used as backup solutions for DNS redundancy. Actually, many organizations, especially small business companies, use public DNS servers for DNS services. Two examples of public DNS servers are:

- Cisco OpenDNS: 208.67.222.222 and 208.67.220.220, and
- Google Public DNS: 8.8.8.8 and 8.8.4.4.

But keep in mind that these public or open DNS services may not always be reliable. They may close these services anytime. For example, as a previously well-regarded free public DNS provider, Norton ConnectSafe closed their public DNS services in November 2018.

Redundancy of File and Database Servers

Mission-critical file and database servers require full redundancy to ensure uninterrupted operation and avoid loss of valuable data. Typical examples include those in banks and other financial organizations. For such critical file and database servers, mirrored servers are highly recommended with separate power supplies and separate networks. They hold the same data so that if a server is down or physically damaged, the mirrored server can become active. To ensure data consistency and integrity, mirrored servers maintain instant synchronization for time-sensitive applications like stock exchanges. For non-real-time applications such as student grade databases, data synchronization can be performed in a batch processing manner, typically overnight.

For non-critical files and databases, if full redundancy is not feasible due to reasons such as budget constraints and other limitations, consider mirrored or duplexed hard drives of the servers, allowing for redundancy at the storage level. Many enterprise networks and data centers use Storage Area Networks (SANs) to enhance the reliability and availability of data. SANs are designed for highly reliable access to large amounts of stored data. Therefore, they provide an option for the redundancy of file and database servers.

Web Server Redundancy

Redundant web servers are useful for minimizing the website downtime and ensuring continuous service availability. When the primary web server fails or requires maintenance, the redundant web server takes over seamlessly, thus providing uninterrupted services to website visitors. The failover process is transparent to users.

An easy way to implement web server redundancy is to use a load balancer, which distributes incoming traffic across multiple web servers. The load balancer can be a hardware device or a software package. In the event of a failure, the load balancer automatically redirects traffic to the operational server, ensuring uninterrupted service. The failover process is automatic and almost instant. It is important to consider redundancy for the load balancer itself to avoid a single point of failure.

If the redundant web servers are located at two different sites, e.g., one on the enterprise network and the other in an external data center, consider placing the load balancer off-site. This can be accomplished by one of the following two options:

- Collocate a redundant hardware load balancer in a high up-time data center to maintain its high reliability and availability, or
- Use a cloud service to deploy a redundant software load balancer.

The choice between these two options depends on specific application requirements and constraints.

Another way to implement web server failover is through DNS settings. If the primary web server becomes inaccessible for a certain duration, it will fail over to the IP address of the secondary web server. When the primary web server becomes responsive again, it resumes its role as the main server.

In cases where the websites of an organization are hosted on an infrastructure provided by an external web hosting service provider, the responsibility for ensuring the reliability and availability of the web servers lies with the service provider. This should be clearly defined in the Service Level Agreements (SLAs) between the organization and the service provider.

Mail Server Redundancy

Mail services are an integral part of fundamental IT services. Modern organizations rely heavily on mail services for their business. In the higher education sector, which we are most familiar with, mail services, along with other IT and Internet services, are essential for normal teaching and learning, research, and other activities. Consequently, ensuring the redundancy of mail servers is a crucial requirement in an enterprise network. Mail servers are also known as mail exchange (MX) servers.

There are several options for redundant or backup MX mail servers, such as store mail solution, shared storage solution, and server replication solution. They are briefly discussed below:

- Store mail solution: This solution stores mail on the secondary server while the primary server is down. If the primary server becomes unavailable, the secondary server will receive and store mails on behalf of the primary server until the primary server is restored. However, this solution does not allow users to access and send mails through the secondary server because the primary server is the only one configured with the authentication details.
- Shared storage solution: This solution shares mail storage between the primary and secondary servers. Only one server is active at a time, and the other remains standby. Thus, the MX servers work in an active-standby mode. The MX-A record in the DNS settings points to the IP address of the active MX server.
- Server replication solution: This solution involves a cluster of mail servers each with independent local mail storage. All mail servers within the cluster are active and automatically synchronize their mail data. Therefore, they work in an active-active mode, and thus are capable of handling both failover and load balancing scenarios.

It is worth mentioning that an MX-record is a type of resource record in the DNS. It specifies the host name of the MX server that handles emails for a domain and a preference code. The lower the value of the preference code is, the higher the priority of handling emails for the domain is. Moreover, an MX A-record (address record) determines which IP address belongs to a domain name. Emails will be routed to the designated IP address set in the A-record of the host using the DNS.

To check MX records, we can use the utility `nslookup` in a terminal window. Type in `nslookup` to execute the utility. Default server and its address will display. Then, issue an command `set type=mx` to set the query type. After that, type in the

```
>nslookup
Default Server:  UnKnown
Address:  fe80::1

> set type=mx
> gmail.com
Default Server:  UnKnown
Address:  fe80::1

Non-authoritative answer:
gmail.com  MX preference = 20, mail exchanger = alt2.gmail-smtp-in.l.google.com
gmail.com  MX preference = 5, mail exchanger = gmail-smtp-in.l.google.com
gmail.com  MX preference = 40, mail exchanger = alt4.gmail-smtp-in.l.google.com
gmail.com  MX preference = 10, mail exchanger = alt1.gmail-smtp-in.l.google.com
gmail.com  MX preference = 30, mail exchanger = alt3.gmail-smtp-in.l.google.com
>
> set type=CNAME
> gmail.com
Default Server:  UnKnown
Address:  fe80::1

gmail.com
primary name server = ns1.google.com
responsible mail addr = dns-admin.google.com
serial  = 467166862
refresh = 900 (15 mins)
retry   = 900 (15 mins)
expire  = 1800 (30 mins)
default TTL = 60 (1 min)
>
```

Fig. 5.18 Examples of MX records and CNAME records looked up by using `nslookup`

domain name to be looked up and press Enter. The MX records of the domain will appear. Figure 5.18 shows Gmail's MX records and CNAME records as examples.

As for mail servers, it is also an option to use mail services from an external mail service provider. In this case, the mail service provider assumes responsibility for the reliability and availability of the mail servers in accordance with the SLAs.

A final note on the redundancy of web and mail servers is about the configuration of *DNS round robin*. DNS round robin configures a list of addresses in a circular mander for redundant web or mail servers. For each request, it responds with a different address from the circular list, thus providing a certain degree of redundancy. However, because other DNS servers on the Internet cache previous name-to-address mapping information, DNS round robin may not be able to provide fast failover of the servers. Nevertheless, as it is a simple configuration, DNS round robin can be an option when other solutions are not readily available for the network.

5.6.4 Route and Media Redundancy

On the Internet, multiple routes exist from one router to another. Therefore, in planning a network, there is typically no need to consider route redundancy on the Internet. Once network traffic reaches the Internet, routing protocols can effectively find a path to route the traffic on the Internet.

However, route and media redundancy should be considered within a network and from the network to the Internet. The connections of hosts to switches, switches to routers, and routers to other routers and WANs within the network are all essential to maintain network connectivity.

For WAN links, it is recommended to have a primary link and a backup link to establish a reliable connection to the Internet. These two links can be provided by two different ISPs, and may employ different technologies. For example, the primary WAN link could be a leased line, while the backup link could be an ISDN connection. Redundant WAN links can be configured in either an active-standby mode, or an active-active mode with load balancing.

Within a network, backup paths can be considered for connections to routers, switches, and other network devices. For critical routers and switches, meshed or partially meshed connections are recommended in order to minimize the impact of link failures on network performance. When desining backup paths, two key considerations are:

- The capacity of the backup paths, and
- The time required to switch to backup paths should the primary path fail.

It is noted that in comparison with the primary path, backup paths may use different technologies and may have lower capacity. Therefore, it is important to test the backup solution to ensure that it meets the required backup requirements. Tested backup paths can also be combined with load balancing techniques.

Meshed or partially meshed network connections can lead to loops. Loops may cause communication failure or performance degradation. For example, a router queries its neighboring routers for a path to an intended destination. If a neighboring router also does not know such a path, it will query its neighboring routers. Such a query may come back from a loop to the router that originally sent out the query.

Loops can be avoided by using the Spanning Tree Protocol (STP) specified in IEEE 802.1D. STP dynamically prunes a meshed or partially meshed topology of connected layer-2 switches into a spanning tree. The resulting topology is loop-free, which spans the entire switched domain with branches spreading out from a stem without loops or polygons.

In cases where the physical topology of a network changes, the established spanning tree may fail to work. STP will respond to the topological change of the network and build a new spanning tree. As this process takes time, some upper-layer network services may be timeout during the convergence period. If this happens, reconnection to these upper-layer services will be required. To speed up the reconstruction of the spanning tree, Rapid Spanning Tree Protocol (RSTP) has been developed in

IEEE 802.1w to supplement IEEE 802.1D. Building upon STP, RSTP provides rapid convergence of the spanning tree by

- Assigning port roles, and
- Determining the active topology.

This allows for faster recovery in case of topological changes.

5.7 Integration of Various Architectural Models

To fully capture the functions and features of a network, multiple architectural models discussed in previous sections need to be integrated, forming an overall architecture of the network. This integration process typically begins with a hierarchical topological model, such as the core/distribution/access model or LAN/MAN/WAN model. Then, incorporating other functional, component-based, and flow-based models into the base model as needed. Once all main functions and features of the network are characterized, a complete reference architecture is established.

Figure 5.19 shows the use of the core/distribution/access model as the base model for developing a complete network architecture. Various architectural models such as flow-based models, redundancy models, functional models, enterprise edge, and the LAN/MAN/WAN model are added to the base model. This provides a top-level view of the entire network architecture.

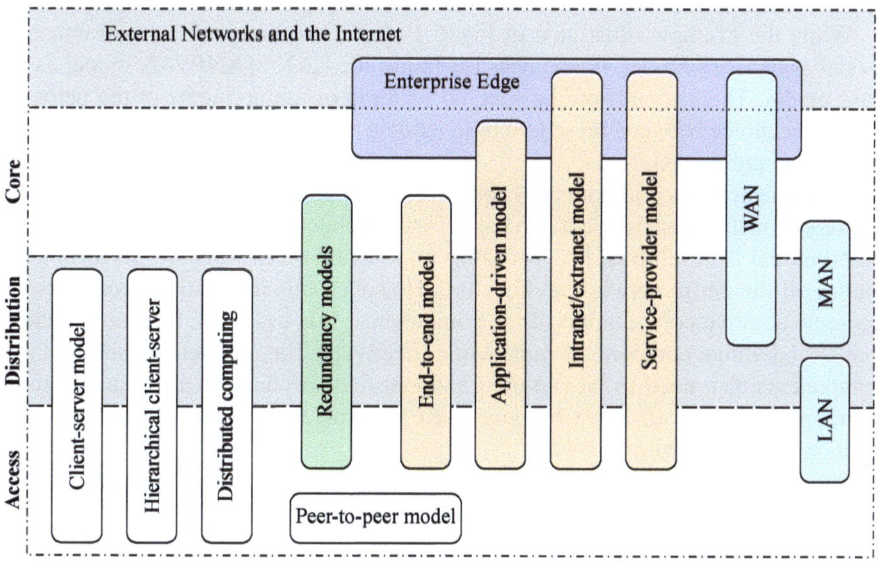

Fig. 5.19 Integration of network architectural models

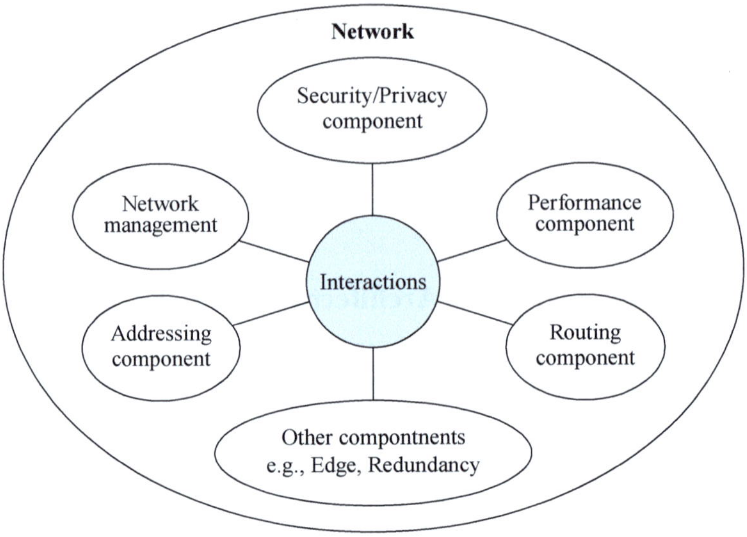

Fig. 5.20 Network architecture with integrated components

It is worth mentioning that depending on specific network requirements, the interactions between each integrated model and the core, distribution, and access layers may be different from those shown in Fig. 5.19. For example, if all applications operate independently without relying on external networks or service providers, the application-driven model may not interact with the edge model.

While the example illustrated in Fig. 5.19 has used the core/distribution/access model as the base model, it also possible to use the LAN/MAN/WAN model as the base model. This depends on the specific scenario or requirements of the network. Making a choice between these two base models requires a thorough analysis of the network requirements.

From the network component perspective, Fig. 5.20 depicts the integration of various component-based models into the network architecture. After each component-based model is developed, it is necessary to assemble them together to form a big picture of the entire network architecture. However, this integration process is not a simple additive combination of the components. It is essential, but not sufficient, that all individual components must work effectively. The interactions among these components also need to be captured and clarified. If conflicts arise among these components, trade-offs must be developed to resolve the conflicts and achieve a harmonious architecture.

The edge and redundancy components shown in Fig. 5.20 have already been discussed previously in detail. The other components in this figure have been introduced briefly in this section. Their detailed descriptions will be provided in the upcoming chapters.

5.8 Summary

As a high-level view of network structure, network architecture can be approached from various perspectives. Overall, it is typically developed with a hierarchical topology. The widely used core/distribution/access and LAN/MAN/WAN architectural models are hierarchical, each of which has three layers. The core/distribution/access model considers network architecture from the functional perspective, while the LAN/MAN/WAN model deals with network architecture from the physical perspective. These two architectural models are generally used as the starting point for network architecture planning.

With the considerations of traffic flows from various network services and applications, flow-based architectural models apply architectural features, functions, and characteristics at the locations of data sources, data sinks, data access interfaces, and data transport in the involved networks. Four typical flow-based architectural models have been discussed in this chapter: peer-to-peer, client-server, hierarchical client-server, and distributed computing architectural models. They are useful when specific considerations are required to meet the traffic flow requirements for network QoS, management, security, and other aspects.

Functional architectural models discussed in this chapter include application-driven, end-to-end service, intranet-extranet, and service-provider architectural models. These models describe functional requirements of the network in the development of network architecture. Critical applications may demand special attention in network architecture design to meet their specific QoS, timeliness, security, and other performance requirements. As network performance is measured end-to-end, ensuring end-to-end performance of network services is essential for services with QoS requirements. The design of intranet-extranet is closely linked to security architecture, which will be discussed later in detail. The service-provider architectural model emphasizes the provisioning of services from either internal or external providers.

Component-based architectural models are essentially functional models, which describe network architecture from individual network components. This chapter has introduced several essential components for network architecture design: addressing, routing, network management, performance, security, redundancy, and enterprise edge. The architecture design for each component is to identify its capacity, choose and design appropriate mechanisms to support the capacity, and investigate its interactions with other components. These components and their interactions determine the functional behaviors of the overall network.

References

1. McCabe, J.D.: Network Analysis, Architecture, and Design (3rd ed.). Morgan Kaufmann Publishers, Burlington, MA 01803, USA (2007). ISBN 978-0-12-370480-1
2. Ersue, M., Claise, B.: An overview of the IETF network management standard. RFC 6632, RFC Editor (2012). https://doi.org/10.17487/RFC6632
3. Union, I.T.: X.700: Management framework for open systems interconnection (OSI) for CCITT applications. CCITT X.700, ITU (1992)
4. for Standardization, I.O.: Information processing systems-open systems interconnection-basic reference model-Part 4: Management framework. ISO/IEC 7498-4:1989 (1st ed.). ISO (1989)
5. Deering, S.: ICMP router discovery messages. RFC 1256, RFC Editor (1991). https://doi.org/10.17487/RFC1256
6. Li, T., Cole, B., Morton, P., Li, D.: Cisco hot standby router protocol (HSRP). RFC 2281, RFC Editor (1998). https://doi.org/10.17487/RFC2281
7. Nadas, S.: Virtual router redundancy protocol (VRRP) version 3 for IPv4 and IPv6. RFC 5798, RFC Editor (2010). https://doi.org/10.17487/RFC5798

Chapter 6
Network Addressing Architecture

A network address is used to identify a network or network device on the Internet for data communication. It is an IP address in TCP/IP networks, such as the Internet. IP addressing largely determines the format of the packet transferred over a network or the Internet. It also determines how to route and deliver traffic from end to end. Therefore, network addressing architecture is an important component in the planning and design of the overall network architecture. It characterizes how IP addressing resources can be better used, how traffic routing can be improved, and how the hierarchy, isolation, and grouping of users and devices can be supported through various addressing mechanisms and strategies.

This chapter will begin with a review of the fundamentals of IPv4 addressing. Then, it will discuss some effective IPv4 addressing mechanisms, including classful versus classless addressing, fixed-length versus variable-length subnetting, supernetting, and private addressing. This will be followed by discussions of effective IP addressing strategies. In general, hierarchical IP addressing allocation is recommended. Both dynamic and static settings of IP addresses have their application scenarios. Regarding IPv6 addressing, after a brief introduction to IPv6, a few of the most significant features of IPv6 addressing will be discussed in more detail, such as autoconfiguration, built-in security, and built-in true Quality of Service (QoS). Furthermore, techniques for the coexistence of IPv4 and IPv6 are highlighted, such as dual stack, tunneling, and translation.

6.1 Review of IPv4 Address Representation

IPv4 is a layer-3 protocol specified in the IETF RFC 791 (September 1981) [1]. To extend its functionality, it has been updated in the IETF the IETF RFC 2474 (December 1998) [2] for the Differentiated Service field, and RFC 6864 (February 2013) [3] for the IPv4 Identification field. This section provides a brief review of IPv4 address representation.

Y.-C. Tian and J. Gao, *Network Analysis and Architecture*, Signals and Communication Technology, https://doi.org/10.1007/978-981-99-5648-7_6

6.1.1 IPv4 Address Structure

An IPv4 address is a 32-bit number, which serves as an identifier for a network or network device. It is represented by four octets, which are expressed in decimal digits and separated by dots. For example, the following address is a valid IPv4 address: 192.168.1.10. The corresponding binary representation for each of the four octets is shown in Table 6.1. Therefore, IPv4 addressing has an address space ranging from all 0s to all 1s across the eight bits of each octet. In decimal representation, the entire IPv4 address space spans from 0.0.0.0 to 255.255.255.255.

6.1.2 Binary and Decimal Representations

Both binary and decimal representations are used for IP addresses. Binary numbers are intrinsic to hardware machines while decimal values are more human-readable. The binary representation is particularly helpful for comprehending address classes, subnetting, and supernetting. Therefore, it is important to understand the relationship between these two representations. Figure 6.1 shows this relationship.

Table 6.1 Decimal and binary representations of IPv4 address 192.168.1.10

	1st Octet	2nd Octet	3rd Octet	4th Octet
Decimal	192	168	1	10
Binary	11000000	10101000	00000001	00001010

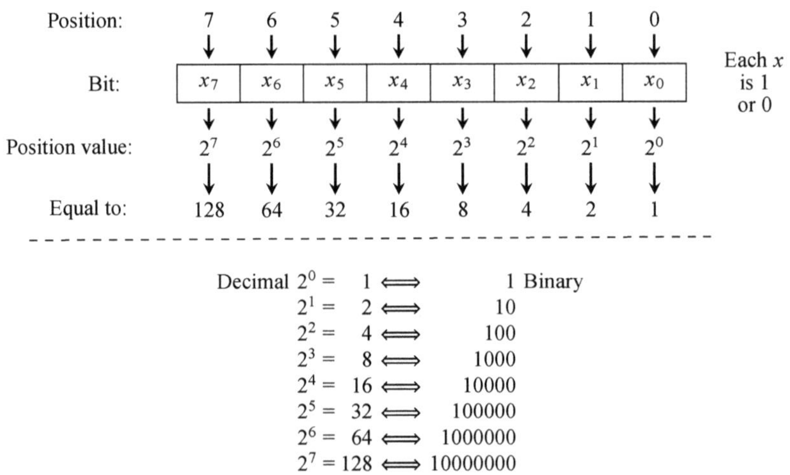

Fig. 6.1 Relationship between binary and decimal representations

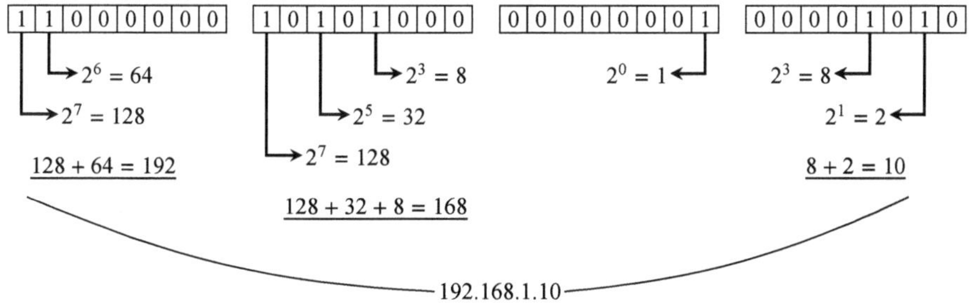

Fig. 6.2 An IPV4 address (Table 6.1): from binary to decimal format

From the relationship shown in Fig. 6.1, it is straightforward to convert between binary and decimal representations. For instance, let us consider the binary IP address 11000000.10101000.00000001.00001010 from the example depicted in Table 6.1. The process of converting this binary IP address to its corresponding decimal representation is illustrated in Fig. 6.2. The resulting decimal representation is 192.168.1.10.

Similarly, by using the relationship shown in Fig. 6.1, converting a decimal representation into the corresponding binary format is also easy. For example, for IP address 158.134.34.120, the following operations are carried out:

(1) For the first octet, we have

$$\begin{aligned}158 &= 128 + 16 + 8 + 4 + 2 \\ &= 2^7 + 2^4 + 2^3 + 2^2 + 2^1 \longleftrightarrow 10011110\end{aligned}$$

(2) For the second octet, it is seen that

$$\begin{aligned}134 &= 128 + 4 + 2 \\ &= 2^7 + 2^2 + 2^1 \longleftrightarrow 10000110\end{aligned}$$

(3) The third octet gives

$$\begin{aligned}34 &= 32 + 2 \\ &= 2^5 + 2^1 \longleftrightarrow 00100010\end{aligned}$$

(4) It follows from the fourth octet that

$$\begin{aligned}120 &= 64 + 32 + 16 + 8 \\ &= 2^6 + 2^5 + 2^4 + 2^3 \longleftrightarrow 01111000\end{aligned}$$

Finally, we have 158.134.34.120 $\Longleftrightarrow$ 10011110.10000110.00100010.01111000.

6.1.3 Static, Dynamic, and Automatic IP Addressing

A static IP address is a fixed IP address assigned to a device for consistent and permanent network communication. Once a device is configured with a static IP address, it typically retains that address permanently until it is decommissioned or there is a network architecture change that necessitates an address change. Static addresses are typically used for routers, servers, and other important devices.

In general, when a communicating node, such as a computer, boots up, it seeks a DHCP server to obtain an IP address. If the node successfully connects to a DHCP server, it dynamically acquires an IP address lease from the server. Before the lease expires, the node can renew the lease by communicating with the DHCP server. Once the lease expires, the DHCP server reclaims the address and can assign it to another node.

However, what happens if there is no reachable DHCP server? In such cases, the operating system can trigger the use of Automatic Private IP Addressing (APIPA) to allow the computer to self-configure an IP address and subnet mask automatically. The block of IP addresses reserved for APIPA is 169.254.0.0/16. This means that the usable range of APIPA addresses is from 169.254.0.1 to 169.254.255.254, with a subnet mask of 255.255.0.0.

While a host can use an autoconfigured IP address from APIPA when no DHCP server is reachable, it does not obtain a default gateway configuration in this scenario. Therefore, APIPA addresses are limited to use within a LAN. Devices configured with APIPA addresses follow a peer-to-peer communication model within the LAN. Furthermore, the activation of APIPA autoconfiguration often indicates a network connectivity problem on the host. It is recommended to troubleshoot the network connectivity issue and check the status of the DHCP server.

When an APIPA address is autoconfigured, the APIPA service periodically checks for the presence of a DHCP server every three minutes. If a DHCP server is detected on the network, the server replaces the APIPA address with a dynamically assigned address.

6.2 IPv4 Addressing Mechanisms

This section discusses several well-known IPv4 addressing mechanisms, which can be considered in network planning and design. The main IPv4 addressing mechanisms to be discussed below include classful and classless addressing, private addressing, subnetting, and supernetting.

6.2.1 Classful IPv4 Addressing

Classful addressing is an IPv4 address architecture, which is specified in the IETF RFC 791 (September 1981) [1]. To enhance its functionality, updates have been made to classful addressing in later RFCs including the IETF RFC 2474 (December 1998) [2] on the Differentiated Service field, and RFC 6864 (February 2013) [3] on the IPv4 Identification field.

In IPv4 addressing, five classes of IP addresses are defined. Figure 6.3 provides an overview of these five classes of IP addresses [1]. Among these classes, Classes A, B, and C are designated for **public addressing**. Class D is reserved for multicast addresses. Class E is reserved for future use. The number of networks and hosts that each of the Classes A, B, and C can represent is tabulated in Table 6.2.

It is seen from Fig. 6.3 and Table 6.2 that an IPv4 address is divided into three fields: Class ID, Network ID, and Host ID. Classes A, B, and C have class IDs of binary 0, 10, and 110, respectively, located at the beginning of the first octet. The network IDs for Classes A, B, and C have sizes of 7 bits, 14 bits, and 21 bits, respectively. As a result, the host IDs in Classes A, B, and C are represented using 24 bits, 16 bits, and 8 bits, respectively.

In an IP address, the Class ID and Network ID together form the *network portion* of the address, while the host ID forms the *host portion* of the address. Therefore, IP addresses in Classes A, B, and C,

- Class A (/8): 8 bits for the network portion, and 24 bits for the host portion,

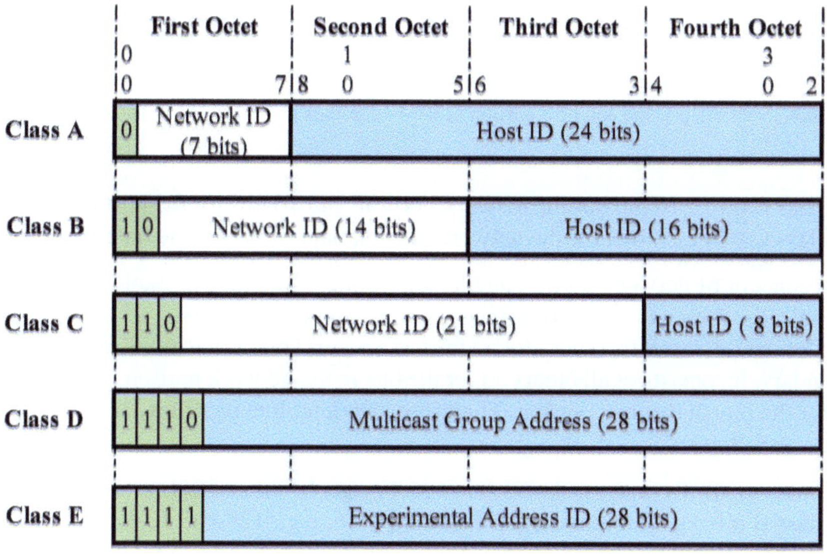

Fig. 6.3 Five classes of IPv4 addresses each begins with a Class ID followed by a network ID and then a host ID

Table 6.2 IPv4 address classes illustrated in Fig. 6.3

Class	Network portion			Host portion	
	Class ID	Network ID		Host ID	
		#bits	#networks	#bits	#hosts/network
A	0	7	2^7	24	2^{24}
B	10	14	2^{14}	16	2^{16}
C	110	21	2^{21}	8	2^8
$2^7 = 128$;		$2^8 = 256$;		$2^{14} = 16{,}384$;	
$2^{16} = 65{,}536$;		$2^{21} = 2{,}097{,}152$;		$2^{24} = 16{,}777{,}216$	

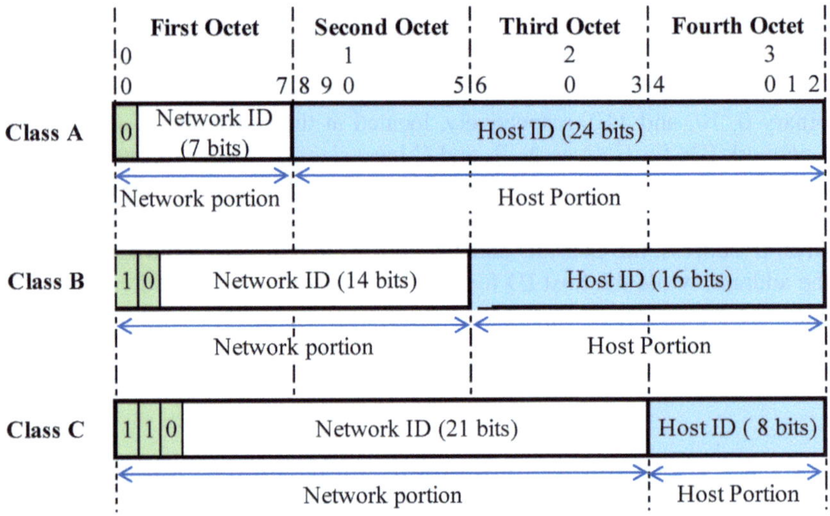

Fig. 6.4 Network portion and host portion in IPv4 addresses in Classes A, B, and C

- Class B (/16): 16 bits for the network portion, and 16 bits for the host portion, and
- Class C (/24): 24 bits for the network portion, and 8 bits for the host portion.

The concept of network portion and host potion is illustrated in Table 6.2. It is also graphically shown in Fig. 6.4 for Classes A, B, and C addresses.

The default mask for a classful address is a natural mask that aligns with the class boundary. It consists of all binary 1s for the bits in the network portion, and all binary 0s for the bits in the host portion. Therefore, the default masks for Classes A, B, and C are as follows:

- Class A (/8): Default mask 255.0.0.0 or /8, e.g., 10.90.20.30/8,
- Class B (/16): Default mask 255.255.0.0 or /16, e.g., 129.100.10.1/16,
- Class C (/24): Default mask 255.255.255.0 or /24, e.g., 192.168.1.10/24.

The biggest limitation of classful IPv4 addressing is its inefficiency in utilizing IP address resources. In classful addressing, each device requires a classful IP address, resulting in a significant number of classful addresses being necessary even for a small network. However, this is not practical in reality due to the scarcity of existing IPv4 address resources, which cannot adequately support the rapid growth of networks and the Internet. This has motivated the development of classless IPv4 addressing, and more recently IPv6 addressing.

6.2.2 Private Addressing

In each of the three Classes A, B, and C, there is a specific range designated for private IP addresses. The ranges of private IP addresses are as follows:

- Class A: 10.0.0.0 through 10.255.255.255,
- Class B: 172.16.0.0 through 172.31.255.255, and
- Class C: 192.168.0.0 through 192.168.255.255.

Private IP addresses are commonly used for LANs in residential, office, and enterprise environments. In an enterprise network, which serves numerous departments and workgroups, many of these segments can be designated as private networks and thus can use private IP addresses. For example, a computer laboratory in a university can be designed as a private network.

The use of private IP addresses offers several advantages. For example, it conserves public IP address resources, solving the problem of insufficient IP address resources. This is particularly beneficial for IPv4 networks where address resources are constrained.

Private IP addresses are routable within the private network where they are used. However, they are not visible to, and directly accessible from, public networks or the Internet. This contrasts with public IP addresses, which must be unique and globally identifiable to ensure proper communication over the Internet.

To enable communication between a network node with a private IP address and other network devices on the Internet, the technique of NAT is employed. NAT maps an IP address space into another, for example, from a private address apace to a public one. This mapping is achieved by modifying the network address information in the IP header of packets as they traverse a traffic routing device. NAT allows multiple devices within a private network, such as a home network, to share a single Internet-routable IP address assigned to a NAT gateway. This enables the devices to access the Internet using a single public IP address, optimizing the utilization of limited public IP addresses.

The concept of NAT, which includes basic NAT and Network Address Port Translation (NAPT), is formally specified in the IETF RFC 3022 (January 2001) [4]. This RFC defines the mechanisms and behavior of NAT. Additionally, the IETF RFC 2663 (August 1999) [5] describes the operation of NAT devices and associated considerations in general. It also defines the terminology used to identify different flavors of NAT.

6.2.3 Classless IPv4 Addressing

Classless addressing is a concept used for Classless Inter-Domain Routing (CIDR), which is specified in the IETF RFC 1518 [6] and RFC 4632 [7]. It eliminates the fixed natural boundary between the network portion and host portion of classful IP addresses. Specifically, classless addressing uses the most significant bits from the host portion of an address to divide a classful address into smaller subnets. This process in known as *subnetting*, which is defined in the IETF RFC 950 [8] and RFC 1878 [9]. With subnetting, the length of the host portion in an address is reduced, causing the boundary between the network portion and host portion to shift further into the host portion of the original classful address. This subnetting process is graphically shown in Fig. 6.5 for an illustrative example using a class B address 129.80.0.0.

It is seen from Fig. 6.5 that subnetting extends the original network portion by appending an subnet ID, which is represented by the bits borrowed from the original host portion, thus forming a new network portion for subnect addresses. The host portion of the subnet addresses is derived from the original portion by excluding the bits that have been used for the subnet ID. In the example provided in Fig. 6.5 for a class B address 129.80.0.0, borrowing 5 bits from the host portion results in a new network portion of /21 subnet addresses, where the original network portion of 16 bits are augmented by the borrowed 5 bits. If these 5 bits are set to 10001, it corresponds to the subnet address 129.80.136.0/21 with a subnet mask 255.255.248.0. The value 248 in the subnet mask is derived from the binary value 11111000 of the third octet of the IP address.

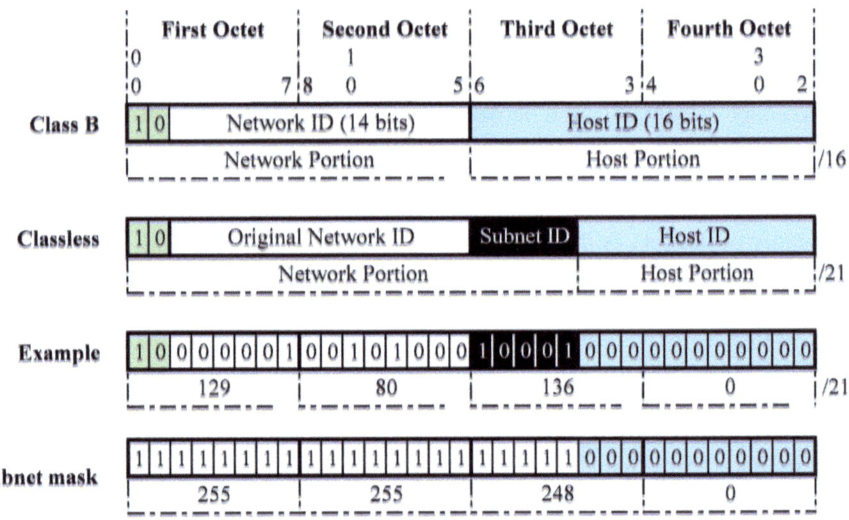

Fig. 6.5 Subnetting that uses one or more significant bits from the original host portion to form subnets

Table 6.3 Calculating the first address of a subnet

Subnet address	129.80.140.2	10000001.00101000.10001010.00000010
		AND
Subnet mask	255.255.248.0	11111111.11111111.11111000.00000000
First address	129.80.136.0	100000001.00101000.10001000.00000000

Given a subnet address and its subnet mask, it is easy to calculate the first address of the subnet. For example, if the given subnet address is 129.80.140.2 and subnet mask 255.255.248.0, we have the calculation process shown in Table 6.3.

6.2.4 *Fixed-Length Subnetting*

As specified in the IETF RFC 950 [8], fixed-length subnetting uses a fixed number of bits from the host portion of an address to represent the subnet ID. It segments an address into smaller addresses of equal size. Using a class B address as an example, Table 6.4 shows how fixed-length subnetting works. It demonstrates various scenarios where up to eights bits from the network portion of the address are used for subnetting. It is worth noting that since a class B address has 16 bits in its host portion, it is possible to segment the address into subnets by borrowing more than eight bits from the host portion of the address.

A class B address represents a /16 address, such as 129.80.0.0/16 (or denoted as 129.80.0.0 with subnet mask 255.255.0.0). If one bit is borrowed from host portion of the address, i.e., from the third octet of the address, to represent the subnet ID, the address can be segmented into two equal-size /17 subnets as the bit can have a binary value of either 0 or 1. These two /17 subnets are 129.80.0.0/17 and 129.80.128.0, respectively. The subnet mask for these two subnets is 255.255.128.0.

Table 6.4 Fixed length subnetting of 129.80.0.0/16 for equal-size subnets

#bits from 3rd octet	3rd octet value range			#subnets	Slash notation	Subnet mask
1	0 0000000	–	1 0000000	$2^1 = 2$	/17	255.255.128.0
2	00 000000	–	11 000000	$2^2 = 4$	/18	255.255.192.0
3	000 00000	–	111 00000	$2^3 = 8$	/19	255.255.224.0
4	0000 0000	–	1111 0000	$2^4 = 16$	/20	255.255.240.0
5	00000 000	–	11111 000	$2^5 = 32$	/22	255.255.248.0
6	000000 00	–	111111 00	$2^6 = 64$	/22	255.255.252.0
7	0000000 0	–	1111111 0	$2^7 = 128$	/23	255.255.254.0
8	00000000	–	11111111	$2^8 = 256$	/24	255.255.255.0

Similarly, if two bits are borrowed from the third octet of the address, four equal-size /18 subnets can be created from the four combinations of the binary values of the two bits: 00, 01, 10, and 11. Thus, the addresses of the four subnets are:

$$129.80.\boxed{00}000000.00000000/18 \rightarrow 129.80.0.0/18$$
$$129.80.\boxed{01}000000.00000000/18 \rightarrow 129.80.64.0/18$$
$$129.80.\boxed{10}000000.00000000/18 \rightarrow 129.80.128.0/18$$
$$129.80.\boxed{11}000000.00000000/18 \rightarrow 129.80.192.0/18$$

The subnet mask of these four subnets is:

$$255.255.\boxed{11}000000.00000000 \rightarrow 255.255.192.0.$$

Fixed-length subnetting can be visualized as cutting a pie into 2^i pieces of equal size, where i is an integer denoting the number of bits used from the host portion of the address for subnetting. This concept is illustrated in Fig. 6.6 for subnetting a Class B address.

- Using one bit for subnetting implies to cut the pie into two halves, i.e., two /17 subnets.
- With two bits for subneting, cut the pie into four quarters, i.e., four /18 subnects.
- If three bits are used for subnetting, the pie is cut to four quarters first, and then each of the four quarters is further cut into two halves, giving eight equal-size pieces, i.e., eight /19 subnets.

It is seen from this process that when one more bit is used for subnetting, simply further cut each of the pieces previously obtained into two halves, and meanwhile the value of the slash notation is incremented by 1.

6.2.5 *Variable-Length Subnetting*

In practical networking, different networks or user groups within an enterprise network often have very different sizes. Fixed-length subnetting, which divides the address space into equal-sized subnets, can result in inefficient utilization of address resources. For example, consider a network with 2,600 nodes. It would require a ./20 address, which provides an address space of $2^{32-20} = 2^{12} = 4,096$. But for a network with 400 nodes, a /23 address is sufficient, which offers an address space of $2^{32-23} = 2^9 = 512$. Therefore, assigning a /20 address to a network of 400 nodes would result in a significant wastage of address resources. The limitation of fixed-length subnetting has motivated the development of variable-length subnetting [8, 9].

Variable-length subnetting allows each segmented network to use an address of an appropriate size based on its actual requirements. This flexibility avoids the wastage of address resources that may occur when using fixed-length subnetting. For the

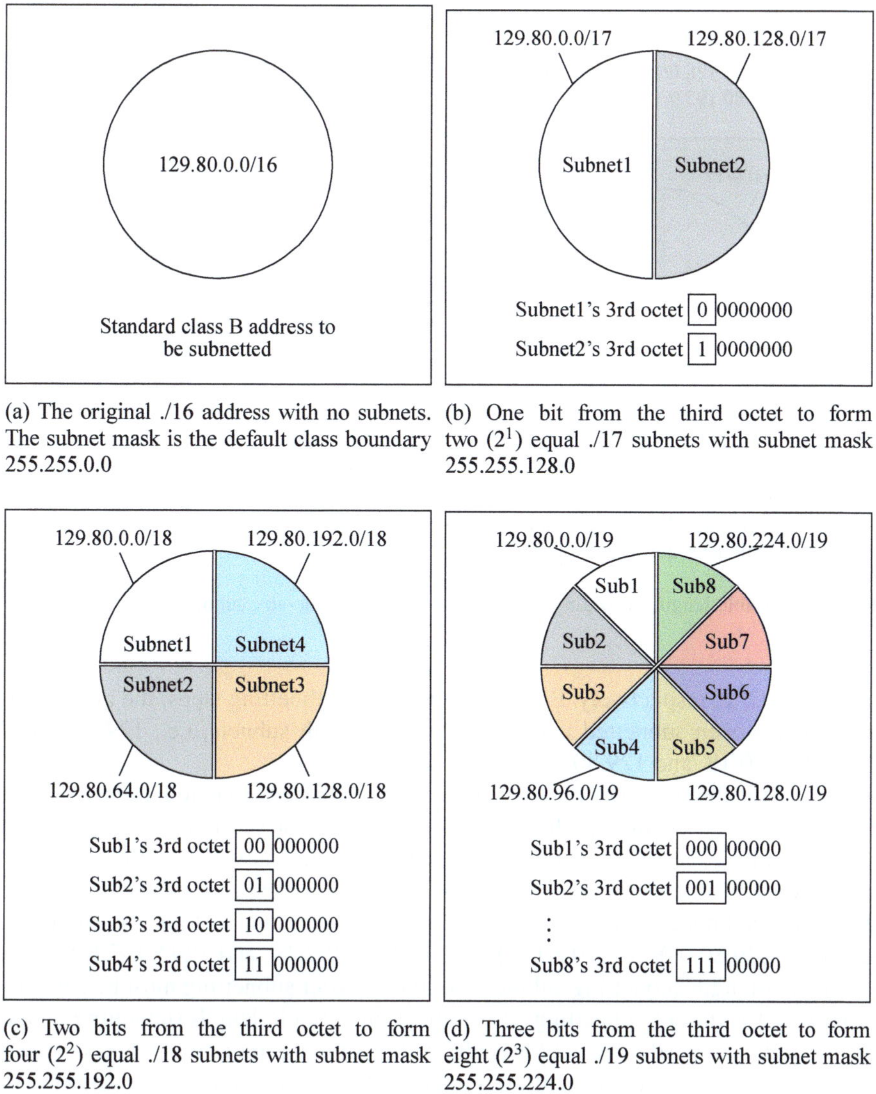

Fig. 6.6 Fixed-length subnetting of a /16 address for equal-size subnets

example mentioned above, a network of 2,600 nodes may use a ./20 subnet, while a network of 400 nodes may use a /23 subnet.

Let us discuss the principle of variable-length subnetting through the visualized example in Fig. 6.7. In the left subfigure of Fig. 6.7, we cut the pie, i.e., a /16 address space, into two equal halves, and assign the first half (129.80.0.0/17) to Subnet1. Next, we further cut the second half of the pie, i.e., a /17 address space, into two equal halves, and allocate them (129.80.128.0/18 and 129.80.192.0/18) to Subnet2_1

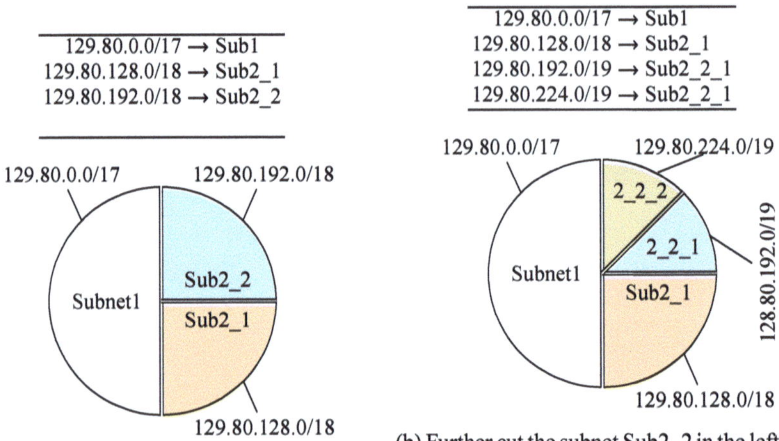

(a) Cut the pie to two halves. The first half is assigned to Subnet 1. The other half is further cut into two halves, which are allocated to Subnet2_1 and Subnet2_2, respectively.

(b) Further cut the subnet Sub2_2 in the left subfigure into two halves, which are then assigned to Subnet2_2_1 and Subnet2_2_2, respectively. Now, there are four variable-size subnets altogether.

Fig. 6.7 Variable-length subnetting of a /16 address for variable-size subnets

and Subnet2_2, respectively. By performing these subnetting steps, the original /16 address has been subnetted into three variable-length subnets, i.e., 129.80.0.0/17, 129.80.128.0/18, and 129.80.192/18.

Any of the three subnets obtained above can be further subnetted. Let us consider Subnet2_2 as an example, which is shown in the right subfigure of Fig. 6.7. Cut this piece of pie, i.e., a /18 address space, into two equal halves, resulting in two /19 addresses: 129.80.192/19 and 129.80.224.0/19. Assign these two addresses to Subnet 2_2_1 and Subnet 2_2_2, respectively. Now, we have a total of four variable-length subnects: 129.80.0.0/17, 129.80.128.0/18, 129.80.192/19, and 129.80.224.0/19.

In fixed-length subnetting, all pieces of the pie after subnetting must be cut in the same way for further subnetting. By contrast, in variable-length subnetting, not all pieces of the pie need to be cut for further subnetting. Keep in mind that if the pie or any piece of the pie after subnetting is to be cut for further subnetting, it must be cut into two halves or more generally 2^i equal-size pieces, where i is an integer valid for further subnetting. Cutting a piece of the pie into other number of equal-size pieces, e.g., 3 or 5 equal-size pieces, is invalid in subnetting.

6.2.6 *Supernetting*

Supernetting is the opposite of subnetting. Conceptually, subnetting segments a network into multiple smaller subnets, whereas supernetting aggregates multiple networks into a bigger one, known as a supernetwork or simply a supernet.

Motivation of Supernetting

Networks are primarily aggregated for the purpose of route summarization. To explain route summarization more clearly, let us recall that routers store known routes in their routing tables and advertise these routes to other routers for packet routing. The size of the routing table significantly impacts the convergence, stability, and overhead of the routing algorithm running on the router. Therefore, it is advantageous to reduce the size of the routing table without compromising timely routing decisions. Route summarization serves as an effective tool for achieving this goal. It consolidates routes to multiple networks with similar network prefixes into a single routing entry. This routing entry points to a supernetwork while encompassing all the individual networks.

Supernetting with Examples

Technically, subnetting shifts the boundary between the network portion and host portion of an address towards the host portion. This process involves 'borrowing' one or more most significant bits from the host portion to form a subnet ID, which becomes part of the network portion in the resulting subnet addresses. By contrast, supernetting operates in the opposite direction by moving the boundary towards the network portion. In supernetting, one or more least significant bits from the network portion are 'given away', resulting in a shorter network portion and a larger network address. This is graphically shown in Fig. 6.8.

Let us demonstrate how supernetting works. Consider a simple example of two addresses: 129.80.192.0/19 and 129.80.224.0/19, which are shown in Fig. 6.7. These two addresses can be aggregated into a single address: 129.80.192.0/18. Why? We have the process shown in Fig. 6.9. In these two addresses, the 19th bit in the network portion has a full range of binary values, i.e., 0 and 1, and all other more significant bits in the network portion do not change. This means that this bit can be dropped

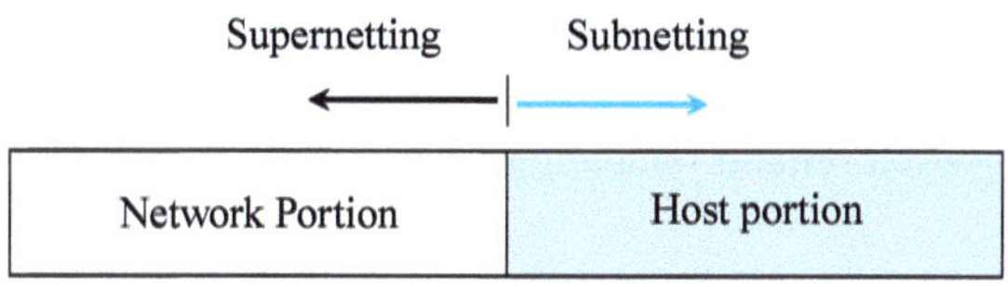

Fig. 6.8 Illustration of supernetting and subnetting

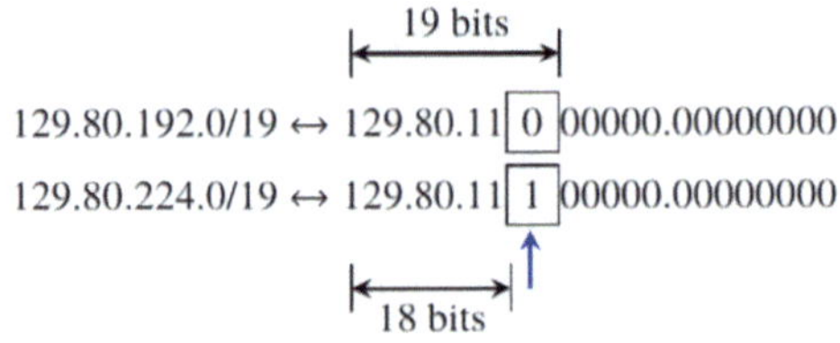

Fig. 6.9 Aggregation of two addresses

from the network portion to form a bigger network 129.80.192/18 without affecting the delivery of traffic to either of the two smaller networks 129.80.192.0/19 and 129.80.224.0/19. Any traffic sent to 129.80.192.0/19 or 129.80.224 can be simply routed to 129.80.192.0/18.

Now, let us consider a more complicated example of six addresses: 129.80.144.0/27, 129.80.144.32/27, 129.80.144.64/27, 129.80.144.96/27, 129.80.144.128/25, and 129.80.145.0/24. We have the following representations of these addresses:

129.80.144.0 /27 ↔ 129.80.10010000.00000000/27
129.80.144.32 /27 ↔ 129.80.10010000.00100000/27
129.80.144.64 /27 ↔ 129.80.10010000.01000000/27
129.80.144.96 /27 ↔ 129.80.10010000.01100000/27
129.80.144.128 /25 ↔ 129.80.10010000.10000000/25
129.80.145.0 /24 ↔ 129.80.10010001.00000000/24

For the first four /27 subnets, the two least significant bits in the network portion have a full range of binary values 00, 01, 10, and 11. All other bits in the network portion remain unchanged. Therefore, the last two bits in the network portion can be dropped to get a supernet 129.80.144.0/25 from the four /27 addresses. This leads to the following three networks:

129.80.144.0 /25 ↔ 129.80.10010000.00000000/25
129.80.144.128 /25 ↔ 129.80.10010000.10000000/25
129.80.145.0 /24 ↔ 129.80.10010001.00000000/24

In these three addresses, the first two /25 addresses have a network portion of 25 bits. Their first 24 bits are in common and the 25th bit has a full range of binary values 0 and 1. Thus, the 25th bit can be given away from the network portion without affecting traffic routing. This results in the following two supernets:

129.80.144.0/24 ↔ 129.80.10010000.00000000/24
129.80.145.0/24 ↔ 129.80.10010001.00000000/24

For these two /24 addresses, the least significant bit in the network portion, i.e., the 24th bit, covers the full range of binary values 0 and 1. The first 23 bits in the network portion remain unchanged. Therefore, these two addresses can be supernetted into a single /23 address 129.80.144.0/23. The corresponding mask in binary format is 11111111.11111111.11111110.00000000, which translates to 255.255.254.0 in decimal format.

Rules for Supernetting

For a block of addresses to be supernetted, several conditions must be satisfied. They are often summarized as *three specific rules* in many references:

(1) Contiguity: The addresses to be aggregated must be contiguous without any gaps.
(2) Same size and power of 2: The address to be aggregated must have the same size, and the number of the addresses must be a power of 2, i.e., 2^i, where i is an integer.
(3) Divisibility: The first address must be exactly divisible by the number of networks to be supernetted. Therefore, it must be either zero, or an even number that is a multiple of the block size.

While these rules are correct, they can be simplified into *a single general rule* as follows:

> *The non-common bits in the network portion of the addresses to be supernetted must have a full range of binary values from all 0s to all 1s.*

This general rule captures the essence of the three specific rules mentioned above and provides a more concise and comprehensive guideline for supernetting.

Let us examine a few examples shown in Table 6.5. We consider example (E1) first with the following four addresses:

129.80.144.0 /27 $\leftrightarrow$ 129.80.10010000.00000000/27
129.80.144.32 /27 $\leftrightarrow$ 129.80.10010000.00100000/27
129.80.144.64 /27 $\leftrightarrow$ 129.80.10010000.01000000/27
129.80.144.96 /27 $\leftrightarrow$ 129.80.10010000.01100000/27

There are 27 bits in the network portion of these four addresses. Among these 27 bits, the first 25 bits are common while the 26th and 27th bits are non-common. The non-common 26th and 27th bits have a full range of binary values from all 0s to all 1s, i.e., 00, 01, 10, and 11. Therefore, these four addresses can be supernnted into a single address, which is 129.80.144.0/25, as discussed previously.

In this example, The non-common bits in the network portion of the addresses cover the full range from all 0s to all 1s. It is easy to show that this block of four addresses meets the three supernetting rules discussed above: (1) Contiguous, (2) Same size and power of 2, and (3) Exactly divisible by the block size as the fourth octet of the first address is zero.

Similarly, for example (E2) shown in Table 6.5, the two addresses 129.80.144.64/27 and 129.80.144.96/27 are two consecutive /27 addresses, implying that there is no gap between these two addresses. The number of same-size (/27) addresses is 2, which is a power of 2. The fourth octet of the first address is 64, which is exactly divisible by the block size (2). Therefore, the two addresses meet the three

Table 6.5 Supernetting examples

Example	Addresses	Supernet
(E1)	129.80.144.0/27 ↔ 129.80.10010000.00000000/27 129.80.144.32/27 ↔ 129.80.10010000.00100000/27 129.80.144.64/27 ↔ 129.80.10010000.01000000/27 129.80.144.96/27 ↔ 129.80.10010000.01100000/27	129.80.144.0/25
(E2)	129.80.144.64/27 ↔ 129.80.10010000.01000000/27 129.80.144.96/27 ↔ 129.80.10010000.01100000/27	129.80.144.64/26
(E3)	129.80.144.0/27 ↔ 129.80. 10010000.00000000/27 129.80.144.32/27 ↔ 129.80. 10010000.00100000/27 129.80.144.96/27 ↔ 129.80. 10010000.01100000/27	Invalid (not contiguous)
(E4)	129.80.144.32/27 ↔ 129.80.10010000.00100000/27 129.80.144.64/27 ↔ 129.80.10010000.01000000/27	Invalid
(E5)	129.80.144.0/27 ↔ 129.80.10010000.00000000/27 129.80.144.128/25 ↔ 129.80.10010000.10000000/25	Invalid (not same size)

supernetting rules and can be supernetted into a single address 129.80.144.64/26. If we examine the network portion of these two addresses, we can observe that the the 27th bit is a non-common bit, which has a full range of binary values 0 and 1. According to the simplified single suppernetting rule, these two address can be aggregated into a single address.

Next, consider example (E3) illustrated in Table 6.5 for the following three addresses:

129.80.144.0/27 ↔ 129.80.10010000.00000000
129.80.144.32/27 ↔ 129.80.10010000.00100000
129.80.144.96/27 ↔ 129.80.10010000.01100000

Three same-size addresses are not in a power of 2 and thus cannot be summarized into a single address. The non-common bits of the network portion of these addresses are the 26th and 27th bits, which do not have a full range of binary values from all 0s to all 1s because the binary value 10 is missing. This missing value means a gap 129.80.144.64/27 ↔ 129.80.10010000.01000000. Using a pie to visualize these addresses as we did previously, we will see that a quarter of the pie is missing.

For the example (E4) demonstrated in Table 6.5, the two addresses 129.80.144.32/27 and 129.80.144.64/27 have their 26th and 27th bits non-common in their network portion. These two bits have binary values 01 and 10, which do not cover the full range of all 0s and all 1s because the values 00 and 11 are missing. Therefore, these two addresses cannot be summarized into a single address.

Finally, we examine example (E5) shown in Table 6.5. The two addresses 129.80.144.0/27 and 129.80.144.128/25 do not have the same size: one is a /27 address and the other is a /25 address. Therefore, they cannot be suppernetted into a single address.

6.3 IPv4 Addressing Strategies

All addressing mechanisms discussed in the previous sections are useful in the development of addressing strategies for network planning and design. In most cases, many of them are used together in a network. For example, dynamic addressing is used for end hosts, while static addressing is used for routers and some servers, such as web servers. Public addresses are always used, while private addressing is almost ubiquitous in residential, office, and enterprise networks. Variable-length subnetting has been a standard technique in networking to conserve address resources, while supernetting is a standard tool for efficient routing.

In order to use addressing mechanisms in a systematic manner, addressing strategies should be planned, developed, documented, and managed, particularly for large-scale networks. Their necessity is also justified by the fact that no mechanisms exist for automatic and dynamic subnetting and subnet allocation. Ad-hoc subnetting and subnet allocation will cause difficulties in network management, troubleshooting, and future upgrades.

6.3.1 Effective Addressing Strategies

Many addressing strategies have been shown to be effective in practical networking in terms of the scalability and manageability of the resulting network architecture. Some of these strategies are listed below:

- Design a structured model for addressing prior to address assignment.
- Assign blocks of addresses hierarchically to enhance network scalability and manageability. Thus, the aforementioned structured model for addressing follows a hierarchical approach. When referring to a *block* of addresses, it means a power of 2, such as 2^2, 2^3, 2^4, and 2^5 addresses.
- Allocate blocks of addresses based on the physical locations of networks rather than user group membership. This prevents issues when individuals or groups of users relocate from one physical location to another. If users from a work group are located in multiple physical networks, a VLAN can be created for them.
- Allow for future network growth by reserving a certain number of addresses in each network. The number of addresses to be reserved depends on the network size and projected future growth, and there is no standard answer.
- Use private addresses with NAT where possible to conserve address resources within residential, office, and enterprise networks.
- Implement dynamic addressing for end systems to achieve network flexibility and minimal configuration. However, routers and certain servers such as web servers and email servers require static addressing for the benefits of routing.
- Addressing and addressing-related management are generally centralized in a network. However, for large-scale networks that span multiple cities, states, or even countries, addressing authority can be delegated to regional networks.

While the main focus of this section is on IPv4 addressing strategies, it is worth noting that the addressing strategies listed above are also applicable in IPv6 addressing, with the exception of NAT. NAT is not required or recommended in native IPv6 networks. The topic of IPv6 addressing will be discussed later in this chapter.

6.3.2 *Hierarchical Address Assignment*

Among various addressing strategies, an important aspect is to establish network hierarchy, design a structured hierarchical addressing model, and assign addresses hierarchically. Let us use an example to discuss how to deal with all these tasks required for hierarchical address assignment.

Network Hierarchy from Requirements Analysis

Network hierarchy can be established through requirements analysis. It shows the organizational structure, physical locations of organizational departments or subnetworks, and the number of users or hosts in each subnetwork. An example of a network hierarchy and requirements obtained from requirements analysis is shown in Table 6.6.

In this example, the network consists of two Autonomous Systems: AS1 and AS2. AS1 is located in City 1 (C1), while AS2 is spread across four cities: City 2 (C2) through City 5 (C5). The organizational departments are shown as follows:

- D1 through D3 in three different physical buildings within City 1 of AS1;
- D4 through D6 in three different physical buildings within City 2 of AS2; and
- D7, D8, and D9 in Cities 3, 4, and 5, respectively, of AS2.

The number of hosts in each organizational department is also clarified, as shown in Table 6.6.

Table 6.6 Network hierarchy and requirements for IP address allocation

AS	Location	Department	#Hosts
AS1	C1: City 1	D1: HQ in A Block	2,600
		D2: Finance in B Block	400
		D3: HR in C Block	400
AS2	C2: City 2	D4: Engineering in Building A	150
		D5: Sales in Building B	
		D6: Customer Service in Building C	40
	C3: City 3	D7: Sales in a building	55
	C4: City 4	D8: Sales in a building	23
	C5: City 5	D9: Sales in a building	36

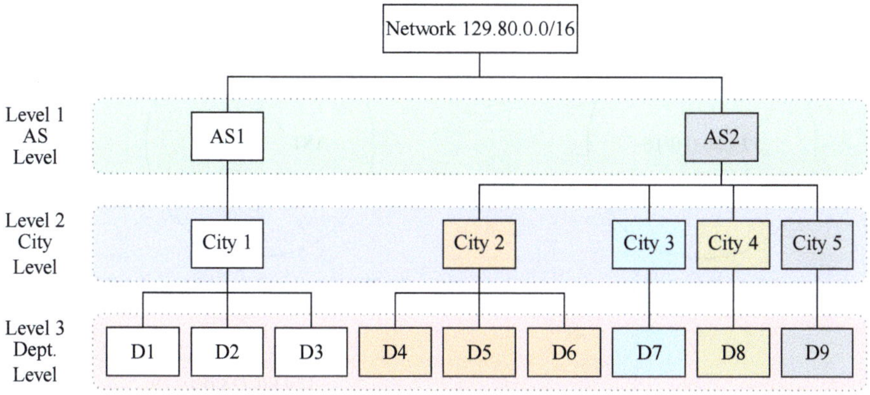

Fig. 6.10 Three-level hierarchical IP addressing for the requirements shown in Table 6.6

Structured Hierarchical Model for Addressing

For the requirements listed in Table 6.6, there are many different ways to solve the IP address assignment problem, leading to different solutions. Nevertheless, building a structured model for IP address assignment is highly recommended, particularly for large-scale networks. The model can be built with a layered hierarchical structure using a top-down approach.

From a top-down perspective, the network hierarchy shown in Table 6.6 motivates us to use a three-layer hierarchical model for addressing. The top level is the AS level, which deals solely with AS1 and AS2. The second level deals with cities, and the bottom level is dedicated to the department level. This hierarchical model with three levels for addressing is demonstrated in Fig. 6.10, in which the IP address allocation will be discussed below.

Variable-Length Subnetting and Hierarchical Address Assignment

Using the hierarchical model illustrated in Fig. 6.10, we now subnet the network address 129.80.0.0/16 with variable-length subnetting and assign addresses hierarchically. For clear visualization, we follow the previously discussed procedures of cutting a pie into pieces of different sizes for variable-length subnetting. This process is demonstrated in Fig. 6.11.

For the AS level, we cut the pie of a /16 address into two halves, which represent two /17 addresses of equal size. Then, assign one half of the pie to AS1 and the other half to AS2. Therefore, we have 129.80.0.0/17 for AS1, and 129.80.128.0/17 for AS2, respectively.

Then, we move down to the city level. The subnetting process is as follows:

- As AS1 is located in City 1, we simply allocate the corresponding half of the pie to that city, implying that there is no need for further subnetting.
- However, AS2 is spread across four cities C2 through C5. We will need to cut the corresponding half of the pie into four equal-size pieces for the four cities. This

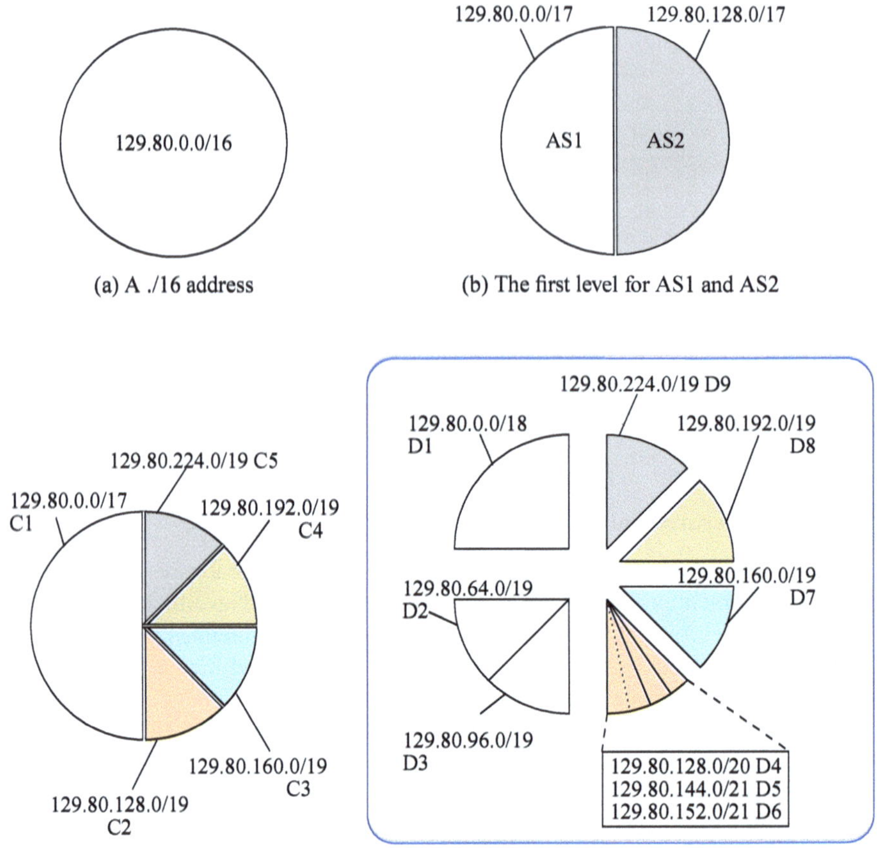

(a) A ./16 address

(b) The first level for AS1 and AS2

(c) The second level for cities

(d) The third level for departments

Fig. 6.11 Hierarchical allocation of a Class B IP address 129.80.0.0/16 for the requirements shown in Table 6.6

results in four /19 addresses: 129.80.128.0/19, 129.80.160.0/19, 129.80.192.0/19, and 129.80.224.0/19. Then, allocate these four pieces (/19 addresses) to the four cities C2 through C5.

Next, we consider the department level. We design subnetting and addressing for this level using the following procedure:

- In City 1, there are three departments D1, D2, and D3, each located in a separate physical building. Therefore, the first half of the pie (129.80.0.0/17) is cut into two equal halves, giving two subnets 129.80.0.0/18 and 129.80.64.0/18.
  - We could assign 129.80.0.0/18 to D1.
  - We further cut 129.80.64.0/18 into two equal halves, yielding two equal-size addresses 128.80.64.0/19 and 129.80.96.0/19. Assign these two /19 addresses to D2 and D3, respectively.

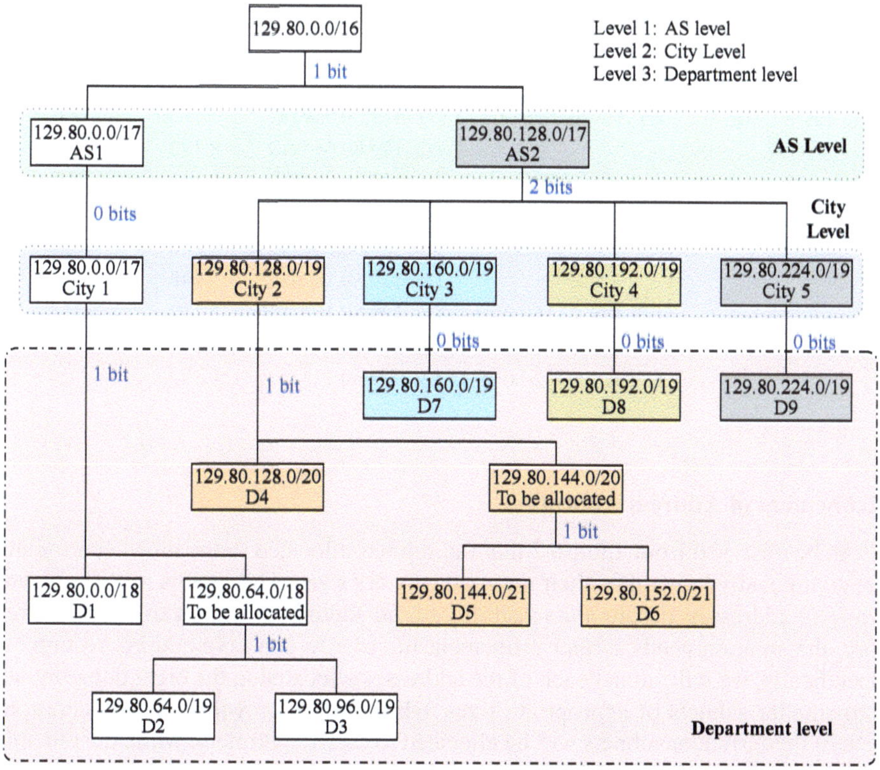

Fig. 6.12 Block diagram of three-level hierarchical IP addressing for the requirements shown in Table 6.6

- In City 2, there are three departments D4, D5, and D6, each located in a separate physical buildings Therefore, we divide the corresponding address 128.80.128.0/19 into two equal halves, resulting in two equal-size /20 addresses: 129.80.128.0/20 and 129.80.144.0/20.
  - Assign 129.80.128.0/20 to D4.
  - Further cut 129.80.144.0/20 into two equal halves, giving two equal-size addresses: 128.80.144.0/21 and 129.80.152.0/21. Assign these two /21 addresses to D5 and D6, respectively.
- For each of Cities 3, 4, and 5, there is only one department. Thus, there is no need for further subnetting. Simply assign the corresponding address to the department.

The resulting hierarchical address allocation is depicted in the block diagram shown in Fig. 6.12, which implements the hierarchical model designed previously in Fig. 6.10. The address allocation is summarized in Table 6.7. This solution represents a feasible allocation for the address assignment problem.

Table 6.7 Feasible address allocation for the requirements specified in Table 6.6. It can be further refined to save address resources

AS	City	Department	#Addr	#Hosts
AS1 129.80.0.0/17	C1: 129.80.0.0/17	D1: 129.80.0.0/18	>16 k	2,600
		D2: 129.80.64.0/19	8,192	400
		D3: 129.80.96.0/19	8,192	400
AS2 129.80.128.0/17	C2: 129.80.128.0/19	D4: 129.80.128.0/20	4,096	150
		D5: 129.80.144.0/21	2,048	70
		D6: 129.80.152.0/21	2,048	40
	C3: 129.80.160.0/19	D7: 129.80.160.0/19	8,192	55
	C4: 129.80.192.0/19	D8: 129.80.192.0/19	8,192	23
	C5: 129.80.224.0/19	D9: 129.80.224.0/19	8,192	36

Refinement of Address Allocation

It can be observed from Table 6.7 that the subnets allocated to the nine departments are significantly larger than their required network sizes. This results in a significant waste of addresses, despite the feasibility of the address allocation solution. Therefore, the solution needs further refinement in order to conserve address resources. Specifically, we will subnet each of the addresses allocated to the eight departments into smaller subnets of appropriate sizes, taking future growth into consideration. One of these smaller subnets will be allocated to each department, while the remaining subnets will be reserved for future use. A refined solution to the address allocation problem is illustrated in Table 6.8 for AS1 and Table 6.9 for AS2, respectively.

Table 6.8 Refined address allocation to AS1 129.80.0.0/17 with subnets of appropriate sizes meeting the requirements in Table 6.6

AS	City	Subnet → Department	#Addr	#Hosts
AS1 129.80.0.0/17	C1 129.80.0.0/17	129.80.0.0/20 → D1	4,096	2,600
		129.80.16.0/20 (Reserved)	4,096	
		129.80.32.0/19 (Reserved)	8,192	
		129.80.64.0/23 → D2	512	400
		129.80.66.0/23 (Reserved)	512	
		129.80.68.0/22 (Reserved)	1,024	
		129.80.72.0/21 (Reserved)	2,048	
		129.80.80.0/20 (Reserved)	4,096	
		129.80.96.0/23 → D3	512	400
		129.80.98.0/23 (Reserved)	512	
		129.80.100.0/22 (Reserved)	1,024	
		129.80.104.0/21 (Reserved)	2,048	
		129.80.112.0/20 (Reserved)	4,096	

Table 6.9 Refined address allocation to AS2 129.80.128.0/17

AS	City	Subnet → Department	#Addr	#Hosts
AS2 129.80.128.0/17	C2 129.80.128.0/19	**129.80.128.0/24 → D4**	256	150
		129.80.129.0/24	256	Reserved
		129.80.130.0/23	512	Reserved
		129.80.132.0/22	1,024	Reserved
		129.80.136.0/21	2,048	Reserved
		129.80.144.0/25 → D5	128	70
		129.80.144.128/25	128	Reserved
		129.80.145.0/24	256	Reserved
		129.80.146.0/23	512	Reserved
		129.80.148.0/22	1,024	Reserved
		129.80.152.0/26 → D6	64	40
		129.80.152.64/26	64	Reserved
		129.80.152.128/25	128	Reserved
		129.80.153.0/24	256	Reserved
		129.80.80.154/23	512	Reserved
		129.80.80.156/22	1,024	Reserved
	C3 129.80.160.0/19	**129.80.160.0/26 → D7**	64	55
		129.80.160.64/26	64	Reserved
		129.80.160.128/25	128	Reserved
		129.80.161.0/24	256	Reserved
		129.80.162.0/23	512	Reserved
		129.80.164.0/22	1,024	Reserved
		129.80.168.0/21	2,048	Reserved
		129.80.176.0/20	4,096	Reserved
	C4 129.80.192.0/19	**129.80.192.0/27 → D8**	32	23
		129.80.192.32/27	32	Reserved
		129.80.192.64/26	64	Reserved
		129.80.192.128/25	128	Reserved
		129.80.193.0/24	256	Reserved
		129.80.194.0/23	512	Reserved
		129.80.196.0/22	1,024	Reserved
		129.80.200.0/21	2,048	Reserved
		129.80.208.0/20	4,096	Reserved
	C5 129.80.224.0/19	**129.80.224.0/26 → D9**	64	36
		129.80.224.64/26	64	Reserved
		129.80.224.128/25	128	Reserved
		129.80.225.0/24	256	Reserved
		129.80.226.0/23	512	Reserved
		129.80.228.0/22	1,024	Reserved
		129.80.232.0/21	2,048	Reserved
		129.80.240.0/20	4,096	Reserved

6.4 IPv6 Addressing

As an improved version of IPv4, IPv6 is defined in the IETF RFC 4291 (February 2006) [10] and RFC 8200 (July 2017) [11]. This section gives a brief introduction to IPv6 addressing, discusses why IPv6 is developed to improve IPv4, and reviews some unique features of IPv6 addressing.

IPv6 is designed to solve the problem of IPv4 address exhaustion. It allows for a significant larger number of public IP addresses on the Internet. In comparison with an IPv4 address that is 32-bit long, an IPv6 address is defined with 128 bits. Thus, IPv6 is four times as long as IPv4 in size, giving a vastly expanded address space.

In recent years, the adoption of IPv6 has been steadily increasing. As of the time of writing this section, global IPv6 deployment has reached 35.6% of overall networks. This adoption rate is monitored by organizations such as Google, Facebook, and Asia Pacific Network Information Centre (APNIC). Figure 6.13 depicts the global IPv6 deployment presented by the Internet Society, based on the statistics from Google, Facebook, and APNIC. It clearly shows a significant trend of increasing IPv6 adoption.

6.4.1 *IPv6 Notation*

As IPv6 addresses are long, it is inconvenient to represent them in decimal numbers as done for IPv4. Instead, IPv6 addresses are written in a shortened format consisting of eight blocks of hexadecimal numbers separated by colons. Each block, also known as a quartet, represents two hexadecimal numbers. The general form of an IPv6 address is as follows:

```
x:x:x:x:x:x:x:x
```

where each 'x' separated by colons represents a quartet. For example, the following representation is a valid IPv6 address:

```
2001:0db8:3c4d:0015:0000:0000:1a2f:1a2b
```

The range of IPv6 address space spans from all 0s to all 1s across all 128 bits in binary format. In hexadecimal format, the range can be represented as:

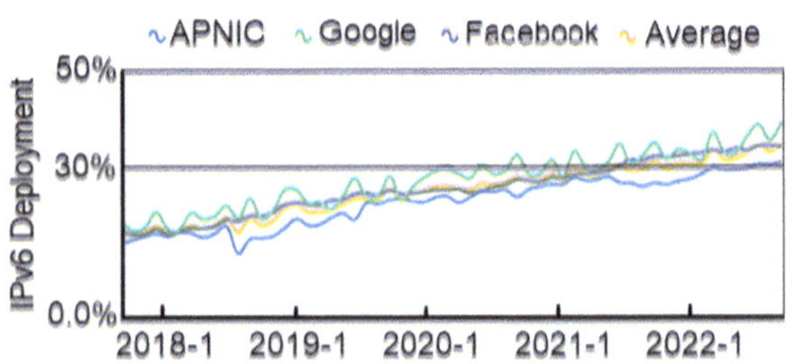

Fig. 6.13 Global IPv6 deployment (https://pulse.internetsociety.org/technologies, accessed 6 Aug 2022)

From `0000:0000:0000:0000:0000:0000:0000:0000`
To `ffff:ffff:ffff:ffff:ffff:ffff:ffff:ffff`

Even with the hexadecimal representation, IPv6 addresses can still appear quite long. To improve readability, a shortened notation for IPv6 addresses is preferred. There are a few rules for shortening IPv6 addresses:

Rule 1: Leading zeros with a quartet are omitted. For example,
`2000:0000:0000:0005:0000:0000:0000:006b`
→ `2000:0:0:5:0:0:0:6b`

Rule 2: A series of consecutive quartets consisting of only hexadecimal 0s is replaced by a double colon '::'. For example,
`2000:0:0:5:0:0:0:6b`
→ `2000 :: 5:0:0:0:6b`
or `2000:0:0:5::6b` (preferred: shortest and valid)
In this example, the notation `2000:0:0:5::6b` is the preferred representation because it is the shortest and valid. Also, avoid using ambiguous representations like ~~`2000::5::6b`~~ (because it is not clear how many 0s each ::represents).

Rule 3: For an IPv4-mapped IPv6 address, embed the IPv4 address into the last four octets and use a representation of combined colon and dotted notations. This enables direct communication between IPv6 and IPv4 applications. For example,
`0:0:0:0:0:ffff:192.1.56.10`
→ :: `ffff.192.1.56.10/96` (shortened format)
where '/96' means that the first 96 bits are fixed `0:0:0:0:ffff`.

6.4.2 *IPv6 Site Prefix, Subnet ID, and Interface ID*

The eight quartets of an IPv6 address are divided into three groups, as shown in Fig. 6.14:

- The leftmost three quartets form the *site prefix*. The site prefix describes the *public topology* that is typically allocated to the site by an ISP or Routing Information Protocol (RIP).
- The next quartet represents the *subnet ID* of the network site. The subnet ID defines the *private topology* or *site topology* of the network.
- The rightmost four quartets make up the *interface ID* of the network. The interface ID serves as a unique identifier for the interface on the local link. It can be automatically configured from the interface's MAC address or manually configured.

In IPv6, the 48-bit site prefix and 16-bit subnet ID together form a 64-bit network prefix. A link, also known as a local link, refers to any LAN bounded by routers. It is worth mentioning that in comparison with IPv4, there are some changes in the terminology used in IPv6 to describe TCP/IP communication.

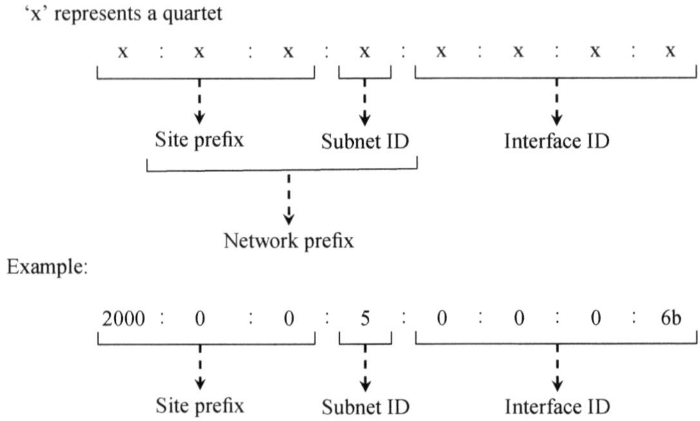

Fig. 6.14 IPv6 site prefix, subnet ID, and interface ID

6.4.3 IPv6 Anycast Addresses

IPv6 classifies IP addresses differently from IPv4. Depending on their usage, IPv6 supports three types of IP addresses: unicast addresses, multicast addresses, and anycast addresses. Unicast and multicast addresses are also supported in IPv4. However, IPv6 no longer supports broadcast addresses but introduces a new type of IP address called anycast addresses.

Anycast addresses are structurally identical to unicast addresses, making them indistinguishable in syntax. They share the same address scopes as unicast addresses. The key distinction between anycast and unicast addresses is their administrative purpose. An anycast address identifies multiple destinations, and packets addressed to an anycast address are delivered to the closest destination. For instance, a DNS server may send a DNS request to multiple DNS servers assigned with the same anycast address. When a router receives such a DNS request, it forwards it to the nearest DNS server rather than broadcasting it to all DNS servers sharing the anycast address.

6.4.4 IPv6 Unicast Addresses

An IPv6 unicast address specifies a single node on a network. IPv6 defines four main types of unicast addresses: loopback address, global addresses, link local addresses, and unique local addresses, as shown in Fig. 6.15. These types of unicast addresses are described below.

Loopback address ::1/128

127 bits	1 bit
0000 0000 0000	1

Global address 2000::/3

3 bits	45 bits	64 bits
001	Global routing prefix	Interface ID

Link local address FE80::/10

64 bits	64 bits
1111 1110 1000 0000 .... 0000 FE80::/10 (also written as FE80::/64)	Interface ID

Unique local addresses FC00::/7 (FC00::/8, i.e., L=0, not specified)

7 bits	1	40 bits	16 bits	64 bits
1111110	L	Global ID	Subnet ID	Interface ID

FD00::/8 resulting from setting L = 1 above

8 bits	40 bits	16 bits	64 bits
11111101	Global ID	Subnet ID	Interface ID

Fig. 6.15 IPv6 unicast addresses. The global ID in unique local addresses is random with a high probability of global uniqueness [12]

Loopback address:

The IPv6 loopback address functions similarly to the IPv4 loopback address. Any packets sent to the loopback address are routed back internally without being transmitted over the network. In binary format, it has 127 leading 0s followed by a 1. Thus, it is written as ::1/128.

Global address:

Similar to public IPv4 addresses, global unicast addresses in IPv6 are unique and globally routable addresses used on the public Internet. The currently released global addresses begin with the prefix `2000::/3`, with the first three bits fixed as `001`. The notation /3 indicates that these three bits are constant and do not change. Additionally, a 16-bit subnet ID can be used to identify a specific segment or subnet within a network.

Link local address:

A link local address is used for communication between nodes on the same link, similar to an autoconfigured APIPA address in IPv4. Link local addresses have a prefix of `FE80::/10`, which means that the first 10 bits are fixed and followed by 54 zeros. In a more concise representation, link local addresses are written as `FE80::/64`. It is worth noting that link local addresses are restricted to the local link and cannot be routed on the Internet.

Unique local address:

A unique local address in IPv6 functions similarly to a private address in IPv4. Just like IPv4 private addresses, IPv6 unique local addresses can be used within a single site or organization without the requirement of centralized registration. However, these addresses are only routable within the scope of the private network and cannot be used on the global Internet.

As shown in Fig. 6.15, unique local addresses are specified as `FC00::/7`, with the first 7 bits fixed as `1111110`. The eighth bit $L = 0$ has not been defined yet but may be specified in the future. Therefore, the use of `FC00::/8` should be avoided before it is formally defined.

The setting of $L = 1$ is defined for locally assigned addresses. Effectively, it gives `FD00:: /8` for locally assigned unique local addresses. Therefore, locally assigned unique local addresses will follow the format `fdxx:xxxx:xxxx::/48`, where 'x' represents a hexadecimal digit. The 10 hexadecimal digits (`xx:xxxx:xxxx`), or 40 bits, following `fd` in the unique local addresses form the global ID. The global ID is generated pseudo-randomly with a high probability of global uniqueness. The IETF RFC 4193 (October 2005) has recommended an algorithm to generate a pseudo-random global ID [12, pp. 4–6],

Considering the 16-bit subnet ID in unique local addresses (Fig. 6.15), the range of unique local address is as follows

from : `fdxx:xxxx:xxxx:0 :: /64`
through : `fdxx:xxxx:xxxx:ffff :: /64`

This range represents the blocks of routable addresses within private /48 networks.

6.4.5 IPv6 Multicast Addresses

A multicast address in IPv6 serves as an identifier for a group of interfaces, typically belonging to different nodes. These interfaces can be part of any number of multicast groups. When a packet is sent to a multicast address, it is delivered to all interfaces associated with that address.

Over time, a comprehensive list of IPv6 multicast addresses has been registered with the Internet Assigned Numbers Authority (IANA). To access the most up-to-date list of registered IPv6 multicast addresses, refer to IANA's IPv6 Multicast Address Space Registry [13].

General Format of IPv6 Multicast Addresses

The general format of IPv6 multicast addresses is shown in Fig. 6.16. It is seen from Fig. 6.16 that multicast addresses have the prefix `1111 1111`, indicating that they fall within the `FF00::/8` range. This format is originally defined in the IETF RFC 4291 [10]. Later, it is updated in a few other RFCs. For example, the IETF RFC 7371 provides updates to the IPv6 multicast addressing architecture [14], while the

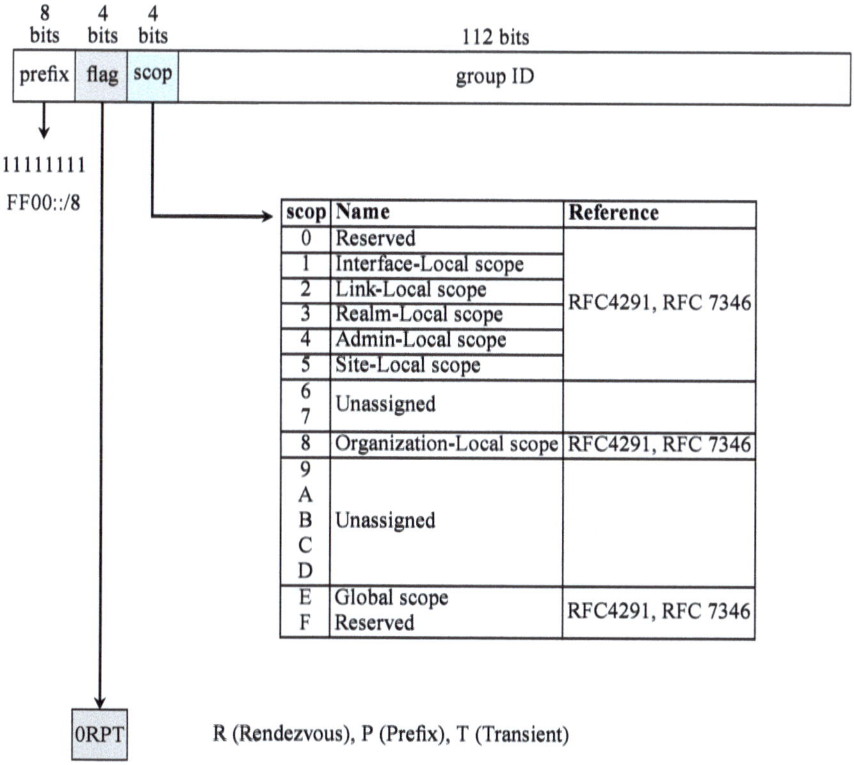

scop	Name	Reference
0	Reserved	RFC4291, RFC 7346
1	Interface-Local scope	
2	Link-Local scope	
3	Realm-Local scope	
4	Admin-Local scope	
5	Site-Local scope	
6 7	Unassigned	
8	Organization-Local scope	RFC4291, RFC 7346
9 A B C D	Unassigned	
E F	Global scope Reserved	RFC4291, RFC 7346

Bit	0	1
R	Rendezvous point not embedded.	Rendezvous point embedded
P	Without prefix information.	Address based on network prefix
T	Well-known multicast address.	Dynamically assigned multicast address

Fig. 6.16 General format of IPv6 multicast addresses [10, 15]

IETF RFC 7346 discusses the IPv6 multicast address scopes [15]. The assignment of new IPv6 multicast addresses is governed by the rules outlined in the IETF RFC 3307 [16].

In Fig. 6.16, the 4-bit flag field is specified as `0RPT` in RFC 4291, where R, P, and T represent Rendezvous, Prefix, and Transient, respectively. Each of these bits can be set to 0 or 1. The settings of these three bits and their meanings are defined in RFC 4291 [10]. They are shown in the lower table within Fig. 6.16. The embedding of the Rendezvous Point (RP) address in an IPv6 multicast address was originally introduced in the IETF RFC 3956, which updates the IPv6 multicast addressing format presented in the IETF RFC 3306 [17]. The encoding of the Rendezvous Point address in an IPv6 multicast address is a specification of a group-to-RP mapping mechanism. It not only allows for easy deployment of scalable inter-domain multicast but also simplifies intra-domain multicast configuration [18, p. 1].

The scope of IPv6 multicast addresses is coded in the scop bits, as defined in the IETF RFC 7346 [15]. The scop coding is demonstrated in Fig. 6.16. It is seen from this figure that certain scop values are currently unassigned, which include scop values 6, 7, and 9 through D in hexadecimal format.

Pre-defined IPv6 Multicast Addresses

There are a few pre-defined multicast addresses. For these addresses, the bits in the group ID are assigned specific scope values. However, when the T bit in the flag field is set to 0, the use of the group ID for any other values is not allowed. Table 6.10 lists the multicast addresses that are pre-defined in the IETF RFC 4291 [10, pp. 16–17].

Unicast-Prefix-based IPv6 Multicast Addresses

The general format shown in Fig. 6.16 for IPv6 multicast addresses does not support the dynamic allocation of IPv6 multicast addresses. Therefore, the IETF RFC 3306 [17] is developed to specify unicast-prefix-based IPv6 multicast addresses. It introduces encoded information in the multicast addresses, allowing for dynamic allocation of IPv6 multicast addresses and IPv6 source-specific multicast addresses. By delegating multicast addresses simultaneously with unicast prefixes, network operators can identify their multicast addresses without the need to run an inter-domain allocation protocol.

Later, the IETF RFC 3956 [18] extends the unicast-prefix-based IPv6 multicast addresses defined in RFC 3306 [17] by encoding the RP in the IPv6 multicast group

Table 6.10 Pre-defined IPv6 multicast addresses in RFC 4291 [10, pp. 16–17]

Description	Multicast addresses

Reserved multicast addresses	`FF00:0:0:0:0:0:0:0`
	through
	`FF0F:0:0:0:0:0:0:0`

All nodes addresses	`FF01:0:0:0:0:0:0:1` (interface local) `FF02:0:0:0:0:0:0:1` (link local)
All routers addresses	`FF01:0:0:0:0:0:0:2` (interface local) `FF02:0:0:0:0:0:0:2` (link local) `FF05:0:0:0:0:0:0:2` (site local)

Solicited-node addresses*	`FF02:0:0:0:0:1:FFXX:XXXX`
	Address range:
	`FF02:0:0:0:0:1:FF00:0000`
	through
	`FF02:0:0:0:0:1:FFFF:FFFF`

*A solicited-node address is formed by taking the last 24 bits of a unicast or anycast address and then appending those bits to the prefix `FF02:0:0:0:0:1:FF00::/104`. IPv6 addresses that differ only in the higher-order bits will map to the same solicited-node address, thus reducing the number of multicast addresses that a node must join.

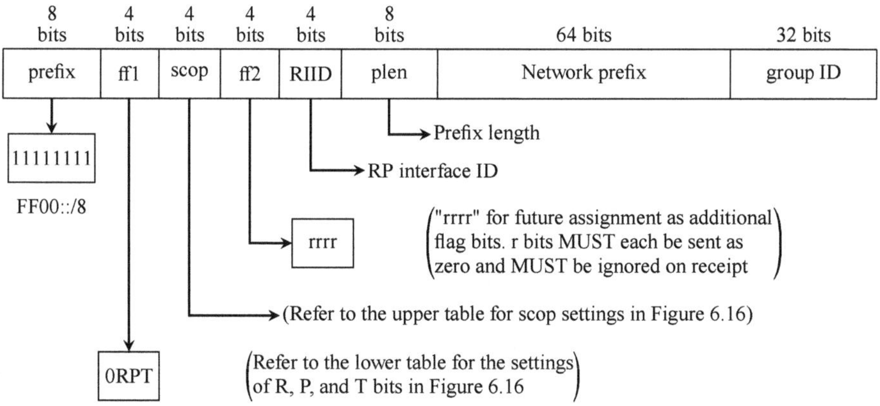

Fig. 6.17 Unicast-prefix-based IPv6 multicast addresses specified in RFC 7371 [14], which updates RFC 3306 [17] and RFC 3956 [18]

address. This provides a simple solution for both IPv6 inter-domain Any Source Multicast and IPv6 intra-domain Any Source Multicast with scoped multicast addresses.

The specifications in RFC 3306 [17] and RFC 3956 [18] involve a reserved field. They are updated in the IETF RFC 7371 [14] by redefining the reserved bits as generic flag bits. RFC 7371 also provides some clarifications related to the use of these flag bits.

In comparison with the general format of IPv6 multicast addresses shown in Fig. 6.16, unicast-prefix-based IPv6 multicast addresses depicted in Fig. 6.17 have two flag fields: ff1 and ff2. The ff1 field in Fig. 6.17 is renamed from the flag in Fig. 6.16. As the scop field is already defined, the second flag field ff2 has to be placed after the scop field. It is actually defined in RFC 7371 [14] from the first four bits of the reserved eight bits specified in RFC 3306 and RFC 3956. In RFC 7371 [14], multicast flag bits denote both ff1 and ff2.

As indicated in Fig. 6.17, the ff2 field reserves four bits as "rrrr" for future assignment as additional flag bits. RFC 7371 states that the "r bits MUST each be sent as zero and MUST be ignored on receipt" [14].

6.4.6 Multicast Flooding

Multicast can cause traffic congestion known as multicast flooding. This is because a multicast group is assigned a single IP address, which means that no specific MAC address of the multiple nodes in the multicast group can be associated with that IP address. As a result, when a switch receives a multicast message, it must flood all its interfaces with the transmission. Effectively, this appears similar to broadcast in IPv4 networks. Multicast flooding is a common issue in both IPv4 and IPv6 networks.

In IPv4 networks, the multicast flooding problem can be addressed by enabling the Internet Group Management Protocol (IGMP) on the switch that handles multicast traffic. IGMP, as a network-layer protocol, is used by IPv4 networks to report their IP multicast group memberships to neighboring multicast routers so that multicast traffic can be directed to the correct devices. Since switches are layer-2 devices, they lack the layer-3 IGMP information that identifies multicast group members. IGMP snooping enables switches to detect IGMP messages, extract information from these messages, and add accurate entries in their MAC address tables.

The first version of IGMP (IGMPv1) is specified in the IETF RFC 1112 [19], which is the first Internet Standard for multicast membership management. The second version of IGMP (IGMPv2) is specified in the IETF RFC 2236 [20], which adds support for *low leave latency*, i.e., a reduction in the time it takes for a multicast router to learn that there are no longer any members of a particular group present on an attached network. The current version of IGMP is version 3, known as IGMPv3, which is specified in the IETF RFC 3376 [21]. As an improved version of IGMPv2, IGMPv3 adds support for *source filtering*. As stated in RFC 3376, source filtering enables a system to report interest in receiving packets *only* from specific source addresses, or from *all but* specific source addresses, sent to a particular multicast address [21, p. 1]. This avoids delivering multicast packets from specific source addresses to networks where there are no interested receivers.

Similar to IGMP in IPv4 systems, the Multicast Listener Discovery (MLD) protocol is developed for IPv6 networks to manage multicast group memberships. It is used by an IPv6 router to discover the presence of multicast listeners on directly attached links and to determine which multicast addresses are of interest to those neighboring nodes. The current version of MLD is version 2, known as MLDv2, which is specified in the IETF RFC 3810 [22]. MLDv2 implements the functionality of IGMPv3. It is an improved version of MLD version 1 (MLDv1), which implements the functionality of IGMPv2.

It is seen from the above discussions that IGMPv3 and MLDv2 allow a host to inform its neighboring routers of its interest in receiving IPv4 and IPv6 multicast transmissions, respectively. For Source-specific Multicast (SSM), a receiver is required to specify both the source's network-layer address and the destination's multicast address to be able to receive the multicast transmission. To make a router and host SSM-aware, both IGMPv3 and MLDv2 are updated in the IETF RFC 4604 [23]. More specifically, RFC 4604 defines the concept of SSM-aware routers and hosts. It also clarifies and modifies the behavior of IGMPv3 and MLDv2 on SSM-aware routers and hosts to accommodate SSM [23, p. 1].

6.4.7 Assignment of IPv6 Addresses to Interfaces

As specified in the IETF RFC 4291 (February 2006) [10], All types of IPv6 addresses are assigned to interfaces, rather than nodes. This is different from IPv4 addressing that assigns IPv4 addresses to nodes. As each interface belongs to a single node, the unicast address assigned to the interface of a node can be naturally used as an identifier for the node. It is interesting to note that unlike IPv4 that assigns a single IP address to a node, all interfaces in IPv6 are required to have at least one link local unicast address even when globally routable addresses are also assigned [10]. This means that a single interface may have multiple IPv6 addresses of any type (unicast, anycast, and multicast) or scope. The link local address of an interface is required for IPv6 sublayer operations of the Neighbor Discovery Protocol, as well as for some other IPv6-based protocols, such as DHCPv6.

Another interesting feature in IPv6 addressing is that a unicast address or a set of unicast addresses may be assigned to multiple physical interfaces if the multiple physical interfaces are treated as one interface when presenting it to the Internet layer. This is advantageous for load-sharing over multiple physical interfaces [10].

In IPv6 networks, a host is required to recognize the following addresses as identifying itself [10, p. 17]:

- Its required link-local address for each of its interfaces,
- Any additional unicast and anycast addresses manually or automatically configured for all interfaces of the node,
- The loopback address,
- The all-nodes multicast addresses `FF01:0:0:0:0:0:0:1` (interface local) and `FF02:0:0:0:0:0:0:1` (link local),
- The solicited-node multicast address for each of its unicast and anycast addresses, and
- Multicast addresses of all other groups to which the node belongs.

Also, as specified in the IETF RFC 4291 [10, p. 17], a router is required to be able to recognize all addresses that a host is required to recognize, and the following addresses as identifying itself:

- The subnet-router anycast addresses for all interfaces for which it is configured to act as a router,
- All other anycast addresses with which the router is configured, and
- The all-router multicast addresses `FF01:0:0:0:0:0:0:2` (interface local), and `FF02:0:0:0:0:0:0:2` (link local), and `FF05:0:0:0:0:0:0:2`(site local).

6.4.8 IPv6 Header Structure

In comparison with IPv4, IPv6 has not only a much larger address space, as discussed previously, but also a much simplified header structure. The improved header structure results in less processing overhead. Figure 6.18 shows a comparison between the IPv4 and IPv6 headers.

IPv6 Fixed Header

It is seen from Fig. 6.18 that several fields in IPv4 header do not exist anymore in IPv6 header. For example,

- The IHL (Internet Header Length) field is no longer needed in IPv6 because the IPv6 header has a fixed length of 40 bytes.
- The fields of *Identification*, *F (Fragment)*, and *fragment offset* no longer exist in the IPv6 header. This is because there is no fragmentation anymore in IPv6 routing, and the Identification field in IPv4 is normally used for fragmentation and reassembly. The elimination of fragmentation in IPv6 allows more efficient routing. In IPv6 networks, fragmentation is done by the sender rather than by routers and other network devices.
- There is no header *checksum* anymore in the IPv6 header. This results from the consideration that the checksum functionality provided in lower-layer protocols (e.g., link-layer PPP and Ethernet) and upper-layer protocols (e.g., transport-layer TCP and UDP) should be sufficient.
- The fields of *options* and *padding* fields in IPv4 header are no longer needed in IPv6 due to the fixed length of the IPv6 header.

It is also noticed from Fig. 6.18 that in comparison with the IPv4 header, the IPv6 header has a few fields that are similar to, or used to replace, the corresponding IPv4 fields:

- The *Traffic Class* filed in IPv6 is similar to the *ToS* field in IPv4.
- The IPv6 *Payload Length* field replaces the IPv4 *Total Length* field.

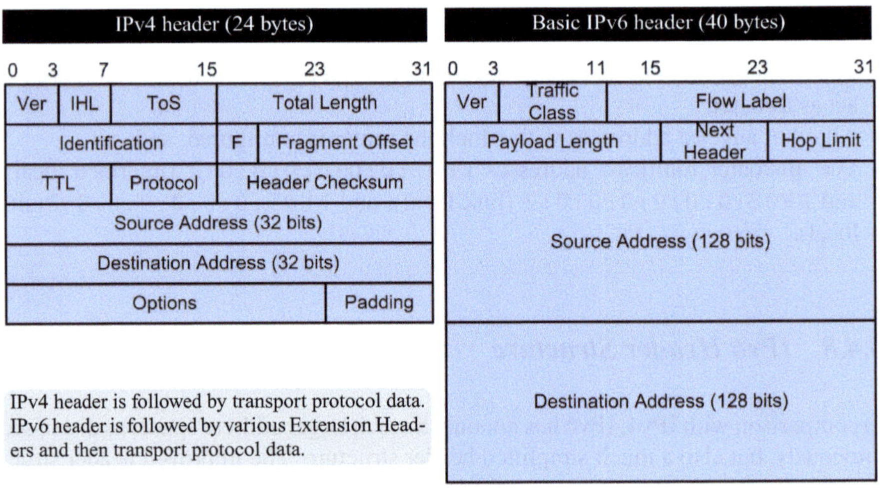

Fig. 6.18 Comparison between IPv4 and IPv6 headers

- The IPv6 *Next Header* field replaces the *Protocol* field in IPv4. It indicates the type of header that immediately follows the IPv6 header. This allows IPv6 extension headers to be embedded into an IPv6 packet.
- The *Hop Limit* field in the IPv6 header is similar to the TTL (Time to Live) field in the IPv4 header. It indicates the maximum number of links over which the IPv6 packet is allowed to travel before being discarded. It is decremented by one by each node that forwards the packet. The packet is discarded if the Hop Limit is decremented to zero.

Moreover, IPv6 has a new field called *Flow Label* in the header. This field is used for QoS support through flow labeling to distinguish delay-sensitive packets. Therefore, IPv6 is said to have built-in true QoS.

IPv6 Extension Headers

In IPv4, the IPv4 header is followed directly by the transport protocol data. However, in IPv6, the IPv6 header is followed by various Extension Headers and then the transport protocol data. These extension headers serve as essential components of IPv6 and provide additional information used by routers, switches, and hosts to determine how to handle or process an IPv6 packet. For example, security mechanisms can be incorporated into IPv6 through extension headers.

Unlike the fixed-length IPv4 header, an IPv6 extension header can be of arbitrary length. However, to improve the processing of extension headers and the subsequent transport protocol data, an IPv6 extension header is always an integer multiple of eight octets in size. This will retain the alignment of subsequent headers.

Most IPv6 extension headers are not examined or processed by routers along the delivery path of a packet until the packet reaches its final destination. This is because routers typically only need to inspect the main IPv6 header and the next header field to determine the appropriate forwarding path for the packet. By bypassing the examination of extension headers, routers can improve their performance and forwarding efficiency. This is particularly beneficial for packets that contain options in IPv6. The options in IPv6 provide additional functionality and flexibility but are not required to be examined by routers during the normal forwarding process. Instead, the destination node is responsible for processing the extension headers and options, allowing for more efficient routing and forwarding in intermediate routers.

Six extension headers have been defined in IPv6. They are listed in Table 6.11 in the preferred order for the case where there is more than one extension header following the fixed header.

Some examples of IPv6 packets with and without extension headers are illustrated in Fig. 6.19. It is seen from Fig. 6.19 that

- When no extension header is included. as shown in the upper part of Fig. 6.19, When no extension header is included, the transport protocol data (e.g., TCP) immediately follows the IPv6 header.
- If only one extension header is included, it is inserted between the fixed header and the transport protocol data. This is shown for the routing header in the middle part of Fig. 6.19.

Table 6.11 IPv6 Extension Headers

No.	Extension Header
(1)	**Hop-by-Hop Option Header:** It provides a set of options that can be used by routers to perform certain management or debugging functions
(2)	**Routing Header:** It is used to mandate a specific routing (similar to source routing options in IPv4)
(3)	**Fragmentation Header:** It is defined for fragmentation and reassembly (similar to fragmentation options in IPv4)
(4)	**Authentication Header (AH):** It provides authentication and integrity
(5)	**Encapsulating Security Payload (ESP) Header:** It provides authentication, integrity, and encryption
(6)	**Destination Option Header:** It defines a set of options that are intended to be examined only by the destination node. These options provide additional functionality or features specific to the destination node. An example of a Destination Option Header is Mobile IPv6, which allows for seamless mobility of a node across different networks

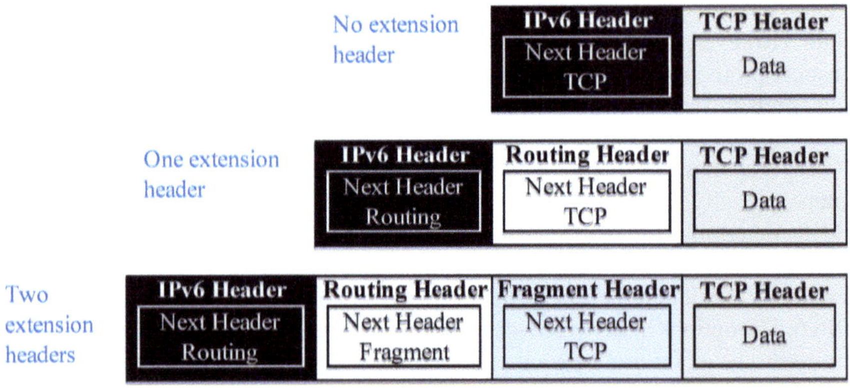

Fig. 6.19 Examples of IPv6 packets with and without extension packets

- For packets with more than one extension header, such as a routing header and a fragment header as depicted in the lower part of Fig. 6.19, sort these extension headers in the preferred order mentioned earlier, and then insert them between the fixed header and the transport protocol data. In this example, the routing header appears before the fragment header. The fragment header is then followed by the transport protocol data.

6.4.9 Key Benefits of IPv6 Addressing

After the brief introduction to IPv6 addressing in previous discussion, it is now ready to discuss the advantages of IPv6 over IPv4.

Key Benefits of IPv6

Key benefits of IPv6 include, but are not limited to:

- Much larger address space, resulting in no need for NAT anymore,
- Elimination of broadcast, completely solving the broadcast storm problem that could occur in IPv4 networks,
- Simpler header structure with a fixed header size of 40 bytes, resulting in less processing overhead,
- No requirement for fragmentation, leading to more efficient routing,
- Autoconfigurarion, making IPv6 nodes to behave like plug-and-play devices,
- Built-in true QoS through flow labeling that distinguishes delay-sensitive packets,
- Built-in security with Internet Protocol Security (IPsec) for authentication and privacy support, and
- Embedded mobility and interoperability support.

Among these features, autoconfiguration, built-in true QoS and security, and embedded mobility and interoperability will be discussed later in the next few sections. Other features are discussed below.

Much Enlarged Address Space

Using 128 bits to represent an address, IPv6 has a much larger address space than IPv4, which uses 32 bits to identify an address. A noticeable benefit of such a vast number of IPv6 addresses is that NAT is no longer needed in IPv6 networks. With the elimination of NAT in IPv6, many NAT-induced problems simply disappear. For example, some security protocols and QoS protocols require knowledge of the network's actual structure to be effective, but they encounter insurmountable difficulties with NAT. It is worth mentioning that there has been a misconception that NAT in IPv4 increases security. However, NAT is not designed as a security solution and does not offer meaningful security.

No More Broadcast

There is no longer any broadcast in IPv6 networks because IPv6 does not provide broadcast addresses. This means that the broadcast storm problem, which can occur in IPv4 networks, is eliminated completely in IPv6 networks. While broadcast traffic is necessary and useful in IPv4 networks, excessive broadcast traffic can lead to network performance degradation. In an IPv4 network, a broadcast storm may occur when there is a high volume of requests for an IP address, the broadcast domain is too large, or there is a switching loop in the Ethernet network topology. For large broadcast domains, physical segmentation of the domains is an effective method to reduce broadcast storms. This can be achieved through the use of either layer-3 routers or layer-2 logical VLANs.

Simplified Header Structure

IPv6 has a simplified header structure compared to IPv4, resulting in reduced processing overhead in IPv6 network communication. In particular, it features a fixed header size of 40 bytes. This has been discussed previously in detail.

No More Fragmentation in Routing

It has been established through previous discussions that fragmentation is not required in IPv6 routing, leading to increased routing efficiency. This is due to the fact that fragmentation is the responsibility of the sender in IPv6 network communication. In IPv6, a minimum Maximum Transmission Unit (MTU) of 1, 280 octets is mandated, but hosts are strongly recommended to use Path MTU Discovery for possible MTUs greater than the minimum [11]. Path MTU Discovery in IPv6 enables a host to dynamically discover and adapt to variations in the MTU size along a given data path. This functionality is specified in the IETF RFC 8201 [24]. It is worth mentioning that Path MTU Discovery relies on ICMPv6 for its operation. But the delivery of ICMPv6 messages may not be guaranteed for some paths. Therefore, an extension to Path MTU Discovery has been defined in the IETF RFC 8899 [25]. RFC 8899 specifies a technique called Packetization Layer Path MTU Discovery, which does not rely on ICMPv6.

6.5 IPv6 Autoconfiguration

One of the most interesting and valuable addressing features implemented in IPv6 is the ability to enable devices to configure themselves automatically and independently. This feature is known as IPv6 autoconfiguration. In contrast to IPv4, where hosts are allocated IP addresses through DHCP, IPv6 takes a step further by introducing a method for certain devices to automatically configure IP addresses for interfaces and other settings without relying on DHCP. This is a new IPv6 feature, which does not exist in IPv4. It is particularly advantageous in various networking scenarios, such as a home network connecting TVs, mobile phones, air conditioners, fridges, and numerous other devices that would be inconvenient to depend on a DHCP server for address assignment. Therefore, IPv6 autoconfiguration facilitates the plug-and-play functionality of network devices.

IPv6 knows both stateful and stateless autoconfiguration. Stateful autoconfiguration, akin to DHCP in IPv4 or DHCPv6 in IPv6, relies on a DHCPv6 server for operation. By contrast, stateless autoconfiguration is a unique feature of IPv6 that operates independently without the need for a DHCPv6 server. It is a unique IPv6 feature, which does not exist in IPv4, as mentioned earlier. In the following, both stateless and stateful autoconfiguration in IPv6 will be explored in more detail.

6.5.1 *IPv6 Stateless Autoconfiguration*

StateLess Address AutoConfiguration (SLAAC) is an IPv6 function for a host to configure its IPv6 address automatically. It is specified in the IETF RFC 4862 [26], which is later updated in the IETF RFC 7527 with enhanced duplicate address detection [27].

Main Features of SLAAC

The design goals of SLAAC include the following four main aspects [26, pp. 7–8], which can be interpreted as four features of SLAAC:

- Automatic Configuration: SLAAC eliminates the need for manual configuration of IPv6 addresses for individual hosts. It provides a mechanism for hosts to create their own unique IPv6 addresses for each interface. This is achieved by combining a prefix with a unique interface identifier, often derived from the link-layer address.
- Plug-and-Play Communication for Small Sites: Small sites with a group of hosts connected to a single link can achieve communication without relying on a DHCPv6 server or router. Link-local addresses are used in such scenarios, which have a well-known prefix identifying the link. Hosts form link-local addresses by combining the link-local prefix with their interface identifier.
- Prefix Discovery for Large Sites: In larger sites with multiple networks and routers, the reliance on a DHCPv6 server is also eliminated. Hosts in such environments determine the prefixes that identify the subnets they are connected to by relying on the routers on the site. The routers periodically send Router Advertisement (RA) messages that optionally list the active prefixes on the link.
- Graceful Renumbering: Address autoconfiguration should enable the smooth renumbering of hosts on a site. This is often necessary when an organization changes its ISP. IPv6 allows interfaces to be assigned multiple addresses, and addresses are leased in both IPv4 and IPv6 networks. Renumbering is achieved by assigning multiple addresses to the same interface, and the lease lifetimes of IPv6 addresses facilitate the gradual phasing out of old prefixes.

It is worth mentioning that SLAAC is performed exclusively for hosts and not for routers. In general, router and servers should be configured manually or through alternative methods.

Steps of SLAAC

ALAAC is performed only on multicast-capable links. It begins when a multicast-capable interface is enabled, for instance, during host bootup. On network initialization, a node can obtain through SLAAC the following information or parameters:

- IPv6 prefix(es),
- Default router address(es),
- Hop limit,
- Link local MTU, and
- Validity lifetime.

The operation of SLAAC involves the following three steps to establish a network connection in IPv6:

(1) Create an IPv6 link-local address.
(2) Check the uniqueness of the created address, if the address ia a duplicate one, authconfiguration fails and stops.
(3) Complete the autoconfiguration.

Let us discuss these three steps below in detail.

Creation of IPv6 Address

In the above Step (1), in order to create an IPv6 address, the first 64 bits (prefix) and last 64 bits (interface identifier) are created separately. Then, they are combined to form a complete IPv6 address. The detailed process is summarized as follows:

- The first 64 bits are created as the link prefix represented by `FE80::/64`. This is shown in Fig. 6.15 for link local address.
- The last 64 bits are created either randomly or by using its 64-bit Extended Unique Identifier (EUI), known as EUI-64. The random generation is the default method in Windows 10.
  The EUI-64 address is specified in the the IETF RFC 4291 [10, pp. 20–21]. It is formed through the following steps: take the 48-bit MAC address, insert a fixed 16-bit value 0xFFFE in the middle of the 48 bits, and invert the value of the seventh bit from the left.
  For example, consider a 48-bit MAC address
  `00:21:2F:B5:66:1A`,
  Insert 16 bit 0xFFFE in the middle of the MAC address, yielding
  `00:21:2F:`<u>*FF : FE*</u>`:B5:66:1A`.
  Then, invert the value of the seventh bit from the left, resulting in the following EUI-64 address
  <u>`02`</u>`:21:2F:`<u>*FF : FE*</u>`:B5:66:1A`.
- Combine the first 64-bit prefix and the EUI-64 address to form the IPv6 address. For the example discussed above, the created IPv6 address is:
  `FE80 :: 0221:2FFF:FEB5:661A`.

Duplicate Address Detection

An address should not be assigned to an interface before it is confirmed to be unique. To check the uniqueness of an address, simply send a Neighbor Solicitation message to the address. Then, follow the following process:

- If no response is received to say that the address is a duplicate one, the address is considered unique. Therefore, the address can be assigned to the interface.
- If the address is already used by another node, a Neighbor Advertisement message will be returned. In this case, the autoconfiguration stops. To recover from an address conflict, there are a couple of options:

- An alternative interface identifier may be provided by the administrator to override the default one, thus allowing the autoconfiguration to start again with this new (presumably unique) interface identifier.
- Alternatively, use manual configuration instead.

As the link-layer address is used as part of the link-local address, the created IPv6 address is supposed to be unique. However, by default, all unicast addresses, regardless of how they are obtained, should be tested for their uniqueness prior to their assignment to an interface. This precaution is necessary because address conflicts can occur due to various factors such as the presence of duplicate MAC addresses.

If an address duplicate does happen, it may indicate that duplicate hardware addresses are in use. In such cases, IPv6 operation on the affected interface should be disabled to prevent further issues. Consequently, the node associated with that interface will no longer send or receive IPv6 packets. This means that:

- Outgoing IPv6 packets are not sent: Any IPv6 packets will not be transmitted from the affected interface.
- Incoming IPv6 packets are silently dropped: Any IPv6 packets received by the affected interface will be discarded without any notification or response.
- No forwarding of IPv6 packets: The node will not forward any IPv6 packets to the affected interface.

Disabling IPv6 on an interface is a protective measure taken to avoid address conflicts and ensure the stability and integrity of the network.

Prefix Discovery and Finalizing Autoconfiguration

After an address is confirmed to be unique and can be assigned to an interface, ask if a router on the network can provide autoconfiguration information. This is performed by sending out a Router Solicitation (RS) message. If a router responds with an RA message, use whatever information that might be, such as the IP address of a DNS server or network prefix, to complete the autoconfiguration. This process is known as *Prefix Discovery*.

In cases where no router is found or the RA message enables the use of DHCPv6, stateful autoconfiguration can be used. It is worth mentioning that "the DHCPv6 service for address configuration may still be available even if no routers are present" [26, p. 8].

Prefix discovery is part of the link-layer *Neighbor Discovery* protocol specified in the IETF RFC 4861 [28]. In IPv6 networks, hosts on the same link use Neighbor Discovery to discover the presence of each other, determine link-layer addresses of each other, find routers, and maintain reachability information about the paths to active neighbors [28, p. 1]. The Neighbor Discovery protocol defines five ICMPv6 packet types, which follow ICMPv6 message format in general, to perform various neighbor discovery functions [28, pp. 11–12]:

- RS (Type 133), which is used by hosts to find routers on an attached link.
- RA (Type 134), which is used by routers to advertise their presence together with various link and Internet parameters either periodically or in response to an RS

message. It is noted that there are two specific flags defined in the RA message format: one-bit ManagedFlag M and one-bit OtherConfigFlag O. With these flags, RA messages allow routers to inform hosts how to perform Address Autoconfiguration, e.g., through DHCPv6 or autonomous (stateless) address configuration [28, p. 12].
- Neighbor Solicitation (Type 135), which is used by nodes to determine the link-layer address of a neighbor or verify that a neighbor is still reachable via a cached link-layer address.
- Neighbor Advertisement (Type 136), which is used by nodes to respond to a Neighbor Solicitation message, or send unsolicited Neighbor Advertisement messages to provide new information quickly without being prompted by a Neighbor Solicitation message.
- Redirect (Type 137), which routers may use to inform hosts of a better first-hop router for a destination.

These ICMPv6 messages contain one or more options. Five options are defined altogether in the IETF RFC 4861 [28, p. 28]:

- Source Link-Layer Address,
- Target Link-Layer Address,
- Prefix Information,
- Redirected Header, and
- MTU.

The flags of the prefix information option have been updated in the IETF RFC 8425 [29]. These updated flags provide hosts with more control and information about the advertised prefixes, allowing for improved stateless autoconfiguration and better management of network connectivity in IPv6.

6.5.2 IPv6 Stateful Autoconfiguration

As mentioned previously in the discussion of SLAAC for node configuration, if SLAAC fails to find a router or an RA message directs SLAAC to use DHCPv6, stateful DHCPv6 will be used for node configuration. Specified in the IETF RFC 8415 [30], stateful DHCPv6 provides an IPv6 address assignment approach that allows hosts to obtain interface addresses or configuration information and parameters from a DHCPv6 server, which maintains a database that checks which addresses have been assigned to which hosts. The parameters used for node configuration can be provided statelessly or in combination with stateful assignment of one or more IPv6 addresses and/or IPv6 prefixes.

Stateful DHCPv6 can operate either instead of, or in addition to, SLAAC. Therefore, stateful autoconfiguration and stateless autoconfiguration complement each other in IPv6 networks. For example, a host may use stateless autoconfiguration to configure its own addresses, but use stateful autoconfiguration to obtain other information.

It is worth mentioning that there is a concept of "stateless DHCPv6", which is discussed in IETF RFC 8415 [30, p. 18]. Stateless DHCPv6 does not aim to provide an address lease to a node. It is used for a node to obtain other configuration parameters, such as a list of DNS recursive name servers or DNS domain search lists. It is stated in RFC 8415 [30, p. 18] that stateless DHCPv6 may be used when a node initially boots or when the software on the node requires some missing or expired configuration information that is available from a DHCPv6 server.

DHCPv6 differs significantly from DHCPv4. Extending DHCPv6 to carry IPv4 address information and configuration information is still an open issue as of the time of writing this book. It is mentioned in RFC 8415 [30, p. 8] that DHCPv4 should be used instead of DHCPv6 when conveying IPv4 configuration information to nodes. In IPv6-only networks, a transport mechanism is developed in the IETF RFC 7341 [31] to carry DHCPv4 messages by using DHCPv6 for dynamic provisioning of IPv4 address and configuration information.

6.6 Built-in Security in IPv6

IPv6 is said to have native or built-in security for authentication and privacy. This claim is reasonable in the sense that IPsec is included as a mandatory feature in IPv6, and extension headers are also designed in IPv6 for security enhancements. The security enhancements in IPv6 are discussed below in comparison to IPv4.

6.6.1 IPsec in IPv6 and IPv4

As a set of security specifications, IPsec is originally designed as part of the IPv6 protocol suite. It is formally specified in the IETF RFC 4301 [32]. This RFC is later updated by two other RFCs: RFC 6040 on tunneling of explicit congestion notification, and RFC 7619 on the NULL authentication method in the Internet Key Exchange Protocol Version 2 (IKEv2).

In contrast to IPv6, IPv4 is designed without security considerations in mind. As a result, there is a significant need for security measures in IPv4 networks. Consequently, IPsec has also been adopted for use in IPv4. However, unlike IPv6, where IPsec is mandatory, support for IPsec in IPv4 is optional. Nevertheless, IPsec is available for use in IPv4 networks although its implementation is generally more complex compared to IPv6.

As mentioned previously, the inclusion of IPsec is mandatory in IPv6. However, this does not necessarily mean that IPsec must be used in IPv6. According to current IPv6 standards, the use of IPsec in IPv6 is not mandatory. If IPv6 networks do not actually use IPsec, they cannot be claimed to be inherently more secure than IPv4 networks.

In IPv4 networks, IPsec is commonly employed for VPNs. In this scenario, IPsec is terminated at the network edge due to the prevalent use of NAT, which alters IPv4 headers and disrupts IPsec functionality. Consequently, IPsec is rarely used for securing end-to-end traffic within IPv4 networks.

However, IPsec in IPv6 provides end-to-end security, ensuring that data transmission is protected from the source node to the destination node. Since NAT is no longer necessary in IPv6, it becomes practical to use IPsec for securing end-to-end IPv6 traffic. For example, IPv6 IPsec can be employed to safeguard all traffic within a data center. As another example, IPv6 deployment can leverage IPsec-based end-to-end security, allowing the decommissioning of existing VPN connections.

6.6.2 Extension Headers for Security and Privacy

IPsec defines cryptography-based security. It introduces two security headers:

- Authentication Header (AH), which ensures the integrity and data origin authentication of IP packets and also provides protection against replay attacks, and
- Encapsulating Security Payload (ESP), which ensures the confidentiality of IP packets through payload encryption and optionally provides integrity and authentication.

Both AH and ESP provide connectionless security services. They can be used independently or in combination. As part of the IPsec protocol suite, AH and ESP extension headers can be implemented in both IPv4 and IPv6 networks.

AH Extension Header

The AH Extension Header is defined in the IETF RFC 4302 [33]. It provides data integrity, data origin authentication, and protection against replays at the IP layer in both IPv4 and IPv6.

- Data integrity ensures that the content of the data in an IP packet received by a node remains unchanged from when it was sent by the source node.
- Data origin authentication refers to the fact if a node receives a packet with a given source IP address in the IP header, it can be ensured that the IP packet indeed comes from that IP address.
- Anti-reply protection results from the fact that an IP packet with a sequence number that has already been processed will not be accepted as a valid packet. This prevents attackers from re-transmitting or manipulating previously captured packets to gain unauthorized access.

To use AH header, the protocol header (IPv4, IPv6, or IPv6 Extension) immediately preceding the AH header is required to set a value 51 to the Protocol field for IPv4, or the Next Header field for IPv6 and IPv6 Extension. The format of AH is depicted in Fig. 6.20 [33]. All the fields shown in Fig. 6.20 are mandatory. This means that they are always present in the AH format and are included in the Integrity Check Value (ICV) computation.

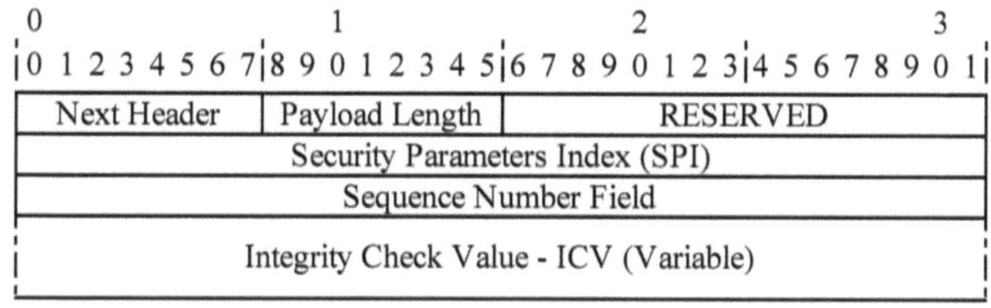

Fig. 6.20 AH format [33, p. 4]

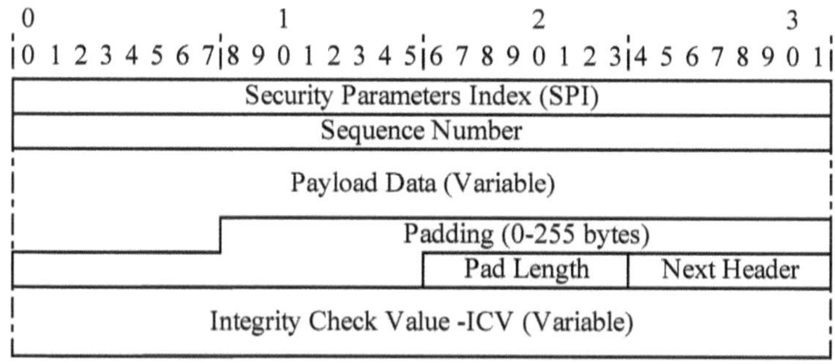

Fig. 6.21 Top-level format of an ESP packet [34, p. 5]

Among all the fields in the AH header, the 32-bit Security Parameters Index (SPI) is used by a receiver to identify the *Security Association* to which an incoming packet is bound. The ICV field has a variable length, which however must be an integral multiple of 32 bits. The details of ICV computation will not be discussed here. For more information about ICV computation, refer to the IETF RFC 4302 [33].

ESP Extension Header

ESP is specified in the IETF RFC 4303 [34] to provide a mixed security in both IPv4 and IPv6. Its main objective is to provide confidentiality through payload encryption. ESP also offers integrity, authentication, and anti-replay service.

To use ESP header, the protocol header (IPv4, IPv6, or IPv6 Extension) immediately preceding the ESP header should set a value of 50 to the Protocol field for IPv4, or the Next Header field for IPv6 and IPv6 Extension. The top-level format of an ESP packet is illustrated in Fig. 6.21 [34]. The packet starts with a 32-bit SPI field, followed by a 32-bit Sequence Number. Then, following the Sequence Number is the Payload data. This is followed by Padding, Pad Length, and Next Header fields. Finally, the optional ICV field completes the packet.

Detailed discussions on the mechanisms of confidentiality, integrity, authentication, and anti-replay protection provided by ESP will not be discussed here. For all these aspects, refer to the IETF RFC 4303 [34].

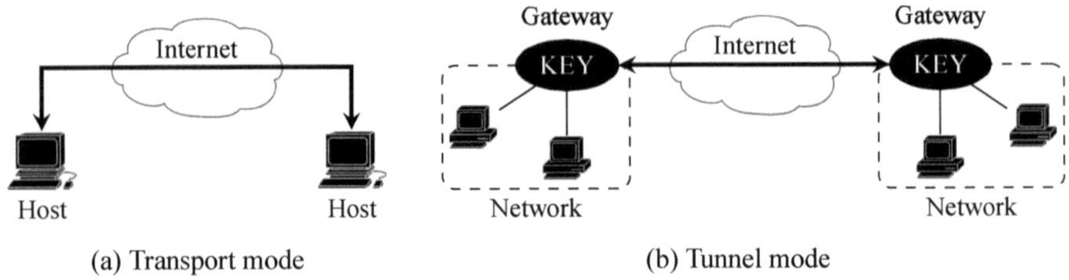

Fig. 6.22 Transport mode and tunnel mode of IPsec

Transport and Tunnel Modes of IPsec

There are two transfer modes for secure connection with IPsec: transport mode and tunnel mode. These two nodes are used in different network scenarios.

- The transport mode connects two end hosts directly.
- The tunnel mode creates a connection between two networks.

These two modes are graphically shown in the logical diagrams of Fig. 6.22.

From the packet format perspective, the transport and tunnel modes in IPsec have different packet structures:

- The transport mode does not change the IP packet header. Only the Protocol field in IPv4 or the Next Header filed in IPv6 is changed to 51 for AH or 50 for ESP. In IPv4, the checksum of the IP packet header needs to be re-calculated. The source and destination addresses of the IPsec tunnel must respectively be the source and destination addresses of the IP packet header. This is why the transport mode is applicable only to communications between two hosts.
- In the tunnel mode, the original IP packet header is encapsulated within a new IP packet. An AH or ESP header is inserted between the original and new IP headers. The original IP address in the original IP packet header is protected and thus hidden by IPsec as part of the payload in the new IP packet. This is the reason that the tunnel model is mainly applicable to communication between VPN gateways or between a host and a VPN gateway.

For the use of AH, Fig. 6.23 shows three scenarios of an IP packet using TCP: with no AH (middle diagram), AH in transport mode (upper diagram), and AH in tunnel mode (lower diagram). It is seen from this figure that in the transport mode, AH is inserted between the original IP header and the original TCP header. In the tunnel mode, the original IP packet is treated as payload, and thus is not changed. A new IP packet is created with a new IP header, which is followed by AH and then the original IP packet as the payload.

For ESP, Fig. 6.24 depicts three scenarios of an IP packet using TCP: with no ESP (middle diagram), ESP in transport mode (upper diagram), and ESP in tunnel mode (lower diagram). Similar to the use of AH discussed above, the use of ESP in the tunnel mode simply treats the original IP packet as the payload and then adds ESP

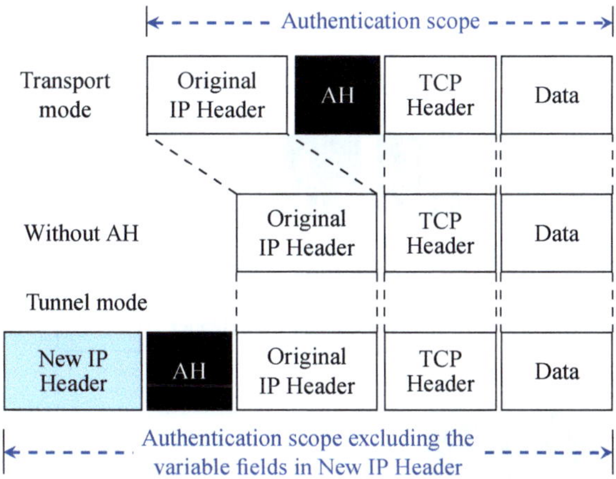

Fig. 6.23 AH in an IP packet using TCP

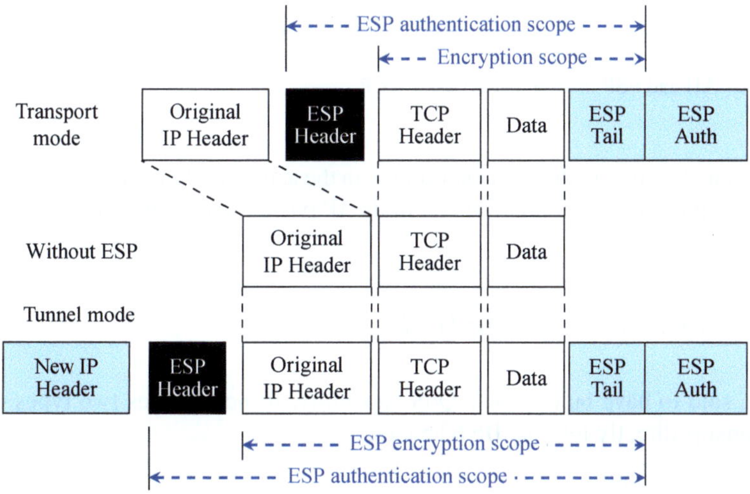

Fig. 6.24 ESP in an IP packet using TCP

and a new IP header in front to form a new IP packet. In the transport mode, insert ESP between the original IP header and the original TCP header.

If both AH and ESP are used, according to the preferred order of IPv6 extension headers, AH precedes ESP. Figure 6.25 shows the use of both AH and ESP in an IP packet.

The scenarios illustrated in Figs. 6.23, 6.24, and 6.25 for the use of AH and ESP headers either individually or in combination are valid for both IPv4 and IPv6. In addition, for IPv6, additional extension headers, if present, can be inserted into an

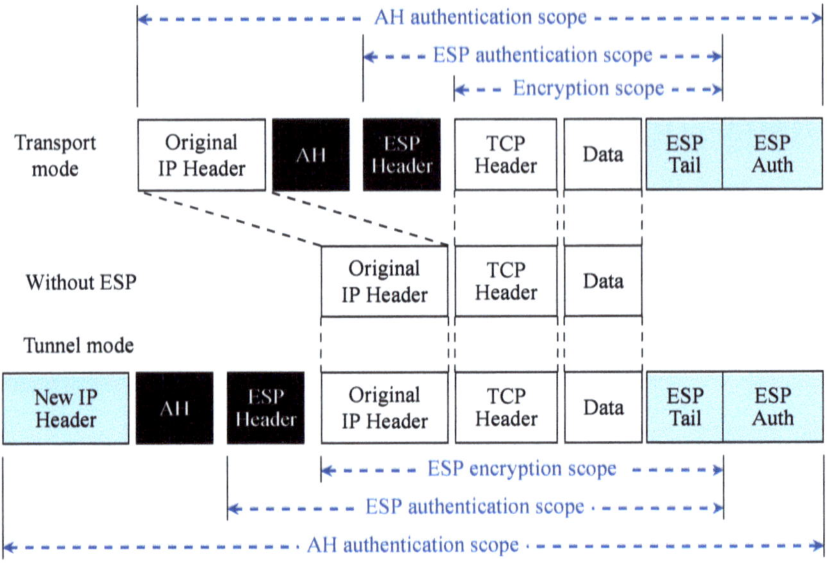

Fig. 6.25 AH and ESP in an IP packet using TCP

appropriate location in the IP packet shown in these figures. For example, if a Routing Header is present, it will precede AH in the IP packet within these figures.

6.7 Built-in True QoS in IPv6

IPv6 is said to have built-in true QoS capability. It incorporates two types of QoS mechanisms directly into the IPv6 header:

- Traffic class, and
- Flow label.

In IPv6, a host can use one or both of these fields to identify traffic packets that require special QoS treatment by IPv6 routers. These fields allow the host to request specific handling and prioritization for the identified packets.

6.7.1 Traffic Class

Similar to the *ToS* field in the IPv4 header, the *Traffic Class* field in the IPv6 header allows hosts to identify different classes or priorities for IPv6 packets. The first 6 bits of the Traffic Class field can be used to set specific precedence or Differentiated Services field CodePoint (DSCP) values, as defined in RFC 2474 [2] and updated in RFC 8436 [35]. The last 2 bits of the Traffic Class field are used for Explicit Congestion Notification (ECN), as defined in RFC 3168 [36]. It is noticed that the routers that forward the packets also use the Traffic Class field for the same purpose.

When the Traffic Class field is employed, the interface to the IPv6 service within a host must provide the Traffic Class bits for an upper-layer protocol, such as TCP. These Traffic Class bits must be present in packets originated by the upper-layer protocol. If there is no intention to use the Traffic Class field, all 8 bits of the Traffic Class field should be set to zero.

Nodes that support some or all of the Traffic Class bits can override the values of these bits. They can only change the values in packets that the nodes originate, forward, or receive, as required for that specific use. Therefore, it is possible that the Traffic Class bits in a received packet differ from those in the packet sent by its source.

However, not all Traffic Class bits in a packet can always be changed along the transmission path. If a node does not support a specific use, it should not make any changes to the values of the corresponding Traffic Class bits.

After packets are marked in the Traffic Class field, various methods can be employed for QoS management. Typical examples include traffic classification, conditioning, shaping, queuing, and scheduling. The packets are processed based on policies that reflect their respective QoS levels.

6.7.2 Flow Labeling

A flow is a sequence of packets that are transmitted from a specific source to a particular, unicast or multicast, destination. Without presence in IPv4, flow labeling is a new feature in IPv6 to label flows. As shown in Fig. 6.18, the IPv6 header includes a 20-bit *Flow Label* field, which enables efficient classification of IPv6 flows at the IP layer based solely on the fixed header of IPv6. More precisely, the flow label enables per-flow processing of packets for differentiation at the IP layer. The IETF RFC 6437 [37] specifies the Pv6 *Flow Label* field. It also clarifies the requirements for flow labeling and flow state establishment.

IPv6 Flow Label Specification

The 20-bit *Flow Label* field is used by a source node, which generates traffic, to label packets belonging to a specific flow. With the flow label as well as the source and destination addresses in the fixed header of IPv6, the flow to which a particular packet belongs could be identified uniquely. Then, packets are processed in a flow-specific manner. Packets that do not belong to any flow carry a flow label of zero.

While the flow label could be used in a stateless or stateful manner, it is most commonly used in stateless scenarios:

- When the flow label is used statelessly, a node that processes the flow label does not need to store any information about the flow either before or after a packet is processed.

- In stateful scenarios, the information about the flow, including the flow label value, needs to be stored for future use. In this case, a signaling mechanism will be useful to notify downstream nodes that the flow label is being used in a specific way. An example of such a signaling mechanism is the IntServ RSVP [38], which can be used for this purpose.

Flow label values should be chosen randomly to enhance security. It is recommended in RFC 6437 [37] that flow labels be chosen from a uniform probability. By doing so, it becomes unlikely for third parties to predict the next flow label value generated by a source. Sequential assignment of flow label values is particularly discouraged in order to avoid potential attacks.

Once set to a non-zero value, the Flow Label of a packet should not be changed along the path of the packet from the source to the destination. The only exception to this rule is for compelling operational security reasons, as indicated in RFC 6437 [37, p. 8].

The *Floe Label* is an unprotected field in the IPv6 header, even when IPsec authentication is employed. No mechanisms exist to verify whether a flow label has been modified or it is chosen from a uniform distribution. Therefore, as a rule, any forwarding nodes such as routers and load distributors MUST NOT rely on the assumption of a uniform distribution of flow label values.

It is a misconception that flow label values can be used to reorder packet flows. The use of the Flow Label field does not affect the general principle that packet flows should not be reordered. Even if the flow label value is set to zero, reordering packet flows is still considered unacceptable. This has been indicated clearly and explicitly in RFC 6437 [37, p. 4].

Requirements for Stateless Flow Labeling

An essential requirement for flow labeling is that source nodes should assign each unrelated transport connection and application data steam to a new flow. As source nodes have easy access to the information of destination address, source address, protocol, destination port, and source port, this information can be used to define a flow.

It is also desirable that flow label values are chosen from a uniform distribution for easy load distribution. While RFC 6437 does not mandate a specific algorithm for choosing flow label values, it specifies requirements for designing such an algorithm. It also provides an example of stateless 20-bit hash function algorithm [37, p. 14].

It is mandatory that if the flow label is not set, it MUST carry a value of zero. For packets with a flow label value of zero, forwarding nodes MAY change the flow label value. They are presumably the first-hop or ingress routers. In such cases, RFC 6437 recommends that the flow label field be set to a uniformly distributed value as for source nodes [37, p. 6].

Requirements for Flow State Establishment

The general requirements for flow state establishment are:

- Source nodes that set flow label values may participate in a flow state establishment method that assigns specific treatments to specific flows.
- It is required that any such method MUST not disrupt the nodes that participate in stateless flow labeling.
- The nodes that set flow label values in a stateful manner MUST choose labels randomly from a uniform distribution, as described previously as a desirable requirement.

6.8 Coexistence of IPv6 and IPv4

IP networks have been evolving significantly in the last few decades, driving the development of IPv6. While Internet Service Providers and enterprises are dealing with the rapid growth of their networks by using IPv6, they also need to serve a large number of existing IPv4 users. However, IPv6 has not been designed to be backward compatible with IPv4. Therefore, IPv6 and IPv4 must coexist for a potentially extended period. It is a general requirement that the coexistence of IPv6 and IPv4 be transparent to end users.

There are generally three main techniques that can help achieve the coexistence of IPv6 and IPv4. They are

- Dual stack, which runs both IPv6 and IPv4 simultaneously,
- Tunneling, which enables the transmission of IPv6 packets over IPv4 networks, and
- IPv4-to-IPv6 translation, known as NAT64 [39], which translates IPv6 packets to IPv4 ones or vice versa. It is primarily used by ISPs.

As NAT64 is primarily used by ISPs, this section focuses more on the dual stack and tunneling. Both dual stack and configured tunneling mechanisms are specified in the IETF RFC 4213 (October 2005) [40].

It is interesting to note that the public side of NAT64 devices generally uses IPv6 rather than IPv4. This is because ISPs cannot further grow their IPv4 networks easily by either assigning public IPv4 addresses to customers or obtaining new public IPv4 addresses for their own networks. However, they must continue to serve both IPv4 customers and new IPv6 customers.

6.8.1 Dual Stack

Dual stack is the most direct technique to achieve the coexistence of IPv6 and IPv4. With dual stack, each network device is configured with the capability to run both IPv6 and IPv4 simultaneously. Figure 6.26 demonstrates the dual stack configuration

Fig. 6.26 Demonstration of dual stack in Windows by using command `ipcinfig` in a command window. Use command `ifconfig` in MacOS and other Linux systems

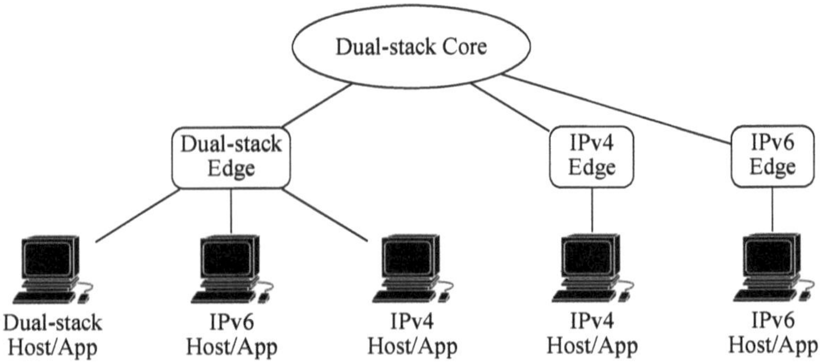

Fig. 6.27 Dual stack configured in a network

of a host. It can be observed from this figure that the host is configured with both IPv6 and IPv4 addresses. The IPv4 address 131.181.33.27/24 is a globally routable public address, while the IPv6 address starts with fe80 and thus is a link-local address. The host runs both IPv6 and IPv4 simultaneously.

Figure 6.27 illustrates the coexistence of IPv4 and IPv6 via dual stack in a network. The core network is configured with dual stack and interconnected with three types of edges: dual-stack edge, IPv6 edge, and IPv4 edge. Hosts and applications with dual stack, pure IPv6, and pure IPv4 can be connected to the dual-stack edge. IPv6-only hosts and applications can be connected to the IPv6-only edge. Similarly, IPv4-only hosts and applications can be connected to the IPv4-only edge.

In dual stack, IPv4 communication will use the IPv4 protocol stack, while IPv6 communication will use the IPv6 protocol stack. In general, the IPv6 protocol stack takes higher priority over the IPv4 protocol stack. But the decision to use IPv4 or IPv6 depends on the response to DNS requests.

As an IPv6 transition technology, dual stack offers several noticeable advantages. A significant advantage is its simplicity and cost-effectiveness. As long as hardware devices support both IPv4 and IPv6, they can be easily configured with dual stack. The configuration process is typically automated and transparent to end users. Another advantage is that dual stack eliminates the need for translation between the IPv4 and IPv6 protocol stacks, resulting in efficient traffic processing. Moreover, dual stack provides the flexibility to discontinue IPv4 in the future when IPv6 becomes fully functional. While IPv4 and IPv6 currently coexist, dual stack is not a permanent long-term solution. If every network device can be configured with both IPv4 and IPv6, why should IPv4 continue to be used at all?

6.8.2 *IPv6-over-IPv4 Tunneling*

Tunneling is also a typical technique used in IPv6 transition. To enable IPv4 routing infrastructure to carry IPv6 traffic, IPv6 datagrams are encapsulated within IPv4 packets.

Encapsulation of IPv6 Datagrams

The encapsulation of an IPv6 datagram within an IPv4 packet is shown in Fig. 6.28. It is seen from this figure that an IPv4 header is added to the IPv6 datagram. Creating and adding the IPv4 header are the tasks of the encapsulator at the entry point of the tunnel. upon reaching the exit node of the tunnel, the decapsulator reassembles the packet if needed, removes the IPv4 header, and processes the received IPv6 packet.

In reality, the encapsulation also needs to deal with more complicated scenarios, particularly those related to fragmentation and ICMPv4 errors on too-big packets. These complicated scenarios are discussed in detail in the IETF RFC 4213 [40, pp. 8–13].

How to use Tunneling

As described in the IETF RFC 4213 [40, p. 6], tunneling can be used in a variety of ways. More specifically, it can be used in the following four scenarios:

- Router-to-Router tunnel: IPv6/IPv4 routers that are interconnected by an IPv4 infrastructure use this type of tunnel to tunnel IPv6 traffic between themselves. In this case, the tunnel spans one segment of the end-to-end path that the IPv6 packet takes.
- Host-to-Router tunnel: IPv6/IPv4 hosts tunnel IPv6 traffic to an intermediary IPv6/IPv4 router reachable via an IPv4 infrastructure. The tunnel spans the first segment of the end-to-end path of the packet.
- Host-to-Host tunnel: IPv6/IPv4 hosts interconnected by an IPv4 infrastructure use this type of tunnel to tunnel IPv6 packets between themselves. In this case, the tunnel spans the entire end-to-end path of the packet.
- Router-to-Host tunnel: IPv6/IPv4 routers use this type of tunnel to tunnel IPv6 packets to their final destination IPv6/IPv4 host. This tunnel spans only the last segment of the end-to-end path.

Fig. 6.28 Encapsulation of IPv6 within IPv4

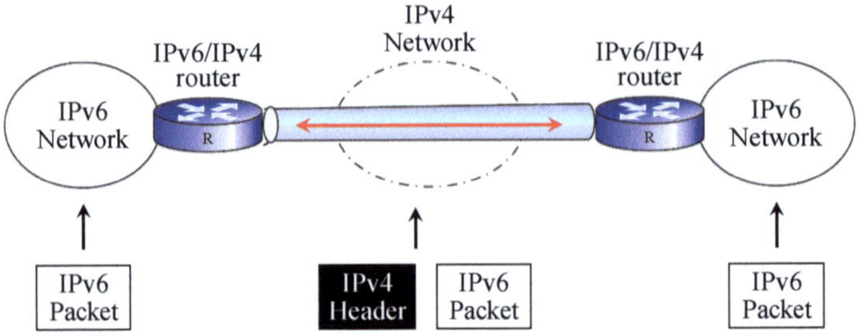

Fig. 6.29 Router-to-router tunneling over IPv4

While configured tunneling can be used in all these four scenarios, it is most likely to be used in router-to-router tunneling. This is due to the requirement of explicit configuration of tunneling endpoints [40, pp. 6–7]. Figure 6.29 demonstrates router-to-router tunneling over an IPv4 network.

6.9 IPv6 Network Planning

Planning IPv6 for a new network or an existing IPv4 network is a challenging task and thus requires a major effort. For an existing IPv4 network, it is unlikely that the network can be converted to an IPv6-only network in one step without significant interruptions to network services. Therefore, it is important to plan for the coexistence of both IPv4 and IPv6. In general, the deployment of IPv6 on an existing IPv4 network should be phased gradually, allowing for a smooth transition and minimizing disruptions to network services.

6.9.1 Checklist for IPv6 Planning

In an Oracle online document [41, Chap. 4], 11 tasks are listed which are considered to be necessary in order to accomplish planning for IPv6 deployment. While this document has been published for over a decade, the 11 tasks listed in the document still form a basic checklist for IPv6 planning. They are summarized in Table 6.12, and should be considered sequentially. Some of the tasks listed above are obvious, e.g., IPv6 support of hardware, an ISP that supports IPv6, and site prefix. Some others require detailed analysis, planning, and development, such as addressing scheme, dual stack or tunneling, and security. A few of these tasks will be discussed briefly in the following subsections.

Table 6.12 Checklist for IPv6 planning [41, Chap. 4]

No.	Task
(1)	Ensure that your hardware is IPv6 ready. Verify that your hardware can be upgraded to IPv6. Specific hardware devices that need to be checked for IPv6 support include routers, firewalls, various servers, switches, and end-user hosts
(2)	Ensure that your ISP supports IPv6. If your current ISP does not support IPv6, find an alternative ISP
(3)	Ensure that your applications are IPv6 ready. Verify that your applications can run in an IPv6 environment
(4)	Obtain a 48-bit site prefix for your site. Obtain it from either your ISP or the nearest Regional Internet Registry (RIR)
(5)	Create a subnet addressing plan. This is to plan the overall IPv6 network topology and addressing scheme before configuring IPv6 on various nodes in the network
(6)	Design a plan for tunnel usage. This determines which routers should run tunnels to other subnets or external networks
(7)	Create an addressing plan for entities on the network. This is to ensure that a plan for addressing servers, routers, and hosts is in place before IPv6 configuration can be performed
(8)	Develop an IPv6 security policy. This requires a plan for security architecture, mechanisms, and protocols for the entire IPv6 network
(9)	Enable the nodes to support IPv6. This requires configuring IPv6 on not only hosts but also routers
(10)	Turn on network services. This requires that existing servers support IPv6
(11)	Update name servers for IPv6 support. More specifically, make sure that DNS servers are updated with the new IPv6 addresses

6.9.2 *Preparing Network Services*

Many network services are already IPv6-ready, such as HTTP and DNS, while others may still be limited to IPv4 in a network. With dual stack, nodes configured for IPv6 can run IPv4 services. However, it is worth noting that not all services accept IPv6 connections when IPv6 is enabled. Generally, services that have been ported to IPv6 will accept IPv6 connections, but those that have not may continue to function using the IPv4 protocol stack.

Special attention should be given to servers during the preparation of network services. In IPv6, servers are typically treated as hosts, and their IPv6 addresses are automatically configured by default. However, automatically configured IPv6 addresses may not always meet the requirements, especially in scenarios where multiple NICs are installed on a server. In such cases, manual configuration of the interface ID portion of the IPv6 addresses for each interface should be considered.

It is worthy to emphasize the need to verify IPv6 support for the most important network services, even if not initially for all services. Services such as DNS, firewalls, mail, and HTTP should be among the top priorities when preparing network services.

6.9.3 Planning for Tunnels

As depicted in Fig. 6.29, tunnels serve as a means for isolated IPv6 networks to communicate with each other over IPv4 routing infrastructure. The IPv6 implementation offers various tunnel configurations, including router-to-router, host-to-router, host-to-host, and router-to-host, serving as transition mechanisms for transitioning to a mixed use of IPv4 and IPv6 [40]. Despite the existence of these tunneling options, global network statistics shown in Fig. 6.13 reveal that approximately two-thirds of networks worldwide are still deployed in IPv4. Consequently, IPv6 traffic originating from an IPv6 or dual-stack network will often need to traverse the Internet through tunnels to reach its destination in an IPv6 or dual-stack network.

When preparing for tunneling, consider which tunnels need to be configured and where they should be deployed. As discussed earlier, each of the four types of tunnels, i.e., router-to-router, router-to-host, host-to-router, and host-to-host tunnels, has its specific application scenarios. Select one or more of these tunnel types to meet the specific tunneling requirements of the network at hand.

6.9.4 Security Considerations

A comprehensive security plan is necessary when building a new IPv6 network or upgrading an existing IPv4 network to an IPv6 or dual-stack network. While IPv6 is known to have built-in security through the mandatory inclusion of IPSec, the use of IPSec is not mandatory in IPv6.

More importantly, IPv6 network security deals with not only IPSec but also many other aspects. The following aspects will need to be considered for IPv6 security [41, p. 90]:

- Both IPv6 packets and IPv4 packets require the same level of filtering.
- IPv6 packets are often tunneled through a firewall. Thus, either of the following scenarios needs to be implemented:
  - Perform content inspection inside the tunnel using the firewall, or
  - Deploy an IPv6 firewall with similar rules at the opposite tunnel endpoint.
- Some transition mechanisms exist that use IPv6 over UDP through IPv4 tunnels. They might be risky by short-circuiting the firewall.
- IPv6 nodes are globally reachable from outside an enterprise network. To restrict public access to the network, stricter firewall rules should be established compared to an IPv4 network. For this purpose, statefull firewalls can be an option.

Overall, when planning for IPv6 network security, it is important to address aspects such as filtering, tunneling through firewalls, risks associated with transition mechanisms, and the need for stricter firewall rules to restrict public access to the network.

6.10 Summary

Layer-3 IP addressing is essential for network devices to be visible and accessible on the Internet. Together with routing, it provides the fundamental functionality for end-to-end delivery of packets from source nodes to destination nodes. A plan for IP addressing in a network will largely determine the connectivity, scalability, and manageability of the network. It will also determine routing efficiency, security design, and the easiness for future growth. Therefore, IP addressing is an important component in the overall network architecture.

Currently, the majority of the networks on the Internet are still IPv4 networks. This implies that planning for any networks must consider IPv4. In IPv4 networks, subnetting and supernetting are widely employed, with subnetting conserving IP address resources and supernetting enhancing routing efficiency. Subnetting borrows one or more most significant bits from the host portion of an IP address to create subnets. This shifts the boundary between the network portion and host portion of an IP address towards the right. By contrast, supernetting gives away one or more least-significant bits from the network portion of an address to aggregate or summarize multiple networks into a single address. As a result, it moves the boundary between the network portion and host portion of the IP addresses to be aggregated towards the left. Following subnetting, IP addresses are allocated hierarchically in blocks. For hosts, IP addresses are typically assigned via DHCP. For routers and some servers, IP addresses will need to be configured manually in general. To save address resources, private IP addresses are also widely used through NAT in residential, office, and enterprise networks.

IPv6 networks are increasingly prevalent, accounting for slightly over one third of all networks. The primary motivation behind IPv6 development is to deal with the issue of IPv4 address exhaustion. However, IPv6 is not merely an extension of IPv4. Actually, it lack backward compatibility with IPv4. In addition to a significantly larger address space and simplified header structure, IPv6 incorporates unique features that are absent in IPv4, such as autoconfiguration, built-in security, and built-in true QoS. It is important to note that while the inclusion of IPsec is mandatory in IPv6, its use in IPv6 is not obligatory. IPsec is designed for both IPv4 and IPv6, making it available for use in both IPv6 and IPv4 networks.

Planning IPv6 will need to consider various factors. Hardware devices and network services must support IPv6. Applications to be deployed must be able to execute in IPv6 environments. More specifically, the coexistence of IPv4 and IPv6 should be considered. There are generally three techniques for transitioning to IPv6: dual stack, tunneling, and NAT64. Among these transition techniques, NAT64 is predominantly used by ISPs. Dual stack is considered to be a temporary solution because it does not save any IPv4 address resources. Tunneling can be configured in multiple ways to accommodate diverse requirements, with router-to-router tunneling being the most common configuration. Other configurations include router-to-host, host-to-router, and host-to-host tunneling.

References

1. Postel, J.: Internet protocol. RFC 791, RFC Editor (1981). STD 5. https://doi.org/10.17487/RFC0791
2. Nichols, K., Blake, S., Baker, F., Black, D.: Definition of the differentiated services field (DS field) in the IPv4 and IPv6 headers. RFC 2474, RFC Editor (1998). https://doi.org/10.17487/RFC2474
3. Touch, J.: Updated specification of the IPv4 ID field. RFC 6864, RFC Editor (2013). https://doi.org/10.17487/RFC6864
4. Srisuresh, P., Egevang, K.: Traditional IP network address translator (traditional NAT). RFC 3022, RFC Editor (2001). https://doi.org/10.17487/RFC3022
5. Srisuresh, P., Holdrege, M.: IP network address translator (NAT) terminology and considerations. RFC 2663, RFC Editor (1999). https://doi.org/10.17487/RFC2663
6. Rekhter, Y., Li, T.: An architecture for IP address allocation with CIDR. RFC 1518, RFC Editor (1993). https://doi.org/10.17487/RFC1518
7. Fuller, V., Li, T.: Classless inter-domain routing (CIDR): The Internet address assignment and aggregation plan. RFC 4632, RFC Editor (2006). BCP 122. https://doi.org/10.17487/RFC4632
8. Mogul, J.C., Postel, J.: Internet standard subnetting procedure. RFC 950, RFC Editor (1985). STD 5. https://doi.org/10.17487/RFC0950
9. Pummill, T., Manning, B.: Variable length subnet table for IPv4. RFC 1878, RFC Editor (1995). https://doi.org/10.17487/RFC1878
10. Hinden, R., Deering, S.: IP version 6 addressing architecture. RFC 4291, RFC Editor (2006). https://doi.org/10.17487/RFC4291
11. Deering, S., Hinden, R.: Internet protocol, version 6 (IPv6) specification. RFC 8200, RFC Editor (2017). STD 86. https://doi.org/10.17487/RFC8200
12. Hinden, R., Haberman, B.: Unique local IPv6 unicast addresses. RFC 4193, RFC Editor (2005). https://doi.org/10.17487/RFC4193
13. IANA: IPv6 multicast address space registry. Online reference (2022). https://www.iana.org/assignments/ipv6-multicast-addresses/ipv6-multicast-addresses.xhtml. Accessed 6 July 2022
14. Boucadair, M., Venaas, S.: Updates to the IPv6 multicast addressing architecture. RFC 7371, RFC Editor (2014). https://doi.org/10.17487/RFC7371
15. Droms, R.: IPv6 multicast address scopes. RFC 7346, RFC Editor (2014). https://doi.org/10.17487/RFC7346
16. Haberman, B.: Allocation guidelines for IPv6 multicast addresses. RFC 3307, RFC Editor (2002). https://doi.org/10.17487/RFC3307
17. Haberman, B., Thaler, D.: Unicast-prefix-based IPv6 multicast addresses. RFC 3306, RFC Editor (2002). https://doi.org/10.17487/RFC3306
18. Savola, P., Haberman, B.: Embedding the rendezvous point (RP) address in an IPv6 multicast address. RFC 3956, RFC Editor (2004). https://doi.org/10.17487/RFC3956
19. Deering, S.E.: Host extensions for IP multicasting. RFC 1112, RFC Editor (1989). https://doi.org/10.17487/RFC1112
20. Fenner, W.: Internet group management protocol, version 2. RFC 2236, RFC Editor (1997). https://doi.org/10.17487/RFC2236
21. Cain, B., Deering, S., Kouvelas, I., Fenner, B., Thyagarajan, A.: Internet group management protocol, version 3. RFC 3376, RFC Editor (2002). https://doi.org/10.17487/RFC3376
22. Vida, R., Costa, L.: Multicast listener discovery version 2 (MLDv2) for IPv6. RFC 3810, RFC Editor (2004). https://doi.org/10.17487/RFC3810
23. Holbrook, H., Cain, B., Haberman, B.: Using Internet group management protocol version 3 (IGMPv3) and multicast listener discovery protocol version 2 (MLDv2) for source-specific multicast. RFC 4604, RFC Editor (2006). https://doi.org/10.17487/RFC4604
24. McCann, J., Deering, S., Mogul, J., Hinden, R.: Path MTU discovery for IP version 6. RFC 8201, RFC Editor (2017). STD 87. https://doi.org/10.17487/RFC8201

25. Fairhurst, G., Jones, T., Tüxen, M., Rüngeler, I., Völker, T.: Packetization layer path MTU discovery for datagram transports. RFC 8899, RFC Editor (2020). https://doi.org/10.17487/RFC8899
26. Thomson, S., Narten, T., Jinmei, T.: IPv6 stateless address autoconfiguration. RFC 4862, RFC Editor (2007). https://doi.org/10.17487/RFC4862
27. Thomson, S., Narten, T., Jinmei, T.: Enhanced duplicate address detection. RFC 7527, RFC Editor (2015). https://doi.org/10.17487/RFC7527
28. Narten, T., Nordmark, E., Simpson, W., Soliman, H.: Neighbor discovery for IP version 6 (IPv6). RFC 4861, RFC Editor (2007). https://doi.org/10.17487/RFC4861
29. Troan, O.: Iana considerations for IPv6 neighbor discovery prefix information option flags. RFC 8425, RFC Editor (2018). https://doi.org/10.17487/RFC8425
30. Mrugalski, T., Siodelski, M., Volz, B., Yourtchenko, A., Richardson, M., Jiang, S., Lemon, T., Winters, T.: Dynamic host configuration protocol for IPv6 (DHCPv6). RFC 8415, RFC Editor (2018). https://doi.org/10.17487/RFC8415
31. Sun, Q., Cui, Y., Siodelski, M., Krishnan, S., Farrer, I.: DHCPv4-over-DHCPv6 (DHCP 4o6) transport. RFC 7341, RFC Editor (2014). https://doi.org/10.17487/RFC7341
32. Kent, S., Seo, K.: Security architecture for the Internet protocol. RFC 4301, RFC Editor (2005). https://doi.org/10.17487/RFC4301
33. Kent, S.: IP authentication header. RFC 4302, RFC Editor (2005). https://doi.org/10.17487/RFC4302
34. Kent, S.: IP encapsulating security payload (ESP). RFC 4303, RFC Editor (2005). https://doi.org/10.17487/RFC4303
35. Fairhurst, G.: Update to IANA registration procedures for pool 3 values in the differentiated services field codepoints (DSCP) registry. RFC 8436, RFC Editor (2018). https://doi.org/10.17487/RFC8436
36. Ramakrishnan, K., Floyd, S., Black, D.: The addition of explicit congestion notification (ECN) to IP. RFC 3168, RFC Editor (2001). https://doi.org/10.17487/RFC3168
37. Amante, S., Carpenter, B., Jiang, S., Rajahalmer, J.: IPv6 flow label specification. RFC 6437, RFC Editor (2011). https://doi.org/10.17487/RFC6437
38. Zhang, L., Berson, S., Herzog, S., Jamin, S.: Resource reservation protocol (RSVP) – version 1 functional specification. RFC 2205, RFC Editor (1997). https://doi.org/10.17487/RFC2205
39. Bao, C., Huitema, C., Bagnulo, M., Boucadair, M., Li, X.: IPv6 addressing of IPv4/IPv6 translators. RFC 6052, RFC Editor (2010). STD 5. https://doi.org/10.17487/RFC6052
40. Nordmark, E., Gilligan, R.: Basic transition mechanisms for IPv6 hosts and routers. RFC 4213, RFC Editor (2005). https://doi.org/10.17487/RFC4213
41. Oracle: System administration guide: Ip services. Online document (2011). Part No: 816–4554–2. https://docs.oracle.com/cd/E18752_01/pdf/816-4554.pdf. Accessed 6 Aug 2022

Chapter 7
Network Routing Architecture

Routing is a network-layer (layer-3) function in ISO's seven-layer network architecture. It involves the process of finding a communication path, typically from multiple options, to deliver network traffic from its source to the intended destination. Routing decisions are made within a network, such as an enterprise network, spanning multiple networks like campus networks in a university, or across the Internet. Through these decisions, network packets are routed from their source nodes to their destination nodes, passing through intermediate nodes like gateways, and routers. These intermediate nodes have routing protocols installed and executed to handle traffic routing and packet forwarding. Each routing device responsible for traffic routing and packet forwarding maintains a dynamic routing table, which keeps a record of routes to various network destinations. If a specific destination is not listed in the routing table, the routing protocol installed on the device will search for it by communicating with other routers, typically neighboring ones.

7.1 Categories of Routing Protocols

From different perspectives of discussions, routing protocols can be classified in different ways. For example, according to where they are used logically and physically within networks, they can be categorized as Interior Gateway Protocol (IGP) or Exterior Gateway Protocol (EGP). In terms of the metrics used to determine the optimal routing paths, routing protocols can be grouped as distance-vector routing protocols, link-state routing protocols, path-vector routing protocols, or variations thereof. Moreover, routing protocols can be classified as either classful or classless, depending on whether they use classful or classless information for routing decisions. Figure 7.1 illustrates the classification and relationships of routing protocols in computer networks.

Y.-C. Tian and J. Gao, *Network Analysis and Architecture*, Signals and Communication Technology, https://doi.org/10.1007/978-981-99-5648-7_7

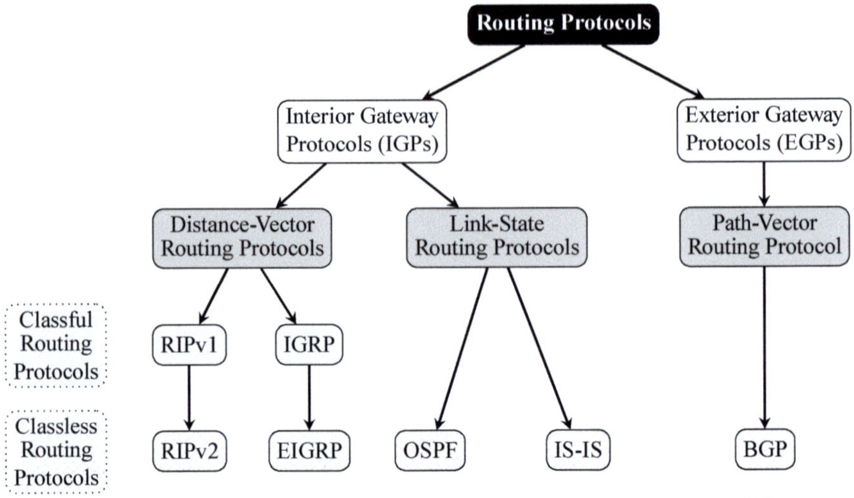

RIPv1	Routing Information Protocol version 1	Broadcast UDP
RIPv2	Routing Information Protocol version 2	Multicast UDP
IGRP	Interior Gateway Routing Protocol	Encapsulated in IP
EIGRP	Ehnanced Interior Gateway Routing Protocol	Encapsulated in IP
OSPF	Open Shortest Path First	Encapsulated in IP
IS-IS	Intermediate System to Intermediate System	On data-link layer
BGP	Border Gateway Protocol	Over TCP

Fig. 7.1 Classification of routing protocols

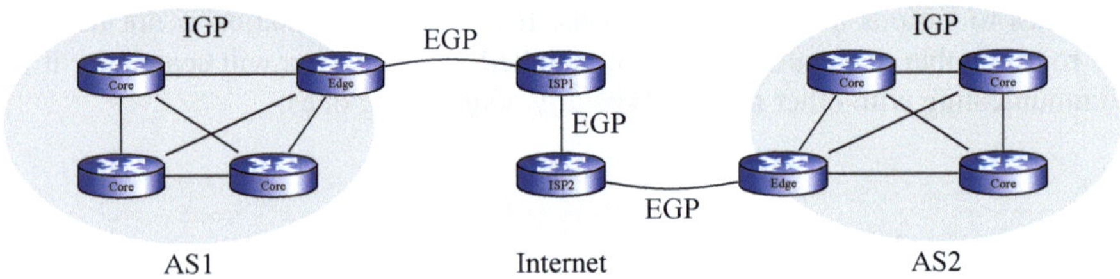

Fig. 7.2 IGPs and EGP in networks

The use of IGPs and EGPs in networks is depicted in the diagram of Fig. 7.2. IGPs exchange routing table information and find network paths between routers within a routing domain, such as an Autonomous System (AS) comprising multiple LANs. Both distance-vector routing protocols and link-state routing protocols fall under the category of IGPs. Therefore, IGPs include protocols like Routing Information Protocol (RIP) (RIPv1 and RIPv2), Interior Gateway Routing Protocol (IGRP) and Enhanced Interior Gateway Routing Protocol (EIGRP), Open Shortest Path First (OSPF), and Intermediate System to Intermediate System (IS-IS).

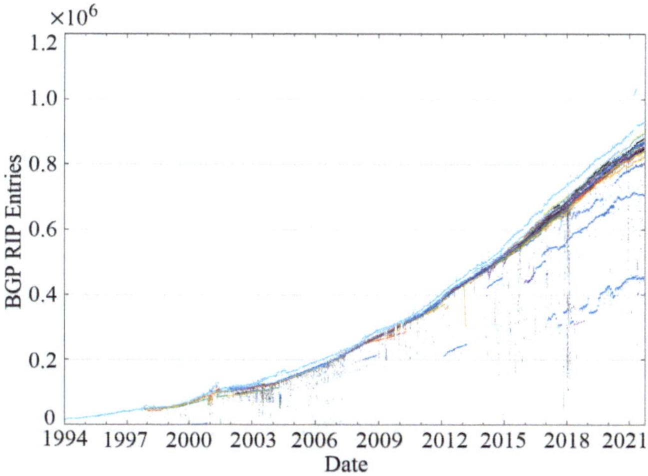

Fig. 7.3 Growth of the BGP table—1994 to present as of 15 Sep. 2021 [3]

By contrast, EGP exchanges routing information and finds network paths between different ASs. It is employed by edge routers and exterior routers to route traffic outside of the ASs.

7.1.1 Path-Vector Routing Protocol

Under the category of EGP, a path-vector-based routing protocol is designed to function across multiple ASs. The only EGP currently in use is the Border Gateway Protocol (BGP), whose version 4 (BGP4) is specified in the IETF RFC 1771 [1]. The application of BGP on the Internet is discussed comprehensively in the IETF RFC 1772 [2].

The primary objective of BGP is to enable routers to exchange information regarding paths to destination networks. By calculating loop-free network paths across the Internet, BGP uses a path-vector routing algorithm and, therefore, is a path-vector routing protocol. This means that BGP keeps track of the path in terms of the AS it traverses rather than the route through individual routers within an AS. For this reason, BGP is commonly used by network service providers or companies with multiple Internet Service Providers (ISPs). It is widely supported by most router manufacturers. Moreover, BGP exchanges routing information through a TCP connection, making it a reliable routing protocol in terms of TCP network communications. Impressively, as of the 15th of September 2021, the active BGP Forwarding Table entries in the Forwarding Information Base (FIB) have reached nearly 10^6. The underlying BGP Routing Table entries in the Routing Information Base (RIB) are approximately 30 times as large as those in the FIB. This is illustrated in Fig. 7.3 [3].

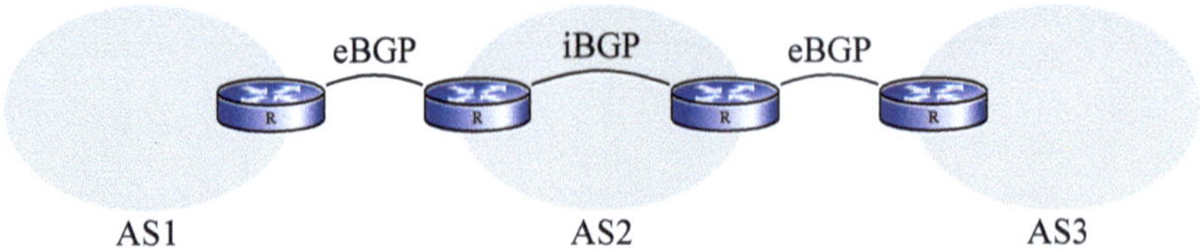

Fig. 7.4 eBGP and iBGP

In the discussions on the application of BGP from the IETF RFC 1772, traffic flows are described as local and transit traffic. In terms of traffic flows, ASs can be classified into stub ASs, multihomed ASs, or transit ASs. A stub AS has only one connection to another AS and only carries local traffic. A multihomed AS is connected to more than one other AS. But it does not carry transit traffic. With connections to more than one other AS, a transit AS carries both local and transit traffic flows. BGP aims to control transit traffic flows and thus primarily operates on transit ASs.

While BGP exchanges routing information externally between ASs, there are scenarios in which several border routers belonging to the same AS are connected to external networks. This requires BGP to have the capability to function internally, allowing these border routers to exchange routing information. Therefore, both *external BGP (eBGP)* and *internal BGP (iBGP)* have been developed. eBGP operates between ASs to exchange information on paths to destination networks. In comparison, iBGP can be used to route traffic between routing domains, such as between border routers, within an AS. Figure 7.4 illustrates an example of the use of eBGP and iBGP.

7.1.2 Distance-Vector Routing Protocols

Distance-vector routing protocols are IGPs, which operate within an AS. They use distance or hop count as the primary metric for determining the best forwarding path. Therefore, a shorter distance indicates a better path. As shown in Fig. 7.1, two widely used distance-vector routing protocols are RIPv2 and EIGRP.

Distance-Vector Routing Operations

A distance-vector routing protocol periodically informs its neighboring routers about its routing table information. By learning from the up-to-date network topology of its neighbors, each router maintains a distance-vector table that describes the distance between itself and all possible destinations.

Using the distance-vector algorithm, a router calculates distance values based on the distance vectors provided by the neighboring routers. Let us consider a scenario with three routers as depicted in Fig. 7.5, where each router has its initial routing table. Each router shares its routing table with the other two routers. Upon receiving

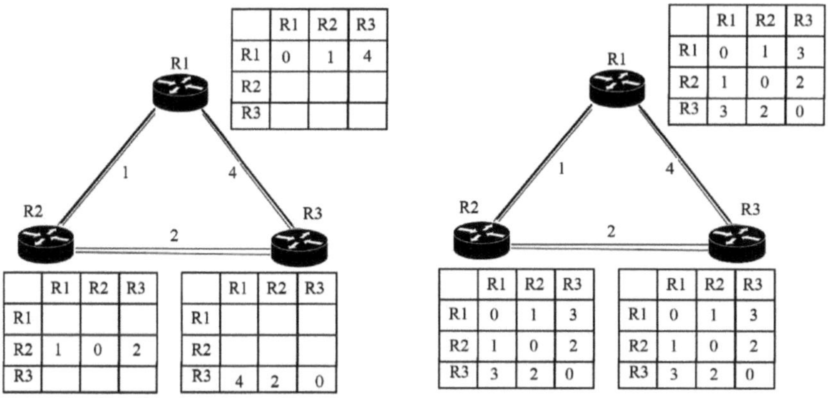

(a) Initial routing tables (b) Updated routing tables

Fig. 7.5 Updating routing tables in distance vector routing protocols

the shared routing table from R2, R1 recognizes that the distance D(R1,R2)= 1 is already the shortest path to R2. The distance D(R1,R3) can be refined by evaluating $\min(D(\text{R1,R2}) + D(\text{R2,R3}), D(\text{R1,R3})) = \min(1 + 2, 4) = 3$. Consequently, R1 updates its routing table with a new distance of 3 to R3. Similarly, through running its own distance-vector algorithm, R3 updates its routing table with a new distance of 3 to R1. Eventually, all three routers possess routing tables with the shortest paths to each other.

To accommodate the dynamic changes in network topology, routers need to periodically exchange their routing tables with their respective neighbors. A shorter update period for routing tables leads to increased overhead, while a longer period results in a slower response to network topology changes. RIPv2 updates routing information every 30 s, whereas EIGRP updates its routing table every 90 s.

The routing table in a router is typically extensive, containing a significant amount of information. When calculating the best paths for routing table updates, the process of periodically exchanging large routing tables with neighboring routers may require multiple update rounds to reach convergence. However, it is important to note that the network topology can change during this convergence process, potentially rendering the calculated paths outdated. Therefore, the convergence speed is a critical factor in routing protocols and should be taken into account when determining the update period for routing tables.

Distance routing protocols possess both advantages and disadvantages. The primary advantage of distance-vector routing protocols is their simpler configuration and maintenance compared to link-state routing protocols, which will be discussed later. However, these protocols also exhibit certain disadvantages when compared to link-state routing protocols. They have slower convergence, are susceptible to the count-to-infinity problem, generate more network traffic, and result in larger routing tables for extensive networks. The count-to-infinity problem arises from network loops. It can be mitigated by setting a maximum network diameter, such as 15 hops, with which any 16-hop routes will be recognized unreachable. Nevertheless, the slow convergence of routing table updates remains an issue. To provide a quantitative measurement, updating routing tables among four routers with a 30 s update period may typically take almost 10 minutes to converge. Two methods have been developed to expedite convergence: triggered updates for new path changes, and holddown timers for erroneous routing information.

RIPv2

As an IGP, RIP is the first standard routing protocol developed for IP networks. It employs a simple routing metric, i.e., the hop count, and thus is a distance-vector routing protocol. Because of its simplicity, RIP has been popularly adopted in IP networks for four decades and is still being used in many networks. As an improved version of RIPv1, RIPv2 is specified in the IETF RFC 2453 (November 1998) [4] and later updated with authentication enhancement in the IETF RFC 4822 (February 2007) [5]. The support of RIP for IPv6, known as RIPng (RIP next generation), is specified in the IETF RFC 2080 (January 1997) [6]. While RIPng is developed based on RIPv1, its concepts are applicable in RIPv2.

From RIPv1 to RIPv2, several enhancements have been introduced. For example, the original classful routing in RIPv1 evolves to Classless Inter-Domain Routing (CIDR) [7, 8] in RIPv2. Also, the broadcast mechanism in RIPv1 (via address 255.255.255.255) for exchanging information is replaced by multicast in RIPv2 (via address 224.0.0.9), reducing routing traffic effectively. Moreover, RIPv1 does not support authentication of update messages, but RIPv2 has introduced this feature. It is worth mentioning that RIPng does not support updates authentication because it is developed based on RIPv1.

Different routing protocols have different ways to exchange routing information. Basically, RIP uses the layer-4 UDP transport protocol to send routing updates. In comparison, OSPF uses layer-3 communications on top of IP to exchange routing information. IS-IS uses layer-2 functions to update routing information.

The main advantage of RIP is its simplicity, making it easy to configure and use. However, RIP sends the whole routing table to neighboring routers, leading to several issues, such as traffic overhead and slow convergence. To address these issues, RIP limits the maximum hop count to 15. But this gives rise to further complications. For example, any destinations beyond 15 hops will be identified as being unreachable. Moreover, RIP allows a maximum of 25 RouTe Entries (RTEs) in each datagram. If there are more than 25 RTEs, more than one datagrams must be sent. No defined limit exists for the number of datagrams comprising a complete routing table update, further slowing down the convergence of the routing protocol. A general suggestion is to consider using RIP in small-scale networks with low hierarchy.

EIGRP

Within the category of IGPs, EIGRP is an advanced distance-vector routing protocol evolved from predecessor IGRP. It was initially developed as a proprietary routing protocol by Cisco for use exclusively on Cisco routers. Later, EIGRP became an open standard in 2013 and was formally specified in the IETF RFC 7868 (May 2016) [9].

As an advanced version of IGRP, EIGRP maintains backward compatibility with a mechanism to import IGRP routes to EIGPR, or export EIGRP routes to IGRP. With the widespread adoption of EIGPR, IGRP has become a legacy routing protocol. Moreover, EIGRP possesses the capability to redistribute routes for RIP, OSPF, IS-IS, and BGP. This makes EIGRP flexible to work with other routing protocols.

EIGRP inherits the load balancing capability from IGRP, supporting load balancing over both equal-metric and non-equal-metric paths. This feature, known as the *variance feature*, enables a path that is twice as good as another to be used twice as frequently. A variance, i.e., a multiplier, defines which paths are included for load balancing. It has a default value of 1, indicating load balancing over equal-metric paths. The maximum variance value allowed is 128. By multiplying the minimum metric of a route by the specified variance value, a threshold is established. Paths with metrics lower than this threshold are included in load balancing. For example, if the minimum metric of a route is 10 and the variance value is 5, then all paths with metrics less than $10 \times 5 = 50$ would be included in load balancing. It is worth mentioning that as a popularly used link-state routing protocol, OSPF lacks the capability of non-equal-metric load balancing.

Unlike RIP, which uses a single hop count as the routing metric, EIGRP and its early version IGRP use a composite metric derived from bandwidth, delay, reliability, and load. The bandwidth metric considers the lowest bandwidth segment along the routing path. The delay metric is the sum of all delays across outgoing interfaces on the path. It is a fixed value and not dynamically calculated. Reliability is based on the reported interface reliability by routers in the path, while load is calculated based on the reported interface load. The use of reliability and load is optional. It is not enabled by default unless explicitly configured. The formula for calculating a single metric in EIGRP is a simple weight expression based on the so-called five K values (K_1 through K_5) as follows [9, p. 41]:

$$\text{metric} = 256\left[\left(K_1 \cdot \text{BW} + \frac{K_2 \cdot \text{BW}}{256 - \text{Load}} + K_3 \cdot \text{Delay}\right)\frac{K_5}{K_4 + \text{REL}}\right] \tag{7.1}$$

where the value 256 scales the 24-bit IGRP metric up to the 32-bit EIGRP metric. Bandwidth (BW) is the lowest interface bandwidth along the path. Delay is the sum of all outbound interface delays on the path. Load and reliability (REL) are expressed as percentages ranging from 1 to 255.

By default, the five K values in Eq. (7.1) are set as $K_1 = K_3 = 1$ and $K_2 = K_4 = K_5 = 0$. The bandwidth is measured in 10^7 (i.e., 10 Gbps) divided by the lowest interface bandwidth. On Cisco routers, the interface bandwidth is expressed in kbps. Therefore, the bandwidth term in Eq. (7.1) is actually a weighted bandwidth term. The delay is the interface delay measured in 10 μs. As $K_3 = 1$ is set, the delay term is used without scaling into the weighted formula. For example [9, pp. 41–42], for an Ethernet interface of 10 Mbps bandwidth and 1 ms delay, the calculated EIGRP bandwidth metric is $256 \times 10^7/10{,}000 = 256,000$, and the calculated EIGRP delay metric is $256 \times 100 = 25{,}600$.

It is understandable that each of the five K values must be set the same value on all routers in an EIGRP network. Failure to do so could result in routing loops. It is the responsibility of network administrators to verify these settings on all EIGPR routers as EIGRP routers do not perform this check automatically.

While EIGRP remains a distance-vector routing protocol like IGRP, there are several differences between them. For example, IGRP is a classful routing protocol, while EIGRP allows for the use of CIDR. Unlike IGRP, EIGRP employs sporadic, partial, and bounded features when sending updates. Updates are only sent when metric changes are detected, making them sporadic rather than periodic. Moreover, only the routes that have changed are updated during these sporadic updates, resulting in partial updates rather than complete updates of the entire routing table. Furthermore, updates are only sent to affected routers, rather than all routers within the network, reducing update traffic overhead significantly. These features contribute to a more efficient utilization of network resources.

EIGRP relies on the diffused update algorithm (DUAL) for loop-free shortest path calculation with small overhead. As a result, the convergence rates of EIGRP become comparable with those of link-state routing protocols. Unlike basic Bellman-Ford distance-vector protocols without coordination, DUAL performs coordinated updates that target only the affected parts of the network. This feature is known as diffusing computation. DUAL operates based on the states of a route (Active or Passive) and also uses the concept of a *feasible successor*. A route in the active state means that neighbors that do not pass the Feasibility Condition check provide the lowest-cost path. Thus, the path is not guaranteed to be loop-free. In comparison, a route is in the passive state when at least one neighbor that provides the current least-total-cost path passes the Feasibility Condition check. Therefore, the path is guaranteed to be loop-free. A neighboring router with the least-cost path to a destination is known as a *feasible successor*. In other words, a path with feasible successors is in the passive state, while a path without feasible successors is in the active state.

Within DUAL, the feasibility condition is tested for loop freedom of a path to destination. It is a sufficient condition, implying that a path that meets this condition is guaranteed to be loop-free. However, the feasibility condition is not a necessary condition, implying that not all loop-free paths meet this condition. The feasibility condition states that when a neighbor's advertised distance to a destination is strictly less than the current *feasible distance* to that destination, then the neighbor is on a loop-free path to the destination. In other words, if a neighboring router is closer to the destination in terms of cost, it lies on a loop-free route to the destination. The *feasible distance* corresponds to the lowest distance to the destination since the route last entered to the passive state from the active state.

When a router detects a link failure, it immediately switches to an alternative route for packet forwarding if a feasible successor exists, thereby avoiding any routing traffic overhead. In the absence of a feasible successor, the router queries neighboring routers to find an alternative loop-free route. This query propagates throughout the network until an alternative route is discovered.

Overall, EIGRP is suitable for large-scale networks with a relatively low hierarchy. It is available on all Cisco routers. However, the compatibility of EIGRP with non-Cisco vendors remains an issue. While EIGRP is claimed to be an open standard through the IETF RFC 7868 [9] as part of Cisco's effort to facilitate interoperability with non-Cisco routers, certain core details are omitted from the RFC specifications. This makes it difficult to configure non-Cisco routers to work with EIGRP for seamless interoperability.

7.1.3 Link-State Routing Protocols

In computer networking, link-state routing protocols belong to the class of IGPs. They are dynamic routing protocols in which each router shares knowledge of its neighbors with every other router in the network. As shown earlier in Fig. 7.1, OSPF and IS-IS are two link-state routing protocols.

Link-State Routing Operations

Similar to distance routing protocols, link-state routing protocols share routing information with other routers. But they differ from distance-vector routing protocols in several aspects. The first difference is that a link-state router does not send its entire routing table to other routers. Instead, it only shares information about its neighborhood, which is known as an *adjacency* with its neighboring routers. The message sent out by a router about its adjacency with its neighbors is a *link state advertisement (LSA)*.

The second difference is that a router running a link-state routing protocol does not have regular communications with other routers for LSAs. Instead, it only communicates when it detects a change in the network.

The third difference is that link-state routing protocols use a flooding process to distribute LSAs. This means that each router sends LSAs to all routers in the entire internetwork. An LSA is passed around from one router to another, with each router making a copy of the LSA without modifying it.

The fourth difference is that each link-state router builds a complete link topological graph of the entire internetwork. By collecting LSAs from all other routers in the internetwork, each router maintains a *link state database*, which describes a graph of the internetwork. The link topological graph of the network is calculated on each router by using the link-state routing algorithm, which is known as Dijkstra's algorithm. Dijkstra's algorithm operates in iteration. The least cost paths derived after k iterations are known as k-destination nodes.

How does Dijkstra's algorithm work in link-state routing protocols? Let N and N_{all} denote the number and set of all available nodes in the network, respectively, i.e., $N_{\text{all}} = \{n_1, n_2, \cdots, n_N\}$. Also, let $N_{\text{spt}} \subseteq N_{\text{all}}$ represent the set of nodes that have already been included in the shortest path tree. Furthermore, the link cost from node

n_i to node n_j, where $n_i \in N_{\text{all}}$ and $n_j \in N_{\text{all}}$, is denoted as $c(i, j)$. If nodes n_i and n_j are not adjacent, $c(i, j)$ is set to ∞. The current least cost of the path from the source node $n_s \in N_{\text{all}}$ to the destination node $n_v \in N_{\text{all}}$ is denoted by $D(s, v)$, which may or may not be further improved. The variable $P(v)$ represents the previous node, which is a neighbor of n_v, along the current least cost path from the source to n_v.

Dijkstra's algorithm can be logically described by Algorithm 7.1. The algorithm begins with an initialization step, followed by an iterative loop. The number of iterations executed in the loop is equal to the total number of nodes available in the network.

Algorithm 7.1: Dijkstra's algorithm in link-state routing protocols

Input: N, N_{all}, $c(i, j)$ for all adjacent nodes, source node n_s
Output: $D(s, v)$, $\forall n_v \in N_{all}$
1 Initialization: Initialize an empty N_{spt};
2 Add source node n_s to N_{spt};
3 **for** *all nodes* $n_v \in N_{all}$ **do**
4   **if** n_v *is adjacent to* n_s **then**
5   Set $D(s, v) \leftarrow c(s, v)$;
6   **else**
7   Set $D(s, v) \leftarrow \infty$
8 **do**
9   Find a node $n_w \notin N_{spt}$ such that $D(s, w)$ is a minimum;
10   Add n_w to N_{spt};
11   Update $D(s, v)$ for all n_v adjacent to n_w and not in N_{spt}:
12         $D(s, v) \leftarrow \min(D(s, v), D(s, w) + c(w, v))$;
13 **until** *all nodes are on shorted path tree, i.e.,* $N_{spt} = N_{all}$;
14 **return**;

Consider an example shown in Fig. 7.6. Initially, add the source node n_0 onto an empty N_{spt} (lines 1 and 2 of Algorithm 7.1). In lines 3 through 5 of Algorithm 7.1, the distance values from the source n_0 to all its adjacent nodes that are not in N_{spt} (i.e., n_1, n_2, and n_3) are calculated as:

$$D(0, 1) \leftarrow 4, \ D(0, 2) \leftarrow 6, \ D(0, 3) \leftarrow 8 \tag{7.2}$$

From lines 6 and 7 of Algorithm 7.1, set the distance values of all other nodes that are not adjacent to the source node n_0 to ∞, i.e.,

$$D(0, w) \leftarrow \infty \ \forall w \text{ not adjacent to } n_0. \tag{7.3}$$

Then, iterations are executed as described below.

In the first iteration, from all nodes that are not in N_{spt}, find a node w with the least distance $D(0, w)$ (line 9 of Algorithm 7.1). From Eqs. (7.2) and (7.3), node n_1 with a distance 4 is selected. It is then added to N_{spt} (line 10 of Algorithm 7.1).

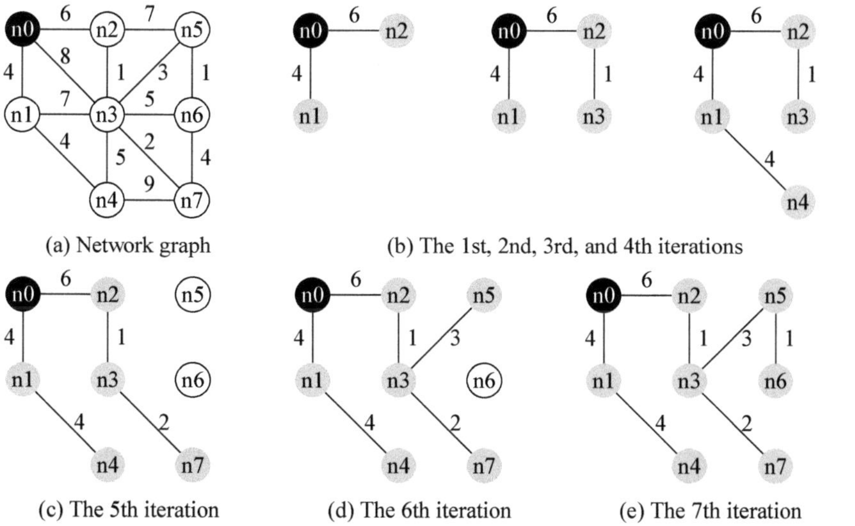

Fig. 7.6 Dijkstra's algorithm for a graph with source node n_0

According to lines 11 and 12 of Algorithm 7.1, for all nodes that are adjacent to n_1 but not in N_{spt} (i.e., n_3 and n_4), update

$$\begin{aligned} D(0,3) &\leftarrow \min(D(0,3), D(0,1) + D(1,3)) = \min(8, 4+7) = 8 \\ D(0,4) &\leftarrow \min(D(0,4), D(0,1) + D(1,4)) = \min(\infty, 4+4) = 8 \end{aligned} \tag{7.4}$$

This completes the first iteration, giving the following results:

$$\begin{cases} N_{spt} = \{n_0, n_1\}, \\ D(0,1) = 4, \\ D(0,2) = 6,\ D(0,3) = 8,\ D(0,4) = 8, \\ D(0,v) = \infty \text{ for all other nodes} \end{cases} \tag{7.5}$$

For the second iteration, scan all nodes that are not in N_{spt} to find a node with the least distance from the source, giving node n_2 according to Eq. (7.5) derived from the first iteration. Then, add n_2 to N_{spt}, and update the distance values from the source to all nodes that are adjacent to n_2 but not in N_{spt}. This yields:

$$\begin{cases} N_{spt} = \{n_0, n_1, n_2\}, \\ D(0,1) = 4,\ D(0,2) = 6, \\ D(0,3) = 7,\ D(0,4) = 8,\ D(0,5) = 13, \\ D(0,v) = \infty \text{ for all other nodes} \end{cases} \tag{7.6}$$

The second iteration is now completed. The resulting shortest path tree is shown in the left diagram of Fig. 7.6b.

After the third iteration is executed, the following results are obtained:

$$\begin{cases} N_{spt} = \{n_0, n_1, n_2, n_3\}, \\ D(0,1) = 4,\ D(0,2) = 6,\ D(0,3) = 7, \\ D(0,4) = 8,\ D(0,5) = 10,\ D(0,6) = 12,\ D(0,7) = 9 \end{cases} \tag{7.7}$$

The resulting shortest path tree is depicted in the middle diagram of Fig. 7.6b.

This process repeats until all nodes are added to N_{spt} (line 13 of Algorithm 7.1). The resulting shortest path tree is shown in Fig. 7.6e. The corresponding distance values from the source n_0 to all other nodes are:

$$\begin{cases} D(0,1) = 4, \quad D(0,2) = 6, \quad D(0,3) = 7,\ D(0,4) = 8, \\ D(0,5) = 10,\ D(0,6) = 11,\ D(0,7) = 9. \end{cases} \tag{7.8}$$

To automate the above computation process, a C program named `sla.c` is provided . It is illustrated below:

```
/* lsa.c for Dijkstra's shortest path algorithm */

#include <limits.h>
#include <stdio.h>
#include <stdbool.h>

#define N 8 /* #nodes in network graph */

/* From the nodes not in Nspt, find the node with min dist */
int minDistance(int D[], bool Nspt[]){
          int minD = INT_MAX, minD_index;
          for (int v = 0; v < N; v++)
            if ((!Nspt[v]) && (D[v] <= minD)){
                  minD = D[v];
                  minD_index = v;
            }
          return minD_index;
    }

    /* Dijkstra's single source shortest path algorithm */
    void dijkstra(const int graph[N][N], const int src, int D[N]){
          bool Nspt[N];/*true in shortest path tree, false otherwise*/

        /* initially, for all i, false for Nspt[i] and D(i)=infinity */
          for (int i = 0; i < N; i++){
                D[i] = INT_MAX;
                Nspt[i] = false;
          }
          D[src] = 0; /* src to src */

          /* Find a node with shortest path for src & not in Nspt */
          for (int count = 0; count < N - 1; count++) {
```

```
            int w = minDistance(D, Nspt);
            Nspt[w] = true; /* add to Nspt */

          /* Update dist of adjacent to the node & not in Nspt*/
          for (int v = 0; v < N; v++)
            if ((!Nspt[v]) && (graph[w][v]) && (D[w] != INT_MAX)
                && (D[w] + graph[w][v] < D[v]))
              D[v] = D[w] + graph[w][v];
      }
      return;
}

int main(){
      /* graph adjacency matrix */
      int graph[N][N] = { { 0, 4, 6, 8, 0, 0, 0, 0 },
                          { 4, 0, 0, 7, 4, 0, 0, 0 },
                          { 6, 0, 0, 1, 0, 7, 0, 0 },
                          { 8, 7, 1, 0, 5, 3, 5, 2 },
                          { 0, 4, 0, 5, 0, 0, 0, 9 },
                          { 0, 0, 7, 3, 0, 0, 1, 0 },
                          { 0, 0, 0, 5, 0, 1, 0, 4 },
                          { 0, 0, 0, 2, 9, 0, 4, 0 }};
      int D[N]; /* array of shortest distance from src to i */

      dijkstra(graph,0,D); /* Dijkstra's algorithm */

      printf("Src node \t Distance\n");
      printf(" -> Node \t from Source\n");
      for (int i = 0; i < N; i++)
        printf("    %d \t\t %   d\n", i, D[i]);

      return 0;
}
```

Compile the C code `sla.c` by using the gcc compiler in a terminal window. Then, run the resulting executable. The program produces the following results:

```
$ gcc lsa.c
$ ./a.out
Src node            Distance
-> Node            from Source
0                    0
1                    4
2                    6
3                    7
4                    8
5                    10
6                    11
7                    9
```

As expected, the results obtained from executing the C program match the values derived manually in Eq. (7.8) and the corresponding diagram shown in Fig. 7.6e. This verifies the correctness of the C program.

OSPF Routing Protocol

Falling into the group of IGPs, OSPF is a link-state routing protocol designed for IP networks. It operates within a single AS and is based on the Shortest Path First technology. The OSPF Version 2 is specified in the IETF RFC 2328 (April 1998) [10] for IPv4 networks. Its extension to IPv6 networks is defined in the IETF RFC 5340 (June 2008) [11].

OSPF is designed with explicit support for CIDR and the tagging of externally-derived routing information. As a dynamic routing protocol, OSPF quickly responds to any topological changes within the AS while generating only small amounts of routing protocol traffic.

An important concept in OSPF is to structure or subdivide networks into routing *areas*. This concept should be considered in conjunction with well-structured IP address allocation and subnetting. By using IP address summarization and supernetting, not only OSPF administration becomes easier, but also OSPF traffic and resource consumption could be significantly reduced. Each routing area in OSPF is identified by a 32-bit number. It consists of groups of contiguous networks and attached hosts. The topology of a routing area remains invisible to the outside of that area. Consequently, routers internal to an area have no knowledge of the network topology external to the area. This isolation of internal and external knowledge enables a significant reduction in routing traffic. Routers within the same area have an identical link-state database. However, due to the introduction of multiple routing areas, the entire AS is unable to maintain an identical link-state database.

In the OSPF architecture, routers are classified into four types: Internal Routers (IRs), Area Border Routers (ABRs), Backbone Routers (BRs), and AS Boundary Routers (ASBRs). IRs and all networks directly connected to them belong to the same area. They run a single instance of the basic routing algorithm. In comparison, ABRs are attached to more than one area. They execute multiple instances of the basic routing algorithm, with each instance for an attached area. BRs have an interface to the backbone area of the network. They also include ABRs that interface with routing areas. However, according to the specifications of OSPF version 2 in the IETF RFC 2328, BRs do not have to be ABRs. ASBRs routers are responsible for exchanging routing information with routers belong to other ASs. They advertise AS external routing information throughout the AS, and the routing paths to each ASBR are known to all routers within the AS. It is worth mentioning that in OSPF, the functions of ASBRs are specified differently from those in other routing architectures, in which ASBRs may be internal or area border routers and may or may not participate in the backbone.

When an OSPF network is designed with multiple routing areas, a backbone area is required, which is also known as *Area 0* or *Area 0.0.0.0*. The backbone area must have contiguous IP addresses, although its physical connectivity can be configured via virtual links. Other areas are connected to the backbone through ABRs. The backbone area is responsible for distributing routing information between non-backbone areas. Figure 7.7 illustrates a simple OSPF architecture with a backbone connected to multiple areas through ABRs.

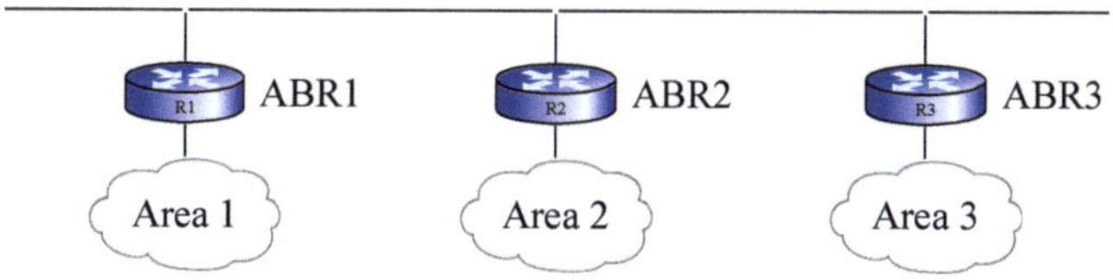

Fig. 7.7 An OSPF network with a backbone connected with multiple routing areas via ABRs

Inter-area routing in OSPF is performed through the backbone area. The backbone area facilitates the exchange of routing information between different areas, enabling communication and connectivity between the non-backbone areas. When a packet needs to be forwarded from a source in one non-backbone area to a destination in another non-backbone area, the packet follows a specific path through the backbone area. Initially, the packet reaches the ABR of the source area. From there, it is forwarded through the backbone to the ABR of the destination area. Finally, the packet is delivered to the destination within that area.

Physically, this inter-area routing process establishes a star topology within the AS, where the backbone area acts as the central hub, and the non-backbone areas function as the spokes. This design ensures scalability and allows the AS to expand by incorporating additional areas, while still maintaining efficient routing and reduced routing traffic.

IS-IS Routing Protocol

As a type of IGPs, IS-IS is a link-state routing protocol developed for use with the Open System Interconnection (OSI) protocol suite. It is formally specified in ISO/IEC 10589:2002 [12], which is an refinement of an earlier version known as ISO/IEC 10589:1992. It is worth mentioning that ISO uses different terminology from what is commonly used today. For example, "intermediate system" and "end system" from ISO actually refer to "router" and "host", respectively. Interestingly, a draft version of ISO/IEC 10589 was republished by IETF in RFC-1142, which was frequently referenced for discussions of IS-IS even after ISO published its second version, ISO/IEC 10589:1992. Therefore, RFC-7142 was published in February 2014 to formally reclassify RFC-11142 as Historic. It explicitly stated that "All references to IS-IS should be to the latest edition of the IS-IS standard (currently ISO/IEC 10589:2002, Second Edition), and RFC 1142 is only of historic interest." Nevertheless, there were a number of RFCs discussing the use of IS-IS for routing in IP environments, such as RFC 1195 (December 1990) [13], entitled "Use of OSI IS-IS for Routing in TCP/IP and Dual Environments".

An IS-IS router exchanges network topology information with its neighboring IS-IS routers. This topology information is then flooded throughout the AS, ensuring that every router in the AS has a global view of the AS's topology. By using a link-state routing algorithm, such as Dijkstra's algorithm, this topology information is used to derive end-to-end paths for packet forwarding within the AS. When making

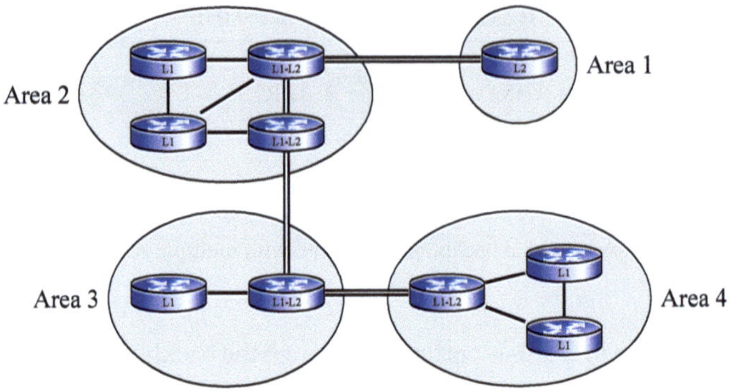

Fig. 7.8 Interconnection of IS-IS networks, in which the string of the L2 system in Area 1 and all L1-L2 systems in the other three areas form the backbone

routing decisions, a link-state routing protocol selects the next-hop address for packet forwarding based on the best path towards the destination.

Similar to OSPF, IS-IS can be deployed in a hierarchical topology. However, unlike OSPF, IS-IS does not have a designated backbone area. In IS-IS, there are three types of routers: *Level 1 System*, *Level 2 System*, and *Level 1–2 System*:

- A Level 1 system is an intra-area router that only has knowledge of the local area and learns prefixes specific to that area. It creates and maintains a level 1 link-state database and SPF tree for the local area.
- A Level 2 system is a backbone router that has knowledge of both intra-area and inter-area routers. It creates and maintains a level 2 link-state database and SPF tree for the backbone.
- A Level 1–2 system is a router that performs the functions of both a Level 1 system and a Level 2 system. It creates and maintains two separate link-state databases: one for the Level 1 system and the other for the Level 2 system. For each link-state database, an SPF tree is also created.

Figure 7.8 illustrates the internetworking of IS-IS networks with different types of routers. It is seen from this figure that in Area 1, there is a Level 2 router but no Level 1 router, eliminating the need for any Level 1–2 routers in that area. In Area 2, there are two Level 1 routers and two Level 1–2 routers. These two Level 1–2 routers form a Level 1 adjacency and a Level 2 adjacency with each other. The Level 2 router in Area 1, and all the Level 1–2 routers from Areas 2 to 4, form a continuous string of backbone routers, as indicated by the double lines in Fig. 7.8.

For a comprehensive understanding of IS-IS, refer to the original ISO standard ISO/IEC 10589:2002 [12]. To better understand IS-IS, some comparisons are given below between IS-IS and OSPF.

Both IS-IS and OSPF are link-state routing protocols, which use Dijkstra's algorithm to compute the best paths through the network. They both support variable-length subnets, multicast for discovering neighboring routers, and authentication of routing updates. Additionally, they both build a topological view of the network.

However, IS-IS differs from OSPF in several aspects. While OSPF is a layer-3 protocol built natively on top of IP networks, IS-IS is an OSI layer-2 protocol. As IS-IS does not rely on IP, it can conceptually support both IP networks and non-IP networks. This feature makes it easier to extend IS-IS from IPv4 to IPv6 networks. By contrast, as OSPF was originally designed for IPv4 networks (RFC 2328 [10]), it required the development of a new standard to handle IPv6 networks, such as RFC 5340 [11]. Actually, various RFCs were developed to provide extensions or enhancements to OSPF.

As discussed earlier, IS-IS defines and uses *areas* differently from OSPF. IS-IS includes Level 1 systems, Level 2 systems, and Level 1–2 systems, whereas OSPF defines areas in a way that an ABR is present in two or more areas simultaneously. Therefore, in OSPF, the area borders exist within the ABR, whereas in IS-IS, the area borders are located between Level 1–2 routers, Level 2 routers, and/or Level 1–2 and Level 2 routers.

7.1.4 Classful and Classless Routing Protocols

Depending on whether or not the subnet mask information is included in routing updates, routing protocols can be divided into classful and classless routing protocols. Classful routing protocols do not include the subnet mask information in their routing updates, while classless routing protocols update their routing tables with subnet mask information.

In the early stages of networking, routing protocols were classful. The only two primary classful routing protocols were RIPv1 and IGRP. RIPv1 is the first-generation routing protocol, while IGRP is the first-generation Cisco proprietary protocol. IGRP has since been obsoleted and replaced by EIGRP in networks. Both RIPv1 and IGRP are considered legacy routing protocols.

Modern routing protocols are classless, meaning that they include the subnet mask information in their routing updates. This shift is primarily due to the widespread adoption of classless IP addressing in modern networks. Classless IP addressing is used together with subnet mask information. The IPv4 routing protocols RIPv2, EIGRP, OSPF, and IS-IS are all classless. They use both IP addresses and subnet masks for routing updates.

It is worth mentioning that the concept of classful routing is specific to early IPv4 networks, it does not apply to IPv6 networks. IPv6 routing protocols are inherently classless as they include the prefix length along with the IPv6 address.

7.1.5 Comparisons of Routing Protocols

A summary of the main features and differences of widely used routing protocols is provided in Table 7.1. It highlights the key aspects of each protocol, and is thus helpful for planning routing architecture in computer networks.

7.2 Routing Architecture and Strategies

The previous section has emphasized that the use of routing protocols is closely related to network architecture and routing performance requirements. This section will briefly discuss routing architectural considerations and the selection of routing protocols in practical network systems.

7.2.1 Architectural Considerations

A router is not needed for communications within a LAN on the same subnet. Network devices on the same subnet are directly interconnected through a layer-2 switch. Mechanisms exist for these devices to communicate with each other at layers 1 and 2 through the layer-2 switch.

However, when it comes to communication between networks within the same enterprise network or over the Internet, a router becomes essential. It serves as the interconnection point between the communicating network and other networks or the Internet. This enables layer-3 communications between networks. The place-

Table 7.1 Comparisons of routing protocols

Protocol type	Distance-vector		Link-state	
Protocol	$\underline{RIPv2}$	$\underline{IGRP, EIGRP}$	$\underline{OSPF}$	$\underline{IS-IS}$
Metric	Hop	Bandwidth, Delay	Bandwidth	By administrator
Periodic	30 s, or 90 s (EIGRP)		Use triggered updates	
What to send	Entire routing table		Link-state information	
Routing loops	Susceptible to routing loops		No risk of routing loops	
Network topology	Do not know network topology		Know the entire topology	
Convergence	Slow		Fast	
Scalability	Small networks		Large networks	
Easiness	Simple		Complex	

ment of routers within a network largely determines the network's hierarchy and interoperability.

In every enterprise network, connectivity to the Internet is established through one or more border routers. Therefore, a hard boundary exists between the enterprise network and the Internet. If there is only one ISP, a border router is placed on the hard boundary, separating internal and external networks. In scenarios involving multiple ISPs, multiple border routers can be installed on the boundary. Each border router connects the internal network to a specific ISP. Figure 7.9 provides a conceptual representation of this configuration.

For the internal network depicted in Fig. 7.9, various architectural topologies can be designed based on the scale and performance requirements of the network. In the case of a small-scale network with only a few tens of network devices, network segmentation may not be necessary. A flat topological network, which directly connects to the border router through a few switches, might suffice. This means that the distribution layer is omitted in the three-layer core/distribution/access architecture.

For medium-scale networks with several hundred network devices, a two-level routing hierarchy can be a suitable choice. The top level comprises the border router as the core router, while the bottom level consists of a few distribution routers. Each of these bottom-level routers connects upwards to the border router and downwards to several access-layer switches.

Large-scale networks require a medium to high hierarchy with multiple levels interconnected through routers. Cisco's three-layer architectural model is often used in this scenario, signifying the functions of the core layer, distribution layer, and access layer. Each of these layers can be organized in one or two levels interconnected through routers. This results in a total network hierarchy of three to over five levels. A hierarchy of 3 to 5 levels is considered medium, while a hierarchy of over 5 levels indicates a high-hierarchy architecture. Figure 7.10 visually illustrates this concept.

With hierarchical routing architecture in mind, it is important to identify routing flows, determine hard and soft routing boundaries, and design policies to manipulate routing flows [14, pp. 269–277] when developing routing architecture. Routers are placed at routing boundaries, and the designed policies for manipulating routing flows are implemented on these routers. Other policies, such as security policies and management policies, can also be implemented on the routers.

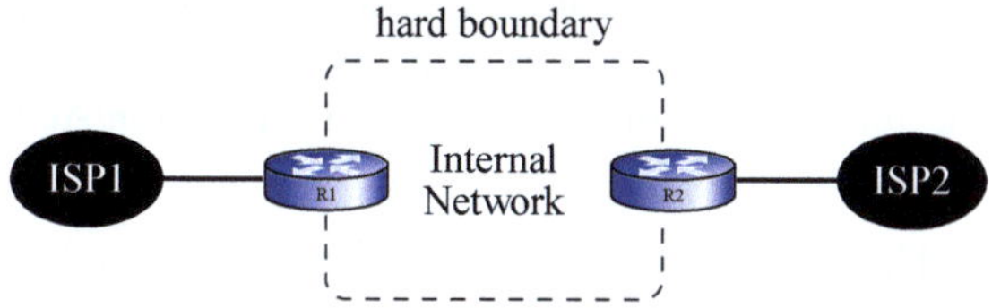

Fig. 7.9 Conceptual interconnection with ISPs via border routers (firewall and other security components are omitted here)

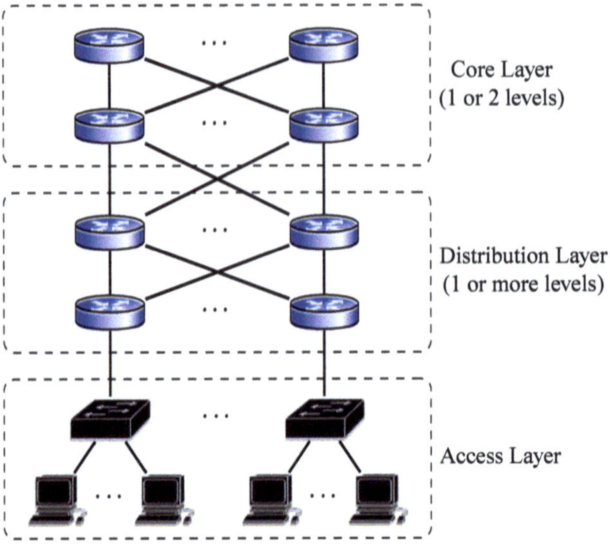

Fig. 7.10 Routing in multiple levels

7.2.2 Choosing Routing Protocols

It has been mentioned previously that a network with a flat topology does not require a router for communications within the network. However, for communications with the Internet, a border router is necessary to handle all inbound and outbound traffic. The selection of a routing protocol compatible with the ISP's routing protocol is important for the border router.

There is another scenario where a routing protocol is not needed for packet forwarding, known as static routing. In static routing, routers simply forward packets directly to static addresses without the need to determine routes. While static routing is efficient and does not involve overhead in discovering forwarding paths, it limits network management and maintenance flexibility. Specifically, static routing does not react to network changes, such as link failures, which can disrupt routing in the network. Therefore, it is generally suggested to avoid static routing unless there are no better design options available for a specific application scenario.

For dynamic routing, what routing protocols are likely to be used in each of the three layers of Cisco's Core/Distribution/Access architecture? This is briefly discussed below:

- **The Core layer** requires high reliability and strong redundant links with good fault tolerance and load sharing. EIGRP, OSPF, and IS-IS provide such support. Due to its slow convergence, RIPv2 is not generally recommended for the core layer.

- **The Access layer** connects hosts to the network. When choosing a routing protocol, factors such as available equipment, IP addressing scheme used, topology, and access network size should be considered. Options could include OSPF, EIGRP, and RIPv2. For some hosts, static routing may be the best option. IS-IS is generally not recommended for the access layer as it requires more configuration knowledge.
- **The Distribution layer** acts as an intermediate layer between the Core and Access layers. Therefore, a routing protocol should be chosen with consideration for both the core and access layers. Depending on the routing protocols selected for the core and access layers, either the core-layer routing protocol (EIGRP, OSPF, or IS-IS) or the access-layer routing protocol (OSPF, EIGRP, or RIPv2) could be run at the distribution layer.

As discussed previously, there are five well-defined and popularly used routing protocols: RIPv2, EIGRP, OSPF, IS-IS, and BGP4. BGP4 is a path-vector routing protocol based on TCP for routing between ASs. It is used for packet forwarding between ASs. RIPv2 and EIGRP are distance-vector routing protocols, while OSPF and IS-IS are link-state routing protocols. The considerations for choosing these protocols are described as follows:

- **RIPv2:** Due to its distance-vector nature and slow convergence, RIPv2 can be considered for small-scale networks with low to medium hierarchy and diversity.
- **EIGRP:** With native support by Cisco's routers and an advanced distance-vector algorithm, EIGRP is suitable for large-scale networks with relatively low hierarchy. However, compatibility with non-Cisco routers can be an issue, potentially causing interoperability problems.
- **OSPF:** The link-state OSPF requires a well-designed hierarchy of the network with a backbone (Area 0) and a number of routing areas. Thus, it is suitable for large-scale networks with relatively high hierarchy.
- **IS-IS:** Initially developed by ISO, IS-IS is also suitable for large-scale networks with relatively high hierarchy. However, it is a layer-2 protocol compared to the layer-3 OSPF. Also, it uses the concept of "Areas" differently from OSPF. Configuring IS-IS in Cisco's routers is not straightforward. Therefore, our general suggestion is to use OSPF unless you well understand what you aim to achieve by choosing IS-IS.
- **BGP4:** BGP operates between ASs, making it suitable for networks of ISPs. When the backbone of an enterprise network interconnects ASs or when organizations require autonomy in interfacing with the backbone, BGP can be considered as the routing protocol for the backbone.

A general guideline in choosing routing protocols is to use the least number of them. This would require a simplified routing architecture. Ideally, using only one routing protocol for the entire enterprise network is the best solution. Therefore, the following scenarios explain how to minimize the number of routing protocols:

- If static routes were previously chosen for certain areas of the network and RIPv2 has now been chosen as the routing protocol for another area, RIPv2 replaces static routes for those areas.

- If RIPv2 was previously chosen for certain areas of the network and OSPF has now been chosen as the routing protocol for another area, OSPF replaces RIPv2 for those areas.
- If BGP is required for a backbone network, it may replace OSPF or RIPv2 that were previously chosen for the backbone.

If a single routing protocol cannot cover all internetworks in the enterprise network, considering an additional one for some areas of the network becomes necessary. In such cases, it is necessary to consider how these two routing protocols coexist in the enterprise network. The concepts of route redistribution and filtering are important when multiple routing protocols have to be used.

7.2.3 Route Redistribution

Consider a scenario where there are two routing domains or ASs, each running a different routing protocol. For example, one network runs OSPF while the other runs EIGRP. This may occur when two companies merge. Before a single routing protocol can be implemented for the merged network, OSPF and EIGRP need to coexist. In this case, routes known to OSPF must be advertised into the portion of the network that runs EIGRP, and vice versa.

The technique of *route redistribution* enables this functionality. As depicted in Fig. 7.11, Router R1 in network A runs OSPF, and Router R3 in Network B runs EIGRP. Router R2 acts as an intermediary between the two networks for route distribution from OSPF to EIGRP and vice versa. R2 is equipped with two interfaces: one connects to Network A with OSPF, and the other connects to Network B with EIGRP. In order for R2 to redistribute routes, route redistribution must be activated through specific commands, such as the `redistribution` command on a Cisco router.

As different routing protocols operate with different metrics, redistributing routes from one routing protocol to another requires careful configuration. One crucial aspect of setting up route redistribution is assigning the metric to routes originating from one network and being redistributed into another network. Table 7.2 presents the default seed metrics for certain routing protocols. These metrics serve as a starting point for the redistributed routes. They can be further adjusted based on specific network requirements and performance considerations.

It is seen from Table 7.2 that when redistributing a route to OSPF, it will be assigned a default metric of 20 (or 1 if the redistribution is from BGP). Any route redistributed to RIP and EIGRP will receive a default metric of *infinity*, indicating that these routes will be considered unreachable by default and will not be advertised. However, if there is a need to advertise such routes, it is important not to rely on the default *infinity* metric. Instead, manually configuring an appropriate metric is recommended using commands such as the `default-metric` command on a Cisco router.

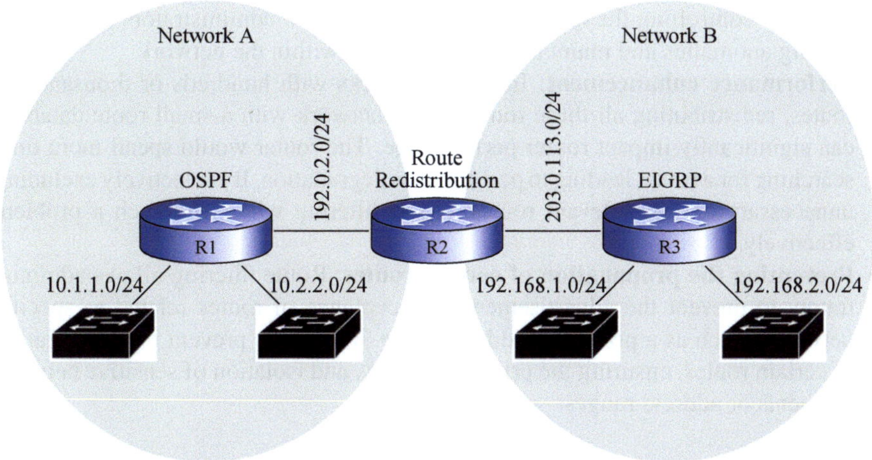

Fig. 7.11 Route redistribution through router R2

Table 7.2 Default seed metrics for route distribution

Routing protocol	Default seed metric
RIP	Infinity
EIGRP	Infinity
OSPF	20 (or 1 when redistributing BGP routes)
BGP	Use the IGP metric value

7.2.4 Router Filtering

In practice, not all routes from one routing domain are typically redistributed to another domain. It is generally not suggested to inject routes into a routing protocol that already has a better path to reach the advertised networks. Also, routing loops can occur when injecting a route into a routing protocol that will subsequently advertise the same route back. Therefore, the technique of route filtering becomes useful to prevent routing loops and ensure security, availability, and performance. Route filtering enables network administrators to selectively exclude certain routes from the local route database, prevent their advertisement to neighboring routers, or avoid their redistribution into another routing protocol. This capability allows for greater control and optimization of route propagation within the routing domain.

There are a few scenarios where route filtering is useful:

- **Preventing suboptimal routing and routing loops**: Route filtering is used with route redistribution, primarily to prevent suboptimal routing and routing loops that may occur when routes are redistributed at multiple redistribution points. By carefully controlling the redistributed routes, network administrators can prevent routing anomalies and maintain optimal routing within the network.

- **Performance enhancement**: In large networks with hundreds or thousands of routes, redistributing all these routes into a network with a small route database can significantly impact router performance. The router would spend more time searching for a route, leading to performance degradation. By selectively excluding unnecessary or less relevant routes, router filtering will avoid such a problem effectively.
- **Preventing the propagation of certain routes**: Route filtering allows administrators to prevent the advertisement or acceptance of routes related to specific networks, such as a private IP address space. This helps prevent the propagation of certain routes, ensuring the privacy, security, and isolation of sensitive network segments or address ranges.

7.3 Software Defined Networks

Software Defined Networking (SDN) is developed from the programmable networks paradigm. It is a network management architecture approach that provides network programmability with the capacity to initialize, control, change, and manage network behavior dynamically via open interfaces. The primary focus of SDN is on making the networking software-defined with centralized packet forwarding decision in a control plane, thus separating it from actual packet forwarding in a data plane. This separation facilitates faster innovation cycles at both planes and enables flexible traffic routing on the fly.

The overall goal of SDN is to leverage this separation and the associated programmability to reduce complexity and accelerate innovation of both the control and data planes. The layered architecture and interfaces of SDN are formally specified in the IETF RFC 7426 [32]. Earlier RFCs, such as RFC 7149 [15], have also addressed various aspects and concepts of SDN.

Let us now delve into the motivations behind SDN, explore the layered architecture of SDN, and subsequently investigate the interfaces developed for SDN.

7.3.1 Motivations Behind SDN

Consider the traditional networking architecture in a cloud data center, as depicted in Fig. 7.12. Cloud services are delivered through Virtual Machines (VMs), which are managed by a hypervisor. In the decentralized environment of the network, various network devices like routers, switches, and firewalls independently make routing decisions and perform data forwarding. Each device is stacked with a control plane and a data plane, each serving different purposes.

Let us consider a router as an example. In its control plane, the router runs a routing protocol, such as OSPF, to establish and update its routing table. The routing table is then used to determine the appropriate forwarding path for packets. Additionally, the router also maintains an ARP table and a switch MAC address table. These functions are essential components of the control plane in a traditional router.

For actual data forwarding in the data plane, packets undergo encapsulation or de-encapsulation by adding or removing headers. IP and MAC addresses are matched for data forwarding. Various management policies, like access lists, are also enforced in this stage. Finally, the data is forwarded to the next hop in the network.

The functions of centralized routing decision-making and actual data forwarding have long been fundamental in traditional networking. However, traditional networking lacks programmability support, meaning that network device configuration and management of network behavior cannot be dynamically controlled and adjusted in a centralized manner. This limitation is the major motivation for the development of SDN.

Let us analyze a typical scenario in a cloud data center, as presented in Fig. 7.12. When an end user initiates a request using the Host in the figure, a new VM_{n+1} is created, along with the deployment of any requested platforms and applications. These tasks can be easily accomplished through the hypervisor. However, more complex tasks follow the creation of VM_{n+1} and the deployment of requested platforms and applications. For example, setting up a VLAN on all switches is necessary to enable

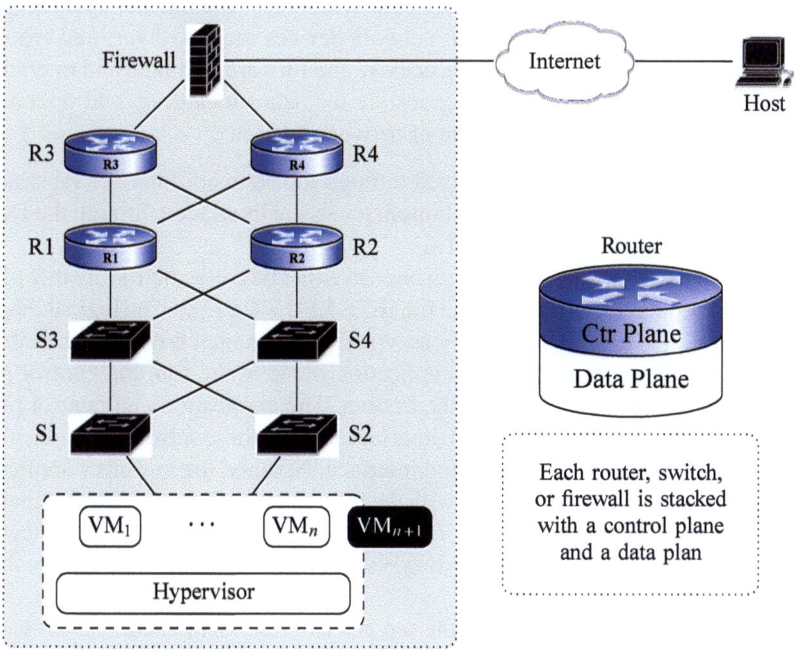

Fig. 7.12 Networking in a cloud data center

the Host's access to VM_{n+1}. A new subnet must be assigned to facilitate this. On all switches, new interfaces with IP addresses should be created, and router management protocols like HSRP need to be configured. Firewall configurations must also be updated to allow access to the new subnet and VM_{n+1}. All these tasks are time consuming and cannot be efficiently performed within a centralized environment in traditional networking. This motivates the development of a new network architecture featuring a centralized software control plane to manage all control plane tasks within the network, as indicated by the shaded area in Fig. 7.12. Meanwhile, the data plane tasks of the network are separated from the control plane tasks to form a separate data plane.

7.3.2 Layered Architecture of SDN

SDN networks are built upon a layered architectural model, encompassing multiple planes that are classified differently from different perspectives. The original RFC 7426 [32] provides a summary of SDN architecture abstractions using a three-level schematic:

- At the top level is the application plane where various applications and services that define network behavior reside.
- The middle level consists of the control plane and management plane. The control plane makes decisions regarding packet forwarding, while the management plane monitors, configures, and maintains network devices such as routers and switches.
- Within the bottom level, two planes coexist: the forwarding plane and operational plane. The forwarding plane is responsible for data forwarding, and operational plane manages the operational state of network devices.

The top and middle levels are interacted through the Network Services Abstraction Layer (NSAL), while the middle and bottom levels are interfaced through the Device and resource Abstraction Layer (DAL).

In practical implementation, SDN planes can either be collocated with other planes or physically separated, as indicated in the IETF RFC 7426 [32]. The logical diagram shown in Fig. 7.13 illustrates a widely accepted SDN architectural model. featuring three planes at different layers: the application plane at the top, the control plane in the middle, and the data plane at the bottom. The application and control planes communicate through a northbound interface, while the control and data planes exchange information via a southbound interface. Notably, the top-layer application plane does not directly communicate with the bottom-layer data plane. In comparison to the architecture described in RFC 7426 [32], the management plane is integrated into the control plane, and the forwarding plane and operational plane are combined into the data plane.

The application plane, located at the top layer in Fig. 7.13, encompasses various application tasks as described in RFC 7426 [32]. These tasks include end-user appli-

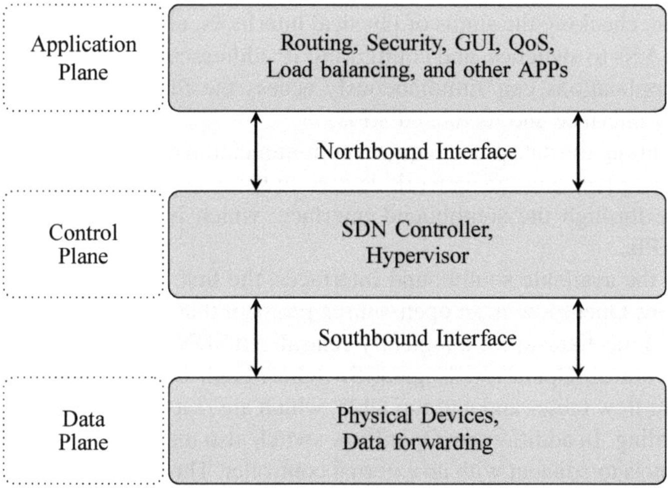

Fig. 7.13 SDN architecture

cations, learning networks for routing, security protection, GUIs, QoS management, load balancing, and other system services and applications.

As illustrated in Fig. 7.13, in the control plane of SDN, a central controller is employed to perform a range of tasks, either fully or partially. It assumes responsibility for topology discovery and maintenance, packet route selection and installation, and the provisioning of path failover mechanisms. In addition, the SDN controller also feeds the data plane with information required for data forwarding. With complete access to comprehensive network information, the controller facilitates easy configuration of the entire network. The SDN controller can be implemented as a physical hardware device or a VM. To manage and configure the underlying VMs associated with network resources, a hypervisor is also placed in the control plane.

With the separation from the control plane, the data plane in Fig. 7.13 is designed to perform several tasks including actual data forwarding, fragmentation and reassembly, and replication to support multitasking on network devices. Depending on the underlying network implementation, the network devices can be virtual, physical, or a combination of both.

7.3.3 *SDN Interfaces*

To enable access to the SDN controller from the application plane, the northbound interface is essential. This interface allows for the configuration of the SDN controller and retrieval of information from the SDN. It uses APIs to provide services and enables other applications to interact with the SDN controller. Tasks that can be accomplished through the northbound interface include retrieving network device information, checking the status of physical interfaces, obtaining network topology, adding VLANs to switches, and configuring IP addresses and routing for new VMs. Multiple applications can simultaneously access the SDN controller through the northbound interface and its related APIs.

Programming the data plane requires communication between the SDN controller in the control plane and the network devices in the data plane. This communication takes place through the southbound interface, which is a software interface that employs APIs.

Among the available southbound interfaces, the first and most widely used one is OpenFlow. OpenFlow is an open-source protocol that is actively developed [16]. It provides a mechanism for a logically centralized SDN controller to communicate with, and control, OpenFlow Logical Switches. Each OpenFlow switch maintains one or more flow tables and a group table, which are responsible for packet lookups and forwarding. In addition, an OpenFlow switch also maintains one or more OpenFlow channels to connect with an external controller. Through the OpenFlow switch protocol, the SDN controller can add, delete, and update flow entries in the flow tables. It can also perform various other tasks for managing OpenFlow switches.

7.3.4 SDN in a Service Provider Environment

SDN has been extensively investigated from various perspectives. For example, the IETF RFC 7149 [15] has discussed the use of SDN within the service provider environment. Service providers face the challenge of managing complex networking environments that involve multiple services, protocols, technologies, and dynamic adaptations in the near future.

The separation of the forwarding and control planes has been a common practice in router implementations for decades. Routing processes have traditionally been performed in software, while data forwarding has been handled in hardware. Therefore, the notion of separating the forwarding and control planes in SDN can be seen as a tautology. However, the concept of a centralized control plane in SDN introduces a new approach to routing decisions.

Enhanced operational flexibility is a key objective for service providers in adopting SDN. However, flexibility goes beyond the separation of control and forwarding planes. There are other requirements for flexibility, such as handling burst traffic or providing elevated QoS for a short period of time based on end-user demands. Therefore, it is important for service providers to predict network dynamics and traffic behavior and understand their impact on the network services to be delivered. From this understanding, the exposure of programmable interfaces is not a goal per se, but rather a means to achieve improved flexibility.

From the service provider perspective, SDN can be defined as "a set of techniques to facilitate the design, delivery, and operation of network services in a deterministic, dynamic, and scalable manner" [15]. These techniques can be categorized into several functional meta-domains including those described in the following:

- Dynamic discovery of network topology, devices, and capabilities;
- Exposing network services and their characteristics, and negotiating the set of service parameters for measuring the quality of service provisioning;
- Dynamic resource allocation and policy enforcement schemes derived from service requirements; and
- Dynamic feedback mechanisms to assess the fulfillment and assurance of policy efficiency.

It is worth mentioning that the IETF RFC 7149 [15] also identifies a few scenarios where the deployment of SDN may not be justified. These include the following cases where:

- Fully flexible software implementations are difficult due to software and hardware constraints,
- Fully modular implementations are challenging due to potentially extra complexity, and
- Fully centralized control raises scalability constraints.

7.3.5 Key Features of SDN

While SDN offers numerous advantages, its key features include network programmability, logically centralized network intelligence and control, abstraction of networks, and openness for interoperability.

As SDN is rooted in the concept of programmable networks, the core idea of SDN is to make the networks under management dynamically programmable. Therefore, it is designed to enable the behavior of the networks to be managed and controlled by software executed outside networking devices, such as routers and switches, which provide physical network connectivity. This decoupling of control software from network hardware brings significant flexibility in customizing network behaviors, deploying new network services, and virtualizing network functions. Provisioning differentiated network services to different types of traffic and end users becomes easier in SDN compared to traditional networking.

The separation of network control software and network hardware leads to a logically centralized network topology for network intelligence and control. This sets SDN apart from traditional networking, which is both logically and physically decentralized and distributed. In decentralized network management and control, individual network devices, such as routers or switches, make their own data forwarding decisions without considering the states of other network devices. This lack of a global view hinders efficient network management. By contrast, SDN possesses comprehensive knowledge of the entire network and its devices, enabling centralized management, control, and optimization of data forwarding, bandwidth performance, security protection, and various policies.

Traditional networking tightly couples services and applications with network hardware. For instance, Differentiated Services (DiffServ) and Integrated Services

(IntSev) are configured through direct interactions with routers. By contrast, SDN heavily relies on network abstraction through virtualization in software. Services and applications running on SDN are abstracted from the underlying network hardware and technologies. They interact with networks through various SDN APIs.

The openness of SDN for interoperability stems from its standards-based and vendor-independent nature. With open standards for SDN, the design and development of SDN become vendor-neutral. More specifically, interactions with SDN control plane and data plane are performed by using standard SDN protocols and interfaces. This fosters an interoperable SDN ecosystem that accommodates multi-vendor products.

7.4 Routing in Publish-Subscribe Networks

With the idea of separating routing decision and packet forwarding, routing topology and routing paths can be calculated in the application plane. This section introduces multicast routing based on publish-subscribe networking for low-latency communications in large-scale network systems. The main content is derived from our recent research and development on low-latency communications for Wide Area Control (WAC) in smart grid [17, 18].

7.4.1 Wide Area Networks in Smart Grid

Data communication networks are an integral part of smart grid infrastructure. They are essential to the measurements, control, and operations of power systems. As shown in Fig. 7.14, smart grid communication networks are composed of Wide Area Networks (WANs), Neighborhood Area Networks (NANs), and Home Area Networks (HANs). WANs are fiber optical networks that interconnect power generation facilities, substations, and control centers. Establishing communications between WANs and HANs, NANs are formed from a large number of smart meters and other devices within power distribution networks, typically spanning from substations to the end users of electricity. Connecting to the smart meters, HANs are built within end user residences to monitor and control energy usage.

It is worth mentioning that when discussing networks with power engineers, it is necessary to explicitly indicate whether communication networks or electricity networks are being referred to. Otherwise, power engineers may default to interpreting "networks" as "electricity networks".

WAC is highly desirable for maintaining the stability of large-scale smart grid. A WAC system consists of Phasor Measurement Units (PMUs), Phasor Data Concentrators (PDCs), and associated WAN for communications. A typical WAC system with WAN support is shown in Fig. 7.15. In general, WAC will not function effectively without the support of low-latency communications in WAN. Low-latency communications are also an essential requirement in NANs and HANs to support real-time applications.

As summarized from the literature review in [17], multicast communications are a viable solution for achieving low latency in WAN communications. In addition to IP multicast, there are basically two WAN architectural models that effectively support multicast WAN communications: SDN architecture and Publish-Subscribe Network (PSN) architecture. The concepts and applications of SDN have been comprehensively discussed in the previous section although the discussions are not specifically for smart grid environments. This section focuses on PSN-based multicast routing in WANs specifically for smart grid environments.

PSNs and SDNs exhibit certain similarities but also possess notable differences. They share the common characteristic of separating routing control from data for-

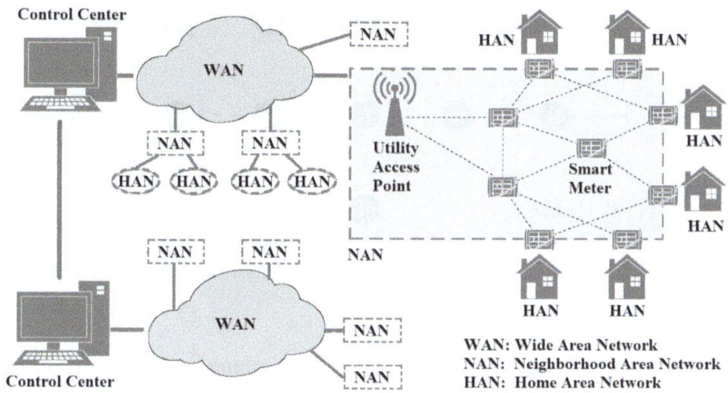

Fig. 7.14 Communication networks in smart grid

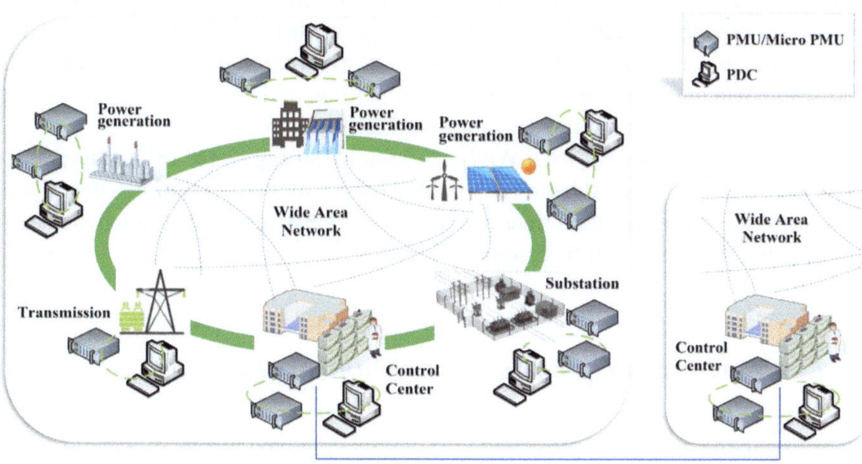

Fig. 7.15 A typical WAC system with WAN support [17, 19]

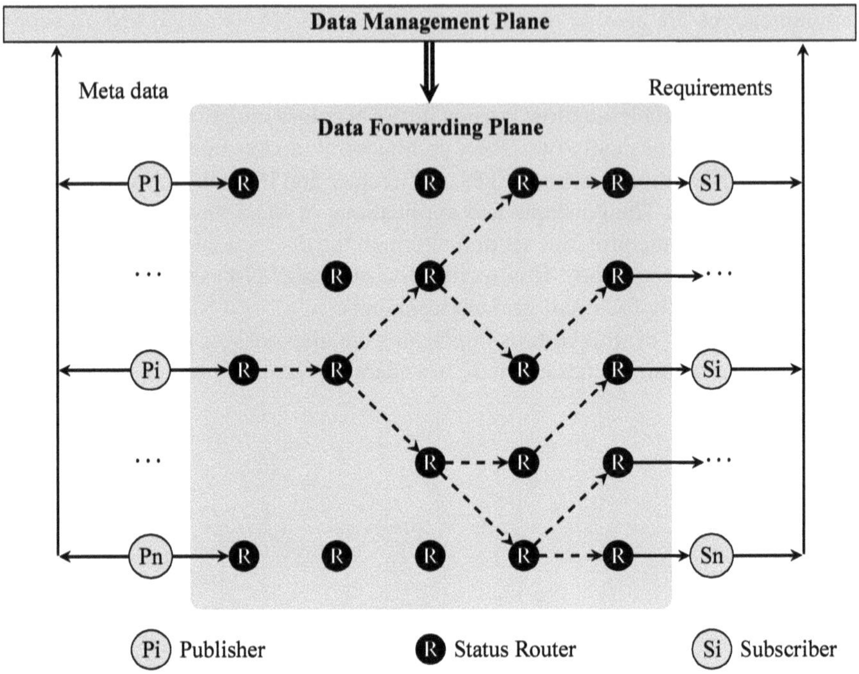

Fig. 7.16 Logical diagram of a PSN with a multicast tree from Pi to multiple subscribers [17, 19]

warding. However, unlike SDNs where the data plane resides in routers and the control plane is centralized in one or more servers, PSNs consist of publishers, subscribers, and status routers as part of their data plane. These components facilitate data forwarding from each publisher to potentially multiple subscribers. The data forwarding process is guided by the data management plane, which is responsible for link maintenance rather than multicast routing control. In the event of a link failure, the data management plane detects the failure, updates the network topology information, and triggers a new calculation of multicast routing.

A logical diagram of PSNs in WANs is depicted in Fig. 7.16. In this diagram, a multicast tree is established from each publisher to potentially multiple subscribers. The creation of these multicast trees can be achieved through a constrained optimization framework, which will be discussed below.

7.4.2 *Constrained Optimization of Multicast Routing*

According to the modelling presented [17], a PSN deployed in a WAC system can be described as a un-directed graph (i.e., network) $G(V, E)$, where V and E represent the sets of Vertices (i.e., nodes) and Edges (i.e., links) in the graph, respectively. An edge $e \in E$ in the graph is characterized by delay $d(e) > 0$ and bandwidth $b(e) > 0$. A source node s and a set of destination nodes $\mathbb{D} \subseteq V - \{s\}$ form a multicast group. A multicast tree T is rooted at s and spans all nodes in $\mathbb{D}$ without any loops. A path $P(T, v) \subseteq T$ is the set of tree links that connect s to $v \in \mathbb{D}$. Then, the end-to-end delay along the path from s to v is expressed as:

$$D(P(T, v)) = \sum_{e \in P(T,v)} d(e). \tag{7.9}$$

The upper bound of the end-to-end delay of the multicast tree is derived from:

$$D^{max}(T) = \max_{v \in \mathbb{D}}(D(P(T, v))). \tag{7.10}$$

For low-latency multicast communications, construct the multicast tree such that the upper bound of the end-to-end delay of the multicast tree is minimized, i.e., $\min D^{max}(T)$. As indicated in [17], this is a classical Steiner tree problem, which is NP-complete. The constraint to this minimization is the bounded bandwidth capacity of the multicast tree,

$$B(T) = \sum_{e \in T} b(e) \leq B^c, \tag{7.11}$$

where B^c is the upper limit of the bandwidth.

Then, the bandwidth-constrained minimum Steiner tree problem for low-latency multicast communications in smart grid WANs is formulated as

$$\begin{cases} \min D^{max}(T) \\ \text{s.t. } B(T) \leq B^c \end{cases} \tag{7.12}$$

A solution to this constrained optimization problem is a multicast tree from the source s to a set of destinations such that the upper bound of the end-to-end delay of the tree is minimized within the bandwidth constraint.

As the constrained optimization problem formulated in Eq. (7.12) is NP-complete, it is not realistic to find an analytical solution, or a numerical solution in a reasonable period of time as the problem size increases. Instead, heuristic methods are more suitable for solving the problem. In [17], the technique of Lagrangian relaxation is adopted to transform the constrained optimization in Eq. (7.12) into an unconstrained problem:

$$\min(D^{max}(T) + \alpha(B(T) - B^c)) \tag{7.13}$$

where α is a Lagrangian multiplier.

Three propositions are established in [17]. They are helpful for further development of an algorithm to solve the low-latency multicast-tree problem. These three propositions are restated below without providing proofs.

Proposition 7.1 *Denote* $L_R(\alpha) = \min(d_T + \alpha(b_T - B^c))$*. Then,* $L_R(\alpha)$ *is a lower bound to Eq. (7.12) for any* $\alpha \geq 0$.

Proposition 7.2 *Let* $T^{min}(w)$ *denote a minimum spanning tree with respect to the aggregate weight* w *between* s *and* $\mathbb{D}$ *found by a Stenier tree construction algorithm, i.e., the prim algorithm. Consider two minimum multicast trees* T_1 *and* T_2*. If* $T_1 = T^{min}(d_{T_1} + \alpha \cdot b_{T_1})$, $T_2 = T^{min}(d_{T_2} + \beta \cdot b_{T_2})$, $\alpha \geq 0, \beta \geq 0$, *and* $\alpha \leq \beta$*, then* $b_{T_1} \geq b_{T_2}$ *and* $d_{T_1} \leq d_{T_2}$ *hold.*

Proposition 7.3 *Consider three spanning trees* T_1, T_2*, and* T_3*. Let* $T_1 = T^{min}(d_{T_1} + \alpha \cdot b_{T_1})$, $T_2 = T^{min}(d_{T_2} + \beta \cdot b_{T_2})$, $T_3 = T^{min}(d_{T_3} + \gamma \cdot b_{T_3})$*, where* $\beta \leq \gamma$, $b_{T_2} \neq b_{T_3}$, $\alpha = \frac{d_{T_3} - d_{T_2}}{b_{T_2} - b_{T_3}}$*. Then,* $d_{T_2} \leq d_{T_1} \leq d_{T_3}$ *and* $b_{T_2} \geq b_{T_1} \geq b_{T_3}$ *hold.*

These propositions provide valuable insights into the constrained optimization problem described in (7.12). Proposition 7.1 establishes a lower bound for the optimization. Proposition 7.2 highlights the impact of the Lagrangian multiplier (α) on bandwidth consumption and end-to-end delay. The higher the value of α is, the smaller the bandwidth consumption is while the larger the end-to-end delay is. Increasing the bandwidth consumption to its upper limit will lead to minimized end-to-end delay. Proposition 7.3 suggests a method for selecting an appropriate value of α to achieve a trade-off between the bandwidth consumption and end-to-end delay.

In Proposition 7.2, a minimum spanning tree, also known as a minimum Steiner tree, of a weighted un-directed graph is a subset of the edges that connect all vertices with no loops and the minimum sum of the edge weights. It is calculated by using the Prim algorithm, which will be discussed below.

7.4.3 Algorithm Design for Problem Solving

From the problem formulation and three propositions discussed earlier, an iterative algorithm is designed in [17] to heuristically solve the constrained optimization problem (7.12. The algorithm is summarized in Algorithm 7.2, which is presented with a slightly modified structure for improved clarity compared to the original version in [17].

The algorithm starts with the shortest delay multicast tree $T^{min}(d)$ and the minimum bandwidth spanning tree $T^{min}(b)$ in lines 1 and 2, respectively. As the minimum delay tree is constructed for the shortest delay from the source node to each of the destination nodes, the total bandwidth consumption, i.e., $T^{min}(d)$, is calculated from unicast communications. In comparison, the minimum bandwidth spanning tree, and later the minimum weight spanning tree, are constructed for multicast communications with the least sum of the bandwidth values from all edges in the tree. Thus, the total bandwidth consumption $T^{min}(b)$ is calculated from multicast communications. The same concept applies to the calculation of the bandwidth consumption $T^{min}(w)$ for the minimum weight spanning tree later.

Algorithm 7.2: Bandwidth-constrained minimum Steiner tree [17, 19]

Input: banwidth b, delay d, B^c, s, $\mathbb{D}$
Output: A multicast tree from s to all nodes in $\mathbb{D}$
1 Minimum delay tree (unicast): $T_2 \leftarrow T^{min}(d)$;
2 Minimum bandwidth spanning tree (multicast): $T_3 \leftarrow T^{min}(b)$;
3 **if** *(Total unicast bandwidth $B(T_2) \leq B^c$)* **then**
4   T_2 is a solution
5 **else if** $(D^{max}(T_2) = D^{max}(T_3))$ **then**
6   T_3 is a solution
7 **else**
8   **loop**
9   Set $\alpha \leftarrow \frac{d_{T_3} - d_{T_2}}{b_{T_2} - b_{T_3}}$;   /* Proposition 7.3 */
10         $T_1 \leftarrow$ min. weight spanning tree $T^{min}(d + \alpha \cdot b)$; /* Proposition 7.2 */
11   **if** $(B(T_1) = B(T_2)$ *or* $B(T_1) = B(T_3))$ *and* $(D^{max}(T_1) = D^{max}(T_2)$ *or* $D^{max}(T_1) = D^{max}(T_3))$ **then**
12   T_3 is a solution;
13   break;   /* terminate loop */
14   **if** $B(T_1) > B^c$ **then**
15   Set $T_2 \leftarrow T_1$;
16   **else**
17   Set $T_3 \leftarrow T_1$;
18   **end-loop**;
19 **return** solution;

If $T^{min}(d)$ meets the bandwidth constraint, it is a solution (lines 3 and 4); otherwise, if the maximum end-to-end delays from $T^{min}(d)$ and $T^{min}(b)$ are the same, use $T^{min}(b)$ as a solution for less bandwidth consumption (lines 5 and 6). If neither $T^{min}(d)$ nor $T^{min}(b)$ satisfies the solution condition (line 7), then enter an iterative loop to find a solution with a trade-off between $T^{min}(d)$ and $T^{min}(b)$ (lines 8 through 18). Upon completion, the algorithm returns the derived solution in line 19.

In the loop, set an initial value α in line 9 according to Proposition 7.3. Then, construct a minimum weight spanning tree in line 10 in terms of the weights calculated from $d + \alpha \cdot b$ (Proposition 7.2). After that, check if the tree meets the solution condition (line 11). If yes, a solution is obtained (line 12), and thus terminate the loop (line 13); Otherwise, update the candidate solution with improved delay and bandwidth performance (lines 14 through 17), and then repeat the loop in line 8.

For the construction of the minimum spanning tree, the Prim algorithm is adopted, which is basically a greedy algorithm. It aims to find a spanning tree with the least sum of the weights of the edges in the tree. The idea of the Prim algorithm is simple with the following steps:

(1) Initialize the tree: choose the source vertex as the root of the tree;

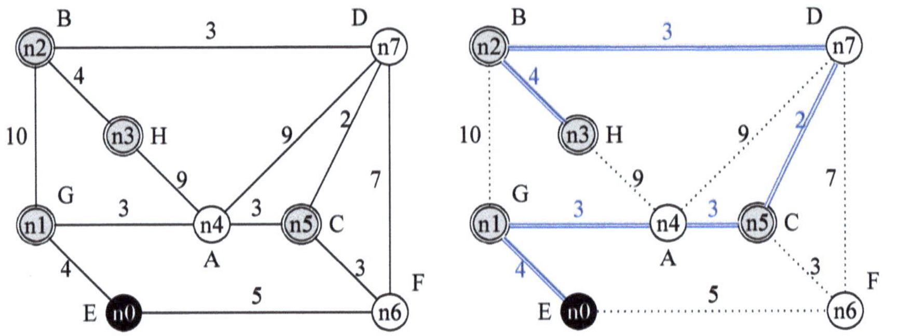

(a) Weighted graph with source n_0 and multicast nodes n_1, n_2, n_3, and n_5.

(b) The minimum spanning tree in double lines (the sum of weights = 19).

Fig. 7.17 The prim algorithm for the minimum spanning tree problem from source n_0 to multicast nodes n_1, n_2, n_3, and n_5

(2) Grow the tree by one edge: from the edges connecting the tree to vertices not yet in the tree, find the edge with the least weight, and add it to the tree. If more than one edge has the same least weight, arbitrarily choose one from these edges.
(3) Repeat step (2) until all vertices are in the tree.

The Prim algorithm for the minimum spanning tree problem is graphically shown in Fig. 7.17 through an example. The source node is n_0, and the multicast nodes from n_0 include n_1, n_2, n_3, and n_5. Starting from n_0, we select the minimum-weight edge from all edges connecting to n_0, giving the edge from n_0 to n_1. Then, pick up the minimum-weight edge from all edges that connect the tree (n_0 and n_1) to the nodes not yet in the tree, giving the edge from n_1 to n_4. Repeat this process until all multicast nodes are in the tree. This leads to the minimum spanning tree in Fig. 7.17b.

A C code is provided below to perform the calculation of the minimum spanning tree:

```
/* prim.c: The prim algorithm for Minimum Spanning Tree (MST)
Use the adjacency matrix representation of the graph */

#include <limits.h>
#include <stdbool.h>
#include <stdio.h>

#define V 8   /* the number of vertices */
#define M 5   /* the no. of multicast vertices including root */

/* The vertex with min key value from those not yet in MST */
int index_minKey(int key[V], bool mstSet[V]){
      int minKey_Val = INT_MAX, minKey_index;
      for (int v = 0; v < V; v++)
        if ((!mstSet[v]) && (key[v] < minKey_Val)){
              minKey_Val = key[v];
```

```
            minKey_index = v;
    }
      return minKey_index;
}

/* Construct MST for graph with adjacency matrix representation. */
void primMST(int graph[V][V],int parentVertex[V]){
      int key[V];      /* key value */
      bool mstSet[V]; /* Set of vertices in MST */

      for (int i = 0; i < V; i++){
            key[i] = INT_MAX;
            mstSet[i] = false;
      }
      key[0] = 0;       /* the 1st vertex as root */
      parentVertex[0] = 0; /* This one is root */

      /* The MST will have V vertices */
      for (int i = 0; i < V - 1; i++){
            int u = index_minKey(key, mstSet);
            mstSet[u] = true; /* add and picked vertex to MST set*/

            /* For vertices not in MST, update key value and parent
               index of the adjacent vertices of the picked vertex. */
            for (int v = 0; v < V; v++)
            if ((graph[u][v])&&(!mstSet[v])&&(graph[u][v]<key[v])){
                  parentVertex[v] = u;
                  key[v] = graph[u][v];
            } /* end-for v */
      }   /* end-for i */
}

/* prune tree brunches not in multicast MST tree */
void pruneMST(int graph[V][V], int multicastSet[M],
              int parentVertex[V], bool childVertex[V]){
      for (int i=0; i<V; i++)    /* Initialize: */
        childVertex[i] = false; /* all not in multicast MST tree*/
      for (int i=0;i<M;i++)
        childVertex[multicastSet[i]] = true; /*in mltcst MST tree*/

  bool changed = true;;
  while (changed){
        changed = false;
        for (int i=0; i<V; i++)
          if (!childVertex[i])
            for (int j=0; j<V; j++)
            if ((i==parentVertex[j]) && childVertex[j]){
                    childVertex[i] = true;
                    changed = true;
              }
      } /* end while */
}

int main(){
      int graph[V][V] = { { 0,  4,  0, 0, 0, 0, 5, 0},
                          { 4,  0, 10, 0, 3, 0, 0, 0 },
                          { 0, 10,  0, 4, 0, 0, 0, 3 },
```

```
                                    { 0,  0,  4, 0, 9, 0, 0, 0 },
                                    { 0,  3,  0, 9, 0, 3, 0, 9 },
                                    { 0,  0,  0, 0, 3, 0, 3, 2 },
                                    { 5,  0,  0, 0, 0, 3, 0, 7 },
                                    { 0,  0,  3, 0, 9, 2, 7, 0 } };
    int parentVertex[V] ={0};
    int multicastSet[M] = {0, 1, 2, 3, 5}; /* 0 as root */
    bool childVertex[V];/*remove vertices not in mltcstMST tree*/

    printf("The Prim algorithm for minimum spanning tree...\n");
    primMST(graph,parentVertex);
    pruneMST(graph,multicastSet,parentVertex,childVertex);

    /* print the constructed multicast MST */
    printf("Edge \tWeight\n");
    for (int i = 1; i < V; i++)
      if (childVertex[i])
        printf("%d->%d \t%d \n",parentVertex[i], i,
               graph[i][parentVertex[i]]);

    return 0;
}
```

Compile the C code with `gcc` compiler in a terminal window. This will generate an executable. Then, running the executable will give the result shown below:

```
$ gcc prim.c
$ ./a.out
The Prim algorithm for minimum spanning tree...
Edge    Weight
0->1    4
7->2    3
2->3    4
1->4    3
4->5    3
5->7    2
```

The sum of the weights of all edges in the resulting tree is 19, which matches the value shown in Fig. 7.17b. This verifies that the C implementation of the Prim algorithm works correctly.

7.4.4 An Illustrative Example

Consider an example graph with 8 vertices shown in Fig. 7.18a and b for the same graph with delay weights and bandwidth weights, respectively. It is the example of Fig. 3 in [17] with the following vertex mapping:

Figure 7.18 : n0 n1 n2 n3 n4 n5 n6 n7
Figure 3 in [17] : E G B H A C F D

Use n_0 through n_7 to name the vertices here for easier programming of the C code of the Prim algorithm given previously.

In this example, n_0 is the source node, and it sends messages to a multicast group formed by four nodes: n_1, n_2, n_3, and n_5. Nodes n_4, n_6, and n_7 are not members of the multicast group. Therefore, our objective is to build a multicast tree from n_0 to the multicast group members n_1, n_2, n_3, and n_5. The tree should have the shortest delay while satisfying the total bandwidth constraint, which is assumed to be bounded by $B^c = 30$. Figure 7.18 illustrates the steps involved in the calculation of the multicast tree.

From line 1 of Algorithm 7.2, the shortest delay tree is illustrated in Fig. 7.18c. From this figure, it is derived that $B(T_2) = 34 > B^c$ (calculated from unicast communications) and $D^{max}(T_2) = 18$. As $B(T_2) > B^c$, the shortest delay tree does not meet the bandwidth constraint.

Then, from line 3 of Algorithm 7.2, the minimum bandwidth spanning tree is obtained (Fig. 7.18d) with $B(T_3) = 19 < B^c$ and $D^{max}(T_3) = 36 > D^{max}(T_2)$. The total bandwidth of the minimum bandwidth spanning tree is within the upper bound, i.e., $B(T_3) < B^c$. However, the delay $D^{max}(T_3) = 36$ is much bigger than that of the shortest delay tree.

Therefore, the maximum delay of the paths in the trees to be checked is between 18 and 36. The total bandwidth of the trees is in the range from 19 to 34. A trade-off could be found between the shortest delay tree and the minimum bandwidth spanning tree. It aims to minimize the maximum delay of the paths in the tree within the bandwidth constraint. So, enter the iterative process from line 8 through line 18.

From Fig. 7.18c and d, calculate α in line 9 of Algorithm 7.2 as:

$$\alpha \leftarrow \frac{36 - 18}{42 - 19} = 18/23. \tag{7.14}$$

Then, derive a weighted graph from Fig. 7.18a and b by calculating the weight of each edge through $d + \alpha b$. This gives Fig. 7.18e, where the minimum weight spanning tree is also illustrated with $B(T_1) = 23$ and $D^{max}(T_1) = 24$. As this minimum weight spanning tree meets the bandwidth constraint, it is a feasible solution to the bandwidth-constrained optimization problem in formulation (7.12).

Now, repeating the process, we have

$$\alpha \leftarrow \frac{24 - 18}{42 - 23} = 6/19. \tag{7.15}$$

Calculate the weight according to $d + \alpha b$ for each of the edges in the graph. This leads to a weighted graph in Fig. 7.18f, which also shows the minimum weight spanning tree. From this figure, it is obtained that $B(T_1) = 26 < B^c$ and $D^{max}(T_1) = 18$. As the obtained $D^{max}(T_1) = 18$ is the same as that of the shortest delay tree (Fig. 7.18c), the best delay performance has been achieved within the bandwidth constraint. Therefore, the tree is the final solution.

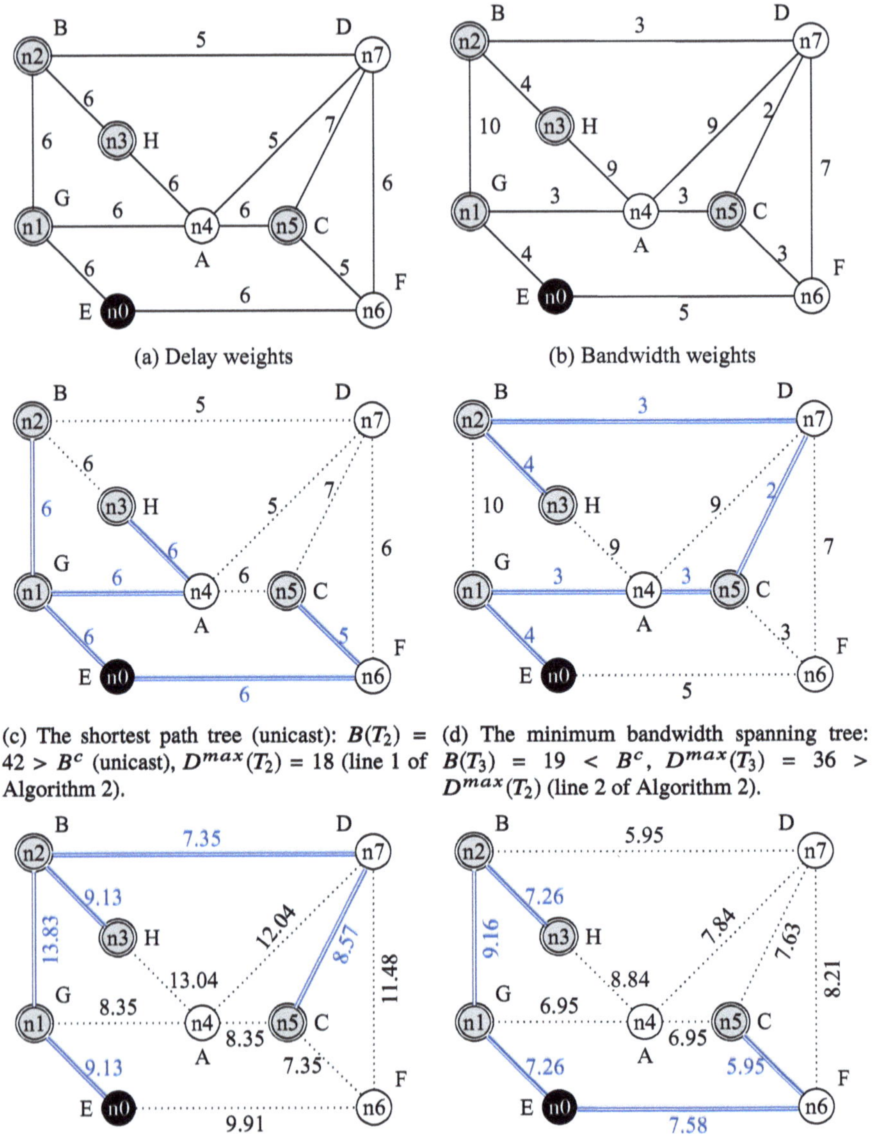

(a) Delay weights

(b) Bandwidth weights

(c) The shortest path tree (unicast): $B(T_2) = 42 > B^c$ (unicast), $D^{max}(T_2) = 18$ (line 1 of Algorithm 2).

(d) The minimum bandwidth spanning tree: $B(T_3) = 19 < B^c$, $D^{max}(T_3) = 36 > D^{max}(T_2)$ (line 2 of Algorithm 2).

(e) Minimum weight spanning tree with $\alpha = (36-18)/(42-19) = 18/23$: $B(T_1) = 23 < B^c$, $D^{max}(T_1) = 24$.

(f) Min. weight spanning tree with $\alpha = (24-18)/(42-23) = 6/19$: $B(T_1) = 26 < B^c$, $D^{max}(T_1) = 18$.

Fig. 7.18 Minimum bandwidth and minimum weight spanning trees from source n_0 to multicast nodes n_1, n_2, n_3, and n_5. Assume that the total bandwidth of the tree is bounded by $B^c = 30$

It is worth mentioning that under $\alpha = 18/23$ (Fig. 7.18e), an alternative minimum weight spanning tree exists. It spans n0-n1-n4-n3-n2 and n4-n5. This tree gives the same results of $B(T_1) = 23$ and $D^{max}(T_1) = 24$.

7.4.5 *Multiple Multicast Trees with Shared Links*

Figure 7.16 has illustrated a single multicast tree from a publisher to multiple subscribers. However, when there are multiple publishers, there will be multiple multicast trees, and these trees may share links. While these shared links may not experience congestion when only one multicast tree is used at a time, they can become congested during the multicast routing of data packets across multiple multicast trees. To address this issue, a Betweenness Centrality to Bandwidth ratio Tree (BCBT) approach has been developed in a constrained optimization framework [18]. The following is a brief discussion of this approach.

Consider a relatively static network, such as those found in WANs in smart grid. Use the same notations as those used previously. The minimum delay tree of the graph is the Shortest Path Tree (SPT), which is defined as:

$$T^{SPT} = \{\min D(P(T, v)), \forall v \in \mathbb{D}\}. \tag{7.16}$$

The delay of each source-destination pair in the SPT represents the shortest possible value. Therefore, it cannot be further reduced.

If all the shared links in the multiple multicast trees constructed from SPT do not experience any congestion, then these multicast trees can be used for multicast routing. However, if the data rate r of a shared link exceeds its bandwidth capacity, new multicast trees, denoted by T^*, need to be constructed to avoid increased communication delay due to traffic congestion on the shared links. This is formally described in [18] as follows:

$$\min D^{max}(T) = \begin{cases} D^{max}(T^{SPT}), \text{ if } \forall e_{ij} \in T^{SPT} \text{ and } r_{ij} \leq b_{ij} \\ D^{max}(T^*), \quad \text{if } \exists e_{ij} \in T^{SPT} \text{ and } r_{ij} > b_{ij} \end{cases} \tag{7.17}$$

Now, two issues need to be investigated. Firstly, how to monitor the switching of trees with shared links from a free flow state to a congestion state, and secondly, how to construct new multicast trees when congestion occurs on shared links of multiple multicast trees. The terminology of "free flow" is taken from the theory of complex networks.

To address the first issue, a data traffic transmission model is presented in [18]. It uses the concept of Betweenness Centrality (BC) from the theory of complex networks. BC quantifies the number of shortest paths passing through an edge e_{ij} in a network. The BC between the i-th and j-th nodes in V is defined as [20]:

$$Bc_{ij} = \sum_{s \neq t} \frac{\sigma_{st}(i, j)}{\sigma_{st}}. \tag{7.18}$$

where $\sigma_{st}(i, j)$ is the number of shortest paths from s to t that pass through the link $e(i, j)$, and σ_{st} is the total number of shortest paths from s to t. The BC indicates the potential traffic volume that an edge needs to handle when using the shortest delay routing.

From Eq. (7.18), for a graph $G(V, E)$ with N_n vertices and packet rate r measured in the number pf packets per unit time, the average number of packets that arrive at edge e_{ij} is estimated in [21] as:

$$\bar{r}_{ij} = \frac{2r}{N_n(N_n - 1)} \cdot Bc_{ij}. \tag{7.19}$$

When the average packet rate $\bar{r}_{ij}$ exceeds the bandwidth capacity b_{ij}, i.e., $\bar{r}_{ij} > b_{ij}$, the link becomes congested. To quantify the congestion likelihood, the BC-to-Bandwidth ratio k_{ij} is defined as:

$$k_{ij} = Bc_{ij}/b_{ij}. \tag{7.20}$$

The largest value of k_{ij} represents the edge where congestion is likely to occur. Therefore, the critical rate r_c at which congestion is likely to occur is estimated as follows:

$$r_c = \frac{1}{2} \cdot \frac{N_n(N_n - 1)}{\max(k_{ij})}. \tag{7.21}$$

For the SPT with N_l leaves and N_g multicast group members in $\mathbb{D}$, the critical data rate is estimated as:

$$r_c^{SPT} = \frac{r_c}{N_l}, \; N_l \in \{1, N_g - 1\}. \tag{7.22}$$

Therefore, if $r < r_c^{SPT}$, congestion is unlikely to occur, indicating a free flow state. Conversely, if $r \geq r_c^{SPT}$, it signifies a congestion state that must be addressed to avoid increased communication delay.

To address the second issues mentioned above for constructing new multicast trees to avoid traffic congestion on shared links, a constrained optimization problem is solved in [18]:

$$\begin{cases} \min \sum\limits_{e(i,j) \in T} k_{ij} \\ \text{s.t. } \max\limits_{v \in \mathbb{D}}(D(P(T, v))) \leqslant \Delta \end{cases} \tag{7.23}$$

where Δ represents the delay tolerance, and k_{ij} is calculated from Eq. (7.20).

However, this constrained optimization problem is NP-complete. To solve it, heuristics are developed in [18] through three steps:

(1) Construct a complete graph,

(2) Construct a constrained spanning tree, and
(3) Construct the final multicast tree.

An algorithm is also design in [18] to implement these steps. The heuristics and algorithm are illustrated with a simple example and also demonstrated through large-scale scenarios [18]. They are omitted here for brevity.

7.5 Routing in Large-Scale Wireless Networks

Routing in large-scale wireless networks poses greater challenges compared to wired networks. This is mainly due to the limited resources in wireless nodes and links. Wireless networks often have constraints such as limited energy, CPU resources, and network resources like frequency spectrum. In ad-hoc mobile networks, where nodes can enter or leave the network randomly within a short period, these constraints are particularly pronounced.

In wireless networks, data communication can be either single-hop or multi-hop to the base station, also known as direct and indirect communication, respectively. The focus of wireless routing is on multi-hop communications, where nodes not only transmit their own data but also act as routers to forward data from other nodes towards the base station.

In wireless sensor networks, two basic techniques are commonly used for routing: flooding and gossiping protocols. Flooding broadcasts a data packet to all neighboring nodes once it is received. This process continues until the packet reaches its destination or the maximum number of hops is reached. Gossiping, on the other hand, is an advanced version of flooding. When a node receives a data packet, it forwards it to a randomly selected neighboring node and repeats this process until the packet reaches its destination or other conditions are met.

Flooding and gossiping protocols are simple and easy to implement. They do not require routing algorithms or network topology maintenance. But flooding and gossiping have their disadvantages. Flooding suffers from excessive resource usage and traffic implosion. Gossiping overcomes the implosion issue but introduces significant delay. Therefore, pure flooding or gossiping protocols are generally not directly employed in practical wireless sensor networks. Instead, wireless routing is designed with specific routing algorithms that incorporate constrained broadcast and the requirement of maintaining network topology to some extent [22, 23].

7.5.1 Classification of Wireless Routing

Wireless routing protocols can be classified in various ways. A basic classification highlights node-centric and data-centric routing protocols.

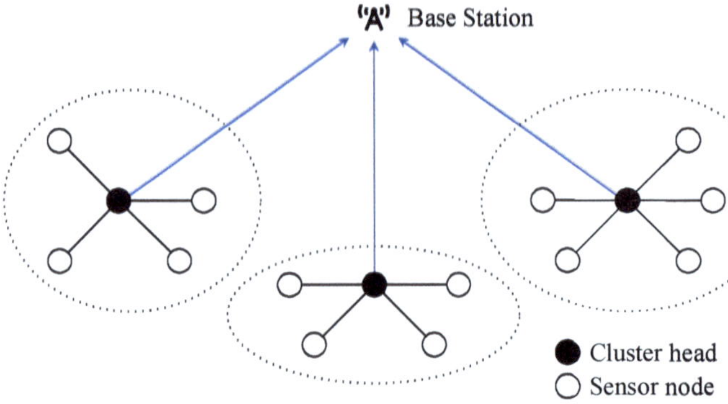

Fig. 7.19 Demonstration of cluster-based LEACH

Node-centric routing protocols specify destination nodes with numeric identifiers. While most ad-hoc routing protocols are node-centric, node-centric communications are not a commonly expected type of wireless sensor network communications. The Low Energy Adaptive Clustering Hierarchy (LEACH) protocol [24, 25] is an example of node-centric routing protocols. In LEACH, sensor nodes are organized into clusters, with one node acting as the cluster head. Most sensor nodes transmit data to cluster heads. The cluster heads aggregate and compress the data and then forward it to the base station, as shown in the diagram of Fig. 7.19. It is interesting to note that an RFC is not found for LEACH. The most similar and relevant IETF working group for this would be Routing Over Low powered and Lossy networks (ROLL), which is related to a number of RFCs [26].

In many wireless sensor network applications, the data sensed by a node is more valuable than the node itself. Therefore, routing protocols designed for wireless sensor networks are more data-centric. Data-centric routing protocols focus on the transmission of information specified by certain attributes rather than collecting data from specific nodes. Therefore, the sink node queries certain areas of the network and waits for relevant data from the sensors with the queried information in the areas. The Sensor Protocols for Information via Negotiation (SPIN) protocol is an example of data-centric routing protocols. SPIN uses three types of messages:

- ADV messages for advertisement,
- REQ messages for data request, and
- DATA messages for data transmission.

As shown in Fig. 7.20, when a node (e.g., Node 1) has new data, it broadcasts an ADV message to its neighbors (i.e., Nodes 2 and 3). After receiving this ADV message, a neighbor (e.g., Node 3) may request data via an REQ message, and the requested data is transmitted via a DATA message (from Node 1). Some comparisons between SPIN and LEACH are tabulated in Table 7.3.

Another classification criterion for wireless routing protocols is whether the routing protocols are Destination-initiated (Dst-initiated) or Source-initiated (Src-

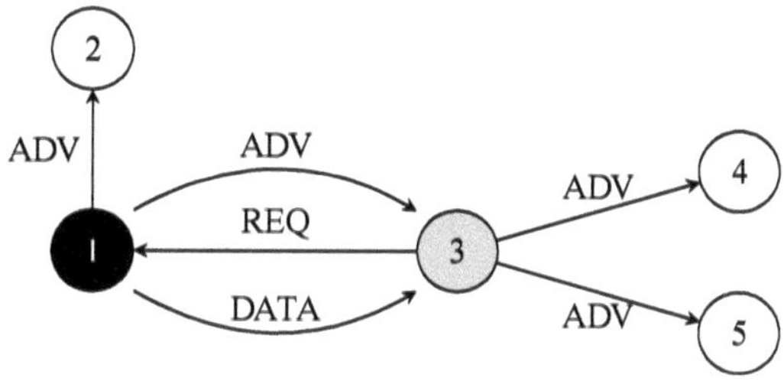

Fig. 7.20 Illustration of SPIN. Node 1 advertises a new data and Node 3 requests the new data

Table 7.3 Comparisons between SPIN and LEACH

Performance Metric	SPIN	LEACH
Routing principle	FLAT	Hierarchical
Energy consumption	Limited	Maximum
Efficiency	Poor	Medium
Delay	Small	Tiny
Packet delivery	Meta-data based	Cluster based
Network lifetime	Good	Good
Mobility	Supported	Fixed base station
Clustering method	N/A	Distributed

initiated). Dst-initiated routing protocols generate traffic paths starting from the destination. In comparison, Src-initiated routing protocols set up routing paths based on the demand of the source node. The source node advertises the data when available and initiates data delivery. LEACH is a Dst-initiated routing protocol, while SPIN is a Src-initiated routing protocol.

Wireless routing protocols are also classified based on the underlying network architecture, distinguishing between flat topology and hierarchical topology. Flat routing treats all nodes equally. There is no hierarchy at all in the organization of the network nodes. By contrast, hierarchical or clustering routing organizes network nodes into clusters, thus introducing hierarchy. Nodes within a cluster communicate with the cluster head, and all cluster heads communicate with each other in higher layers. Hierarchical routing offers scalability, energy efficiency in route discovery, and easier management compared to flat routing.

7.5.2 Proactive and Reactive Routing

In general, wireless routing protocols are either reactive or proactive. A reactive routing protocol creates a routing path only when there is a packet to transmit, triggering routing actions on demand. As a result, a reactive protocol does not maintain the entire network topology. In contrast to reactive routing, a proactive routing protocol establishes routing paths at the start of network operation and continuously updates routing information to maintain reliable paths.

Reactive and proactive routing protocols have their own advantages and disadvantages. Reactive protocols can quickly respond to changes in network topology and link quality by generating routing paths based on local and neighboring node information. However, these protocols need to establish routing paths for each transmission, leading to increased overhead and communication delay.

Proactive routing protocols have the advantage of quick initiation of data transmission, as routing paths are already established. Also, maintaining the entire network topology allows for routing optimization. However, updating routing information in proactive protocols can be slow, especially when there are changes in network topology or wireless link quality. To address these issues, the work presented in [23] has proposed a proactive routing approach that incorporates a hierarchical architecture and an innovative updating process.

7.5.3 Reactive Routing Protocols

Let us briefly discuss three reactive routing protocols developed for large-scale wireless networks: Ad-hoc On-demand Distance Vector (AODV), Dynamic Source Routing (DSR), and Ad-hoc On-demand Multipath Distance Vector (AOMDV).

AODV

AODV is a commonly used reactive routing protocol. It is formally specified in the IETF RFC 3561 [27]. As a distance-vector routing protocol, AODV chooses the shortest and loop-free path from the routing table. Figure 7.21 shows a illustrative diagram of AODV.

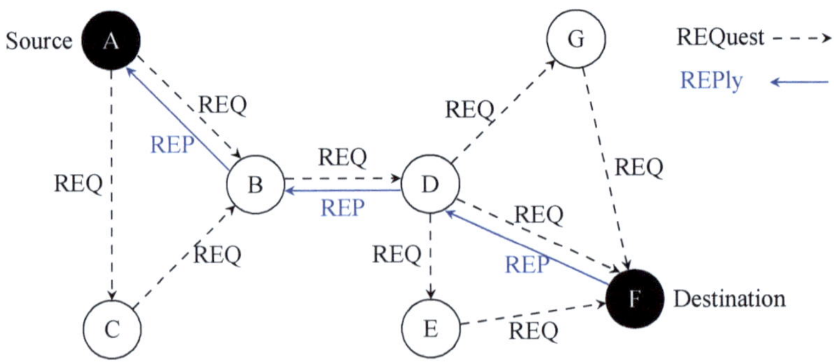

Fig. 7.21 An illustrative diagram of AODV

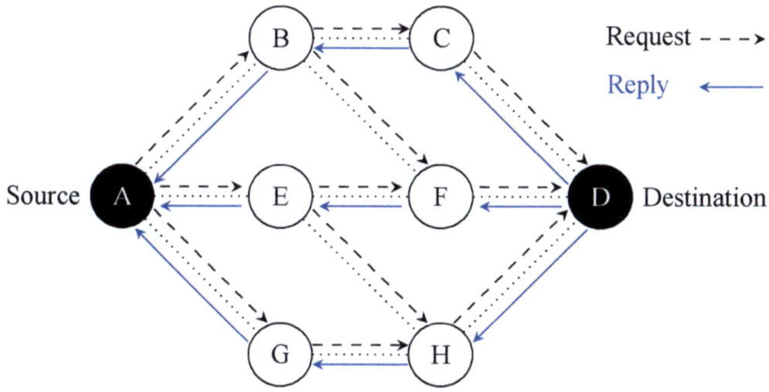

Fig. 7.22 An illustrative diagram of AOMDV that finds multiple paths

AOMDV

The Ad-hoc On-demand Multipath Distance Vector (AOMDV) routing protocol is an extension of the AODV protocol. It is specifically designed to find multiple paths from the source to the destination in wireless networks. By discovering multiple paths, AOMDV provides alternative routes that can be used in case of route failures, thereby improving the reliability of wireless communications and reducing communication delay. The concept of AOMDV routing is graphically shown in Fig. 7.22.

While AOMDV enhances the capabilities of AODV, it does introduce additional overhead during the route discovery process. This overhead is primarily due to the need to find and maintain multiple paths, requiring additional control messages and computational resources. However, the benefits of increased reliability and reduced communication delay outweigh the overhead costs in scenarios where a reliable and efficient routing solution is essential (Fig. 7.22).

DSR

Instead of relying on the routing table at each intermediate node, using source routing for hop-by-hop data delivery can be effective in reducing overhead in route establishment. This approach has led to the development of the DSR protocol, which is defined in the IETF RFC 4728 [28]. In DSR, intermediate nodes use the source route embedded in the packet's header to determine the next node to which the packet should be forwarded. This process is illustrated in the diagram shown in Fig. 7.23.

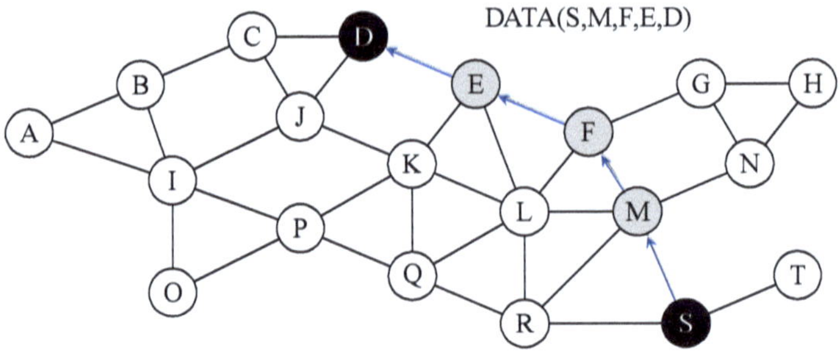

Fig. 7.23 Demonstration of the DSR protocol

7.5.4 Proactive Routing Protocols

Examples of proactive routing protocols include the Optimized Link State Routing (OLSR) protocol, Destination Sequence Distance Vector (DSDV) routing protocol, and Babel. They are briefly discussed in the following.

OLSR

The OLSR protocol is initially defined in the IETF RFC 3626 [29]. Later, it is enhanced in the IETF RFC 7181, known as OLSRv2 [30], which has inspired further developments documented in several RFCs. OLSRv2 retains the basic mechanisms and algorithms of OLSR but introduces a link metric to replace hop count for selecting the shortest routes. It also adopts a more flexible and efficient signaling framework for simplified message exchanges. Figure 7.24 illustrates various OLSR messages for a local node. The resulting network routing topology looks like the one shown in Fig. 7.25.

DSDV

The DSDV routing protocol is a table-driven proactive routing protocol. In DSDV routing, each node in the network maintains a routing table and periodically broadcasts routing updates. The best path is calculated using a distance-vector algorithm modified from the Bellman-Ford mechanism. Figure 7.26 illustrates the structure of a routing table for a specific node. The sequence number shown in Fig. 7.26 is a time indication sent by the destination node. It is set as an even number if the corresponding node is reachable. An odd sequence number indicates that the node is not reachable. When a node (e.g., node 1) finds out that a neighbor node (e.g., node 2) is no longer reachable, it advertises the route to that neighbor node (i.e., node 2) with an infinite metric (distance) and a sequence number one greater than the latest sequence number for the route, forcing any nodes with the node (i.e., node 1) on the

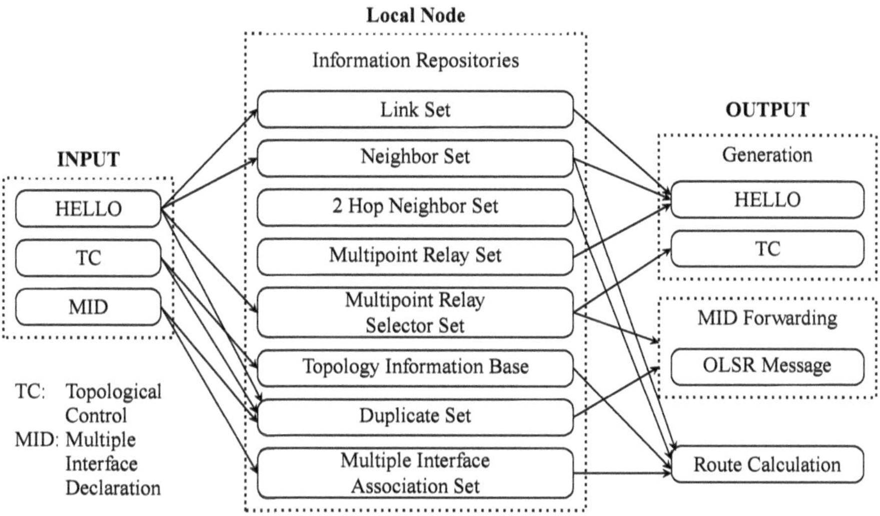

Fig. 7.24 OLSR messages

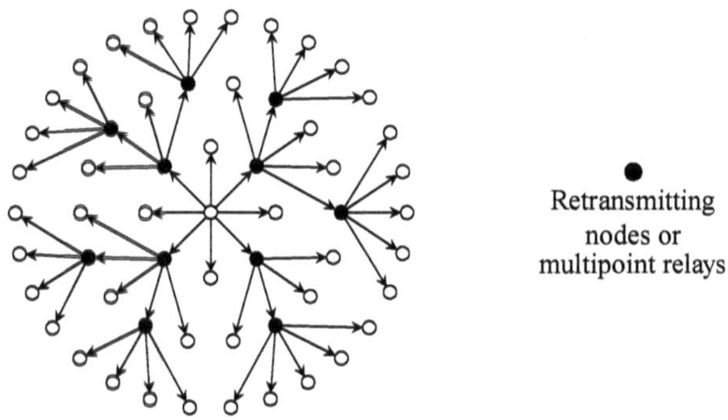

Fig. 7.25 Illustration of OLSR topology

path to the neighbor node (i.e., node 2) to reset their routing tables. Route selection is based on the distance-vector metric and sequence number criteria. The route labeled with the highest sequence number, which indicates the most recent information, is always used. If two routes to the same destination are known with the same distance, the one with a better (higher) sequence number is selected and used.

Babel

The Babel routing protocol is also a distance-vector routing protocol. It is designed to be robust and efficient on both wireless mesh networks and wired networks. Babel is an IETF standard, and the latest version is specified in the IETF RFC 8966 [31].

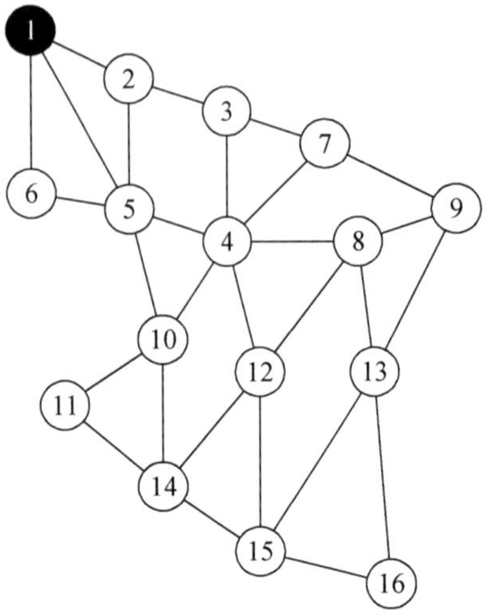

Routing table for node 1

Destination	NextNode	Distance	SeqNo.
2	2	1	22
3	2	2	26
4	5	2	32
5	5	1	134
6	6	1	144
7	2	3	162
8	5	3	170
9	2	4	186
10	6	2	142
11	6	3	176
12	5	3	190
13	5	4	198
14	6	3	214
15	5	4	256
16	5	5	270

Fig. 7.26 The DSDV routing protocol

Basically, Babel incorporates ideas from DSDV, AODV, and EIGRP but employs different techniques to prevent loops.

In Babel, neighbor discovery is performed using the following two messages:

- the hello message, sent every four seconds, and
- the IHU (I Hear You) message, sent every 12 s.

The topology dissemination of the Babel routing protocol is performed via a route update message every 16 s. Babel's updates are exchanged only between neighbors.

In general, topology dissemination is a process used to share information about the network topology among the network nodes. It involves communicating details about the connections and routes within the network. Each node maintains a view of the network topology, which the node uses to determine the best routes for forwarding packets to their destination. When changes occur, such as nodes joining or leaving the network, nodes need to update their view of the network topology to ensure optimal routing decisions. Figure 7.27 provides an illustration of Babel's topology dissemination in a simple network consisting of six nodes and eight edges.

Fig. 7.27 Babel's topology dissemination through communications between neighbors

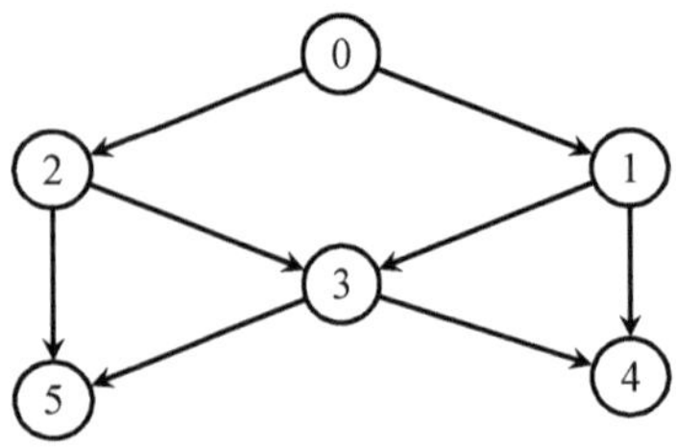

7.6 Summary

Routing plays a significant role in computer networks. It generally includes routing decision and packet forwarding. The routing decision is to find a path to route traffic from the source to the destination, wile the packet forwarding is the actual delivery of data packets along the identified path. Both routing decision and packet forwarding are traditionally implemented in routers in distributed environments, where each router independently determines how to route traffic. Therefore, the implementation of routing heavily relies on routing architecture and routing protocols, and is driven by network scalability, performance, and other considerations.

This chapter has discussed five main routing protocols: RIP, EIGRP, OSPF, IS-IS, and BGP. RIP and EIGRP are distance-vector routing protocols, while OSPF and IS-IS are link-state routing protocols. BGP is a path-vector routing protocol. Each of these protocols has its own strengths and weaknesses, making them suitable for different scenarios. It is important to select the appropriate routing protocols in conjunction with the routing architecture. BGP operates between ASs, RIP is suitable for small-scale networks with low to medium hierarchy and diversity, EIGRP is well-suited for large-scale networks with relatively low hierarchy, and OSPF and IS-IS are suitable for large-scale networks with relatively high hierarchy. Moreover, it is worth noting that while routing is a layer-3 function in the OSI seven-layer reference architecture, some routing protocols are based on layer-4 functions (e.g., UDP-based RIP and TCP-based BGP) or layer-2 functions (e.g., IS-IS).

To enhance network programmability and flexibility, SDN has been developed as an alternative to traditional routing. In SDN, a central SDN controller in the control plane makes routing decisions, while network devices in the data plane perform the actual data forwarding. The application plane is responsible for learning network topology and other network knowledge. It also handles the configuration of the central SDN controller through interfaces like GUIs. SDN employs two interfaces to connect the SDN planes: the northbound interface that connects the application plane and control plane, and the southbound interface that connects the control plane and data plane. With the separation of routing decision and data forwarding in mind, this chapter has extensively discussed publish-subscribe-based multicast routing, with a focus on its application in smart grid communication networks.

Compared to routing in wired networks, routing in wireless networks poses greater challenges primarily due to the limited resources in wireless networks. This chapter has covered several widely used wireless routing protocols and routing architectures. The concepts of proactive routing and reactive routing have been highlighted in this chapter. They are developed with the consideration of routing architectural models suitable for various application scenarios.

References

1. Rekhter, Y., Li, T.: A border gateway protocol 4 (BGP-4). RFC 1771, RFC Editor (1995). https://doi.org/10.17487/RFC1771
2. Rekhter, Y., Gross, P.: Application of the border gateway protocol in the Internet. RFC 1772, RFC Editor (1995). https://doi.org/10.17487/RFC1772
3. Potaroo: Growth of the BGP table - 1994 to present. https://bgp.potaroo.net. Accessed 15 Sep. 2021
4. Malkin, G.: RIP version 2. RFC 2453, RFC Editor (1998). https://doi.org/10.17487/RFC2453
5. Atkinson, R., Fanto, M.: RIPv2 cryptographic authentication. RFC 4822, RFC Editor (2007). https://doi.org/10.17487/RFC4822
6. Malkin, G., Minnear, R.: RIPng for IPv6. RFC 2080, RFC Editor (2007). https://doi.org/10.17487/RFC2080
7. Rekhter, Y., Li, T.: An architecture for IP address allocation with CIDR. RFC 1518, RFC Editor (1993). https://doi.org/10.17487/RFC1518
8. Fuller, V., Li, T.: Classless inter-domain routing (CIDR): the Internet address assignment and aggregation plan. RFC 4632, RFC Editor (2006). BCP 122, https://doi.org/10.17487/RFC4632
9. Savage, D., Ng, J., Moore, S., Slice, D., Paluch, P., White, R.: Cisco's enhanced interior gateway routing protocol (EIGRP). RFC 7868, RFC Editor (2016). https://doi.org/10.17487/RFC7868
10. Moy, J.: OSPF version 2. RFC 2328, RFC Editor (1998). https://doi.org/10.17487/RFC2328
11. Coltun, R., Ferguson, D., Moy, J., Lindem, A.: OSPF for IPv6. RFC 5340, RFC Editor (2008). https://doi.org/10.17487/RFC5340
12. For Standardization, I.O.: Intermediate System to Intermediate System Intra-domain Routing Information Exchange Protocol for Use in Conjunction with the Protocol for Providing the Connectionless-mode Network Service (ISO 8473). ISO/IEC 10589:2002, 2nd edn. ISO (2002)
13. Ginsberg, L.: Use of OSI IS-IS for routing in TCP/IP and dual environments. RFC 1195, RFC Editor (1990). https://doi.org/10.17487/RFC1195
14. McCabe, J.D.: Network Analysis, Architecture, and Design, 3rd edn. Morgan Kaufmann Publishers, Burlington (2007). ISBN 978-0-12-370480-1
15. Boucadair, M., Jacquenet, C.: Software-defined networking: a perspective from within a service provider environment. RFC 7426, RFC Editor (2014). https://doi.org/10.17487/RFC7149
16. Foundation, O.N.: OpenFlow switch specification, version 1.5.1. Published online https://opennetworking.org/wp-content/uploads/2014/10/openflow-switch-v1.5.1.pdf (2015). Accessed 13 Sep. 2021
17. Li, X., Tian, Y.C., Ledwich, G., Mishra, Y., Han, X., Zhou, C.: Constrained optimization of multicast routing for wide area control of smart grid. IEEE Trans. Smart Grid **10**(4), 3801–3808 (2019)
18. Li, X., Tian, Y.C., Ledwich, G., Mishra, Y., Zhou, C.: Minimizing multicast routing delay in multiple multicast trees with shared links for smart grid. IEEE Trans. Smart Grid **10**(5), 5427–5435 (2019)
19. Ding, Y., Li, X.: Low-latency multicast and broadcast technologies for real-time applications in smart grid. In: Tian, Y.C., Levy, D. (eds.) Handbook of Real-Time Computing, pp. 861–892. Springer (2022)
20. Newman, M.E.J.: Scientific collaboration networks. II. Shortest paths, weighted networks, and centrality. Phys. Rev. E 64(1), 016132 (2001)
21. Guimera, R., Diaz-Guilera, A., Vega-Redondo, F., Cabrales, A., Arenas, A.: Optimal network topologies for local search with congestion. Phys. Rev. Lett. **89**, 328170 (2002)
22. Ding, Y., Tian, Y.C., Li, X., Mishra, Y., Ledwich, G., Zhou, C.: Constrained broadcast with minimized latency in neighborhood area networks of smart grid. IEEE Trans. Industr. Inf. **16**(1), 309–318 (2020)
23. Pradittasnee, L., Camtepe, S., Tian, Y.C.: Efficient route update and maintenance for reliable routing in large-scale sensor networks. IEEE Trans. Industr. Inf. **13**(1), 144–156 (2017)

24. Heinzelman, W., Chandrakasan, A., Balakrishnan, H.: An application-specific protocol architecture for wireless microsensor networks. IEEE Trans. Wireless Commun. **1**(4), 660–670 (2002)
25. Fan, X., Song, Y.: Improvement on LEACH protocol of wireless sensor network. In: 2007 International Conference on Sensor Technologies and Applications (SENSORCOMM 2007), pp. 260–264 (2007)
26. Barthel, D., Robles, I., Retana, A., Richardson, M., Cragie, R., Pignolet, Y.A.: Routing over low power and lossy networks (roll). Charter-IETF-ROLL-05, https://datatracker.ietf.org/wg/roll/about/. Accessed on 18 Oct 2021
27. Perkins, C., Belding-Royer, E., Das, S.: Ad hoc on-demand distance vector (AODV) routing. RFC 3561, RFC Editor (2003). https://doi.org/10.17487/RFC3561
28. Johnson, D., Hu, Y., Maltz, D.: The dynamic source routing protocol (DSR) for mobile ad hoc networks for IPv4. RFC 4728, RFC Editor (2007). https://doi.org/10.17487/RFC4728
29. T. Clausen, E., Jacquet, P.: Optimized link state routing protocol (OLSR). RFC 3626, RFC Editor (2003). https://doi.org/10.17487/RFC3626
30. Clausen, T., Dearlove, C., Jacquet, P., Herberg, U.: The optimized link state routing protocol version 2. RFC 7181, RFC Editor (2014). https://doi.org/10.17487/RFC7181
31. Chroboczek, J., Schinazi, D.: The babel routing protocol. RFC 8966, RFC Editor (2021). https://doi.org/10.17487/RFC8966
32. Haleplidis, E., Pentikousis, K., Denazis, S., Salim, J.H., Meyer, D., Koufopavlou, O.: Software-defined networking (SDN): Layers and architecture terminology. RFC 7426, RFC Editor (2015). https://doi.org/10.17487/RFC7426

Chapter 8
Network Performance Architecture

Network performance architecture aims to describe how performance requirements, developed from requirements analysis and specifications, are fulfilled within the network during planning and design. Therefore, the task of architectural design for performance cannot be performed before conducting a comprehensive requirements analysis. To comprehend network performance architecture, it is necessary to understand the concept of performance, the characterization of performance requirements, the focus of performance architecture, and the available mechanisms and techniques that support network performance.

In general, network performance con be considered as a set of performance levels for network bandwidth capacity, latency, jitter, reliability, maintainability, availability, and others. Performance requirements are characterized by specific performance metrics, and their quantitative values, bounds or thresholds, or other statistics. They are expected from both the network and user perspectives. The majority of network services fall under the category of best-effort services, which do not require specific Quality of Service (QoS) support. Consequently, performance architecture does not focus on these best-effort services. Instead, it concentrates on the QoS of non-best-effort services, such as real-time and guaranteed services. Therefore, this chapter on performance architecture will primarily consider QoS support. Various mechanisms and techniques will be introduced to support performance in the planning of performance architecture.

Let us begin with an introduction to QoS control. This will be followed by comprehensive discussions on resource and traffic control, network policies, Differentiated Service (DiffServ), Integrated Service (IntServ), and Service Level Agreements (SLAs).

Y.-C. Tian and J. Gao, *Network Analysis and Architecture*, Signals and Communication Technology, https://doi.org/10.1007/978-981-99-5648-7_8

8.1 Quality-of-Service Control

The concept of QoS has been explained in detail in the flow analysis chapter of this book. The primary objective of QoS is to regulate application-centric and network-centric responses [1, pp. 18–19]. The discussions in this section on QoS control will provide a comprehensive understanding of how QoS operates.

As we have already understood, network services are provisioned as best-effort services by default. Therefore, to implement QoS in a network, the first step is to identify the end users, applications, or devices that require QoS management within the network bandwidth limit. This is part of requirements analysis, which has been discussed previously in a separate chapter. However, having a more detailed analysis of the QoS requirements for users, applications, and devices will help in developing the best QoS management architecture.

In data communications, traffic flows destined for users, applications, and devices with QoS requirements need to be marked prior to their transmission over the network. This is a type of traffic *prioritization*. Mechanisms and protocols have been developed at different layers to mark data packets that require QoS management. They are primarily implemented at Layers 2 and 3.

At Layer 2, the mechanism of Class of Service (CoS) embeds the class of services bits into the frame header to indicate the classification of specific traffic. It can be used to mark outgoing traffic but cannot be used as part of an input traffic policy. CoS is typically offered within an Multi Protocol Label Switching (MPLS) offering. The main types of CoS technologies include IEEE 802.1p Layer 2 Tagging and Type of Service (ToS). They can be set through CLI commands, e.g., on a router. For instance, the following example sets the CoS value to 6 on a router:

```
Router(config)# policy-map policy1
Router(config-pmap)# class class1
Router(config-pmap-c)# set cos 6
Router(config-pmap-c)# end
```

It is worth mentioning that while CoS marks the traffic flows that require QoS management, it does not actually manage the QoS of these flows. The actual manipulation of the traffic relies on *QoS policies*, which will be discussed later.

At Layer 3, there are two standard QoS architectural models for IP networks: DiffServ and IntServ. To mark the data stream that requires DiffServ, a 6-bit Differentiated Services field CodePoint (DSCP) is embedded into the packet header. This 6-bit DSCP is part of the 8-bit Differentiated Service field (DS field) in the packet header. The DS field was originally defined in the IETF RFC 2474 [2] to replace the outdated IPv4 Type of Service (ToS) field. It was later updated in RFC 3260 [3] and most recently in RFC 8436 [4].

After the traffic flows with QoS requirements are marked, the next step in QoS management is to categorize data streams into different groups accordingly. QoS policies are then applied to these different groups of traffic flows to provide preferential treatment of certain data streams over others. Techniques that can implement such QoS treatment include scheduling, queuing, and frame preemption, which are topics of resource management and traffic control:

- Scheduling determines the order in which traffic is processed for transmission over the network.
- Queuing controls the single or multiple queues in a network device to store data packets or frames pending for transmission.
- Frame preemption can interrupt the transmission of preemptable frames to achieve reduced latency for time-sensitive frames in real-time network applications.

For example, if video traffic is tagged for QoS management and a QoS policy is created to allocate more bandwidth to the video traffic, the routing and switching devices will prioritize the video packets or frames by moving them to the front of the queue and transmitting them immediately ahead of other best-effort packets or frames. In comparison, for standard best-effort UDP data transfer without marking QoS requirements, the UDP datagrams and frames will be buffered in the queue until sufficient bandwidth is available for their transmission. In this case, if the queue is full while there are still more UDP datagrams or frames arriving, some datagrams or frames without QoS requirements may be dropped.

8.2 Resource and Traffic Control

QoS provisioning in networking requires service differentiation, which can be achieved by implementing mechanisms for bandwidth management and traffic control. This section discusses four types of such mechanisms: prioritization, traffic management, scheduling and queuing, and frame preemption. These mechanisms can be implemented at Layer 2, Layer 3, or both Layer 2 and Layer 3.

8.2.1 Prioritization

Traffic prioritization is a process that determines which flows should be given higher priority levels than others in traffic management and processing. Traffic flows with higher priority will be served ahead of lower-priority flows. Since network resources are shared among multiple users, applications, and devices, it becomes necessary to prioritize certain network services over others.

A detailed analysis of traffic prioritization is an integral part of requirements analysis for network planning, design, management, and control. This section explores various mechanisms that can be employed to implement effective traffic prioritization.

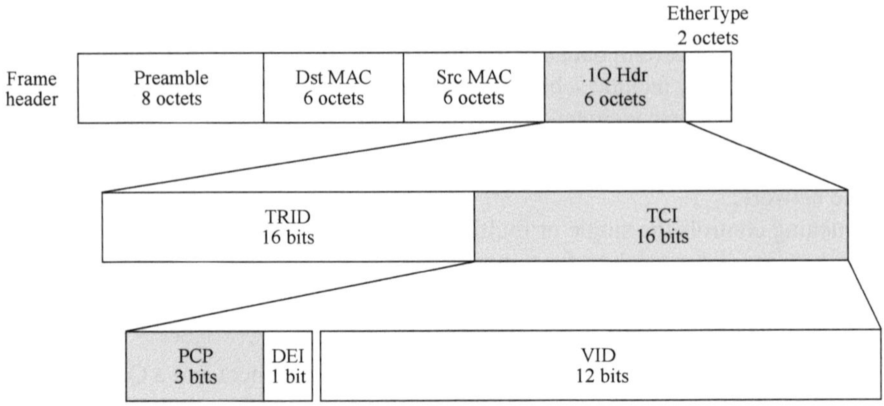

Fig. 8.1 IEEE 802.1p Ethernet frame header

IEEE 802.1p Layer 2 Tagging

As mentioned previously in Sect. 8.1, there are Layer-3 and Layer-2 technologies to mark the outgoing traffic with QoS requirements. At Layer 2, IEEE 802.1Q, often referred to as Dot1q, is developed for bridges and bridged networks to support VLANs on IEEE 802.3 Ethernet networks. Its latest version is ISO/IEC/IEEE 8802-1Q:2020 [5]. IEEE 802.1Q comes with a QoS prioritization mechanism, commonly known as IEEE 802.1p.

IEEE 802.1p inserts a 4-octet field between the source MAC address and EtherType fields of the original frame header. As a result, the maximum frame size is extended from the original 1, 518 bytes to 1, 522 bytes. The minimum frame size remains at 64 bytes, but a bridge may increase it from 64 bytes to 68 bytes to accommodate the additional 4 octets in the frame header. The structure of the IEEE 802.1p frame header is illustrated in Fig. 8.1.

Within the four added octets, two bytes are used as the Tag Protocol Identification (TPID), while the remaining two bytes are defined for the Tag Control Information (TCI). The TCI is composed of a 3-bit Priority Code Point (PCP), a 1-bit Drop Eligible Indicator (DEI), and a 12-bit VLAN Identifier (VID). This structure is also shown in Fig. 8.1.

The 3-bit PCP represents the mapping of the 802.1p CoS to eight levels of frame priority. For the use of the eight priority levels, IEEE has provided some general recommendations as summarized in Table 8.1.

Access Categories in IEEE 802.11e

At Layers 2 and 1, the wireless network standard IEEE 802.11e has been defined as an amendment to IEEE 802.11 wireless networking. A fundamental concept within 802.11e is the provision of differentiated services for network traffic with different priority levels in wireless LAN applications. This enhancement of QoS is particularly beneficial for delay-sensitive applications, including voice, video, and multimedia

Table 8.1 IEEE-recommended use of the eight PCP levels

PCP	Priority (*)	Acronym	Traffic type/Application
000	0	BK	Background/Best-effort data
001	1	BE	Best effort/medium-priority data
010	2	EE	Excellent effort/high-priority data
011	3	CA	Critical applications
100	4	VI	Video, < 10 ms latency and jitter
101	5	VO	Voice, < 10 ms latency and jitter
110	6	IC	Internetwork control
111	7	NC	Network control

(*) The lower the number, the lower the priority level

Table 8.2 Four access categories (ACs)

Access category	Designation	Priority
AC0 (AC_BK)	Background	Lowest
AC1 (AC_BE)	Best Effort	Medium
AC2 (AC_VI)	Video	High
AC3 (AC_VO)	Voice	Highest

services over wireless networks. Considering the numerous amendments to IEEE 802.11, the IEEE has made the decision to consolidate eight of them (802.11a, b, d, e, g, h, i, j) into a unified document, which is now known as the base standard IEEE 802.11-2007.

In IEEE 802.11e, four priority levels have been defined for wireless traffic. They are denoted as four Access Categories (ACs), as illustrated in Table 8.2. The AC0 background traffic (AC_BK) possesses the lowest priority and is therefore handled with best effort without any specific QoS settings. The AC1 best-effort traffic (AC_BE), assigned medium priority, benefits from an increased allocation of wireless bandwidth and other resources compared to AC0 traffic, thereby offering a certain level of QoS. The AC2 video traffic (AC_VI) is granted a higher priority than AC1 (i.e., AC_BE), allowing for more opportunities to contend for wireless channel access. The highest priority is assigned to AC3 voice traffic (AC_VO), indicating that this category of traffic holds the greatest likelihood of successfully contending for wireless channel access.

The management of the four ACs is implemented in 802.11e through the Media Access Control (MAC). The original 802.11 MAC uses the standard Distributed Coordination Function (DCF) or Point Coordination Function (PCF) to control media access. It is worth mentioning that as an optional mode in 802.11, PCF is not widely used or enabled in wireless access points or Wi-Fi adapters. Enhancing the standard DCF and PCF, 802.11e provides a new coordination function, the Hybrid Coordination Function (HCF). HCF uses two methods to control media access: Enhanced Distributed Channel Access (EDCA) and HCF Controlled Channel Access (HCCA),

Table 8.3 Default EDCA settings of Contention Window (CW), AIFNSN, and Transmit Opportunity (TXOP)

AC	Priority	CWmin	CWmax	AIFSN	Max TXOP
AC_BK	Lowest	aCWmin	aCWmax	7	0
AC_BE	Medium	aCWmin	aCWmax	3	0
AC_VI	High	$\frac{\text{aCWmin}+1}{2}-1$	aCWmin	2	>0
AC_VO	Highest	$\frac{\text{aCWmin}+1}{4}-1$	$\frac{\text{aCWmin}+1}{2}-1$	2	>0

both of which have priority differentiation for traffic flows. As HCCA is not mandatory for 802.11e, it is not widely enabled in access points. Therefore, in the following, the focus of discussions would be on EDCA.

EDCA sets the Contention Window (CW) differently for the four ACs according to their priority levels. The AC_BK and AC_BE have the same size of the CW. In comparison, AC_VI has a smaller CW. The CW of the AC_VO is the smallest, indicating its highest priority. The default CW settings for the four ACs are tabulated in Table 8.3.

With EDCA, high-priority traffic experiences a shorter waiting time on a station before being sent out compared to low-priority traffic. Consequently, high-priority traffic has a higher chance of being sent out. This waiting time is controlled by the Arbitration Inter-Frame Space (AIFS), which is characterized by AIFS Number (AIFSN). AIFSN represents the number of slot times used in the AIFS. A lower AIFSN indicates a higher priority and, therefore, a greater chance of being sent out. The default AIFSN values for the four ACs are listed in Table 8.3.

Furthermore, EDCA provides contention-free access to the channel for a specific period known as the Transmit Opportunity (TXOP). During a TXOP, a station is allowed to transmit as many frames as possible. If a frame is too large to fit in a single TXOP, it can be fragmented into smaller frames. Table 8.3 reveals that AC_BK and AC_BE are assigned a maximum TXOP of 0, meaning that they are not granted contention-free access to the wireless channel. By contrast, positive maximum TXOP values are set for AC_VI and AC_VO. The maximum TXOP is set in intervals of 32 μs. In OFDM, the default TXOP values are 94 intervals (3.008 ms) for AC_VO and 47 intervals (1.504 ms) for AC_VI. IEEE 802.11e allows these TXOP values to be configured on access points.

In wireless network applications, various traffic flows are mapped to these four ACs for QoS performance management. Table 8.4 illustrates an example in medical and industrial control systems. In this scenario, the alarm and emergency commands are the most critical and thus are assigned to AC3 (AC_VI). Real-time monitoring traffic is mapped to AC2 (AC_VO) with high priority, which is lower than that of the alarm and emergency commands in AC3 (AC_VI). Non-real-time data transmissions in medical and networked control systems are served with the medium-priority

Table 8.4 Applications of ACs in medical and industrial control systems

Access category	Priority	Medical and ind. control systems
AC0 (AC_BK) Background	Lowest	Non-medical or non-control traffic
AC1 (AC_BE) Best Effort	Medium	Non-real-time data
AC2 (AC_VI) Video	High	Real-time monitoring
AC3 (AC_VO) Voice	Highest	Alarm and emergency commands

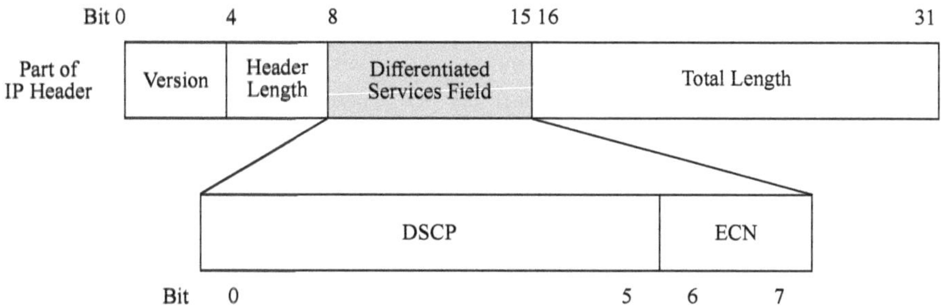

Fig. 8.2 The Structure of the one-octet DS field consisting of 6-bit Differentiated Services Codepoint (DSCP) and 2-bit Explicit Congestion Notification (ECN) in the IP header

AC2 (AC_BE). Moreover, non-medical and non-control application flows over the wireless network are treated as AC0 (AC_BK) without any specific QoS settings.

DS Field in IP

At layer 3, a one-octet DS Field is defined in the IP header for configuring differentiated services. It consists of a 6-bit DSCP and a 2-bit Explicit Congestion Notification (ECN), as specified in detail in the IETF RFC 3260 [3], RFC 3168 [6], and RFC2474 [2]. These RFCs should be read in their entirety because the basic specifications of DS field given in RFC 2474 are updated in the other two RFCs. The structure of the DS field in the IP header is illustrated in Fig. 8.2.

DSCP Codepoints in IP

The 6-bit DSCP in the DS field can accommodate a total of 64 unique codepoints. These codepoints are divided into three distinct pools, which are allocated and managed by the Internet Assigned Numbers Authority (IANA). The three pools of codepoints are shown in Table 8.5.

In old IP header defined in RFC 791, a 3-bit IP Precedence is used in the Type of Service (ToS) byte to differentiate eight priority levels. Later, RFC 2474 [2] redefines the ToS byte as the DS field, where the first six bits define the DSCP and the remaining two bits represent ECN, as shown in Fig. 8.2. To maintain a certain level of backward compatibility with the old IP Precedence, RFC 2474 further assigns

Table 8.5 Three pools of DSCP codepoints specified in RFC 2474 [2, pp. 14–15] and RFC 8436 [4, p. 4] ('x' takes a value of either 0 or 1)

Pool	Codepoint space	Assignment policy
1	xxxxx0	Standards Action (*)
2	xxxx11	Experimental or Local Use (EXP/LU)
3	xxxx01	Standard Action (⋆) [4, p. 4]

(*) xxx000 for backwards compatibility with the old IP Precedence
(⋆) This replaces the original EXP/LU defined in RFC 2474 [2, pp. 14–15]

Table 8.6 Mapping DSCP 'xxx000' to IP Precedence

DSCP			IP Precedence	
xxxx000	Value	Meaning	Value	Description
000 000	0	Best Effort	000 (0)	Routine or Best Effort
001 000	8	CS1	001 (1)	Priority
010 000	16	CS2	010 (2)	Immediate
011 000	24	CS3	011 (3)	Flash (*)
100 000	32	CS4	100 (4)	Flash Override
101 000	40	CS5	101 (5)	Critical (⋆)
110 000	48	CS6	110 (6)	Internet
111 000	56	CS7	111 (7)	Network

(*) mainly for voice signaling or video
(⋆) mainly for voice Real-time Transport Protocol (RTP)

eight RECOMMENDED codepoints ('xxx000) drawn from Pool 1 codepoint space. A typical mapping of DSCP codepoints 'xxx000' to the outdated IP precedence is tabulated in Table 8.6.

ECN Codepoints in IP

In the DS field (Fig. 8.2), RFC 3168 has given clear motivations and specifications for the use the 2-bit ECN. This 2-bit ECN addresses the need for mechanisms that can indicate congestion before a queue actually overflows, such as through the use of Random Early Detect (RED) queuing. This mechanism allows for the marking of packets instead of outright dropping them, providing a congestion indication that can benefit time-sensitive traffic flows.

The two ECN bits are known as ECN-Capable Transport (ECT) bit and Congestion Experienced (CE) bit, respectively. Together, they define four ECN codepoints: 00, 01, 10, and 11. Each codepoint indicates a different ECN scenario, as shown in Table 8.7. The Not-ECT codepoint (00) signifies that a packet is not using ECN. The CE codepoint (11) is set by a router to indicate congestion to the end nodes.

The ECT codepoints ECT(0) and ECT(1) are set by data senders to tell routers that the end-points of the transport protocol are ECN-capable. Both ECT(0) and ECT(1) are treated equivalently by routers, implying that routers do not differentiate between

Table 8.7 ECN codepoints [6, p. 7]

ECT bit	CE bit	Known as
0	0	Not-ECT
0	1	ECT(1)
1	0	ECT(0)
1	1	CE

them. This flexibility allows data senders to choose either ECT(0) or ECT(1) on a packet-by-packet basis to indicate the ECN capability of the end points.

The use of two codepoints ECT(0) and ECT(1) for the same purpose is motivated by two reasons. Firstly, it prevents the CE codepoint from being erased by network elements, ensuring that the indication of congestion is preserved throughout the network. Secondly, it allows data receivers to report the receipt of packets with the CE codepoint set back to the sender, as required by the transport protocol. This reporting enables the sender to be aware of the congestion encountered during packet transmission [6, p. 7].

8.2.2 Traffic Management

Generally speaking, network traffic management is a broad topic that encompasses various techniques for monitoring and managing networks. Its objective is to optimize the performance and security of network applications while operating within the network's available resources. Key network management mechanisms include bandwidth monitoring, packet inspection and classification, marking, policing, shaping, queuing, scheduling, and dispatching.

In this subsection, we specifically focus on three fundamental concepts: *admission control*, *traffic classification and conditioning*, and *traffic shaping*. The concept of prioritization has been discussed previously, which determines the relative importance of network traffic and how each of the traffic flows will be handled. Other concepts related to traffic management will be discussed later, such as queuing and scheduling, network policies, DiffServ, and IntServ.

Admission Control

Network admission control is a mechanism that allows a network to regulate access to its resources. This control can be based on the identity of the user or the priority level of a network flow. When a user is not authorized to access certain network resources, traffic flows from that user attempting to access those resources are blocked. In cases where multiple traffic flows are authorized for access, a best-effort network management approach treats all flows equally, providing them with the same opportunity to acquire network resources. However, when these flows have different levels of priority, the implementation of admission control permits, denies, or delays their access to network resources based on their relative priority.

To illustrate the concept of admission control, let us consider critical real-time traffic with higher priority and general best-effort traffic with the best-effort service level. In the absence of critical real-time traffic, all best-effort traffic flows are served based on the best-effort service, ensuring each flow has an equal chance to obtain network resources such as bandwidth. However, when critical real-time traffic is present, its access takes precedence over other flows, granting it either immediate access or a higher probability of access. If the real-time traffic requires a guaranteed network service, its needs must be fulfilled before serving other flows. For instance, if network resources, such as available bandwidth, have already been fully utilized to accommodate the real-time traffic flows and other flows, any additional requests for network resources will be delayed or denied. If a new request is delayed, network resources that are currently allocated to non-real-time traffic could be reduced in order to admit the new request.

From the implementation perspective, admission control is often applied at the access layer within Cisco's three-layer Core/Distribution/Access architecture. This approach ensures that access to network resources is regulated and managed effectively.

It is worth mentioning that the concept of admission control plays a significant role in IntServ QoS (RFC 1633 [7]). Therefore, it will be discussed more comprehensively when the IntServ QoS technique is introduced.

Traffic Classification and Conditioning

To ensure effective network QoS implementation, traffic classification and conditioning capabilities are essential. Traffic classification aims to prioritize traffic for differentiated services. It has been discussed previously under the topic of prioritization with various techniques such layer-2 tagging and layer-3 DS field composed of DSCP and ECN, and will be further discussed later in the context of DiffServ QoS. Once traffic is classified, traffic conditioning comes into play to enforce the QoS requirements specified in an SLA, e.g., delay requirements.

To enforce the QoS requirements defined in the SLA, packets need to be metered, marked, and shaped/dropped. Thus, a network traffic conditioner consists of four main components: a *meter*, a *marker*, a *shaper*, and a *dropper*. From this perspective, traffic conditioning is sometimes directly referred to as a QoS mechanism that involves packet metering, marking, shaping, and dropping.

Each of the four components of a traffic conditioner functions differently. The *meter* measures the traffic to identify the traffic flows that either conform or do not conform to the SLA traffic profile. For conforming flows, no further action is taken. However, non-conforming flows are marked by the *marker*, allowing for further analysis and potential actions. Some marked packets are forwarded to the receiving device for possible actions, while others may be shaped by the *shaper* for delayed forwarding or even discarded by the *dropper*.

Traffic conditioning plays a significant role in achieving within the DiffServ architecture, as specified in the IETF RFC 2475 [8]). The process and logic flow of traffic conditioning will be discussed later in a dedicated section to DiffServ. In the following, traffic shaping will be examined, which is one of the four components of traffic conditioning.

Traffic Shaping

Traffic shaping is a bandwidth management technique that delays a set of packets so that network traffic on the network can be controlled for optimized and guaranteed performance. It is used for bandwidth throttling, i.e., to control the volume of traffic on the network in a specific period of time to bring the delayed traffic into compliance with a designed traffic profile. It is also used for rate limiting, i.e., to control the maximum rate at which the traffic is sent to the network. Moreover, traffic shaping can be used for other purposes that involve more complex criteria than bandwidth throttling and rate limiting.

There are two main types of traffic shaping: application-based and route-based traffic shaping. Route-based traffic shaping is undertaken based on the route information of previous hop or network hop. In application-based traffic shaping, the applications of interest are identified first. Then, depending on shaping policies, the traffic of these applications may be shaped by using a shaper. However, the use of VPNs that encrypt application traffic can circumvent application-based traffic shaping. Also, many application protocols employ encryption to prevent shaping.

Traffic shaping finds applications in various scenarios. It is widely used by domestic ISPs to manage their networks. For example, ISPs may employ traffic shaping to limit resource consumption by peer-to-peer file-sharing networks such as BitTorrent. Data centers use traffic shaping to maintain SLAs for a wide range of applications and numerous hosted tenants. Furthermore, traffic shaping is also an integral part of traffic conditioning in the DiffServ QoS management (RFC 2475 [8]), which will be discussed later in a separate section.

Traffic shaping is noticeable in the Internet services provided from an ISP to individual customers. During peak times or even at other instances, the data transmission rate over the network can significantly decrease from its normal value. Some customers may complain that their bandwidth is being throttled by their ISPs. However, this can be attributed to traffic shaping management aiming to accommodate more users or applications within the capacity of the ISP networks at that time, or it may be a result of traffic policing management due to various reasons such as the misuse of network resources. To gain clarity, it is advisable to refer to the SLA between the ISP and customers to determine whether traffic shaping and traffic policing are specified within the agreement, often referred to as the Terms and Conditions of the services.

Traffic shaping is often mistakenly conflated with traffic policing, although they are distinct yet related concepts. Traffic shaping adds delays to packets to align them with a desired traffic profile, while traffic policing deals with packet dropping and marking. The concept of traffic policing will be covered later in a separate section.

8.2.3 *Queuing and Scheduling*

Queuing stores layer-3 packets or layer-2 frames in one or multiple queues on a network device, such as a router or switch, before they are processed for transmission over the network. The design of queuing strategies involves two main issues. The first issue pertains to determining the number and order of queues, as well as the placement of traffic within these queues. The second issue focuses on managing queue overflow situations when the queues become fully filled while additional packets or frames continue to arrive. Scheduling addresses the first issue, while queuing strategies tackle the second issue.

Scheduling

For network applications with no QoS requirements, best-effort services are applied to the traffic of these applications, without prioritization. In such cases, a single queue can be used as a buffer to store incoming layer-3 packets for routers or layer-2 frames for switches. The simplest form of a queue is the First In First Out (FIFO) queue. When new data arrives at the device, it is pushed to the back of the FIFO queue for future processing. As the device begins processing the data, the data is popped from the head of the FIFO queue. This ensures that the FIFO queue operates on a First In First Out basis.

When dealing with network traffic that has QoS requirements, multiple queues may be necessary on the network device. These queues are typically configured with different levels of priority to provide differentiated services to various traffic groups. An example of multiple queues is the Class-based Queuing (CBQ). CBQ maintains multiple queues with different priority levels. Traffic is placed into the respective queues based on its priority. Each queue in CBQ can be a FIFO queue. Another example is the IEEE 802.11e standard, which defines four levels of traffic priority known as ACs. Multiple queues are designed to buffer traffic for different ACs, with higher priority queues receiving preferential processing. This allows for differentiated QoS provisioning in the network.

Figure 8.3 illustrates a scenario where multiple queues are maintained on a network device. Each queue is a FIFO queue with an assigned priority level.

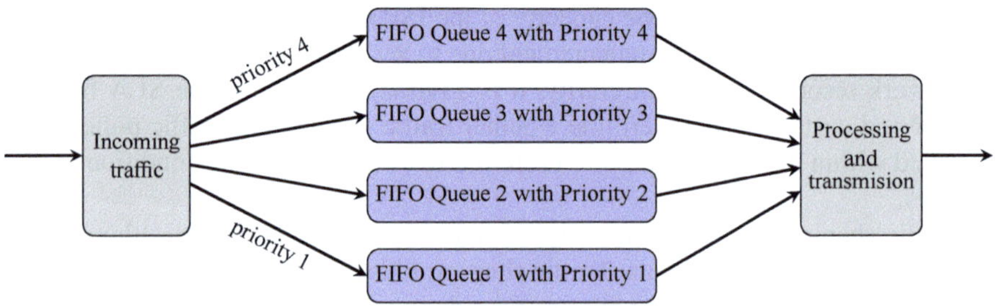

Fig. 8.3 Multiple FIFO queues each with a different priority level

Queuing

When a queue is full and there are still incoming packets, a queuing mechanism is required to decide how to drop packets. Various queuing drop policies are developed to deal with packet drop when a queue is full. The simplest one is the DropTail policy. It simply drops the newly arriving packets until the queue has enough room to accept incoming traffic. In the network simulator `ns2`, the DropTail queue management policy is implemented in the `Queue/DropTail` object.

Another well-known queue management policy is the Random Early Detect (RED) policy. To avoid queue overflow, RED randomly selects a packet from the queue and drops it. This is performed even before the queue becomes full. This is particularly helpful for TCP traffic flows, as early drops of TCP packets force the TCP sender to slow down its transmission rate. While RED is an effective queue management mechanism, dynamically predicting an appropriate set of RED parameters is a difficult task. Therefore, RED is not enabled by default and its use is still limited on the Internet [9, p. 5].

As a variation of RED, Weighted RED (WRED) operates in the same way but supports multiple queues, each corresponding to a different priority level. For a single queue with mixed traffic of different priority levels, different thresholds may be configured for priority levels. For example, a queue may have a lower threshold for low-priority packets, triggering early drops of low-priority packets to protect higher-priority packets in the same queue.

In recent years, the concept of Active Queue Management (AQM) has become increasingly important in queue management for network traffic. The IETF RFC 7567 [9] has presented detailed recommendations for AQM to improve the performance of the Internet. This is basically based on the connectionless IP architecture. While the connectionless nature of IP provides flexibility and robustness, it can lead to congestion collapse under heavy load, causing unacceptably long delays. Internet latency has become a focus of attention to increase the responsiveness of Internet applications and protocols. Queue buildup in network devices is a major source of delay, and an AQM mechanism can provide lower-latency interactive services, reduce packet drops, prevent lock-out behavior, and decrease the probability of control loop synchronization.

Interestingly, RFC 7567 [9] no longer recommends RED or any other specific algorithm as the default. Instead, it recommends processes for selecting appropriate algorithms. Particularly, it emphasizes that a selected algorithm should be able to automate any required tuning for common deployment scenarios. The mechanisms described in RFC 7567 can be implemented in network devices on the path between endpoints or in the networking stacks within endpoint devices.

8.2.4 Frame Preemption

The concept of frame preemption is related to Time-Sensitive Networking (TSN), which is a rich set of standards currently under active development by the Time-Sensitive Networking task group of the IEEE 802.1 working group. The Time-Sensitive Networking task group is renamed from the Audio Video Bridging task group to reflect the expansion of the working area of the standardization group from audio and video applications to general time-sensitive network applications. Overall, TSN standards primarily focus on how time-sensitive data is transmitted over Ethernet networks. Therefore, they are layer-2 mechanisms for time-sensitive frame transmission.

Briefly speaking, the majority of TSN projects define various extensions to the IEEE 802.1Q, often referred to as Dot1q, with its latest version being ISO/IEC/IEEE 8802-1Q:2020 [5], as mentioned earlier. With many extensions to the IEEE 802.1Q, TSN has particularly addressed low-latency and high-availability frame transmission.

One of these extensions forms IEEE 802.1Qbu and 802.3br for Interspersing Express Traffic (IET) and Frame Preemption. The implementation of frame preemption relies on not only bridge management as part of 802.1 but also Ethernet MAC control as part of 802.3. Therefore, frame preemption is defined in two standard documents, i.e., IEEE 802.1Qbu for the bridge management component, and IEEE 802.3br for the Ethernet MAC component.

To explain how frame preemption works, let us consider a simple example in the context of TSN. In TSN, many frames follow a cyclic pattern with QoS requirements for real-time monitoring, control, and other applications. These frames are transmitted periodically in cycles across the network. The cycle time is designed to be a fixed value. Figure 8.4 depicts the transmission of cyclic frames 1, 2, 3, and 4. Alongside cyclic frames, sporadic data frames also exist. For the purpose of demonstrating frame preemption, assume that sporadic frame f5 has no QoS requirement and thus can be transmitted with best effort when spare network resources are available. If f5 can fit within the cycle time, it is simply transmitted within the cycle without impacting the performance of cyclic frames f1 through f4, as shown in Fig. 8.4a.

However, if frame f5 is too long to fit within the cycle time, it will interfere with the normal transmission of cyclic frames f1 through f4. Without frame preemption, this interference may lead to deadline misses for some cyclic frames with QoS requirements. In Fig. 8.4b, frame f5 cannot be transmitted successfully in cycle 2 and thus misses its deadline.

To prevent excessively long or numerous sporadic messages from extending into the next cycle time, the concept of a *guard band* is introduced. The guard band is a configurable period of time at the end of each cycle. The transmission of a message can only be successful when the message transmission finishes before the end of the guard band, in other words, before the end of the cycle.

The guard band mechanism solves the problem of the transmission of sporadic messages that can be transmitted before the end of the guard band. However, the challenge persists for transmitting too long or too many sporadic messages that extend

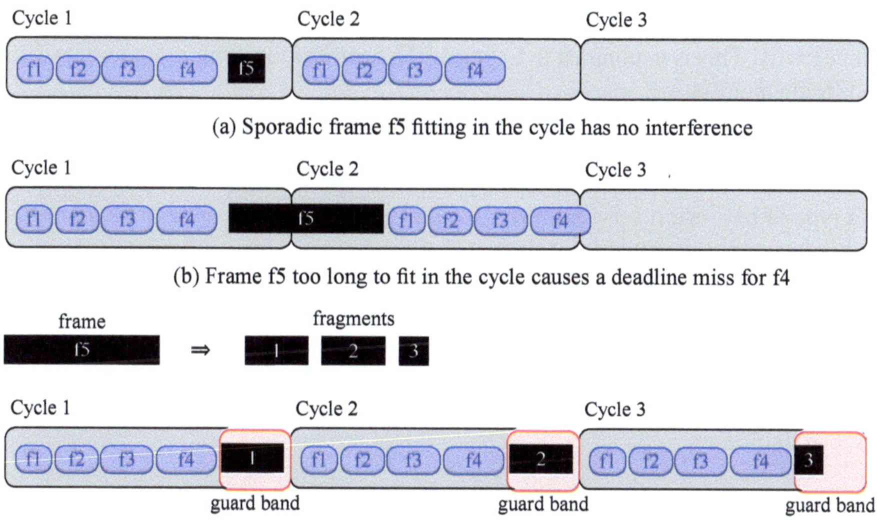

(a) Sporadic frame f5 fitting in the cycle has no interference

(b) Frame f5 too long to fit in the cycle causes a deadline miss for f4

(c) Transmission with frame preemption

Fig. 8.4 Frame preemption

into the next cycle. The mechanisms of *express messages* and *frame preemption* are designed to tackle this problem. Express messages are periodic real-time messages with QoS requirements and should not be preempted. Other messages are considered *normal messages*, which are transmitted with best effort. If these normal messages cannot fit within the guard band, they can be preempted. The process of frame preemption is summarized in Algorithm 8.1.

The technique of *frame preemption* resolves the problem of transmitting messages that are too long or too numerous to fit within a cycle. This is achieved through a few simple steps when sporadic messages need to be transmitted. Figure 8.4c demonstrates the process of frame preemption. In this figure, frame f5 is divided into three fragments. It is preempted twice, eventually completing transmission in Cycle 3.

Algorithm 8.1: Frame preemption

```
for each cycle do
    Transmit express messages from the beginning of the cycle;
    if there are normal messages or their fragments to transmit then
        Resume transmission of the fragments or transmit normal messages;
        if the transmission cannot finish within the guard band then
            Preempt the transmission before the end of the guard band;
```

It is worth mentioning that frame preemption operates purely on a link-by-link basis. This means that it operates from one switch to the next switch, where fragments are reassembled. In other words, frame fragmentation does not provide end-to-end connectivity. This is in contrast to Layer-3 IP fragmentation, which supports end-to-end fragmentation.

In terms of error detection, frame preemption incorporates CRC32 error detection for fragments. It will disregard any errors beyond the CRC32 error detection capacity and thus does not perform error recovery. This is a typical feature of frame preemption as a type of best-effort services. Error handling and recovery can be implemented at higher layers of the networking stack.

Although frame preemption may look appealing, it is not yet widely supported as of the time of writing this book. Its industrial implementation faces significant challenges, including complexity and stringent requirements for accurate time synchronization. Moreover, no standards have been published so far for frame preemption in wireless networks. Therefore, ongoing efforts are being made to further develop TSN technologies.

8.3 Network Policies

Network policies are a broad topic in network architecture, network design, network management, and network operations. This section provides a brief overview of network policies and their impact on network performance. Some of the discussions presented here are taken from Cisco's online documentation, e.g., online reference [10].

8.3.1 *Policies and Their Benefits*

Let us now clarify what policies are, what policies govern, and what benefits that policies can offer.

What are Network Policies?

General network policies are sets of statements and rules that govern the behavior of network devices through management and allocation of network resources among users, applications, links, and devices. Network performance policies, specifically, focus on performance assurance and improvement in network service provisioning. They provide a high-level view of how the network should perform and what behaviors are expected from network components to achieve optimal performance. From this high-level view, various mechanisms are designed and deployed to meet the policy requirements.

Given the close relationship between network performance and various aspects of the network and its components, performance policies are developed in conjunction with other network policies, such as network management policies, network security policies, and privacy policies. Therefore, it is important to understand the impact of network policies on network performance, especially for network applications with QoS requirements.

What do Network Policies Govern?

Network policies govern various aspects of network operations, including user access, device privileges, application prioritization, data management, and location-aware services. For network users, policies should be implemented to recognize and differentiate between different user groups, and particularly grant appropriate rights and privileges accordingly. For example, human resources users may be authorized to access detailed information from HR databases, while other users are restricted from retrieving such data. Similarly, finance officers may have access to finance databases, while this access is restricted for other users. Network administrators typically have more extensive access privileges compared to regular users.

Device access can also be regulated through policies. For example, policies can dictate that only specific devices are allowed to make configuration changes to DNS servers. No modifications to such DNS configurations are enabled from any other devices. Desktop computers and computing nodes may have greater access rights compared to printers and video cameras. Access to finance databases, for instance, may be entirely prohibited for printers and video cameras.

Given that network bandwidth is a shared resource, policies need to prioritize traffic flows from various applications and implement QoS management for high-priority and critical applications. Different network applications have different levels of importance and priority. Therefore, policies should be designed to manage and allocate network resources accordingly, ensuring that critical applications receive the necessary bandwidth and QoS to function optimally.

In the context of QoS management, different types of data are transmitted over networks. Some data can be aggregated for hop-by-hop transmission, while others require individual management to ensure end-to-end QoS. Certain data types, such as financial and medical data, are more sensitive and critical, warranting specialized policies to govern their handling and security.

Moreover, modern network services often incorporate location-aware capabilities. Policies with location awareness need to be established to govern these services effectively. For instance, considerations must be made regarding whether users are allowed to query critical databases using public Wi-Fi networks in locations like coffee shops and airport lounges. Policies should outline the appropriate usage and access privileges in such scenarios to maintain the security and integrity of network services.

Benefits of Policies

Network policies implemented on a network will enable automated management and service provisioning, especially in dynamic and evolving environments. More specifically, facilitate various common tasks such as adding devices and users, launching new applications and services, and managing network resources for QoS. Well-defined policies offer several benefits to a network, for example:

- To align the network with business goals and objectives.
- To provide systematic network management for acceptable use, back-up, archiving, failover, and disaster recovery.
- To ensure consistency of network services across the entire infrastructure.
- To establish dependable and verifiable performance.
- To automate network management processes.
- To provide comprehensive security protection to the network.

Tp enhance the protection of data and network resources, implement security policies that restrict users and devices to the minimum required access level while meeting performance requirements. Any violations to the policies will be promptly detected and then mitigated. Consequently, security risks within the network can be significantly reduced.

In terms of network performance, once business goals and objectives are clarified and well-defined, metrics can be developed to quantify what levels of quality the network is able to deliver for various network applications and services. This facilitates continuous monitoring of network performance, thus ensuring that the deployed policies are followed, network QoS requirements are met, and business goals and objectives are fulfilled.

8.3.2 *Types of Network Policies*

There are different types of network policies. With the understanding discussed above about what policies govern, each type of policy is defined to govern a specific part of the network. Table 8.8 shows some typical types of network policies. It is, however, not an exhaustive list of all types of policies. Part of the table is abstracted from Cisco's online reference [10].

8.3.3 *Simple Yet Effective Performance Policies*

There are a variety of network policies that can impact on network performance. But implementing a few simple yet effective performance policies can significantly improve network performance. Enforcing these policies helps alleviate the burden on the network and optimize resource utilization.

Before the discussions of effective performance policies, it is worth reiterating that network resources are shared among users, applications, and devices. Consequently, the network has a finite capacity to handle traffic. Therefore, the actions of network users and their applications can place a significant burden on the network. Excessive utilization of network resources can lead to excessive load, resulting in substantial performance degradation. Therefore, a key point of enforcing effective performance policies is to reduce unnecessary and potentially detrimental network usage.

Table 8.8 Types of network policies

No.	Policies	Description
1	Access and security	These policies are defined to manage authorization and access control for authorized network access and privacy protection, the security architecture of the network, and security environments. They also determine how policies are enforced in network operations
2	Application and QoS	These policies determine relative importance of various applications, how the traffic should be prioritized for each of the applications, and what mechanisms should be activated for specific applications or QoS requirements
3	Traffic routing	These policies control how traffic from certain types of users or applications should be routed. They should be considered together with routing architecture and traffic manipulation
4	IP-based or role-based	These policies guide the use of static IP-based use of network resources, or dynamic role-based access to network resources. They should be considered when IP-mobility is a requirement or if a user works from different locations
5	Back-up and archiving	These policies decide how databases or other critical data are backed up or archived from time to time. They determine whose job and how often the job are backed up or archived, how the back-up or archiving is made, how the back-up or archiving system is tested
6	Failover	These policies specify how the network keeps running when one or more key pieces of the network are down. They should be considered together with network redundant architecture design. The HSRP is an example that can be used for the management of a router failure
7	Disaster recovery	These policies govern how to make the network running as quickly as possible should a disaster happen such as the damage of the building that hold the network due to, e.g., flood or burning down

Keeping this in mind, three straightforward yet effective performance policies can be implemented: network access management, QoS management, and blocking specific websites and applications. They are discussed in the following.

Access control is previously listed in Table 8.8 as the first type of network policies. It not only determines the level of access granted to users within the network but also governs the devices they are authorized to use. Different users logging into the same devices may have different levels of privileges for network access. Also, organizations may permit users to connect their personal devices, such as laptops and smartphones, to the network. By effectively managing access control, unnecessary utilization of network resources can be minimized.

QoS management has been discussed a lot in this chapter. It is previously listed in Table 8.8 as the second type of network policies. When considering QoS, it is important to recognize that the majority of network services are best-effort services, meaning that they are managed to the best of the network capacity. However, certain critical services, such as voice communications, may require dedicated bandwidth allocation through QoS mechanisms. Mission-critical or safety-critical services should be prioritized with high levels of QoS management, ensuring continuous availability of the required bandwidth and other necessary resources.

Blocking websites and applications belongs to the second type of network policies listed in Table 8.8. In cases where critical services suffer from performance degradation due to insufficient bandwidth or limited network resources, a simple yet effective solution is to identify the sources of excessive network bandwidth consumption, such as specific websites or video-related applications. If a particular website consumes an excessive amount of bandwidth, it can be blocked or subjected to traffic shaping measures for users visiting the site. Similarly, video streaming services, known for their bandwidth-intensive nature, can be restricted if there is insufficient bandwidth available for critical services. Implementing these restrictions can help alleviate network congestion and improve the performance of essential and critical services.

8.3.4 Implementing Network Policies

Network policies should be designed, developed, and formally specified prior to their implementation in a network. To develop and implement effective network policies, it is important to have a clear understanding of the business goals and objectives that the network aims to support. The implemented policies must be closely aligned with the identified goals and objectives. Also, policies must be developed in a systematic manner rather than an ad-hoc approach, especially when dealing with large-scale networks. Furthermore, network policies should include provisions for enforcement and continuous improvement over time.

From the technical perspective, Cisco has summarized seven steps for policy implementation: identification, visualization, defining, modelling, activation, extension, and assurance [10]. These seven steps are tabulated in Table 8.9. They provide a structured framework for successfully implementing network policies.

8.4 Differentiated Services (DiffServ)

DiffServ is one of the two standard layer-3 QoS architectural models (the other being IntServ, which will be discussed later). It approaches QoS from the perspective of aggregating traffic based on Per-Hop Behaviors (PHBs). This means that DiffServ does not aim to provide end-to-end QoS support to individual traffic flows. Instead, it processes packets on a hop-by-hop basis within a DiffServ domain.

When packets enter an ingress node of a DiffServ domain, they are classified and marked with a particular per-hop forwarding behavior. They undergo classification, marking, policing, and shaping/dropping at the boundary node before being

Table 8.9 Seven steps of policy implementation [10]

Step	Description
1. Identification	To identify what users, applications, and devices are on the network
2. Visualization	To understand how users, applications, and users communicate over the network
3. Defining	To define policies that permit, deny, or modify certain traffic flows of network communications
4. Modelling	To conduct model-based assessment of the policies
5. Activation	To activate network devices to enforce the developed policies
6. Extension	To scale up the policy enforcement from devices to the entire network or multiple networks
7. Assurance	To make sure the policies work as expected, and if not, refine and fine-tune the policies as needed

forwarded. Network resources are allocated to them based on service provisioning policies, their PHB classification, and conditioning.

This section will begin with discussions of the DiffServ architecture. Subsequently, other aspects related to DiffServ will be covered, such as the DiffServ domain and region, classification and conditioning, PHBs, and DSCP to service class mapping.

8.4.1 *DiffServ Architecture*

DiffServ follows a simple architectural model recommended in the IETF RFC 2475 [8]. A schematic diagram of the DiffServ architecture is shown in Fig. 8.5. The functions associated with DiffServ are discussed below.

In DiffServ, network traffic enters a *DiffServ domain* through an ingress border router. The traffic is classified and possibly conditioned at the border router, and then assigned to different behavior aggregates, e.g., for queuing and forwarding. Each behavior aggregate has a unique DiffServ codepoint, which is part of the IP header as shown previously in Fig. 8.2. Within the DiffServ domain, core routers forward packets according to the PHBs linked with the codepoint, thus achieving DiffServ QoS. The DiffServ traffic leaves a DiffServ domain through an egress border router.

Depending on the direction of the DiffServ traffic flow as indicated by the dashed arrows in Fig. 8.5, a border router in a DiffServ domain may act as an ingress router or an egress router. It connects the DiffServ domain to other domains, which can be either DiffServ domains or non-DiffServ domains.

It is worth mentioning again that DiffServ does not provide end-to-end QoS support for individual flows. Thus, the DiffServ traffic flow shown in Fig. 8.5 is managed on a hop-by-hop basis from the point at which it enters a DiffServ domain to the

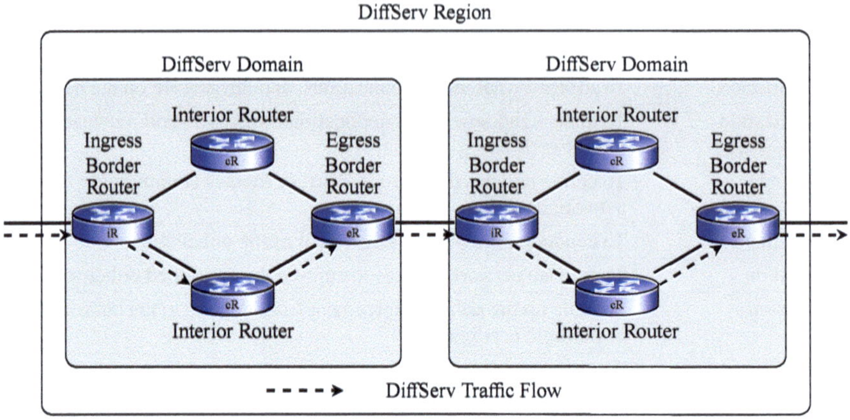

Fig. 8.5 DiffServ architecture

point at which it leaves the same or a different DiffServ domain. Several DiffServ domains can work together to form a *DiffServ region*.

8.4.2 DiffServ Domain and Region

A *DiffServ domain* is a network domain with a contiguous set of DiffServ nodes. It is typically composed of one or more networks under the same network administration. The DiffServ nodes within the same DiffServ domain are governed by a common service provisioning policy and operate with a common set of PHB groups.

Physically, a DiffServ domain consists of boundary nodes and interior nodes. The boundary nodes are located at the well-defined boundary of the DiffServ domain. They interconnect the DiffServ domain with other DiffServ domains and non-DiffServ domains. Residing within the DiffServ domain, the interior nodes are interconnected with boundary nodes and/or other interior nodes within the same DiffServ domain.

A boundary node in a DiffServ domain can be an *Ingress* node if it deals with ingress traffic, or an *Egress* node if it processes egress traffic. It can act as an ingress node for some traffic flows and also an egress node for other traffic flows. In this case, it handles traffic flows in both directions simultaneously. An ingress node classifies and possibly conditions ingress traffic to a specific DiffServ codepoint (Fig. 8.2). Interior nodes within the DiffServ domain forward the traffic according to the behavior indicated in its DiffServ codepoint. Before the traffic leaves the DiffServ domain, an egress node may perform traffic conditioning on the traffic that is being forwarded to a directly connected peering domain.

When two or more contiguous DiffServ domains work together to provide DiffServ QoS support, they form a *DiffServ region*. The simplest DiffServ region configurations require all the DiffServ domains in the same region to follow a common service provisioning policy and support a common set of PHB groups and codepoint mappings. They would function as if they were within the same DiffServ domain. This configuration is feasible when all the DiffServ domains in the DiffServ region are under the same network administration. With such configurations, no traffic conditioning is required between the DiffServ domains.

In general, the DiffServ domains in a DiffServ region may support different PHB groups and different codepoint-PHB mappings. Therefore, for all these domains to work together, the peering domains must each establish a peering SLA that defines a Traffic Conditioning Agreement (TCA). The TCA specifies classifier rules and any corresponding traffic profiles. It also specifies metering, marking, dropping, and shaping rules to be applied to the traffic flows selected by the classifier. In other words, the TCA defines how transit traffic from one domain to another is conditioned at the boundary between the two domains.

8.4.3 Traffic Classification and Conditioning

Traffic flows can be classified in different ways. The IETF RFC 4594 [11] simply classifies traffic flows into two groups: network control traffic and user/subscriber traffic. Each group consists of multiple service classes. The network control traffic group is composed of two service classes:

- Network control class: for routing and network control function.
- Operation, Administration, and Maintenance (OAM) class: for network configuration and management functions.

In the user/subscriber traffic group, ten service classes are defined and recommended in the IETF RFC 4594 [11]. They include an application control class, five media-oriented classes, three data classes, and a best-effort class, as illustrated in Table 8.10.

Aiming to map the received DiffServ traffic to appropriate service classes, the packet classification policy provides guidance to packet classifiers for classifying packets in a traffic stream based on information from the packet header. Two types of packet classifiers have been defined in the IETF RFC 2475 [8, pp. 14–15]: BA (Behavior Aggregate) classifier and MF (Multi-field) classifier. The BA classifier classifies packets based solely on the DiffServ codepoint, while the MF classifier considers additional information from the IP header or the incoming interface. Traffic classifiers at the ingress of the DiffServ domain may choose to honor, ignore, or remark the DiffServ markings of incoming packets to achieve better QoS control within the network's capacity.

Traffic in each service class can undergo further conditioning by subjecting it to rate limiters, traffic policers, or shapers. A traffic conditioner consists of several elements, such as a meter, marker, shaper, and dropper. Figure 8.6 provides a logical view of a packet classifier and traffic conditioner.

Table 8.10 Ten user/subscriber service classes defined and recommended in RFC 4594 [11, p. 15]

Application category	Service class	QoS rating	Flow behavior	Signaled
Application control	Signaling	Responsive	Inelastic	–
Media-oriented	Telephony	Interactive	Inelastic	Yes
	Real-time interactive			
	Multimedia conferencing	Interactive	Rate Adaptive	
	Broadcast video	Responsive	Inelastic	
	Multimedia streaming	Timely	Elastic	
Data	Low-latency data	Responsive	Elastic	No
	High-throughput data	Timely		
	Low-priority data	Non-critical		
Best effort	Standard	Non-critical	Not specified	

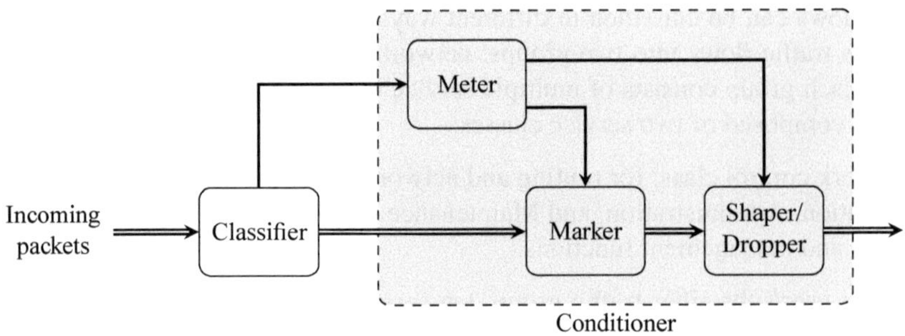

Fig. 8.6 A logical view of DiffServ classifier and conditioner [8, p. 16]

Each element in the conditioner is dedicated to specific functions. A traffic meter measures the properties of packets selected by a classifier and then provides the measured packet information to the packet marker and shaper/dropper, which compare the measurements against a traffic profile specified in a TCA. The traffic marker sets the DS field of a packet to a specific codepoint, thereby associating the marked packet with a particular DiffServ behavior aggregate. With the use of a finite-size buffer, a shaper queues some or all of the packets in a traffic stream to ensure compliance with a specified traffic profile. Conversely, a dropper discards some or all of the packets in a traffic stream in order to bring the stream into compliance with a traffic profile.

8.4.4 Per-hop Behavior (PHB)

DiffServ QoS is managed based on PHBs. In the IETF RFC 2475 [8, pp. 14–15], a PHB is defined as a description of the externally observable forwarding behavior of a DiffServ node applied to a particular DiffServ behavior aggregate. The PHB of a packet is determined by the DS field of the IP header, as previously demonstrated in Fig. 8.2. Theoretically, the six-bit DSCP field has a space of 64 DSCP codepoints. While DiffServ RFCs have recommended certain codepoint encodings, these recommendations are not mandatory. Therefore, network administrators have the flexibility to use the 64 DSCP values.

However, in practice, certain well-defined PHBs are commonly used in most networks. For example, the three pools of DSCP codepoints mentioned in Table 8.5 are generally followed in DSCP configurations. Similarly, the mapping of DSCP codepoints with the pattern 'xxx000' to eight IP precedence values adheres to the RFC 2474 recommendation, as shown in Table 8.6.

More generally, there are four commonly defined PHBs, or PHB groups. They are:

- Default Forwarding (DF) PHB, which is typically associated with best-effort traffic of the CS0 class (Table 8.6, RFC 2474 [2]);
- Expedited Forwarding (EF) PHB (defined in RFC 3246 [12]), which is dedicated to low-loss and low-latency traffic. Within the EF PHB group, RFC 5865 [13] has proposed to add a Voice-Admit class with the DSCP codepoint '101100' (44), which is in parallel with the existing EF codepoint '101110' (46).
- Assured Forwarding (AF) PHB (defined in RFC 2597 [14]), which requires the assurance of delivery under prescribed conditions such as delay and bandwidth requirements; and
- Class Selector PHBs: which aim to maintain backward compatibility with the outdated IP precedence.

In addition, there are some recent developments of new PHBs or DSCP codepoint settings. For example, the IETF RFC 8622 [15] has specified a lower-effort PHB with the recommended DSCP codepoint '000001'.

Within the AF PHB, four classes of DSCP encodings have been defined in the IETF RFC 2597 [14]. They are tabulated in Table 8.11. In all service classes listed in this table, rate-based queuing is adopted.

8.4.5 DSCP to Service Class Mapping

While PHBs specify traffic behaviors or requirements for DiffServ QoS management, they are not uniquely linked to individual applications. Actually, DiffServ does not aim, and it is not possible, to assign a unique DSCP codepoint to every possible application because there are only up to 64 codepoint values. Therefore, DiffServe manages QoS in terms of service classes. Traffic flows from applications belonging to

Table 8.11 Assured Forwarding behavior group defined in RFC 2597 [14]

Class	Drop probability		
	Low	Medium	High
1	AF11: 001010 (10)	AF12: 001100 (12)	AF13: 001110 (14)
2	AF21: 010010 (18)	AF22: 010100 (20)	AF23: 010110 (22)
3	AF31: 011010 (26)	AF32: 011100 (28)	AF33: 011110 (30)
4	AF41: 100010 (34)	AF42: 100100 (36)	AF43: 100110 (38)

the same service class are aggregated on the hop-by-hop basis for packet forwarding. Therefore, it is important to establish a DSCP to service class mapping.

The IETF RFC 4594 [11] has provided detailed recommendations for the use and settings of DSCP codepoints in various service classes. These recommendations are are summarized in Table 8.12. The Telephony EF (46) service class uses priority queuing in traffic processing. All other service classes listed in Table 8.12 use rate-based queuing. Some of these service classes also incorporate AQM.

8.5 Integrated Services IntServ

IntServ is an alternative layer-3 QoS architectural model, originally defined in the IETF RFC 1633 [7]. It operates on a different principle compared to DiffServ. While DiffServ focuses on QoS through traffic aggregation using PHBs, IntServ takes a distinct approach by providing QoS support for individual end-to-end traffic flows. It accomplishes this by defining values and mechanisms to allocate resources along the entire path of the flows, ensuring real-time QoS guarantees. Consequently, IntServ is considered a fine-grained QoS model, while DiffServ is categorized as coarse-grained QoS. This section will discuss the IntServ architecture, flow specifications in IntServ, the Resource Reservation Protocol (RSVP) protocol employed in IntServ, and other related topics concerning IntServ.

8.5.1 Assumptions for IntServ Architecture

IntServ is designed to manage end-to-end QoS for individual traffic flows generated by various applications, particularly real-time applications. When the IntServ architecture is defined in the IETF RFC 1633 [7], several assumptions are made with justifications. The first assumption is that network resources, such as bandwidth, need to be explicitly managed to meet application requirements. This necessitates explicit "resource reservation" and "admission control" for individual traffic flows. This assumption is justified by the fact that real-time services generally require some

Table 8.12 DSCP to service class mapping (sr+bs: single rate with burst size token bucket policer) recommended in RFC 4594 [11, pp. 19–20]

Service class	DSCP name	DSCP value	Conditioning at DiffServ edge	PHB used	AQM
Network control	CS6	110000(48)	(*)	RFC2474	✓
Telephony	EF	101110(46)	Police using sr+bs	RFC3246	–
Signaling	CS5	101000(40)	Police using sr+bs	RFC2474	–
Multimedia conferencing	AF41, AF42, AF43	100010(34) 100100(36) 100110(38)	Using two-rate, three-color marker (e.g., RFC2698)	RFC2597	per DSCP
Real-time interactive	CS4	100000(32)	Police using sr+bs	RFC2474	–
Multimedia streaming	AF31, AF32, AF33	011010(26) 011100(28) 011110(30)	Using two-rate, three-color marker (e.g., RFC2698)	RFC2597	per DSCP
Broadcast video	CS3	011000(24)	Police using sr+bs	RFC2474	–
Low-latency data	AF21, AF22, AF23	010010(18) 010100(20) 010110(22)	Using two-rate, three-color marker (e.g., RFC2697)	RFC2597	per DSCP
OAM	CS2	010000(16)	Police using sr+bs	RFC2474	✓
High-throughput data	AF11, AF12, AF13	001010(10) 001100(12) 001110(14)	Using two-rate, three-color marker (e.g., RFC2698)	RFC2597	per DSCP
Standard	DF (CS0)	000000 (0)	–	RFC2474	✓
Low-priority data	CS1	001000 (8)	–	RFC3662	✓

* Users are not permitted to access to the Network Control service classes
CS6 marked packets from untrusted sources should be dropped/remarked
Otherwise, CS6 marked packets should be policed, e.g., using sr+bs

form of service guarantees. Without dedicated resource reservation, these guarantees cannot be achieved. The nature and requirements of the applications determine whether the service guarantee is interpreted as absolute or statistical, strict or approximate. For example, general real-time video streaming over UDP may be adequately served with a statistical performance guarantee, such as an average datagram loss rate of 0.1%. However, streaming remote surgery data may demand an absolute and strict performance guarantee, such as a latency less than 100 ms.

By reinforcing the importance of the service guarantee, the inappropriateness of a few arguments are discussed in RFC 1633 [7, pp. 4–5], such as:

- "Bandwidth will be infinite." This argument is found to be unrealistic and thus unacceptable. The available bandwidth of a specific link between any two nodes is always shared by multiple users or applications in general. Unless it is well managed, e.g., through allocation and reservation, no one could claim that the required amount of bandwidth is always sufficient anytime we need to serve real-time applications.
- "Simple priority is sufficient." A priority mechanism will help improve the performance of real-time services that are given higher priority levels. It is, however, not a service model, and thus does not provide service guarantee. If there are too many high-priority tasks running concurrently, traffic congestion may occur, causing performance degradation for some real-time services.
- "Applications can adapt." This might be true for some applications, such as adaptive real-time applications. However, it is not true in general for most real-time applications. Thus, this is also an unacceptable assumption.

The conclusion from these discussions is that resource reservation is essential to provide performance guarantee in QoS management for individual end-to-end traffic flows. This would require the routers involved in IntServ QoS management to be capable of reserving resources. This in turn requires flow-specific state in the routers. Thus, it is critical for the routers to manage the flow state in order to be able to reserve resources.

The second assumption made in defining IntServ is that it is desirable to use the Internet as a common infrastructure to support both real-time and non-real-time communication. While it is technically possible to build a separate infrastructure solely for real-time network services, this approach is not a general solution. The Internet is inherently shared by multiple users and applications. It should be used to provide a general solution to both real-time and non-real-time services.

With the Internet as a common infrastructure, it is further assumed that a unified protocol stack model is adopted for both real-time and non-real-time network services. Thus, the layer-3 IP protocol that is popularly used today is considered for real-time data communication. Obviously, it is an option to add a new real-time protocol at layer 3 to serve real-time data. But this would add additional complexity in network design, operation, and management.

A single service model for the Internet would be beneficial for the management of real-time and non-real-time network services. However, it does not necessarily mean a single implementation for packet scheduling and admission control. The IETF RFC

1633 [7] has introduced a reference implementation framework, which can be used to guide different designs.

It is understood from the above discussions and considerations that in order to implement IntServ, it is essential for the routers involved to know the flow state and also have a mechanism to reserve resources. Later, it will be shown that there is also a requirement for signaling across the network.

8.5.2 IntServ Service Model

An IntServ service model is specified in the IETF RFC 1633 [7]. It is designed with the understanding that maintaining compatibility necessitates a relatively stable service interface, regardless of the evolution of network technologies and applications. This model proposes a core set of services that relate most directly to the time-of-delivery of data packets. Services for routing, security, and others that are not directly relevant to data delivery are not included within this service model.

Overall, the service model has described five main aspects relevant to IntServ QoS [7, pp. 11–19]:

(1) QoS requirements and a set of services,
(2) Service requirements and service models,
(3) Packet dropping,
(4) Usage feedback, and
(5) Reservation model.

Let us briefly discuss these five aspects in the following.

A Core Set of Services

Focusing on the time-of-delivery of packets, the core service model of IntServ places its emphasis on per-packet delay, particularly the upper and lower bounds of delay. As a result, network applications are classified into two categories: real-time applications, where late-arriving packets become useless, and elastic applications, which can wait for data to arrive indefinitely. To cater to these application classes, three types of services are specified: fixed-delay real-time service, predictable real-time service, and best-effort service.

There are many examples of real-time applications, including both critical and non-critical ones. For example, networked industrial control systems, air traffic control, and smart grid are safety-critical and mission-critical real-time applications. Playback applications also fall under the real-time category, although their criticality may vary. For playback applications, delay and jitter are two main factors that affect the application performance. For intolerant applications that cannot tolerate delay beyond a specific threshold, a network service with a fixed offset delay must be employed in QoS management.

There are also tolerant applications that can tolerate a certain level of delay variations even beyond the maximum delay bound. For example, missing a few frames in video streaming may not be an issue, but missing too many frames will be a problem. For such tolerant applications, predictive service could be used, which provides a fairly reliable, but not 100% reliable, delay bound.

Elastic applications always wait for the arrival of packets. But a significant increase in delay can result in performance degradation. These applications typically process received packets immediately rather than buffering them for later use. Examples of elastic applications include FTP and email applications. For such applications, the appropriate service model is the 'as-soon-as-possible' (ASAP) service, commonly referred to as the best-effort service.

Resource-Sharing Service Models

While QoS services primarily focus on delay, the sharing of network resources for QoS is centered around the allocation of aggregate bandwidth on individual links. This is referred to as the link-sharing service model.

There are different types of link-sharing approaches, including multi-entity link sharing, multi-protocol link sharing, and multi-service sharing. In multi-entity link sharing, a link is shared among multiple subscribers, organizations, departments, or users. Multi-protocol link sharing allows for fair sharing of a link among multiple protocols, preventing one protocol family from overwhelming the link and excluding other protocol families. In the case of multi-service link sharing, various classes of applications should have shared access to the link, even when coexisting with real-time applications.

Admission control plays an important role in ensuring that the requirements for link sharing are met. It functions similarly to admission control for meeting real-time service commitments. Admission control is a fundamental concept in network traffic control, serving to regulate the admission of traffic into the network based on available resources and QoS constraints.

Packet Dropping

In many applications, not all packets have equal importance. Some packets carry higher priority or significance compared to others. In situations where the network becomes overloaded or there is a risk of failing to meet certain service commitments, it may be necessary to prioritize the handling of packets based on their importance. As a solution, a *preemptable* packet service is proposed to address this requirement. This service allows for the preemption or dropping of packets that are considered less important. Consequently, the queuing delays experienced by other packets that are considered more important can be effectively reduced.

Usage Feedback

The usage feedback service model is also proposed from the network management perspective, rather than the technical perspective. It aims to avoid the abuse of network resources. For example, it can be used to prevent a user from consuming the entire bandwidth of a link in a link-sharing application, ensuring fair resource sharing.

Reservation Model

The reservation model describes the negotiation process for QoS with the network. Various methods exist for negotiating QoS between the application and the network. The IntServ reservation model incorporates two fundamental concepts: receiver initiation, and NONE or ALL service.

In IntServ, the QoS RSVP request is initiated by the receiver host, rather than the sender host. This is well justified in the IETF RFC 1633 [7] and RFC 2205 [16]. A key fact is that a receiver knows exactly what it wants to, or can, receive. Receiver initiation of resource reservation handles heterogeneous receivers more easily than sender initiation. Each receiver simply requests for a resource reservation that is most appropriate to itself. For a receiver to learn the characteristics of the sender's data flow, a high-level mechanism could be used, which is called "out of band" in the RFC 1633 [7, p. 27]. Then, the receiver will generate its own flow specs for its reservation request.

IntServ does not support partial QoS allocation if a router lacks the necessary resources for the requested QoS. The IntServ RSVP protocol makes a binary decision, either accepting or rejecting the requested IntServ QoS. This implies that it can provide either NONE or ALL of the requested resources.

8.5.3 *Overall Architecture*

Before we discuss the IntServ architecture, it is worth emphasizing that the fundamental service model of the Internet, characterized by the best-effort delivery of IP traffic, has remained unchanged for many decades. However, numerous components and mechanisms have been developed to augment the underlying IP service. IntServ is one such addition that intends to supplement, rather than replace, the basic IP service.

Logical View of the Architecture

The IntServ architecture, as proposed in the IETF RFC 1633 [7], consists of two main elements: an extended service model and a reference implementation framework. The service model defines the externally observable behavior for both guaranteed and predictive real-time traffic over shared links, while the reference implementation framework provides guidance on how to implement the service model. As discussed previously, different designs are possible for implementing the defined service model.

To provide a clearer understanding of these architecture elements, let us examine a high-level view of the overall IntServ architecture depicted in Fig. 8.7 for an individual end-to-end traffic flow. In this figure, IntServ is applied to a specific real-time flow between hosts H1 and H2, traversing routers R1, R2, R3, and R4. To ensure resource reservation agreement among all these routers for the flow, certain essential requirements must be met:

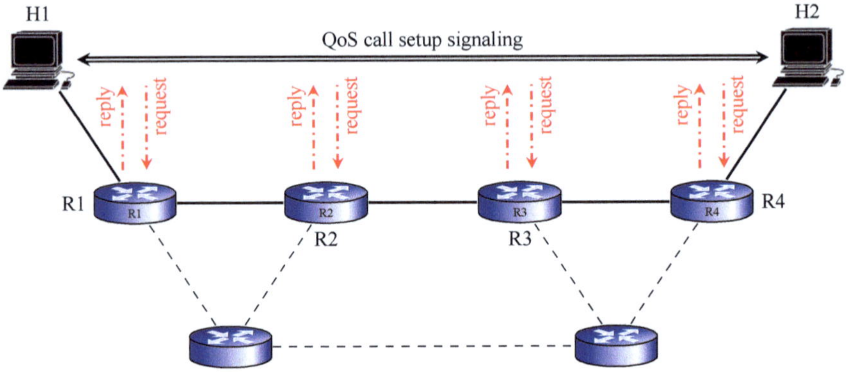

Fig. 8.7 IntServ architecture for end-to-end QoS

- Each router along the path implements IntServ.
- All these routers must possess knowledge about the nature of the traffic flow and be capable of reserving and allocating resources, such as bandwidth, to the flow.
- A signaling system is necessary to communicate flow requirements with all these routers.
- A mechanism is required for establishing and tearing down resource allocations.

Moreover, when IntServ is applied to traffic flows from multiple applications, each application must make an individual IntServ reservation for its own end-to-end flow.

In Fig. 8.7, there is a negotiation process for resource reservation with each of the routers R1 through R4 along the path for the individual end-to-end flow from H1 to H2. In this process, known as 'call setup' or 'call admission', a request of resource reservation is sent to a router. If the router has sufficient resource to reserve for the flow, it responds with a positive 'yes' to the request; otherwise, it replies with a negative 'no'. IntServ QoS can only be applied to the individual end-to-end flow if all the routers along the path respond with 'yes'. If any router responds with 'no', IntServ QoS will not be established for the flow. Therefore, the outcome of the IntServ negotiation is either ALL for providing the requested service or NONE for no service at all. No partial IntServ QoS will be provided.

The steps involved in the call setup process include: (1) flow specs, (2) signaling for call setup, and (3) per-element call admission. Let us now examine these steps below in more detail.

Flow Specs

Flow specs characterize traffic and specify the desired QoS. In order for a router to determine whether or not it has sufficient resources for its participation in an IntServ session, the session must first declare what its QoS requirement is and what the characterization of the traffic has, for which a QoS guarantee is requested. For a flow spec, this is handled from two perspectives: Traffic SPECification (TSPEC) and Request SPECification (RSPEC):

- TSPEC tells what the traffic looks like. In IntServ, every packet that is being sent requires a token. A token bucket is designed, which slowly fills up with tokens. Thus, the rate of token arrivals dictates the average rate of traffic flow. The depth of the token bucket indicates how bursty the traffic is allowed to be. TSPEC includes token bucket algorithm parameters. Typically, it simply specifies the token rate and bucket depth. For example, for a video streaming flow with 50 frames per second and 10 packets per frame, the TSPEC may simply specify a token rate of 500 Hz and a bucket depth of 10.
- RSPEC tells what guarantees the traffic needs. In the simplest case, it can be a normal Internet 'best-effort' service, for which no reservation of resources is needed. However, in general, a 'guaranteed' setting is given for an absolutely bounded service. For example, the latency or packet drop rate should be less than a desired threshold.

In IntServ QoS applications, depending on the service requested, the specific forms of TSPEC and RSPEC may vary. Technical specifications of TSPEC and RSPEC are defined in part in the IETF RFC 2210 [17] and RFC 2215 [18].

Signaling for Call Setup

The TSPEC and RSPEC of an IntServ session must be carried to each of the routers involved in the session. This is achieved through a signaling system for IntServ QoS. A specific protocol, RSVP, has been developed for call setup signaling. The IETF RFC 2210 [17] specifies the use of RSVP with the IntServ architecture. The details of RSVP and its integration with IntServ will be discussed later.

Per-Element Call Admission

Once a router receives the TSPEC and RSPEC for an IntServ QoS guarantee, it checks whether it has sufficient resources to admit the call. The decision for call admission is based on the TSPEC, RSPEC, the resources already committed to ongoing sessions, and the available resources for reservation and allocation. If the router has enough resources to accommodate the request, the decision is positive and the call is admitted. Otherwise, the decision is negative and the call cannot be served.

As mentioned previously, IntServ QoS requires all routers involved along the path of the individual end-to-end traffic flow to satisfy the request. If any of the routers involved is unable to satisfy the request, the IntServ QoS guarantee cannot be provided for the flow.

8.5.4 Controlled-Load Service and Guaranteed QoS

The IntServ architecture defines two major classes of services: controlled-load service and guaranteed service. The controlled-load service is specified in the IETF RFC 2211 [19], while the guaranteed service is defined in the IETF RFC 2212 [20]. These two classes of services are managed differently in IntServ QoS management.

Controlled-Load Service

The controlled-load service, specified in the IETF RFC 2211 [19], provides client traffic flows with "a quality of service closely approximating the QoS that same flow would receive from an unloaded network element". This service uses admission control to ensure that the service is received even when the network element is overloaded. In other words, the packets of a flow from a controlled-load service will pass through the router with a very low drop rate and close-to-zero queuing delay regardless of the traffic load on the router. In real-world networking, the controlled-load service targets real-time multimedia applications.

It is interesting to note that no performance guarantees are quantitatively specified in the controlled-load service. Thus, there are no specifications regarding the approximation of the QoS of an unloaded network element, and the drop rate that is considered to be very low.

Guaranteed QoS

The guaranteed service is defined in the IETF RFC 2212 [20]. It provides "firm (mathematically provable) bounds" on the queuing delays that a packet will experience in a router. By using the guaranteed service, it becomes possible to provide a service that guarantees both delay and bandwidth for a flow. RFC 2212 has discussed three characteristics of the guaranteed service specifications:

- Achieving a guaranteed resource reservation necessitates a setup mechanism. However, RFC 2212 [20] does not specify such a mechanism or a method for identifying flows intentionally, making the service model independent of its implementation. In real-world networking, a call setup mechanism and a signaling system can be adopted, such as flow specs and RSVP (e.g., RFC 2210 [17] and RFC 2215 [18]), as discussed previously.
- To ensure a bounded delay, it is essential that each service element along the path supports guaranteed service or adequately mimics it.
- While applications typically lack control over end-to-end delay because each service element along the path introduces delays that are generally not under the applications' control, the guaranteed service provides considerable control over these delays.

Network delay consists of two parts: a fixed delay (such as transmission delay) and a variable delay (such as queuing delay). The fixed delay is not influenced by the guaranteed service. Instead, it is determined by the setup mechanism, which may choose a specific path from multiple options. On the other hand, the queuing delay is controlled by the guaranteed service because it primarily depends on two parameters: the token bucket and the requested data rate of the application. If the delay exceeds the desired threshold, the application can modify these parameters in a predictable manner to achieve a shorter delay. Examples demonstrating this behavior are provided in the IETF RFC 2212 [20].

8.5.5 Reference Implementation Framework

The IETF RFC 1633 [7] has presented a reference implementation framework to realize the IntServ service model. The framework consists of four basic components:

(1) A packet scheduler,
(2) An admission control routine,
(3) A packet classifier, and
(4) A reservation setup protocol.

The implemented system is placed on each router that participates in IntServ QoS. Figure 8.8 shows the architecture of the reference implementation framework in a router that is interconnected with other routers and participates in an IntServ session [16, p. 5]. Also, IntServ QoS is subject to routing and admission policy control, which are also shown in the figure as part of IntServ's environment in the router.

The functions of the four components of the IntServ architecture shown in Fig. 8.8 can be understood from different perspectives. From the traffic management perspective, the reservation setup protocol component is responsible for IntServ call setup, while the scheduler, admission control, and classifier components work together for IntServ traffic control in the router. From the functional perspective, the classifier and scheduler components form the packet forwarding path, while the reservation setup and admission control components, along with routing and admission policy control, form the background code.

The main functions of the four components of the IntServ architecture depicted in Fig. 8.8 are briefly described as follows. The reservation setup component serves as

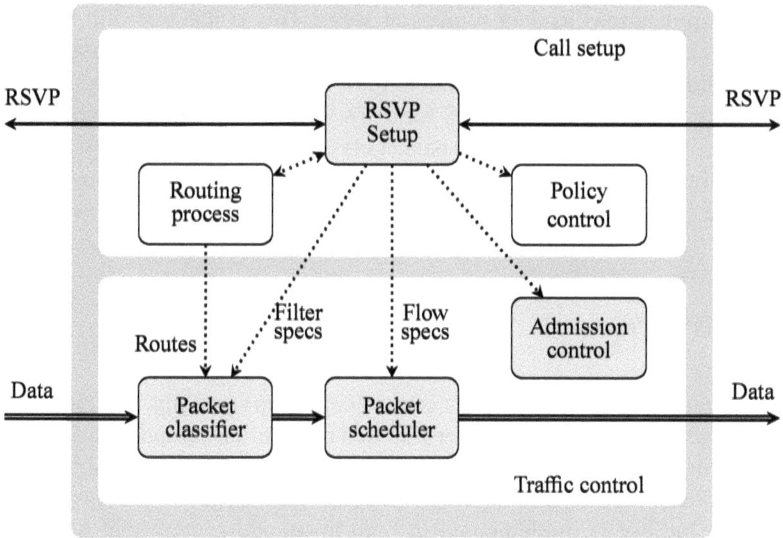

Fig. 8.8 IntServ architecture in a router

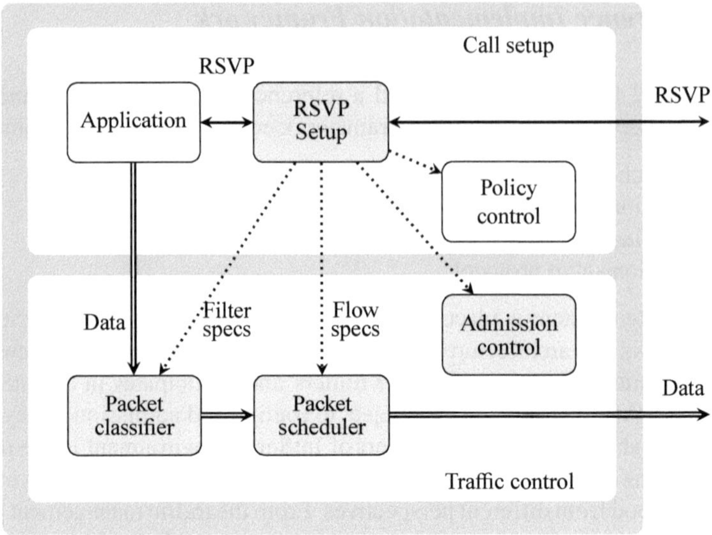

Fig. 8.9 IntServ architecture in a host, which generates data in this example

a signaling system that communicates with other routers or hosts to establish IntServ calls. The packet classifier maps incoming packets into specific classes for traffic control purposes. The packet scheduler manages the forwarding of different packet streams using a set of queues. The admission control component determines whether a new flow is granted the requested QoS.

The implementation framework for a host is similar to that for a router, with the addition of applications as data sources or sinks. This is illustrated in Fig. 8.9 [16, p. 5]. In the case of hosts, data is generated or consumed by applications rather than being forwarded. To enable communication with routers for resource reservation, an application requiring IntServ QoS for a flow must be capable of invoking a local reservation setup agent. However, a packet classifier may not be necessary because the application has knowledge of the data flow, allowing for the assignment of packet classes to be specified through local I/O control associated with the flow.

8.5.6 RSVP Protocol

RSVP is a transport-layer protocol for IntServ call setup signaling. Operating on IPv4 or IPv6 networks, it supports receiver-initiated reservation setup for individual end-to-end data flows of multicast or unicast type. As a signaling protocol, it transports control data only. Application data will be transferred through other protocols such as UDP. As IntServ is implemented in both hosts and routers that participate in IntServ QoS management, RSVP operates on these hosts and routers for IntServ QoS signaling.

RSVP Specifications and Extensions

RSVP was developed and specified through a series of RFCs. Its inclusion as part of the IntServ reference implementation framework was described as early as in 1994 in the IETF RFC 1633 [7]. However, the detailed functional specification of RSVP was described three years later in 1997 in the IETF RFC 2205 [16]. Subsequently, additional RFCs were published to extend or support the RSVP protocol. Notably, as mentioned earlier, RFC 2210 [17] defined the use of two types of services: controlled-load service (RFC 2211 [19]) and guaranteed service (RFC 2212 [20]). Some extensions to the RSVP protocol and related RFCs are tabulated in Table 8.13.

Flow Specs and Filter Specs

It is seen from the IntServ architecture in Fig. 8.8 that RSVP communicates with the classifier and scheduler through filter specs and flow specs, respectively. Filter specs and flow specs are two key concepts in the RSVP reservation model.

RSVP initiates reservation call setup for a flow. It identifies the specific QoS of the flow, in addition to other information of the flow, such as the destination address, the protocol identifier, and optionally the destination port number. The QoS information is specified in a flowspec. RSVP conveys the flowspec from the application to the involved hosts and routers for IntServ QoS management along the flow's path. A

Table 8.13 IntServ RSVP and related RFCs

IntServ RSVP or RSVP support/extension	RFC	Year
Initial description of RSVP as part of an IntServ reference implementation framework	RFC 1633	Jun. 1994
RSVP Version 1 functional specification	RFC 2205	Sep. 1997
The use of RSVP for controlled-load service and guaranteed service	RFC 2210	Sep. 1997
Controlled-load service in IntServ	RFC 2211	Sep. 1997
Guaranteed service in IntServ	RFC 2212	Sep. 1997
RSVP extensions for policy control	RFC 2750	Jan. 2000
RSVP extensions for LSP Tunnels	RFC 3209	Dec. 2001
Generalized Multi-Protocol Label Switching (GMPLS) Signaling Resource reserVation-Traffic Engineering (RSVP-TE) Extensions	RFC 3473	Jan. 2003
Current best practices and procedures for modifying RSVP	RFC 3936	Oct. 2004
RSVP extension for the reduction, instead of tear-down, of bandwidth of a reservation flow	RFC 4495	May 2006
Node-ID-based RSVP hello: A clarification statement	RFC 4558	Jun. 2006
RSVP extensions for path-triggered RSVP receiver proxy	RFC 5946	Oct. 2010
IPv6 flow label specification	RFC 6437	Nov. 2011
RSVP ASSOCIATION object extensions	RFC 6780	Oct. 2012

comprehensive explanation of flow specs has been provided previously in Sect. 8.5.3 when the overall IntServ architecture is discussed. Specifically, the TSPEC within a flowspec describes the characteristics of the traffic flow, while the RSPEC specifies the desired performance guarantees for the flow. RSVP sends a request containing the TSPEC and RSPEC to a router or host, and the response is a binary decision of either accepting or rejecting the request. As previously mentioned, IntServ QoS management does not provide partial performance guarantees.

On the other hand, filter specs specify how data packets under IntServ QoS management are classified for resource reservation. Essentially, a filterspec determines the reservation style for the data packets that receive QoS as defined by the flowspec. Currently, three RSVP reservation styles are defined [16, pp. 11–14]:

(1) Fixed filter. A fixed filter makes a distinct reservation for a specific flow initiated by an explicit sender. The reservation created by the fixed filter is exclusive to that sender and is not shared with any other sender.
(2) Wildcard filter. A wildcard filter makes a single reservation that is shared among all senders in a session. While it does not explicitly specify individual traffic flows, it serves multiple flows within the session, with all of them sharing the allocated resources. The wildcard filter is particularly suitable for applications like video conferencing, where only one or a few senders (speakers) are active at any given time. In such cases, a single reservation is sufficient to cater to all participants.
(3) Shared explicit filter. A shared explicit filter combines elements of both fixed and wildcard filters. It enables the receiving application to establish a shared reservation explicitly among selected senders. Unlike the wildcard filter that allows all participants to share the reservation, the shared explicit filter restricts the sharing to specifically chosen participants.

RSVP Messages

There are two fundamental types of RSVP messages: path messages (*path*) and reservation messages (*Resv*).

Specifically, each RSVP sender host transmits RSVP path messages downstream and hop by hop towards the receiver host along the path determined by the routing protocol. A path message contains the following information in addition to previous hop address:

- Sender template: It describes the format of the data packets that the sender will send out. The sender template is in the form of a *filterspec*, which is used to communicate with the IntServ classifier.
- Sender TSPEC: It describes the traffic characteristics of the data flow initiated by the sender.
- Adspec: It is a package of "One Pass With Advertising (OPWA)" information as described in the IETF RFC 2210 [17]).

The receiver host sends Resv messages upstream towards the sender to request RSVP reservation. The Resv messages include the following data objects:

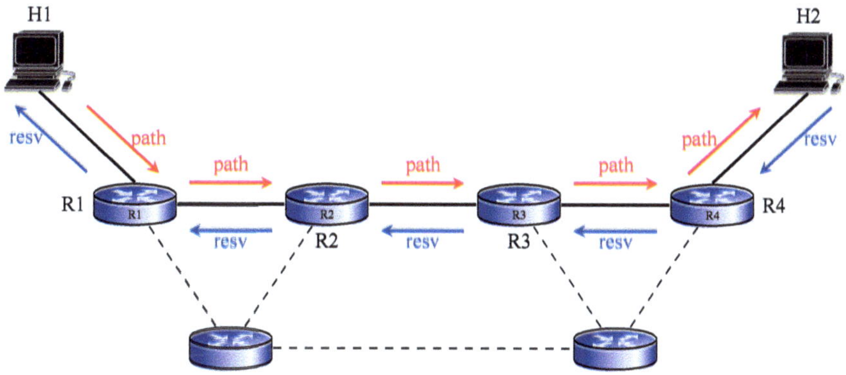

Fig. 8.10 RSVP operation from end to end

- Flowspec: It identifies the resources needed by the flow and specifies the desired QoS parameters.
- Filterspec: It specifies the packets that will receive the requested QoS defined in the flowspec.

Resv messages are forwarded along the exact reverse of the path that the data packets will use. As mentioned earlier, the path is selected by the routing protocol in use.

RSVP Operation

Following the overall IntServ architecture depicted in Fig. 8.7 for end-to-end Quality of Service (QoS), the RSVP operation is illustrated in Fig. 8.10. It is briefly summarized in the following four main steps.

Step1. An RSVP host that needs to send a data flow with specific QoS (i.e., H1 in Fig. 8.10) sends an RSVP *path* message every 30 s to the destination (H2 in Fig. 8.10). The path message will travel along a path selected by the routing protocol towards the destination. Eventually, it will arrive at the destination receiver host H2, and a path from the sender host H1 to the receiver host H2 is established. It is worth mentioning that path messages do not initiate resource reservation.

If a router receives a *path* message but does not understand RSVP, then: 1) it will forward the message without interpreting its contents, and 2) it will not participate in the IntServ RSVP resource reservation.

Step2. A *Resv* message that carries the *flowspec* and *filterspec* information is then sent by the receiver host (i.e., H2 in Fig. 8.10) along the reverse path towards the sender host (i.e., H1 in Fig. 8.10) to request resource reservation. The reservation request specifies the QoS bounds, such as bandwidth and delay, and other requirements, which are defined in the flowspec and filterspec. It is worth reinforcing that a *Resv* message initiates the RSVP resource reservation.

Step3. When a router along the reverse path from the receiver host H2 towards the sender host H1 receives the *Resv* message, the following actions occur:

- If the router is unable to reserve the requested resources, it denies the request and uses RSVP to send a negative reply back to the receiver host H2. The process of negotiation for IntServ QoS fails, and the RSVP call setup session terminates. This means that no IntServ QoS can be provided through this path for the traffic flow. Go back to Step 1.
- Otherwise, if the router is able to reserve the requested resources, it agrees to the request and forwards the message to the next hop. Continue to the next step.

Step 4. If all routers along the path agree to honor the request, RSVP returns a positive reply, and the IntServ RSVP resource reservation is successfully made for the IntServ QoS of the flow.

RSVP Soft State and Interaction with Routing

For general resource reservation, multicast distribution is achieved by using flow-specific state in each of the routers along the path. There are two basic styles of reservation setup:

- The connection-oriented 'hard state' style.
- The connectionless 'soft state' style.

In the hard-state approach, flow-specific state is created and deleted in a deterministic manner through cooperation among the routers along the path. When a host requests an RSVP session, it is the responsibility of the network to create and later delete the necessary state. The hard-state approach demands high reliability of the reservation session.

To mitigate this demand, RSVP adopts the soft-state approach. In the soft-state approach, the reservation state is treated as cached information that is installed and regularly refreshed by the end hosts. Any unused state will be timed out by the routers along the path, simplifying system design and management.

In RSVP operation, RSVP interacts with routing for the installation of flow state along the path of the traffic flow for which the reservation is requested. A basic idea behind the RSVP deign is to enable the coexistence of RSVP with exiting routing protocols, thus eliminating the need to modify the routing protocols when deploying RSVP. In general, the following routing issues have been considered in the development of RSVP setup [7, pp. 28–30]:

(1) Find a route that understands and supports resource reservation.
(2) Find a route with sufficient available resource for the new flow for which a reservation is being requested.
(3) Adapt to a route failure by setting up reservation along a new path.
(4) Adapt to a route change due to reasons other than failure. In this case, a functional route can be 'pinned' to provide consistent support for RSVP reservation. The routing protocol will not change a pinned route as long as it remains functional.

8.6 Service-Level Agreements (SLAs)

An SLA is a formal contract between a service provider and a customer about the service standards that the provider is obligated to meet and the level of accountability if those standards are not met. The customer can be an individual user or an organization, while the service provider can be a traditional network service provider, an ISP, or a cloud service provider. SLAs are now also applied to enterprise network environments to offer not only traditional network infrastructure but also a range of network services to their users.

However, in the network community, the concept of SLAs is sometimes misused or misunderstood for various reasons. For example, SLAs are extensively discussed in the original specifications of the DiffServ architecture in the IETF RFC 2475 [8]. But RFC 2475 primarily addresses SLAs from the technical perspective. As the work on DiffServ progresses, it becomes evident that an 'agreement' implies not only technical aspects but also pricing, contractual, and other business-related considerations. Consequently, in the DiffServ context, the technical aspects of SLAs described in RFC 2475 are further refined and made more restrictive in the IETF RFC 3260 [3]. As a result, new terminology is introduced to describe the elements of service and traffic conditioning in DiffServ:

- A Service Level Specification (SLS) refers to a set of parameters and their values that define the service offered to a traffic stream by a DiffServ domain.
- A Traffic Conditioning Specification (TCS) refers to a set of parameters and their values that define a set of classifier rules and a traffic profile. It is an integral part of an SLS.

Three Types of SLAs

There are three types or levels of SLAs: customer-level SLAs, service-level SLAs, and multilevel SLAs. Let us explain each type below:

- Customer-level SLA: This type of SLA is an agreement between a service provider and its external customers. It focuses on the services provided to external customers and is sometimes referred to as an external SLA. A customer-level SLA typically encompasses all the services that the customer uses. For example, an SLA between an ISP and a large organization may cover services such as web browsing, FTP, email, and VoIP. It establishes the quality of service standards and the corresponding responsibilities of the service provider.
  In contrast to external SLAs, there are also internal SLAs, which are agreements between an organization and its internal customers. Although internal SLAs are not as stringent as external SLAs, they can be considered as a special case of customer-level SLAs. Internal SLAs typically define the services provided by different departments within the organization and the expected levels of service.
- Service-level SLA: This type of SLA is an agreement between a service provider and multiple customers who receive the same service and share a similar agreement. Service-level SLAs focus on providing a consistent level of service to all

customers. For example, a mobile service provider may offer the same quality of service to all of its customers, ensuring a certain level of network coverage, call quality, and data speed. The service-level SLA establishes common service standards applicable to all customers.
- Multilevel SLA: A multilevel SLA divides the service agreement into various levels, each specific to a set of customers or users for the same services within the same SLA. It combines elements of customer-level and service-level SLAs. A multilevel SLA may also include corporate-level SLAs for generic Service Level Management (SLM) applicable to every customer or user throughout the organization. Large companies with significant service offerings typically design multilevel SLAs to address the diverse requirements of their customer base.

Overall, customer-level SLAs concentrate on individual customers, service-level SLAs focus on providing consistent service across multiple customers, and multilevel SLAs combine elements of both to cater to different customer segments within an organization.

Key Components of SLAs

While an SLA may have many components, Table 8.14 lists some key components included in an SLA. This table can be used as a checklist before the SLA is formally signed.

SLA Performance Metrics

Within an SLA, specific performance metrics can be defined, which customers will periodically or irregularly check. Table 8.15 provides examples of SLA performance metrics, although it is not an exhaustive list.

Service levels can be embedded into an SLA, or signed for different users with different performance expectations. For example, a normal user requires only best-effort service from an ISP. But an enterprise SLA may have best-effort, predictable, and guaranteed services from its ISP for different applications. This is shown in the following example of differentiated service levels in an SLA:

- Basic level: All best effort including bandwidth, delay, availability.
- Moderate level: At least 100 Mbps bandwidth in both upstream and downstream, at most 200 ms end-to-end delay, and best-effort availability for a group of applications or services.
- High level: Guaranteed 200 (100) Mbps upstream (downstream) bandwidth, maximum 100 ms delay between two ends, 99.999% uptime for a specific application.

8.7 Summary

Network services are typically provisioned as best-effort services without default QoS control. However, for network services that have real-time or other QoS requirements, specific solutions must be provided for QoS management in the planned

Table 8.14 Key components of an SLA

Component	Description
Agreement overview	The basics of the SLA, such as the parties involved, the commencement date, the end date, and a brief introduction of the services that will be provided
Description of services	Detailed descriptions of each of the services to be offered, how each of the services will be delivered and under what conditions, whether or not maintenance and technical support will be included, what services hours will be, and what processes technologies will be involved, and other related aspects
Exclusions	Clear statements of any specific services that will not be offered but might be otherwise thought as part of standard services
Service performance	Clear and detailed definitions of performance metrics and the performance levels under various categories such as best effort, predicable, and guaranteed services
Redressing	Definitions of any compensation or payment from the service provider to the customer should the SLA cannot be properly fulfilled
Stakeholders	Clear statements of the parties involved in the SLA and their respective responsibilities
Security	Clarifications of all security measures to be taken by the service provider, such as anti-poaching, security, and nondisclosure agreements
Risk management and disaster recovery	Clear statements of risk management processes and a disaster recovery plan
Review and change processes	Statements of how the SLA is regularly reviewed against key performance indicators. Changes may be made as a consequence of a review
Termination	Definitions of the process to deal with SLA termination or expiry. Is a notice period required from the service provider or the customer?
Signatures	The approval of the SLA by all involved parties

network. Network performance architecture focuses on fulfilling QoS requirements that are developed from requirements analysis and clarified in SLAs for network services. The intention of QoS is categorized into application-centric responses and network-centric responses. Application-centric responses deal with the control of network services, while network-centric responses address the control of network resource contention. Effective QoS management requires implementing both types of responses.

To implement QoS management, it is essential to identify the QoS requirements for users, applications, services, or devices within the network capacity. Consequently, the corresponding traffic flows are classified, marked, and prioritized at layer 2 (e.g., in frame headers) or layer 3 (e.g., in IP headers). Various methods are then developed for the control of network resources and network traffic. Examples include traffic management (e.g., admission control and traffic conditioning), queuing and scheduling, and frame preemption.

Table 8.15 Examples of SLA performance metrics

Metric	Description
Abandonment rate	The percentage of queued calls that customers abandon while waiting for answers
Availability	The probability that the services are available as expected when required during the period of a mission, e.g., 99.99% availability of video streaming service in the expected bandwidth and delay. It is related to, but different from, uptime
Average speed to answer	Average time (usually in seconds) it takes for a customer call to be answered by the service desk
Error rate	The percentage of errors in a service, such as coding errors and missed deadlines
First-call resolution	The percentage of incoming customer calls that are resolved straightaway without the need for a callback from the help desk.
Mean time to recovery (MTTR)	The time it takes to recover from a service outage
Specific performance thresholds or benchmarks	Bandwidth, round-trip delay, jitter, packet loss rate, etc., which actual performance will be compared against regularly or irregularly
Time service factor	The percentage of customer calls answered within a given time frame, e.g., 90% in 30 s
Total resolution time	The time it takes for an issue to be resolved after it is lodged by the service provider
Turn-around time	The time it takes for a service provider to resolve a specific issue since it has been received
Uptime	A measure of service reliability to quantify the percentage of time the services are running on average, e.g., 99.999% uptime over a year. Downtime is 100% less uptime, i.e., uptime in % + downtime in % = 100%

Network policies are an integral part of network performance architecture that governs network users, devices, applications, and access to data and other network resources. Typical types of network policies include access and security policies, application and QoS policies, traffic routing policies, IP-based or role-based access policies to resources, backup and archiving policies, failover policies, and disaster recovery policies. Therefore, network policies are related to not only network performance but also many other network aspects such as network management and network security.

In computer networking, there are two standard layer-3 QoS architectural models: DiffServ and IntServ. DiffServ considers QoS from the perspective of aggregating traffic based on PHBs. Without end-to-end QoS, it only provides soft QoS for network applications and services. The DiffServ architecture consists of a traffic classifier and a traffic conditioner. The conditioner is composed of four components: a meter, a marker, a shaper, and a dropper. In comparison with DiffServ, IntServ supports end-to-end QoS for each individual flow that requires it. Therefore, IntServ provides hard

QoS for network applications and services. IntServ is receiver-initiated and requires negotiation and resource reservation for an individual flow. The signaling of IntServ for negotiation and reservation is implemented in the RSVP protocol. The results of the negotiation are binary, either ALL or NONE, implying that no partial IntServ QoS will be supported.

SLAs are an integral part of network performance architecture. They specify detailed performance requirements along with many other aspects. The performance requirements from SLAs are used for QoS management for specific users, organizations, applications, and/or devices. Therefore, they form part of the technical specifications for the planning of network performance architecture, management architecture, and security architecture, all of which fit into the overall network architecture.

References

1. Huston, G.: Next steps for qos architecture. RFC 2990, RFC Editor (2000). https://doi.org/10.17487/RFC2990
2. Nichols, K., Blake, S., Baker, F., Black, D.: Definition of the differentiated services field (DS field) in the IPv4 and IPv6 headers. RFC 2474, RFC Editor (1998). https://doi.org/10.17487/RFC2474
3. Grossman, D.: New terminology and clarifications for Diffserv. RFC 3260, RFC Editor (2002). https://doi.org/10.17487/RFC3260
4. Fairhurst, G.: Update to IANA registration procedures for pool 3 values in the differentiated services field codepoints (DSCP) registry. RFC 8436, RFC Editor (2018). https://doi.org/10.17487/RFC8436
5. Parsons, G.: Telecommunications and exchange between information technology systems–requirements for local and metropolitan area networks–Part 1Q: Bridges and bridged networks. ISO/IEC/IEEE 8802-1Q:2020, 802.1 WG - Higher Layer LAN Protocols Working Group (2020). ISO/IEC/IEEE 8802-1Q:2020
6. Ramakrishnan, K., Floyd, S., Black, D.: The addition of explicit congestion notification (ECN) to IP. RFC 3168, RFC Editor (2001). https://doi.org/10.17487/RFC3168
7. Braden, R., Clark, D., Shenker, S.: Integrated services architecture. RFC 1633, RFC Editor (1994). https://doi.org/10.17487/RFC1633
8. Blake, S., Black, D., Carlson, M., Davies, E., Wang, Z., Weiss, W.: An architecture for differentiated services. RFC 2475, RFC Editor (1998). https://doi.org/10.17487/RFC2475
9. Baker, F., Fairhurst, G.: IETF recommendations regarding active queue management. RFC 7567, RFC Editor (2015). https://doi.org/10.17487/RFC7567
10. Cisco: What is network policy? Online documentation. https://www.cisco.com/c/en/us/solutions/enterprise-networks/what-is-network-policy.html. Accessed on 7 Jan 2021
11. Babiarz, J., Chan, K., Baker, F.: Configuration guidelines for DiffServ service classes. RFC 3168, RFC Editor (2006). https://doi.org/10.17487/RFC4594
12. Davie, B., Charny, A., Bennet, J., Benson, K., Boudec, J.L., Courtney, W., Davari, S., Firoiu, V., Stiliadis, D.: An expedited forwarding PHB (per-hop behavior). RFC 3246, RFC Editor (2002). https://doi.org/10.17487/RFC3246
13. Baker, F., Polk, J., Dolly, M.: A differentiated services code point (DSCP) for capacity-admitted traffic. RFC 5865, RFC Editor (2010). https://doi.org/10.17487/RFC5865
14. Heinanen, J., Baker, F., Weiss, W., Wroclawski, J.: Assured forwarding PHB group. RFC 2597, RFC Editor (1999). https://doi.org/10.17487/RFC2597

15. Bless, R.: A lower-effort per-hop behavior (LE PHB) for differentiated services. RFC 8622, RFC Editor (2019). https://doi.org/10.17487/RFC8622
16. Zhang, L., Berson, S., Herzog, S., Jamin, S.: Resource reservation protocol (RSVP) – version 1 functional specification. RFC 2205, RFC Editor (1997). https://doi.org/10.17487/RFC2205
17. Wroclawski, J.: The use of RSVP with IETF integrated services. RFC 2210, RFC Editor (1997). https://doi.org/10.17487/RFC2210
18. Shenker, S., Wroclawski, J.: General characterization parameters for integrated service network elements. RFC 2215, RFC Editor (1997). https://doi.org/10.17487/RFC2215
19. Wroclawski, J.: Specification of the controlled-load network element service. RFC 2211, RFC Editor (1997). https://doi.org/10.17487/RFC2211
20. Shenker, S., Partridge, C., Guerin, R.: Specification of guaranteed quality of service. RFC 2212, RFC Editor (1997). https://doi.org/10.17487/RFC2212

Chapter 9
Network Management Architecture

The easiest method of managing network devices is in an ad-hoc manner. To check the status of a network device or link, send messages to the device or over the link. Then, observe and analyze the response. If the response shows our expected results, e.g., connectivity and delay, the device or link works well. Otherwise, further identify potential problems and fix them. The operating system has provided some utilities for checking the status of a network device, link, or other aspects. For example, `ping`, `traceroute` (or `tracert` in Windows), and `netstat` are examples of these utilities. By using these utilities, statistics or other specific information can be obtained from remote systems.

With such utilities in ad-hoc operations, network management is often inappropriately considered as an operational issue rather than a design issue. Therefore, it is largely overlooked during network planning. However, as one of the important components of the overall network architecture, network management has an impact on many aspects of the network and should be addressed in advance before network deployment and operation. Considering network management as an integral component during the network planning phase will help avoid many network operational problems in connectivity, scalability, manageability, performance, integrity, and security.

As for the planning of other architectural components, planning network management architecture also requires a good understanding of the customer's requirements and goals through comprehensive requirements analysis. These requirements and goals will drive the planning, design, and implementation of the network management architecture.

Y.-C. Tian and J. Gao, *Network Analysis and Architecture*, Signals and Communication Technology, https://doi.org/10.1007/978-981-99-5648-7_9

9.1 Concepts of Network Management

Network management can be considered as a set of functions that involve planning, monitoring, and controlling network resources to achieve network connectivity, scalability, performance, integrity, and security. It is also a process for managing and implementing network operations, troubleshooting network problems, optimizing network performance, and enhancing network integrity and security. Therefore, various protocols and mechanisms have been developed to monitor and evaluate network events, information flows, integrity and security, and other network performance metrics. These protocols and mechanisms help identify and resolve network problems, thereby improving and enhancing the Quality of Service (QoS) of the network. The IETF RFC 6632 [1] have provided comprehensive descriptions of protocols, mechanisms, and data models for network management.

9.1.1 Network Management Hierarchy

When addressing network management in corporate networks, it is important to consider not only the computer networks themselves but also the business goals, network services, and other aspects that either serve, or are served by, the computer networks. Therefore, a layered architecture is recommended for network management to encompass all these aspects and ultimately serve the business goals. Figure 9.1 illustrates the hierarchy of network management layers [2, p. 300].

In this hierarchy, the higher the layer, the more abstract it becomes. The lower the layer, the more detailed information the layer will provide. Therefore, business management and service management, located at the upper layers, contain more abstract information compared to the three lower layers. Network element management, situated at the bottom layer, involves the most specific and detailed information. The intermediate layers, i.e., network management and element management, are more abstract than network element management but provide more detailed information than the upper layers of business management and service management.

To implement network management, higher layers in the hierarchy are typically managed through policies, while lower layers are monitored and managed through variables and parameters. These management activities are implemented in functional components using various protocols and mechanisms. The ISO CCITT X.700 [3]/ISO/IEC 7498-4 [4] has clearly described a number of functional components that address network management requirements.

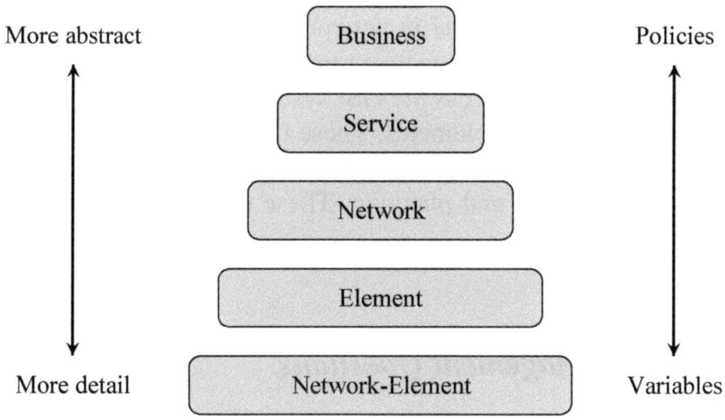

Fig. 9.1 Hierarchical network management

9.1.2 Network Management Framework

There are two main network management frameworks: one defined by the IETF through RFC 6632 [1] and the other specified by the ISO through CCITT X.700 [3]/ISO/IEC 7498-4 [4]. These two frameworks have different views on network management, and thus address network management from different perspectives:

- The IETF standard emphasizes simplicity in network management. It adopts a variable-oriented approach, and the management information exchanges may be unreliable.
- By contrast, the ISO/IEC standard provides a powerful framework for network management. It uses an object-oriented approach, and the management information exchanges are performed reliably.

Focusing on the OSI reference model, the ISO CCITT X.700 [3]/ISO/IEC 7498-4 [4] describes key concepts for OSI management. It provides a structure for OSI management, along with an overview of its objectives and facilities. The standard also details OSI management activities. The main objectives of OSI management are to support:

(a) Planning, organizing, supervising, controlling, and accounting for the use of interconnection services.
(b) Adapting to changing requirements from users and systems;
(c) Ensuring predictable communication behavior.
(d) Ensuring information protection and communication security.

As a subset of the total OSI environment, the OSI management environment includes both the capability to gather information and exercise control, and the capacity to maintain an awareness of, and report on, the status of resources in the OSI environment [3, 4]. The OSI management can be applied to an autonomous OSI system or implemented in cooperation with other OSI systems. In any case, information exchanges are essential in network management. Therefore, from the technical perspective, network management involves two basic functions: transporting management information across the OSI system or systems, and managing network management information elements. These functions are implemented through various network management tasks and protocols. Typical tasks include monitoring, configuring, troubleshooting, and planning. These tasks will be discussed in more detail later.

9.1.3 Network Management Questions

When designing a network management architecture, several key questions need to be addressed clearly and explicitly. The following list provides some of these questions, although it is not exhaustive:

- What architectural model or models are best suitable for the network under consideration? For example, should it be in-band or out-of-band, centralized or distributed, or hierarchical?
- Does the current network or network design need to be reconfigured to meet the network management requirements?
- Which protocols should be chosen for network management?
- Are there specific network locations or portions that require dedicated management for critical network services or applications?
- Will the network management design fulfill all Service Level Agreements (SLAs)?
- What is the impact of network management traffic flows on network performance and capacity planning?

Answering these questions necessitates a detailed analysis, design, and performance evaluation of network management.

9.2 Functional Areas of Network Management

As discussed earlier, network management serves various purposes. The ISO CCITT X.700/ISO/IEC 7498-4 [3, 4] categorizes network management requirements into several functional areas. Specifically, the following functional areas are emphasized: fault management, configuration management, accounting management, performance management, and security management. These five functional areas collectively form the ISO's FCAPS (Fault, Configuration, Accounting, Performance, and Security) model for network management, as shown in Fig. 9.2. According to the ISO CCITT X.700/ISO/IEC 7498-4 [3, 4], specific management functions are provided by OSI management mechanisms within these functional areas. Many mechanisms are used to fulfill the requirements in multiple functional areas. Similarly, multiple

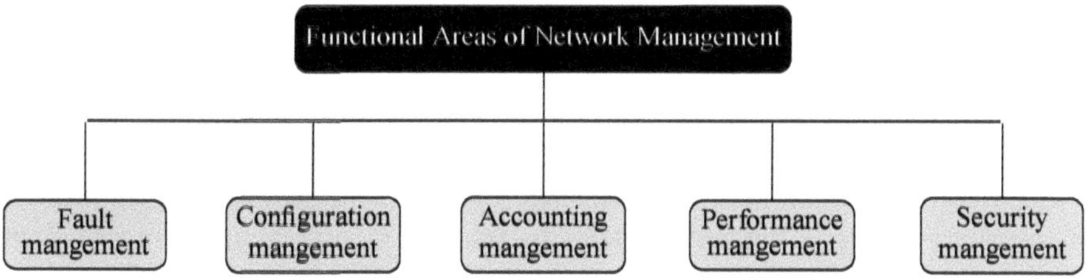

Fig. 9.2 FCAPS model for network management [3, 4]

functional areas may share the same managed objects. Each managed subject is an OSI management view of a resource that is subject to management.

Fault management. In the FCAPS model, fault management focuses on fault detection, isolation, correction, and logging. Fault detection aims to identify faults by recognizing specific events, such as network operation errors. Therefore, fault management encompasses the following main functions [3, 4]: maintaining and examining error information logs, accepting and acting upon error detection notifications, tracing and identifying faults, conducting diagnostic tests, and correcting faulty systems. Fault management relies on network management protocols to transport, report, and record faults and related events.

Configuration management. Configuration management is concerned with monitoring system configuration information and any changes that occur. It is important because many network problems arise from changes in configuration, software, and hardware. Therefore, configuration management aims to identify and track such changes and exert control over the network system accordingly. Typical functions of configuration management include [3, 4]:

- Collecting information on demand about the current condition of the open system.
- Associating names with managed objects and initializing and terminating managed objects.
- Setting network parameters for the routine operation of the open system.
- Changing the configuration of the open system.

Accounting management. Accounting management monitors the usage information of network resources. This enables the establishment of charges for resource usage, allowing individual users, departments, or business units to be appropriately billed for accounting purposes. Accounting management includes functions to inform users of costs incurred or resources consumed, thus enabling charges and billing. In non-billed networks, "accounting" could be replaced by "administration," which manages the resources for authorized user access.

Performance management. Performance management aims to ensure that network performance remains at acceptable levels. It enables the evaluation of (1) the behavior of resources, and (2) the efficiency and effectiveness of communication activities. Typically, performance management includes functions to collect, maintain, and examine network and system performance data; analyze this performance data to determine system performance; and implement performance control by setting system and performance parameters and altering system modes of operation.

Security management. Lastly, security management supports the application of security policies in network systems to enhance Confidentiality, Integrity, and Availability (CIA). It addresses both access control and security data analysis. Data security methods include authentication and encryption coupled with authorization. Overall, security management is achieved through functions to create, delete, and control security services and mechanisms; distribute security-relevant information; and analyze and report security-relevant events.

9.3 Network Management Protocols

Aiming to develop a simple network management framework, the IETF RFC 6632 [1] provides comprehensive discussions on core network management protocols as well as other protocols with specific focuses. The core protocols include Simple Network Management Protocol (SNMP), Syslog protocol, IP Flow Information eXport (IPFIX) and Packet SAMPling (PSAMP), and Network Configuration Protocol (NETCONF).

Serving ISO's OSI-specified networks, Common Management Information Protocol (CMIP) is defined [5]. It is an implementation of the ISO Common Management Information Service (CMIS) for network management [6]. CMIP/CMIS over TCP/IP networks forms CMOT (CMIP over TCP/IP) protocol.

9.3.1 SNMP

The first core protocol outlined in the IETF RFC 6632 for network management is the SNMP [7, 8]. The latest version of SNMP is SNMPv3, which is comprehensively specified from various perspectives, resulting in a series of RFCs. These RFCs include RFC 3410 [7] through RFC 3418. These RFCs have been further updated in subsequent RFCs. For example, RFC 3411 [8] has been updated by RFC 5343 and RFC 5590. These updates aim to enhance and refine the SNMP specifications for efficient network management.

SNMP Components and Architecture

According to RFC 3410 [7], all three versions of SNMP (SNMPv1, SNMPv2, and SNMPv3) share the same basic structure and components and follow a common architecture. Therefore, unless explicitly specified, our discussions on SNMP do not differentiate between the versions of SNMP. However, in contexts where there is no confusion, SNMPv3 may be specifically indicated.

There are four basic components in SNMP:

- Several (typically many) *managed nodes or devices*: These are the entities being managed, each equipped with an SNMP entity called an *agent*. Agents provide access to management instrumentation remotely.
- At least one SNMP *manager*: This is an SNMP entity responsible for management applications.
- A *management protocol*: It is used to convey management information between SNMP entities such as managers and agents.
- *Management information base* (MIB): MIB represents the collection of managed objects that can be accessed and manipulated via SNMP.

These four components form the basic structure of SNMP, which, as mentioned above, is shared by SNMPv1, SNMPv2, and SNMPv3.

The SNMP architecture is designed to be modular, allowing for the evolution of SNMP over time. Actually, SNMP is developed to be much more than just a protocol of moving data over a network. It encompasses various aspects including:

- A data definition language, which defines how data is structured and represented in SNMP.
- Definitions of management information, known as MIB.
- A protocol definition, which specifies the rules and procedures for communication between SNMP entities.
- Security and administration, which address the security measures and administrative controls within SNMP.

The SNMP architecture is developed with a protocol-independent data definition language and MIB, along with a MIB-independent protocol. Interestingly, this separation was originally intended to enable the replacement of the SNMP-based protocol without redefining the management information. However, in practice, it facilitated the transition from SNMPv1 to SNMPv2 and subsequently SNMPv3, rather than transitioning away from MIB. This SNMP architecture design was said to be a right decision with a wrong reason [7, p. 5].

While the three versions of SNMP share the same four components and follow the same architecture, they exhibit some differences. SNMPv2 introduces improvements in data handling and error management compared to SNMPv1, while SNMPv3 offers robust security measures to protect SNMP communication. The security of SNMP will be discussed later.

SNMP Commands

SNMP is widely used for collecting and configuring parameters from network devices, primarily for network management purposes. This functionality is achieved through seven SNMP messaging commands, each associated with a specific type of packet. These commands are as follows: `get`, `set`, `get-next`, `get-bulk`, `response`, `trap`, and `inform`. Table 9.1 provides a summary of these SNMP commands and their corresponding packet types.

Table 9.1 SNMP commands

Command	Function
Manager-to-agent	
`get`	A request to retrieve the value of a variable or list of variables
`set`	A request to issue configurations or commands
`get-next`	A request to find the values of the next record in the hierarchy of the MIB
`get-bulk`	A request to obtain large tables of data by performing multiple `get-next` request commands
Agent-to-manager	
`response`	A reply to a request from the manager
`trap`	Asynchronous trap messages to notify the SNMP manager of the occurrence of a significant event such as an error or failure
Manager-to-agent/manager	
`inform`	An acknowledgment to confirm the receipt of a trap

The SNMP `inform` command, which was introduced in SNMPv2, originally aimed to provide a manager-to-manager notification message regarding information in a remote MIB view, thereby supporting the Manager-of-Managers (MoM) architecture for network management. In fact, manager-to-manager notifications were already achievable in SNMPv1 through the `trap` command. However, the delivery of `trap` and other SNMP messages was not guaranteed due to the underlying UDP transport used by SNMP. To overcome this limitation, the `inform` command was implemented in SNMPv2, enabling manager-to-agent communications to return an acknowledgment upon the receipt of a `trap` notification. This enhancement in SNMPv2 significantly improved the reliability of manager-to-manager notifications and enhanced the overall robustness of SNMP-based network management systems.

SNMP Traps

SNMP traps are the only information sent within an SNMP system without a specific request from the SNMP manager. They are sent from managed devices to the SNMP manager asynchronously when a monitored variable exceeds a predefined threshold, such as packet dropping rate or error rate. However, it may not be necessary for an SNMP agent to send a trap message every time a threshold is crossed in order to avoid potential false positives. Instead, the SNMP agent can be configured to throttle the number of traps sent for the same monitored variable, effectively throttling the trap notifications.

To receive SNMP traps, the SNMP manager simply listens on port number 162 for incoming trap messages from agents under SNMP management. In cases where the SNMP manager requires additional information from a specific agent, it can send a get request to that agent to retrieve the required data.

By employing SNMP traps and using appropriate configurations, SNMP-based systems can effectively monitor and respond to critical events and conditions, ensuring efficient network management and reducing the risk of information overload caused by excessive trap notifications.

MIBs

A MIB within an SNMP agent serves as a repository for information gathered by the agent on the managed device. MIB modules typically include MIB object definitions, and may also define event notifications. These modules may further include compliance statements specified in terms of appropriate object and event notification groups. As described in the IETF RFC 3440 [9, p. 12], MIB modules define the MIBs that are maintained by the instrumentation in managed agents and devices. These MIBs are made remotely accessible by SNMP management agents, conveyed through the SNMP management protocol, and manipulated by SNMP management applications.

In general, management information defined in MIB modules can be used with any version of the SNMP protocol, regardless of the version of the data definition language used. This implies that MIB modules defined in SNMPv1 are generally compatible with SNMPv2 and SNMPv3. Similarly, MIB modules defined in SNMPv3 are also compatible with SNMPv2 and SNMPv1. However, there is an exception to this compatibility. Newer versions of SNMP may introduce new data types that are not recognizable by older SNMP versions. For example, SNMPv1 does not understand the Counter64 data type introduced in SNMPv2.

From the architectural perspective, the MIB is designed with a tree structure. This hierarchical structure is specified in the IETF RFC 1213 [10], which is subsequently updated in RFC 2011, RFC 2012, and RFC 2013 for IP, TCP, and UDP scenarios, respectively. MIB objects with similar characteristics are grouped together under the same branch of the MIB tree.

The separation of MIB from the SNMP protocol allows for independent design and development of MIB modules. Consequently, there is a significant and growing number of standards-track MIB modules. Moreover, there is an even larger and expanding number of enterprise-specific MIB modules that are unilaterally defined by various vendors, research groups, consortia, and other entities. This results in an unknown and conceptually uncountable number of defined MIB objects. The MIBs in an SNMP management system can consist of standard MIB known as MIB-II [10], other standard MIBs specific to device or protocol, Remore MONitoring (RMON) MIBs, and vendor-specific MIBs.

MIB-II specified in the RFC 1213 [10] has defined ten groups of managed objects for TCP/IP networks. These ten groups of managed objects are listed in Table 9.2.

RMON MIBs

MIB-II, as a standard MIB, lacks the capability to provide statistics on data link and physical layers of the OSI's seven-layer network model. To address this issue, the IETF has developed RMON MIBs as an extension to MIB-II. These RMON MIBs are designed to provide Ethernet traffic statistics and fault information, enabling

Table 9.2 Ten groups of managed objects in MIB-II [10]

No.	Group	No.	Group
(1)	System Group	(6)	TCP Group
(2)	Interfaces Group	(7)	UDP Group
(3)	Address Translation Group	(8)	EGP Group
(4)	IP Group	(9)	Transmission Group
(5)	ICMP Group	(10)	SNMP Group

Table 9.3 Ten groups of managed objects from RMON1

No.	Group	Description
1	Statistics	LAN statistics, e.g. utilization, collisions, CRC errors
2	History	Snapshots based on selected LAN statistics
3	Alarm	To set threshold and generate alarms to be sent as RMON SNMP traps
4	Hosts	Host specific LAN statistics, e.g., bytes sent/received, frames sent/received
5	Hosts top N	Record of N most active connections over a given period of time
6	Matrix	The amount of sent/received traffic and the number of errors between pairs of nodes in a LAN
7	Filter	To define packet data patterns of interest, e.g., MAC address or TCP port
8	Capture	To collect and forward packets matching the Filter
9	Event	To send alerts (SNMP traps) for the Alarm group
10	Token Ring	Extensions specific to Token Ring

network administrators to monitor, analyze, and troubleshot a group of LANs and VLANs remotely.

There are several RFCs related to RMON:

- RMON1 is specified in the IETF RFC 2819 [11]. It defines the MIB for monitoring Ethernet-like networks. RMON1 MIB provides ten groups of managed objects, as illustrated in Table 9.3.
- RMON2 is defined in the IETF RFC 4502 [12]. It extends the monitoring capabilities of RMON1 by introducing 10 additional groups of managed objects. These additional groups, as tabulated in Table 9.4, provide enhanced monitoring and analysis features for various network layers and protocols.

There are a few other extensions to RMON. For example, the IETF RFC 3273 [13] introduces HCRMON for high capacity networks. RFC 2613 defines RMON MIB extensions for switched networks [14]. Moreover. RFC 3577 serves as an introduction to the RMON Family of MIB Modules [15]. These RFCs contribute to the development and standardization of RMON MIBs, enhancing the monitoring capabilities of SNMP-based network management systems.

Table 9.4 Additional groups of managed objects from RMON2

No.	Group	Description
1	Protocol Directory	List of protocols the probe can monitor
2	Protocol Distribution	Traffic statistics for each protocol
3	Address Map	To map network-layer (IP) to MAC-layer addresses
4	Network-Layer Host	Layer 3 traffic statistics, per each host
5	Network-Layer Matrix	Layer 3 traffic statistics, per source/destination pairs of hosts
6	Application-Layer Host	Traffic statistics by application protocol, per host
7	Application-Layer Matrix	Traffic statistics by application protocol, per source/destination pairs of hosts
8	User History	Periodic samples of user-specified variables
9	Probe Configuration	Remote configure of probes
10	RMON Conformance	Requirements for RMON2 MIB conformance

Benefits and Disadvantages of RMON

The benefits of using RMON can be summarized from various perspectives. One of the major advantages of RMON is its ability to facilitate remote monitoring and management of networks. This capability provides several benefits. Some benefits are presented below:

- A network administrator can monitor remote offices regardless of their physical location. This is particularly beneficial for small network teams managing large enterprise networks that interconnect several campus networks via WAN. When issues arise with a network device, the network administrator does not necessarily need to travel to the location for troubleshooting. This saves time and resources, increases productivity, and allows for more proactive network management.
- Network administrators in other offices can monitor the network in real time. In large-scale networks, network team members are often spread across different geographical locations and responsible for network management. RMON enables these team members in other offices, cities, or even countries to monitor network performance in real time, even if the management software server is installed at headquarters. If errors occur on a host, a remotely located network administrator can troubleshoot without involving the users of the host.
- RMON allows network users and administrators to be mobile without neglecting the network. They can work from home, coffee shops, hotels, or anywhere on Earth with Internet connectivity, while still having the ability to monitor and manage the network effectively.

- RMON provides the ability to manage access levels with ensured security. For remote access, one way to secure the network is by using encrypted authentication and ensuring that all users log onto the network in the same way. Customizing user access is another approach to enhancing network security.
- With proper connectivity, such as VPN over WAN, RMON is always accessible. This means that critical network devices within the network can be monitored regardless of their physical location or the location of network administrators. This also allows for more frequent monitoring, leading to faster fault diagnosis and troubleshooting.

These benefits make RMON an invaluable tool for network management, enabling efficient remote monitoring and ensuring the smooth operation of networks.

However, RMON also has some observed disadvantages. It is important to consider these limitations when implementing RMON, as they can impact the effectiveness and efficiency of network monitoring and management. Some RMON disadvantages are summarized below:

- The amount of information that RMON MIBs provide is insufficient for network managers and administrators who need to solve complex problems, often at a distance.
- The mechanism employed for data retrieval to a central management console is slow and bandwidth-inefficient.
- RMON values are stored in 32 bit registers, which limit the count value to $2^{32} - 1 = 4,294,967,295$. While this value may seem large, it can be quickly reached in certain RMON application scenarios. For example, in a 100 Mbps fast Ethernet network operating at just 10% loading, the counters will be reset to zero after approximately one hour of activity.
- Full RMON support in hardware typically requires dedicated powerful processor, increasing the cost of devices where RMON is installed. However, as the cost of processors continues to decrease over time, this issue is gradually becoming less significant.

SNMP Security and Administration

The three versions of SNMP differ in their security capabilities. SNMPv1 is known for its fundamental weakness, which is the lack of well-defined authentication schemes. Instead, it relies on a simplistic authentication scheme based on community strings. SNMPv2 without security is referred to as SNMPv2c. Although SNMPv2 incorporates security enhancements compared to SNMPv1, it is considered incomplete in meeting the requirements for "commercial-grade" security.

SNMPv3 addresses these security limitations by introducing significant improvements. It focuses on several key aspects:

- Authentication: SNMPv3 ensures origin identification, message integrity, and some aspects of replay protection.
- Privacy: SNMPv3 includes encryption of SNMP messages to ensure confidentiality.

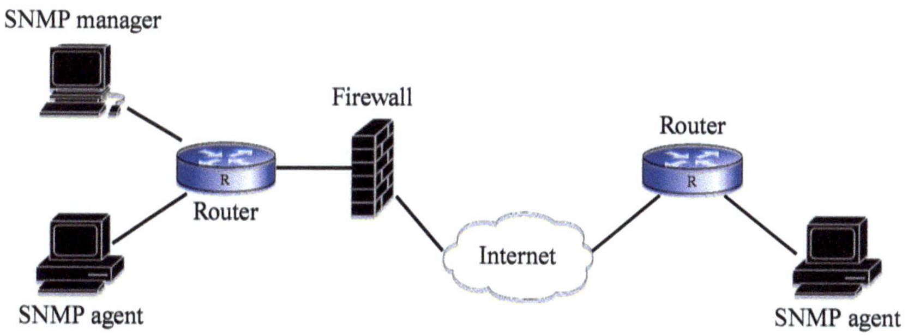

Fig. 9.3 SNMPv3 with security

- Authorization and access control: SNMPv3 effectively manages permissions and access control.
- Configuration and administration: SNMPv3 provides capabilities for suitable remote configuration and administration.

With the enhanced security provided by SNMPv3, remote locations can be inspected securely, going beyond restricting communications solely to the local LANs at those remote locations. This ensures that sensitive data remains protected during transit, allowing for more comprehensive and secure network management. SNMPv3 with security is logically depicted in Fig. 9.3.

SNMP Transport and Transport Security Model

SNMP operates in the application layer of the TCP/IP protocol suite. It is intentionally designed to be a lightweight protocol, adding only minimal overhead to the managed network in terms of bandwidth, memory, and CPU resources of the managed network devices. Therefore, SNMP uses UDP as the underlying transport protocol for message transfer. The SNMP agent listens on port 161 for all get/set request messages, while the SNMP manager listens on port 162 for trap and inform messages.

SNMPv1 and SNMPv2 use a Community-based Security Model. In SNMPv3, a User-based Security model is developed in the IETF RFC 3414 [16]. Later, the SNMP Transport Security Model, defined in RFC 5591 [17], offers an alternative to the User-based Security Model as well as SNMPv1 and SNMPv2 Community-based Security Models.

Briefly speaking, the SNMP Transport Security Model uses lower-layer security mechanisms to provide message-oriented security services such as authentication, encryption, timeliness checking, and data integrity checking. Examples of such mechanisms include the integration of Secure Shell (SSH), Transport Layer Security (TLS), and Datagram Transport Layer Security (DTLS) Transport Models into the SNMP Transport Security Model. When used with TLS or DTLS, SNMP requests are received on port 10161, while SNMP notifications are sent to port 10162.

9.3.2 Syslog Protocol

The second core protocol outlined in RFC 6632 is the Syslog Protocol, with the latest specifications provided in RFC 5424 [18]. The Syslog protocol provides a mechanism for distributing logging information with security considerations. It features a layered architecture that allows the use of reliable and secure transport protocols for transmitting syslog messages. Additionally, it introduces a structured message format to accommodate vendor-specific extensions.

The IETF RFC 5676 formally defines managed objects for mapping SYSLOG messages to SNMP notifications [19]. It specifies a portion of the MIB of SMNP for network management protocols in computer networks.

Moreover, the security enhancement for signed Syslog Messages is also summarized in the IETF RFC 6632 [1], which refers to RFC 5848 [20]. RFC 5848 provides a mechanism for incorporating origin authentication, message integrity, replay resistance, message sequencing, and detection of missing messages into transmitted Syslog messages in conjunction with the Syslog protocol.

Regarding the transport of syslog messages, while UDP can be used as the underlying transport protocol, it is highly recommended to establish a secure connection using TLS. TLS transport mapping for Syslog is formally defined in the IETF RFC 5425 [21].

9.3.3 IPFIX and PSAMP

The third group of core protocols summarized in the IETF RFC 6632 for network management includes the IPFIX and PSAMP protocols. These protocols are developed to meet the requirement of exporting IP flow information in various applications. Several relevant RFCs pertain to IPFIX. For example, the IETF RFC 3917 has listed general IPFIX requirements such as openness, scalability, and collecting processes [22]. In general, many applications necessitate flow export for usage-based accounting, traffic profiling, traffic engineering, attack/intrusion detection, and QoS monitoring and management. Detailed IPFIX requirements derived from these applications can be found in the Appendix of RFC 3917.

The IETF RFC 7011 [23] presents the IPFIX specification for the exchange of IP traffic flow information over the network. It defines a push-based data export mechanism using a simple binary format from an IPFIX Exporter to an IPFIX Collector. More specifically, RFC 7011 describes IPFIX data types, message format, Template Records, data sets, exporting and collecting processes, and transport session. It also specifies how IPFIX Data and Template Records are carried over various transport protocols. In IPFIX, the Stream Control Transmission Protocol (SCTP) defined in RFC 4960 [24] is a mandatory transport protocol to implement, while TCP and UDP are optional.

Furthermore, the IETF RFC 5470 [25] defines an IPFIX architecture for IP traffic flow monitoring, measurement, and exporting. More specifically, it specifies the key IPFIX architectural components, defines the IPFIX architectural requirements (e.g., recovery, security), and describes the characteristics of the IPFIX protocol. The IPFIX architecture includes metering processes, observation points, packet selection criteria, observation domains, exporting processes, and collecting processes. The IPFIX protocol is designed to address these components and provide support for applications. It inherits various existing security mechanisms to ensure different levels of security protection.

Working in conjunction with the IPFIX protocol, the PSAMP protocol and related techniques are comprehensively described in a series of RFCs from RFC 5474 through RFC 5477. RFC 5474 [26] formalizes the general framework of PASAMP, allowing network elements to select subsets of packets and export a stream of reports on the selected packets to a Collector. The set of packet selection techniques supported by PSAMP is described in the IETF RFC 5475 [27]. The PSAMP protocol itself is specified in RFC 5476 [28]. It uses IPFIX protocol to transfer information on individual packets from a PSAMP Exporting Process to a SAMP Collecting Process. Moreover, the IETF RFC 5477 [29] defines an information and data model for PSAMP.

9.3.4 NETCONF for Configuration Management

The last group of core protocols described in RFC 6632 for network management is NETCONF, along with other related supporting protocols for configuration management. To better understand NETCONF, let us summarize the advanced requirements for configuration management as discussed in RFC 3535 [30]. These requirements are tabulated in Table 9.5.

To fulfill these requirements, NETCOME is developed as a protocol to install, manipulate, and delete the configuration of network devices, as specified in RFC 6241 [31]. For Requirement 9) listed in Table 9.5, YANG is developed as the NETCONF data modeling language, as outlined in RFC 6020 [32] and RFC 7950 [33]. YANG provides a simple XML-based syntax and directly addresses Requirements 5) and 7) in Table 9.5. Using the YANG data modeling language, NETCONF offers a basic set of operations to edit and query the configuration on a network device.

To address Requirements 11) listed in Table 9.5, security enhancements for NETCONF are specified in several RFCs, such as NETCONF over SSH in RFC 6242 [34] and NETCONF over TLS with authentication in RFC 7589 [35].

Before NETCONF was developed, the primary approach for making automated configuration changes to the network was CLI scripting. Although CLI scripting is powerful, flexible, and still useful in network management, it lacks transaction management and structured error management. Moreover, CLI scripting becomes difficult to maintain with changes in the structure and syntax of commands over time.

Table 9.5 Advanced NETCONF requirements [30]

Requirement and description	
1	Robustness with minimized disruptions to, and maximized stability of, the network under management
2	A task-oriented view
3	Extensibility for new configuration operations
4	Standardized error handling in configuration
5	Distinction between configuration data and operational state, along with distribution of configurations under transnational constraints
6	Scalability in the number of transactions and managed devices
7	Interoperability of dumping and reloading a device configuration across multiple vendors and device types
8	Both a human interface and a programmatic interface
9	A data modeling language with a human-friendly syntax
10	Easy conflict detection and configuration validation
11	Secure transport, authentication, and robust access control

NETCONF addresses these issues and simplifies network configuration compared to CLI scripting.

SNMP is another approach that could potentially be used for configuration management. However, SNMP lacks a defined discovery process to find appropriate MIBs for configuration management. Due to its lightweight nature, SNMP uses UDP and also lacks useful standard security mechanisms. As a result, SNMP is not commonly used for configuration but is primarily employed for performance and monitoring applications in network management.

9.3.5 CMIP

CMIP is an OSI specified network management protocol. It is defined in ITU-T Recommendation X.711/ISO/IEC International Standard 9596-1 [5] and also in the ITU-T X.700 series of recommendations. CMIP implements CMIS, which is specified in ITU-T Recommendation X.710/ISO/IEC 9595 [6]. Therefore, the terms of CMIS and CMIP are often used interchangeably with the same meaning.

CMIS, defined in ITU-T Recommendation X.710/ISO/IEC 9595 [6], provides management information services used by application processes to exchange information and commands for systems management. There are two types of services available in CMIS for the management of network elements: management operation services and management notification services.

Management operation services are composed of six services, which are listed in Table 9.6. Each service addresses a specific aspect of management operations.

Table 9.6 CMIP management operation services

Service	Function
M-CREATE	Create an instance of a managed object
M-DELETE	Delete an instance of a managed object
M-GET	Request the attributes of a managed object or a set of managed objects
M-CANCEL-GET	Cancel an outstanding GET request
M-SET	Set the attributes of a managed object
M-ACTION	Request an action to be performed on a managed object

Regarding management notification services, CMIS defines the M-EVENT-REPORT service as a common service for management notification. It allows the reporting of management information applicable to the notification. The M-EVENT-REPORT service can be invoked by a CMISE-service-user to report an event about a managed object to a peer CMISE-service-user. It can be requested in a confirmed or non-confirmed mode, with a reply expected in the confirmed mode.

It is worth mentioning that CMIS initially defined management association services in addition to management operation services and management notification services. Later, it was decided that these services could be provided by Association Control Service Element (ACSE), which is used for establishing a call between two application programs. Therefore, these services were removed from ISO/IEC 9595. For complete information, the association services initially defined in, but later removed from, ISO/IEC 9595 are listed below:

- M-INITIALIZE: to create an association with (i.e. connects to) another CMISE
- M-TERMINATE: to terminate an established connection; and
- M-ABORT: to terminate the association in the case of an abnormal connection termination.

These services are formalized as ACSE services in ITU-T Recommendation X.711/ISO/IEC International Standard 9596-1 [5].

CMIP defined in ITU-T Recommendation X.711/ISO/IEC International Standard 9596-1 [5] is a Layer 7 protocol that implements CMIS services. More specifically, it specifies procedures for creating a common Network Management System (NMS) through standardized procedures with clearly specified elements. The defined procedures include Association Establishment, Remote Operations, Event Reporting, Get, Set, Action, Create, Delete, Association Orderly Release, and Association Abrupt Release.

CMIP is implemented in association with two other Layer 7 OSI protocols: ACSE and ROSE (Remote Operations Service Element). ACSE manages associations between management applications (CMIP agents), while ROSE handles data exchange interactions. CMIP assumes the presence of the other six lower layers in the 7-layer OSI model but does not specify the functions provided in those layers to support CMIP functionalities.

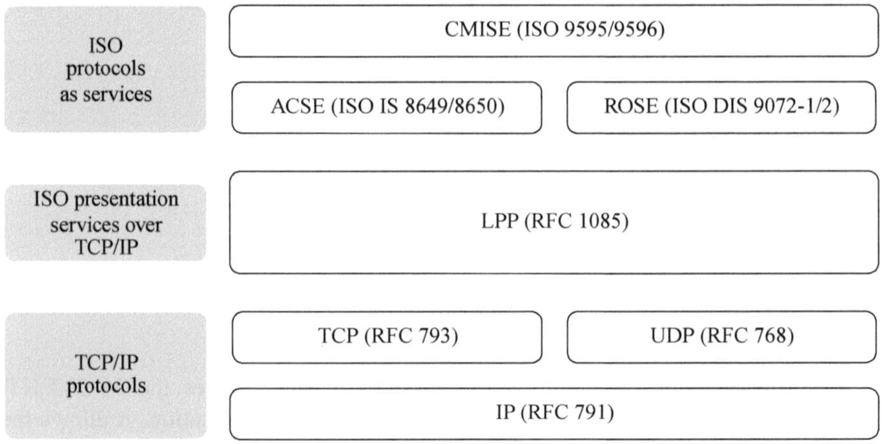

Fig. 9.4 The CMOT protocol suite [36, p. 5]

Moreover, the implementation of CMIP over TCP/IP results in CMOT. The network architecture that uses ISO's CIMS/CIMP over TCP/IP networks is specified in the IETF RFC 1189 [36]. The CMOT protocol suite consists of seven protocols: CMISE, ACSE, ROSE, Lightweight Presentation Protocol (LPP), TCP, UDP, and IP. The ISO protocols CMISE, ACSE, and ROSE are implemented as services on top of the TCP/IP protocol stack. LPP implements ISO's presentation services over TCP/IP. The three TCP/IP protocols TCP. UDP, and IP are core protocols in TCP/IP networks. Figure 9.4 [36, p. 5] shows the complete CMOT protocol suite with the relationships of all included protocols.

It is worth mentioning that both CMIP and SNMP define network management standards, but CMIP is more complex. The ISO aims to provide powerful network management functions, and CMIP is initially designed with the goal of replacing SNMP for more advanced network management operations. This is primarily because SNMP offers only a limited number of such operations. Consequently, the implementation of CMIP is more intricate compared to SNMP. In practice, CMIP is primarily used by some telecommunications service providers for network management, whereas SNMP is an Internet protocol specifically designed for TCP/IP networks and commonly employed in corporate networks.

9.3.6 Protocols and Mechanisms with Specific Focus

In the IETF RFC 6632, ten types of protocols are listed as network management protocols and mechanisms with specific focus. Some of these are still in use, while others are classified as Historic or designated Obsolete. Selected protocols and mechanisms are discussed below, which find wide applications in current networks.

The first type of such protocols is for IP address management. The aim of these protocols is to allocate an IP address to a host automatically when the host joins either an IPv4 network or IPv6 network. Dynamic Host Configuration Protocol (DHCP) has a version for IPv4 networks and another version for IPv6 networks. For a dual-stack network with both IPv4 and IPv6 addressing schemes, a DHCP is able to allocate an IPv4 address and an IPv6 address to a host. Autoconfiguration is a feature in IPv6. Ad-hoc nodes need to configure their network addresses with locally unique addresses and globally routable IPv6 addresses.

The second type of such protocols is for IPv6 operations. As the majority of existing networks are still IPv4 networks, it is essential for IPv6 networks to work with IPv4 networks, implying the co-existence of IPv4 and IPv6 networks. Dual-stack and configured tunneling are two mechanisms to enable IPv6 networks to communicate with IPv4 networks and vice versa.

Policy-based management is also a network management technique with specific focus. A framework has been defined in the IETF RFC 2753 for policy-based admission control [37]. It can be used for managing, sharing, and reusing policies in a vendor-independent, interoperable, and scalable manner.

Diameter base protocol specified in RFC 6733 [38] is a protocol with specific focus on providing an Authentication, Authorization, and Accounting (AAA) framework for applications such as network access or IP mobility. It works in both local AAA and roaming scenarios.

Another AAA protocol designed to manage network access is Remote Authentication Dial-In User Service (RADIUS). With a client-server model, it uses two types of packets to manage the full AAA process: Access-Request and Accounting-Request. The former manages authentication and authorization, while the latter manages accounting.

9.4 Network Management Data Models

Network management data models standardized by the IETF encompass a wide range of components such as MIB modules, IPFIX Information Elements, Syslog Structured Data Elements, and YANG modules. These data models are typically classified into three main categories: generic infrastructure data models, network management infrastructure data models, and data models specific to certain layers. The basic concepts of MIBs have been previously discussed in relation to SNMP, including examples such as MIB-II (Table 9.2) and RMON MIBs (Tables 9.3 and 9.4). A more comprehensive discussion regarding network management data models is presented below.

9.4.1 Generic Infrastructure Data Models

Generic infrastructure data models are core abstractions upon which many other data models are built. Among these models, a significant example is the Interface Groups MIB module, also known as IF-MIB, which is a part of MIB-II (Table 9.2). IF-MIB not only introduces the basic concept of network interfaces but also provides essential monitoring objects widely used for performance and fault management purposes. Another crucial infrastructure data model is the Entity MIB (version 4) specified in the IETF RFC 6933 [39]. This data model defines managed objects that facilitate the management of multiple logical and physical entities controlled by a single SNMP agent. Consequently, it plays a vital role in inventory management activities.

9.4.2 Management Infrastructure Data Models

Several data models are specifically designed for the management of the network management system itself. These data models encompass MIB modules that offer generic functionalities such as calculations involving MIB objects, thresholding and event generation, event notification logging, and system alarms. These models complement the existing SNMP MIB modules used for monitoring and configuring SNMP itself (Table 9.2).

The RMON family of MIB modules, which provide remote monitoring capabilities, can also be used for network management purposes (Tables 9.3 and 9.4). In addition, the IPFIX and Syslog protocols provide standardized information data models, many of which find relevance in network management operations.

9.4.3 Data Models at Specific Layers

Following the four-layer TCP/IP network architecture, data models are developed for specific layers: the data link layer, network layer, transport layer, and application layer. They are briefly described in the following.

At the data link layer, various data models in the form of MIB modules exist for different data link technologies, such as ADSL, ATM, Ethernet, ISDN, and others. These MIBs extend the generic network information data model with interface-specific information. They primarily focus on monitoring capabilities for performance and management tasks, and to some extent, accounting and security management functions.

In the network layer, the MIB-II (Table 9.2) includes IP and ICMP groups of MIBs, which form part of the network layer data models. Also, there are data models defined for routing protocols such as OSPF, IS-IS, BGP-4, and multicast routing. IP performance metrics are also defined as data models for performance management tasks. Furthermore, data models defined in NETCONF also cover the network layer.

For the transport layer, data models are defined in the TCP and UDP groups of MIB-II modules (Table 9.2). Data models are also developed specifically for SCTP at the transport layer.

Since SNMP is an application-layer protocol, the SNMP MIBs defined in MIB-II (Table 9.2) can be generally regarded as application-layer data models. However, in RFC 6632, they are explicitly classified as management infrastructure data models. Some data models are defined for specific application protocols or for instrumenting applications to collect information for performance and fault management. However, compared to the data models in the other three layers mentioned earlier, generic application-layer data models are less widely adopted in deployment.

9.4.4 An FCAPS View of Management Data Models

According to the FCAPS framework, network management data models can be aligned with specific network management functions or tasks for fault management, configuration management, accounting management, performance management, and security management. However, certain data models may not fit into a single category and may span multiple areas. For example, transmission and protocol MIBs are examples of data models that cover multiple FCAPS categories.

Data Models for Fault Management

A significant number of MIB modules are available for fault management purposes. In STD 62 [40], which comprises several RFCs, there is a system-group MIB module (Table 9.2) that is commonly polled to verify the operational status of a device. Another MIB module includes objects for managing notifications, such as tables for addressing, retry parameters, security, lists of targets for notifications, and user customization filters.

The interface group MIB, listed in Table 9.2, is specifically designed to manage and monitor the status of network interfaces, providing valuable information for fault management activities. The RMON MIB modules illustrated in Tables 9.3 and 9.4 can be configured to detect predefined conditions, particularly error conditions, on existing MIB variables. When one of these conditions is met, events can be logged, and depending on the system design, management stations may receive notifications.

Moreover, various other MIB modules defined in different RFCs can be utilized for fault management purposes. Examples of such modules include the Alarm MIB, Alarm Reporting MIB, IPFIX MIB, and MIB for ping/traceroute/lookup operations. These modules offer additional functionality and specific mechanisms to facilitate fault detection, reporting, and resolution in network management operations.

Data Models for Configuration Management

Configuration management encompasses various tasks related to configuring network devices such as physical and logical network topology, inventory management, and software management. A wide range of MIB modules have been defined to support the monitoring of network configuration.

Existing MIB modules such as those in MIB-II (Table 9.2), RMON1 (Table 9.3), and RMON2 (Table 9.4) provide valuable information and capabilities for monitoring network configuration. These modules offer insights into device configuration parameters, performance metrics, and network statistics, enabling effective configuration management.

Moreover, new MIB modules can be designed to cater to specific functionality requirements. These modules can focus on monitoring and modifying operational parameters, such as event reporting thresholds, to provide enhanced configuration management capabilities. By defining custom MIB modules, network administrators can tailor their configuration management processes to meet their specific needs and efficiently manage network devices.

Data Models for Accounting Management

Thus far, a standardized mechanism for collecting usage information of network resources for accounting management has not been established. As a result, accounting data is typically collected or calculated indirectly. For example, RADIUS employs accounting client and server MIB modules that define corresponding objects for IPv6. Moreover, IPFIX and PSAMP use information elements containing IP flow information, facilitating usage-based accounting. Although these approaches provide some solutions, a comprehensive standardized mechanism for direct accounting data collection is still lacking.

Data Models for Performance Management

Performance management is crucial for maintaining the desired level of overall network performance. It monitors network parameters, collects statistical information, and analyzes activity logs. RMON MIBs (Tables 9.3 and 9.4) are commonly used for performance management tasks, providing valuable insights. Moreover, other relevant MIBs can be used, such as those generated for the collection and reporting of metrics that assess the quality of VoIP sessions. These tools aid in effectively managing and optimizing network performance.

Data Models for Security Management

Security management plays a vital role in safeguarding networks and systems against potential attacks, ensuring the preservation of CIA. RADIUS and Diameter are widely recognized as security management protocols. Various data models can be applied to enhance security management practices.

For example, the NETCONF Access Control Model (NACM) defined in the IETF RFC 8341 [41] addresses the requirement of a structured and secure environment for NETCONF protocol or HTTP-based RESTCONF (REpresentational State Transfer

CONFiguration) protocol defined in RFC 8040 [42]. Thus, it provides a standard access control model to restrict NETCONF or RESTCONF protocol access for particular users to a pre-configured subset of all available NETCONF or RESTCONF (REpresentational State Transfer CONFiguration) protocol operations and content. Numerous MIB modules defined in STD 62 [40], which consists of a number of RFCs, can also be used for security management.

9.4.5 Model-Based Network Management

Computers networks are becoming increasingly more complex, posing challenges for network management. While traditional ad-hoc and command-line interface (CLI) approaches still have their uses in specific scenarios, they are not suitable for systematic management of large-scale networks. Protocol-based network management has emerged as the primary means of managing networks. To improve the efficiency of network management, automating all management processes is desired. This has led to the development of model-based network management, a relatively new technique that is being actively developed.

A core concept in model-based network management is the use of a common modeling language for effective communication between network devices and management system modules. The YANG language has been defined as the standard data modeling language for this purpose.

In general, two types of data models have been developed in model-based network management: open models and native models. Open models are designed to be independent of the underlying platform implementation and are implemented in the YANG language. This approach standardizes the configuration of network devices across different vendors. Native models, on the other hand, provide a model-based interface to the existing CLI, enabling a transition from CLI-based device management to a more model-driven approach.

Model-based management offers unique features compared to traditional network management approaches. The data models used in model-based management are programmable, allowing for flexible management operations. New data models can be added to the management system as needed, ensuring scalability. Network management tasks can be automated, e.g., by using Python and other scripting languages, leading to greater efficiency and simplicity. Furthermore, configuration operations can be treated as transactions, where a configuration is either applied successfully or rolled back automatically in the event of a failed attempt.

Let us briefly discuss an example on Cisco's ISO XE platform. A wide range of OS- and platform-specific YANG models have been developed for model-based management. These models provide interfaces in the YANG language to interoperate with existing device CLI, Syslog, and SNMP interfaces. When enabled, they can run natively as Linux processes on the device, translating configuration messages from YANG models into CLI commands that can be executed on the device.

9.5 Network Management Mechanisms

Network management is a broad concept in networking. It involves the management of network devices, monitoring various performance metrics, and handling logical entities such as traffic flows. These aspects are monitored using network management protocols and MIBs, which have been discussed in the last few sections. In this section, we will begin by discussing network devices and their characteristics. This will be followed by introducing network management mechanisms.

9.5.1 Characterizing Network Devices for Management

A *network device* is a network entity that functions as an individual component at a specific layer or across multiple layers. Examples of network devices include NICs at Layers 1 and 2, Layer-2 switches, Layer-3 routers, DHCP servers, DNS servers, and more. Routing switches are devices that operate at both Layers 3 and 2, combining routing and switching functionalities.

While the management of network devices is primarily associated with performance management, other management functions, such as fault management, are also related to network devices. However, measuring and analyzing the performance of network devices is crucial for all these management functions.

In performance management, including network device management, two types of performance should be monitored: *end-to-end management* and *component management*:

- End-to-end management involves measuring, quantifying, and analyzing performance from one end to the other end along a network link that connects multiple network devices. It can measure availability, capacity, data rate, throughput, utilization, latency, jitter, reachability, Round Trip Time (RTT), error, and traffic burstiness. It is worth mentioning that in computer networking, network services are provisioned and managed end-to-end.
- Component management focuses on managing the performance of individual components, links, or devices. This is also known as *per-link, per-element, or per-network* characteristics. For example, data rate and throughput can be measured for a specific interface of a router.

In the practical management of network devices, end-to-end management and per-link/per-element/per-network management can be used separately or combined for specific management tasks. For example, the one-way delay from one end to the other end is the sum of multiple delays across a series of links that form the overall link, as shown in Fig. 9.5.

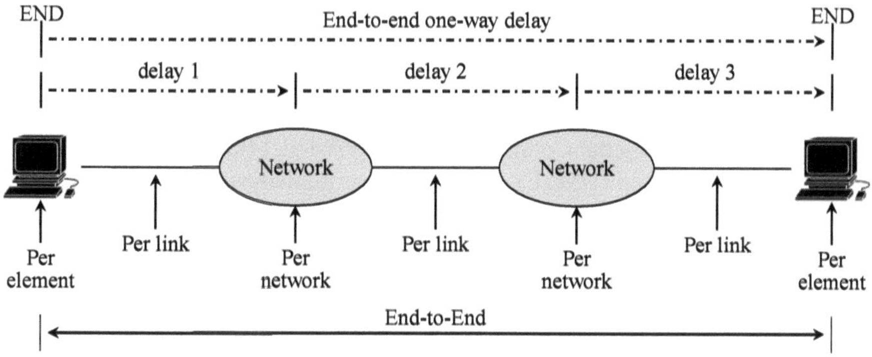

Fig. 9.5 End-to-end and per-link/per-network/per-element management of network devices

9.5.2 *Instrumentation*

Instrumentation is the process of probing the network to gather various management data for network management purposes. It is designed as part of the network management architecture to fulfill the requirements of monitoring network performance metrics. Therefore, instrumentation is planned with considerations for monitoring, visualization, processing, and storage of management data models in the form of various MIBs.

To design effective instrumentation strategies, a detailed analysis should be conducted to understand the requirements of instrumentation for each type of network devices, such as storage, routers, switches, and computing nodes. The analysis can begin by examining the current network settings and configurations and then proceed to identify any existing or potential issues in the current and future networks. It is also important to have a simple, reliable, and scalable design for not only instrumentation but also the entire network management system. When network problems occur, the network management system should not be the first system to fail in the network.

The simplest method of instrumentation is through direct access to network management data. Application-layer protocols such as SSH, FTP, SFTP, and TFTP allow direct access to MIBs from a remote host. SSH, usually running over TCP, enables secure remote login. FTP, which operates over TCP, and its secure version SFTP, are used for file transfers, including MIB files, to and from remote hosts. Trivial File Transfer Protocol (TFTP), as a lightweight file transfer protocol that operates over UDP, can be used to transfer configurations to and from network devices.

Monitoring tools can also be employed for instrumentation. General utilities such as `ping`, `traceroute`, and `tcpdump` are examples of tools used for network performance monitoring and evaluation. At the operating system level, Windows provides Windows Management Instrumentation (WMI), which is a set of extensions to the Windows Driver Model. WMI classes such as `Win32_Process` can gather useful information about processes. The Windows command `wmi` extends WMI for operation from various command-line interfaces. WMI is used to build scripts for web administration tasks, including accessing, reading, and modifying key configuration files.

Windows WMI is an implementation of Web-Based Enterprise Management (WBEM), which defines specifications for discovering, accessing, and manipulating resources modeled using the Common Information Model (CIM) of the Distributed Management Task Force (DMTF). WBEM is also implemented and used in Linux systems. For example, the Subsystem for Linux-based Implementations of WBEM (SBLIM) is a collection of system management tools that enable WBEM on Linux. In Linux systems, the kernel exposes a plethora of information and tunable settings via the `/proc` and `/sys` file systems, which can be accessed for network performance analysis and evaluation.

Network management protocols play an important role in the systematic design of instrumentation. SNMP is the most commonly used network management protocol for this purpose. It provides access to various MIBs, including MIB-II (as shown in Table 9.2), RMON MIBs (as shown in Tables 9.3 and 9.4), and many other specific MIBs discussed extensively in previous sections. Numerous commercial and open-source monitoring software packages use SNMP for accessing MIBs.

In IP networks, the IPFIX and PSAMP protocols can be employed for instrumentation to monitor IP flows and packets. It is worth mentioning that while IPFIX is based heavily on the Cisco-specific NetFlow, it is an open standard that supersedes NetFlow. In the Cisco environment, IPFIX is usually referred to as NetFlow v10. The details of the IPFIX and PSAMP protocols have been discussed previously. In brief, using SCTP as the mandatory transport protocol (with TCP and UDP as optional transport protocols), IPFIX defines how IP traffic flows are exchanged over the network, including the content, format, and structure of the exchanged information. PSAMP, on the other hand, defines how to select subsets of packets and how to send the selected packets to a Collector.

9.5.3 Event Notification and Trend Analysis

Monitoring network parameters can be implemented to enable event notification. When a network parameter exceeds its predefined threshold, a warning or alarm can be triggered to indicate possible or actual performance degradation. For example, an event notifying heavy broadcast traffic on a LAN can be sent to the network administrator and/or network management system. Link failures can be reported through event notification as well. Other examples of events that can be reported include burst traffic originating from a blacklisted address, no response to a critical database query, and SPF failures in email services.

Monitoring for event notification can be designed with periodic polling for real-time monitoring. In this case, it is important to analyze the amount of data generated within a given time period and assess the impact of transporting and storing the data on overall network performance. For example, in a network with 200 devices, each having four interfaces, and monitoring 10 characteristics per interface, the total number of characteristics measured in each polling would be $200 \times 4 \times 10 = 8,000$ characteristics. If 8 bytes of data are generated for each characteristic and the protocol overhead is 60 bytes, the total amount of data generated in each polling period would be $8,000 \times (8 + 60) = 544$ kB or 4.352 M bits. This is a substantial amount of data. With a polling period of six seconds, the required average transmission rate for the generated data would be over 725 kbps. The storage required for the generated data in a day (24 h) would be 921.6 MB, equivalent to over 336.38 GB per year. Therefore, a trade-off must be considered between polling frequency and the amount of generated data. This will be further discussed in Sect. 9.6.4 regarding network management traffic considerations.

Monitoring network parameters can also be designed for trend analysis, aiming to identify network behaviors and dynamics over a period of time, thereby discovering trends or potential issues. The information extracted from trend analysis is valuable for planning network upgrades, growth, or additional resource allocation. By periodically or sporadically polling network parameters, statistics and trends of the parameter measurements or calculated metrics can be obtained for different time intervals, such as an hour, half a day, a day, a week, or a month. For example, Fig. 9.6 shows the measurements of VoIP latency over 24 h and network traffic over a week. The nominal value of VoIP latency shown in Fig. 9.6a is approximately 150 ms. If trend analysis indicates a significant potential increase in VoIP latency to over 300 ms, actions need to be taken to prevent quality degradation of VoIP services. Regarding network traffic, if trend analysis reveals deviations from the pattern shown in Fig. 9.6b, indicating abnormal or unusual behavior. Therefore, further investigations and resolution would become necessary.

9.5.4 Configuration of Network Parameters

Configuration of network parameters is part of configuration management in network management. It can be done through:

- SSH and CLI operations;
- Direct access to configuration files;
- Utilities at the operating system level; and
- Protocols.

For SSH and CLI operations, and direct access to configuration files, discussions have been conducted previously when introducing instrumentation mechanisms. Similar techniques can be used to configure network parameters, for instance, by accessing the Linux `/proc` and `/sys` file systems.

For operating system utilities that allow network parameter configuration, the WBEM technique is useful, which has been discussed when presenting instrumentation mechanisms. It has implementations in both Linux and Windows environments. In Linux, SBLIM is an IBM-initiated open source project that implements the functions of WBEM. In Windows, WMI is an implementation of WBEM.

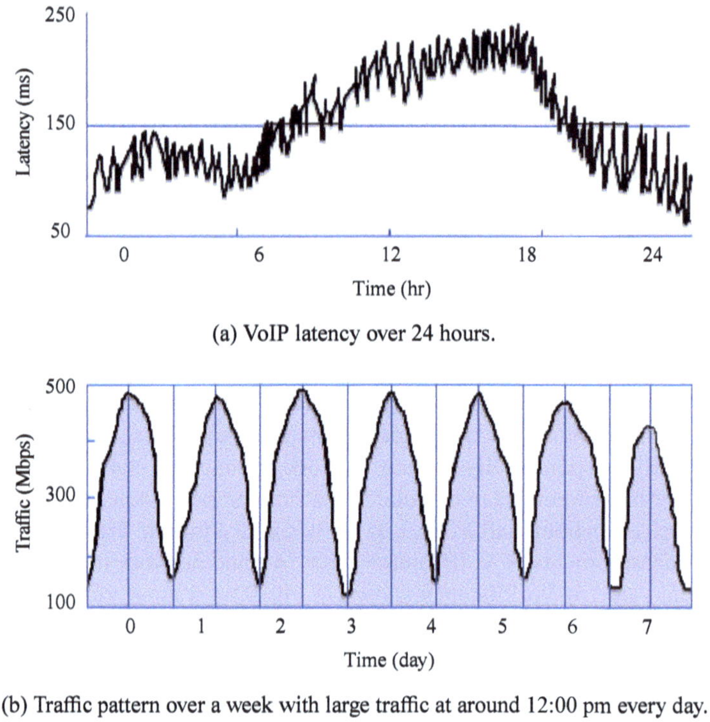

Fig. 9.6 Trend analysis for VoIP latency over 24 h and network traffic over a week

A number of protocols enable network parameter configurations. For example, the SNMP `get` and `set` commands (Table 9.1) request to retrieve and configure a variable or a list of variables, respectively. That being said, SNMP is generally not used for network configuration. The NETCONF protocol discussed previously (RFC 6241 [31]) is designed specifically for configuration management. A number of other protocols have been developed to support NETCONF, such as the data modeling language YANG (RFC 6020 and RFC 7950 [32, 33]).

9.6 Management Architectural Models

In network management architecture, there are three essential components, e.g., managed devices, network management agents residing in the managed devices, and one or more Network Management Systems (NMSs) that manage the network. These components are described below:

- A network device under management is a network node such as a server, router, switch, or an end-user's host, from which management information is collected.

The collected information can be stored and pre-processed locally. The raw or pre-processed information will be sent to NMSs.

- An agent is a network management software installed on a network device. It manages the collection of management information and sends the collected information to NMSs using a protocol such as SNMP.
- An NMS is an application-layer software system that runs various applications for network management. Typically, it communicates with agents; receives, displays, processes, and stores Network Management Data (NMD); and monitors and controls the managed devices. For large-scale computer networks, an NMS runs on a server with powerful capabilities in computing, graphics, memory, and storage.

Different arrangements of managed devices, agents, and NMSs in a network form different architecture models for network management. Typical architectural models for network management include in-band or out-of-band management, and centralized, distributed, or hierarchical management, which will be discussed below in detail.

9.6.1 In-Band and Out-of-Band Network Management

In-band management is a network management model in which the traffic flows for network management follow the same paths as normal traffic flows from network applications. A network administrator may use SSH or SNMP tools to reach network devices over existing networks for network management tasks. For the scenario of SNMP network management shown in Fig. 9.3, a logical diagram of in-band network management is depicted in Fig. 9.7a.

As no additional network paths are required to transfer network management data, it simplifies the design and implementation of network management architecture. Existing network paths are used for the transmission of both network management traffic and normal network communication traffic. However, the overhead introduced by the management to the network paths may affect the performance of normal network communications. It may even overload the paths, thus causing network congestion. Therefore, the implementation of in-band network management in a network needs to be carefully evaluated and closely monitored. Otherwise, critical network monitoring may not be available when it is needed.

For large or business-critical networks, in-band network management is not enough. If the network is down, network devices become unreachable over the network for network management, leading to a big risk for the organization and its business that the network serves. Therefore, an alternate or secondary network path is necessary for the management of the network. The network management model with different network paths for network management traffic flows and normal traffic flows is known as *out-of-band management*. For the topology of the SNMP network management in Fig. 9.3, out-of-band management is shown in Fig. 9.7b. It transfers network management traffic flows over dedicated network paths, thus increasing the

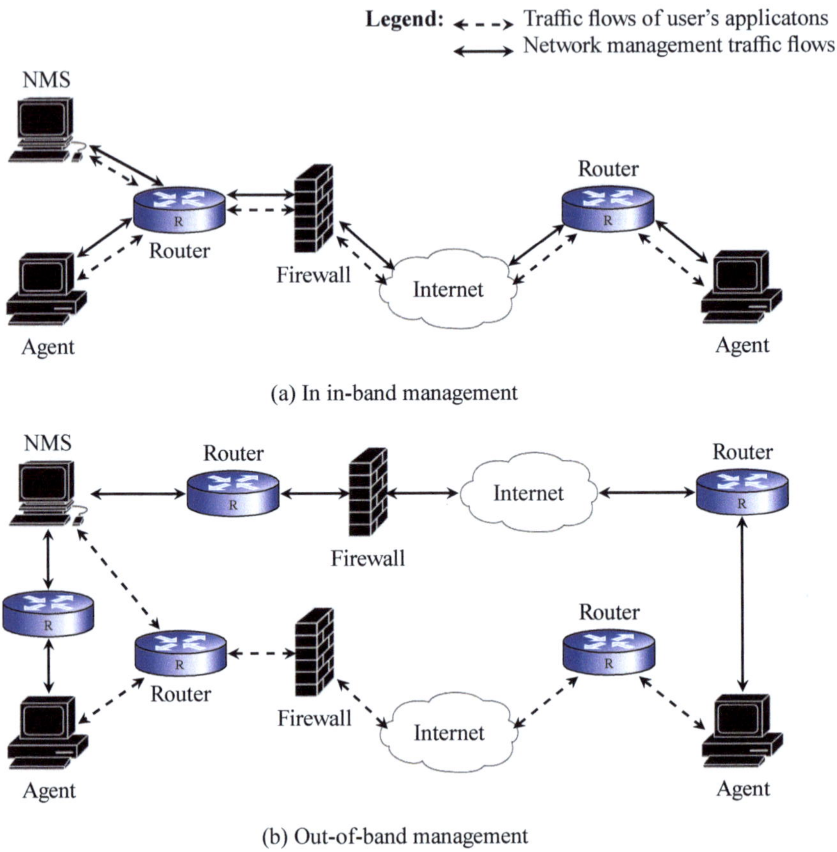

Fig. 9.7 In-band and out-of-band network management. In in-band management, management traffic flows follow the same paths as the traffic flows of user's applications. In comparison, in out-of-band management, separate and dedicated paths are used for management traffic flows

reliability of network management. It allows monitoring the network continuously even if some errors or events occur that disable the connections of normal network paths over which in-band management works.

To make out-of-band network management reliable, it is usually provided via a separate network. Out-of-band network management over a separate network has distinct advantages. If the network under management is down, the devices in the network are still accessible remotely over the separate network for re-configuration, re-setting, and troubleshooting. Also, with a separate network for network management traffic flows in out-of-band management, it becomes easier to integrate enhanced security mechanisms and policies into the network management. As all network devices are accessible remotely through network management, improved security is important for the security protection of the entire network and network devices.

```
Hostname#configure t
Hostname(config)#control-plane
Hostname(config-ctrl)#management-plane
Hostname(config-mpp)#inband
Hostname(config-mpp-inband-int name)#interface{typeinstance|all}
Hostname(config-protocol-peer)#allow{protocol|all} [peer]
Hostname(config-protocol-peer)#address ipv4{peer-ip-address
                                        |peerip-address/length}
Hostname(config-protocol-peer)#Use the commit or end command.
Hostname#show mgmt-plane[inband|out-of-band]
                        [interface{type instance}
```

(a) In-band management

```
Hostname#configure t
Hostname(config)#control-plane
Hostname(config-ctrl)#management-plane
Hostname(config-mpp)#out-of-band
Hostname(config-mpp-outband)#vrf vrf-name
Hostname(config-mpp-outband)#interface{typeinstance|all}
Hostname(config-mpp-outband-int name)#allow{protocol|all} [peer]
Hostname(config-protocol-peer)#address ipv6{peer-ip-address
                                        |peerip-address/length}
Hostname(config-protocol-peer)#Use the commit or end command.
Hostname#show mgmt-plane[inband|out-of-band]
                        [interface{type instance} |vrf]
```

(b) Out-of-band management

Fig. 9.8 Configuration of in-band and out-of-band management by using Cisco's *Management Plane Protection Commands* [43]

In-band and out-of-band management can be configured in various network management environments. For instance, Cisco has developed *Management Plane Protection (MPP) Commands*, which can be used in Cisco IOS XR and Cisco CRS Router [43]. Figure 9.8 shows how these MPP commands are used to configure in-band and out-of-band management. More examples are given in [44] on how to implement MPP.

9.6.2 Centralized and Distributed Network Management

In the *centralized management* architecture, a single NMS manages all network devices either directly for small-scale networks or indirectly through intermediate management nodes for large-scale networks. Agents are distributed across the network under management. They send management information data such as `ping` responses to the centralized NMS. From the traffic flow perspective, the central-

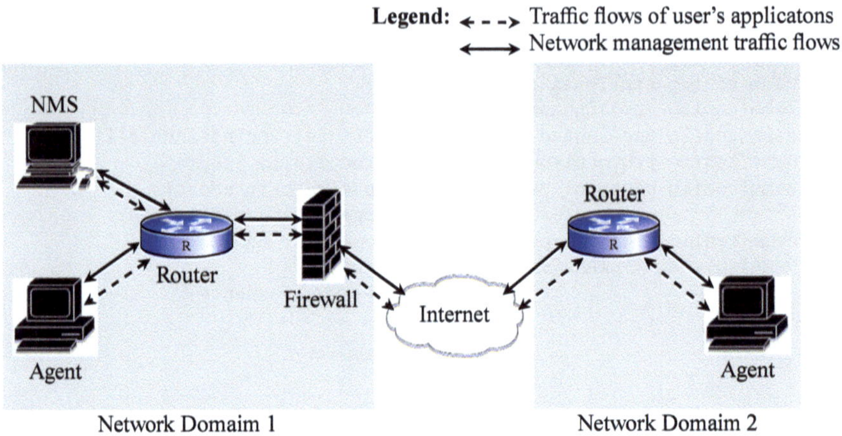

Fig. 9.9 Centralized network management, in which a centralized NMS manages all network domains across the Internet

ized management behaves like a client-server system. For the network considered in Figs. 9.3 and 9.7a, the architecture of centralized network management is shown in Fig. 9.9. It is shown in Fig. 9.9 that a single NMS manages the entire network with two network domains across the Internet. The two domains could be on two geographical locations, such as two campuses, two cities, or even two countries.

Centralized management has all management functions and NMS on the single NMS. Therefore, it is simple and relatively easy to implement and maintain particularly for small-scale networks. Also, a variety of software tools are available for use on a centralized NMS.

However, the centralized NMS is not always reliable. It is a single point of failure. If it fails, the network management becomes dysfunctional. Therefore, for the management of critical networks or critical components of networks, when a centralized NMS is used, consider other management measures as well, e.g., distributed management components or nodes, to enhance the reliability of the network management.

Distributed management refers to the arrangement where multiple NMSs or multiple components of one or more NMSs are deployed across the network under management. This concept is depicted in Fig. 9.10 through two different scenarios. Distributed network management localizes the management traffic within each managed domain. In comparison to centralized management, distributed management enhances reliability by eliminating the single point of failure issue and improves scalability as it can easily extend to larger networks.

In the case of distributed network management with multiple NMSs, the architecture is illustrated in Fig. 9.10a. Here, a local Element Management System (EMS) performs the tasks of an NMS. Since full management functionalities for the entire network across the Internet may not be required in a localized management domain, a local EMS may operate with fewer management functions compared to those for

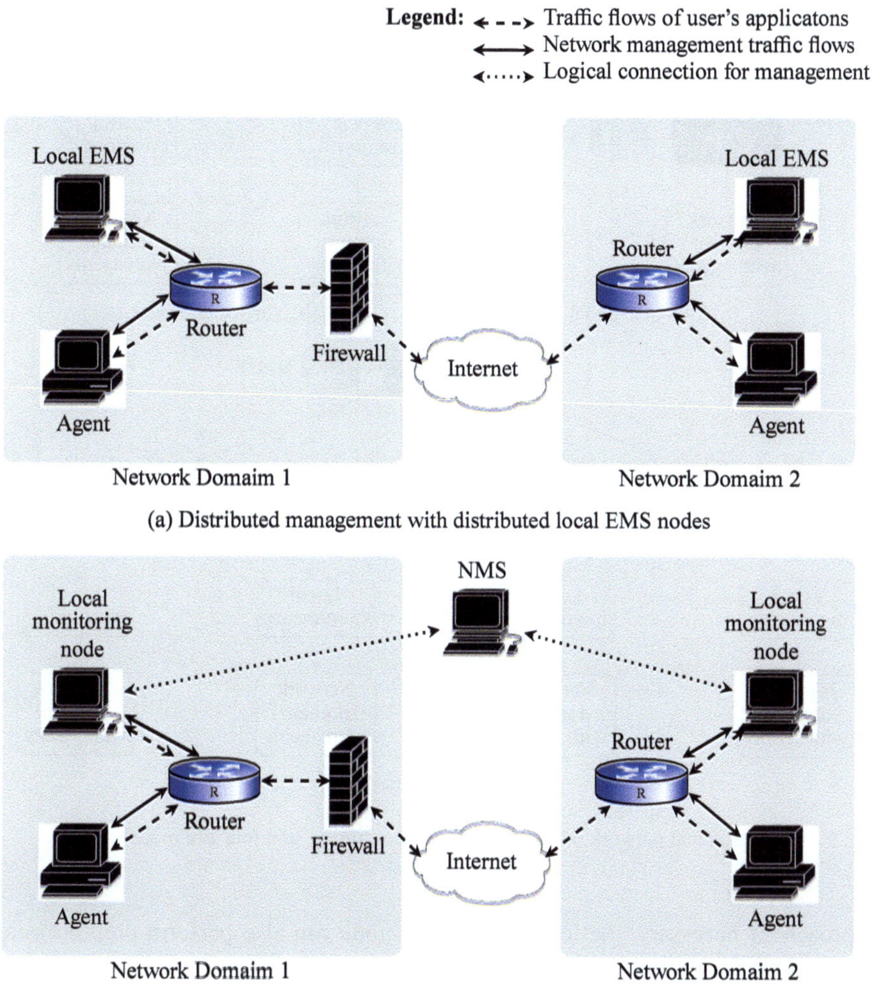

(a) Distributed management with distributed local EMS nodes

(b) Distributed management with local monitoring nodes managed by an NMS

Fig. 9.10 Distributed network management with distributed local EMS nodes or distributed local monitoring nodes

the overall network. Each local EMS operates independently and makes autonomous decisions regarding the management of its network domain.

On the other hand, Fig. 9.10b presents distributed network management with a single NMS and multiple monitoring components distributed throughout the network. Each local monitoring node is responsible for monitoring the network devices within its vicinity, collecting management information from these devices, and subsequently forwarding the information to the NMS through either an in-band or out-of-band

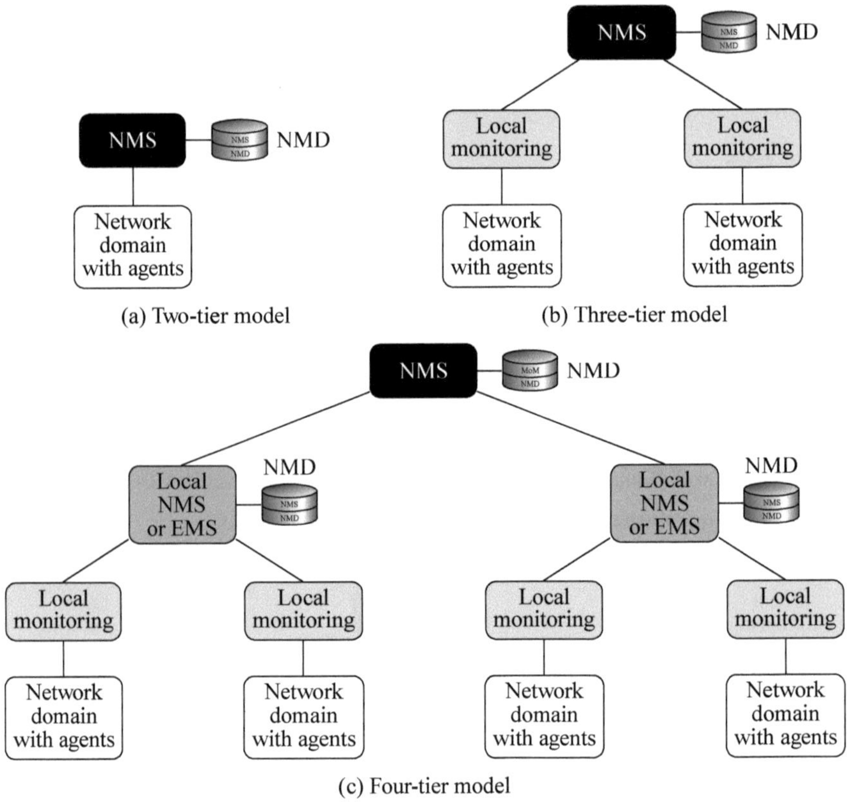

Fig. 9.11 Hierarchical network management with two-, three-, and four-tire models

approach. If necessary, the local monitoring node can also perform preprocessing and/or store the collected management information locally.

9.6.3 Hierarchical Network Management and MoM

Network management is typically implemented with at least two tiers: an NMS tier and a tier consisting of managed devices installed with network management agents. However, in practical network management, especially in large-scale networks, three or more tiers are commonly utilized. When certain functions of an NMS are distributed among multiple regional NMSs or local EMSs, a hierarchical model with three or four tiers is formed. This hierarchical model is depicted in Fig. 9.11 with two-tier, three-tier, and four-tier representations. In these hierarchical management models, each NMS, local NMS, or EMS has its own database for NMD.

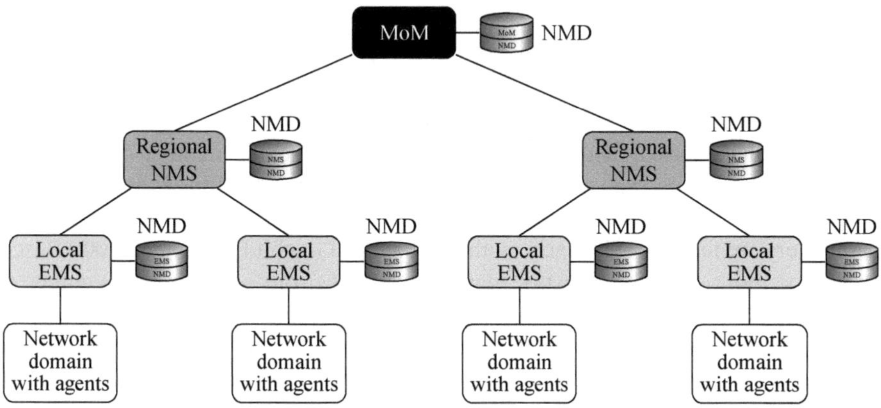

Fig. 9.12 Manager of managers (MoM), in which each manager has its own database NMD

Furthermore, for managing large-scale networks, an NMS can be deployed to oversee multiple regional NMSs. Each regional NMS manages one or more local EMSs, which in turn monitor and manage network domains embedded with management agents. This creates a Manager of Managers (MoM) architectural model, as illustrated in Fig. 9.12. In the MoM model, each manager maintains its own database for NMD. It may also transmit raw or pre-processed data to the manager at its immediate upper tier.

Distributed and hierarchical management models offer improved reliability and scalability. However, compared to centralized management, they are more complex to design, implement, and manage. Security becomes a critical concern in these models and must be addressed with care. It is essential to consider network security architecture in conjunction with these management models.

9.6.4 Consideration of Network Management Traffic

Network management has a significant impact on the capacity requirements of network traffic within the managed network. Therefore, it is crucial to carefully evaluate the overhead generated by network management and incorporate it into flow analysis and capacity design considerations. For the traffic generated from periodic polling of network parameters, a simple example has been discussed previously in Sect. 9.5.3 on monitoring for event notification and trend analysis.

Several factors should be taken into account when estimating management traffic. These include the number of devices to be managed, the number of interfaces on each device, the number of parameters to be monitored, the data collection interval, and the transmission of raw or pre-processed management information. From these factors, an estimation of the average data rate of management information can be derived.

As a general guideline, it is recommended to keep the management traffic within 2% to 10% of the capacity of a LAN. However, this range largely depends on the overall bandwidth capacity of the LAN. For example, in the case of 10 Mbps Ethernet on 10 BaseT networks, 5% of the bandwidth capacity corresponds to a 500 kbps data rate. In such scenarios, distributed management or a smaller number of monitored parameters would be preferred. On the other hand, for 100 Mbps Fast Ethernet on 100 BaseT networks, centralized management and a larger number of monitored parameters are feasible. Similarly, with 1000 Mbps Gigabit Ethernet on 1000 BaseT networks, more complex network management functions and a higher number of monitored parameters with shorter intervals can be accommodated.

It is worth noting that 10 BaseT Ethernet networks are outdated in enterprise systems. Since 2002, 10 Gbps Ethernet over fiber (IEEE 802.3ae) has been available for LANs. From 2006, 10G BaseT, which offers 10 Gbps Ethernet over copper twisted pair cables (IEEE 802.3an), has been deployed in LANs. Currently, there are ongoing developments of 100 Gbps Ethernet, 200 Gbps Ethernet, and 400 Gbps Ethernet under the IEEE 802.3 umbrella (Reference: IEEE 802.3 Task Force, Study Group, and Ad Hoc Officers, accessed on 25 Nov. 2021). For a 10 Gbps Ethernet network, 2% of the 10 Gbps bandwidth would equate to 200 Mbps, which is more than sufficient to fulfill the general requirements of management traffic.

If the management traffic exceeds 10% of the LAN capacity, it may be necessary to segment the LAN or redesign the network management to reduce the management traffic. Conversely, if the management traffic is less than 2%, particularly less than 1%, more parameters can be monitored at a higher frequency, or multiple LANs can be merged into a single LAN.

In general, WAN connections also require management to ensure connectivity and performance across the entire managed network. WAN connections can be monitored per WAN-LAN interface. Often, a WAN monitoring device on a WAN-LAN interface is also part of a LAN and can be used to monitor all devices on the LAN as well as network services across the WAN.

9.7 Summary

Network management focuses on the management of network devices and resources for the provisioning of various network services. It is an integral part of the overall network architecture. This chapter has introduced two main frameworks for network management: the IETF standard and the OSI FCAPS framework. While both frameworks adopt a hierarchical architecture for network management, their specific components within the hierarchy differ. The IETF standard outlines a hierarchical structure encompassing business, service, network, element, and network management, arranged from the highest level to the lowest. In contrast to the IETF standard, the ISO FCAPS framework defines five management components or functions: fault, configuration, accounting, performance, and security.

The IETF and ISO frameworks embody different philosophies. The IETF standard emphasizes the simplicity of network management, employing a variable-oriented approach. By contrast, the ISO FCAPS framework prioritizes powerful management capabilities, utilizing an object-oriented approach. Consequently, specific protocols have been developed by the IETF and ISO for network management purposes. The IETF protocol family encompasses SNMP, Syslog, IPFIX and PSAMP, as well as NETCONF and RESTCONF. On the other hand, the ISO protocol family includes CMIP/CMIS along with various supporting protocols. The implementation of CMIP/CMIS over TCP/IP gives CMOT.

A significant part of network management is MIBs, which range from standard MIB-II, RMON1 MIBs, RMON1 MIBs, to a large number of other specific MIBs. To support effective and efficient management of these MIBs, various models are designed in a number of RFCs as standards. Furthermore, a standard data modeling language, YANG, is also developed for formalizing data model modules. It facilitates model-based network management, which automates the processes of network management and interoperates with existing device CLI, Syslog, and SNMP interfaces.

The mechanisms of network management include instrumentation, monitoring for event notification, monitoring for trend analysis, configuration of network parameters, and troubleshooting. They are applied for end-to-end network management based on per link, per network, or per element.

Depending on the requirements of specific application scenarios, either in-band or out-of-band management can be designed for a network. It is also popular to combine both in-band and out-of-band management in the same network. The network paths dedicated to out-of-band management ensure the reliability of network management with improved timeliness and responsiveness. From the architectural perspective, network management can be centralized, distributed, or hierarchical. Hierarchical management inherits the advantages of centralized and distributed management. Thus, it is widely applied in networks. Eventually, the network management architecture will be integrated into the overall network architecture.

References

1. Ersue, M., Claise, B.: An overview of the IETF network management standard. RFC 6632, RFC Editor (2012). https://doi.org/10.17487/RFC6632
2. McCabe, J.D.: Network Analysis, Architecture, and Design, 3rd Edn. Morgan Kaufmann Publishers, Burlington, MA 01803, USA (2007). ISBN 978-0-12-370480-1
3. Union, I.T.: X.700: Management framework for open systems interconnection (OSI) for CCITT applications. CCITT X.700, ITU (1992)
4. for Standardization, I.O.: Information processing systems - open systems interconnection - basic reference model - Part 4: Management framework. ISO/IEC 7498-4:1989, First edn. ISO (1989)
5. Union, I.T.: ITU-T Recommendation X.711: Information technology – open systems interconnection – common management information protocol: Specification. ISO/IEC 9596-1:1998, ITU-T (1998)
6. Union, I.T.: ITU-T Recommendation X.710: Information technology - open systems interconnection - common management information service. ISO/IEC 9595:1998, ITU-T (1997)
7. Case, J., Mundy, R., Partain, D., Stewart, B.: Introduction and applicability statements for Internet standard management framework. RFC 3410, RFC Editor (2002). https://doi.org/10.17487/RFC3410
8. Harrington, D., Presuhn, R., Wijnen, B.: An architecture for describing simple network management protocol (SNMP) management frameworks. RFC 3411, RFC Editor (2002). https://doi.org/10.17487/RFC3411
9. Ly, F., Bathrick, G.: Definitions of extension managed objects for asymmetric digital subscriber lines. RFC 3440, RFC Editor (2002). https://doi.org/10.17487/RFC3440
10. McCloghrie, K., Rose, M.: Management information base for network management of TCP/IP-based internets: MIB-II. RFC 1213, RFC Editor (1991). https://doi.org/10.17487/RFC1213
11. Waldbusser, S.: Remote network monitoring management information base. RFC 2819, RFC Editor (2000). https://doi.org/10.17487/RFC2819
12. Waldbusser, S.: Remote network monitoring management information base version 2. RFC 4502, RFC Editor (2006). https://doi.org/10.17487/RFC4502
13. Waldbusser, S.: Remote network monitoring management information base for high capacity networks. RFC 3273, RFC Editor (2002). https://doi.org/10.17487/RFC3273
14. Waterman, R., , Lahaye, B., Romascanu, D., Waldbusser, S.: Remote network monitoring MIB extensions for switched networks version 1.0. RFC 2613, RFC Editor (1999). https://doi.org/10.17487/RFC2613
15. Waldbusser, S., Cole, R., Kalbfleisch, C., Romascanu, D.: Introduction to the remote monitoring (RMON) family of MIB modules. RFC 3577, RFC Editor (2003). https://doi.org/10.17487/RFC3577
16. Blumenthal, U., Wijnen, B.: User-based security model (USM) for version 3 of the simple network management protocol (SNMPv3). RFC 3414, RFC Editor (2002). https://doi.org/10.17487/RFC3414
17. Harrington, D., Hardaker, W.: Transport security model for the simple network management protocol (SNMP). RFC 5591, RFC Editor (2009). https://doi.org/10.17487/RFC5591
18. Gerhards, R.: The syslog protocol. RFC 5424, RFC Editor (2009). https://doi.org/10.17487/RFC5424
19. Schoenwaelder, J., Clemm, A., Karmakar, A.: Definitions of managed objects for mapping SYSLOG messages to simple network management protocol (SNMP) notifications. RFC 5424, RFC Editor (2009). https://doi.org/10.17487/RFC5676
20. Kelsey, J., Callas, J., Clemm, A.: Signed syslog messages. RFC 5848, RFC Editor (2010). https://doi.org/10.17487/RFC5848
21. Miao, F., Ma, Y., Salowey, J.: Transport layer security (TLS) transport mapping for syslog. RFC 5425, RFC Editor (2009). https://doi.org/10.17487/RFC5425
22. Quittek, J., Zseby, T., Claise, B., Zander, S.: Requirements for IP flow information export (IPFIX). RFC 3917, RFC Editor (2004). https://doi.org/10.17487/RFC3917

23. Claise, B., Trammell, B., Aitken, P.: Specification of the IP flow information export (IPFIX) protocol for the exchange of IP traffic flow information. RFC 7011, RFC Editor (2013). https://doi.org/10.17487/RFC7011
24. Stewart, R.: Stream control transmission protocol. RFC 4960, RFC Editor (2007). https://doi.org/10.17487/RFC4960
25. Sadasivan, G., Brownlee, N., Claise, B., Quittek, J.: Architecture for IP flow information export. RFC 5470, RFC Editor (2009). https://doi.org/10.17487/RFC5470
26. Duffield, N., Chiou, D., Claise, B., Greenberg, A., Grossglauser, M., Rexford, J.: A framework for packet selection and reporting. RFC 5474, RFC Editor (2009). https://doi.org/10.17487/RFC5474
27. Zseby, T., Molina, M., Duffield, N., Niccolini, S., Raspall, F.: Sampling and filtering techniques for IP packet selection. RFC 5475, RFC Editor (2009). https://doi.org/10.17487/RFC5475
28. Claise, B., Johnson, A., Quittek, J.: Packet sampling (PSAMP) protocol specifications. RFC 5476, RFC Editor (2009). https://doi.org/10.17487/RFC5476
29. Dietz, T., Claise, B., Aitken, P., Dressler, F., Carle, G.: Information model for packet sampling exports. RFC 5477, RFC Editor (2009). https://doi.org/10.17487/RFC5477
30. Schoenwaelder, J.: Overview of the 2002 IAB network management workshop. RFC 3535, RFC Editor (2003). https://doi.org/10.17487/RFC3535
31. Enns, R., Bjorklund, M., Schoenwaelder, J., Bierman, A.: Network configuration protocol (netconf). RFC 6241, RFC Editor (2011). https://doi.org/10.17487/RFC6241
32. Bjorklund, M.: YANG - a data modeling language for the network configuration protocol (NETCONF). RFC 6020, RFC Editor (2010). https://doi.org/10.17487/RFC6020
33. Bjorklund, M.: The YANG 1.1 data modeling language. RFC 7950, RFC Editor (2016). https://doi.org/10.17487/RFC7950
34. Wasserman, M.: Using the NETCONF protocol over secure shell (SSH). RFC 6242, RFC Editor (2011). https://doi.org/10.17487/RFC6242
35. Badra, M., Luchuk, A., Schoenwaelder, J.: Using the NETCONF protocol over transport layer security (TLS) with mutual X.509 authentication. RFC 7589, RFC Editor (2015). https://doi.org/10.17487/RFC7589
36. Warrier, U.S., Besaw, L., LaBarre, L., Handspicker, B.D.: Common management information services and protocols for the Internet (CMOT and CMIP). RFC 1189, RFC Editor (1990). https://doi.org/10.17487/RFC1189
37. Yavatkar, R., Pendarakis, D., Guerin, R.: A framework for policy-based admission control. RFC 2753, RFC Editor (2000). https://doi.org/10.17487/RFC2753
38. Fajardo, V., Arkko, J., Loughney, J., Zorn, G.: Diameter base protocol. RFC 6733, RFC Editor (2012). https://doi.org/10.17487/RFC6733
39. Bierman, A., Romascanu, D., Quittek, J., Chandramouli, M.: Entity MIB (version 4). RFC 6933, RFC Editor (2013). https://doi.org/10.17487/RFC6933
40. Editor, R.: Std62. RFC Editor Online Documentation. https://www.rfc-editor.org/info/std62. Accessed on 7 Dec 2021
41. Bierman, A., Bjorklund, M.: Network configuration access control model. RFC 8341, RFC Editor (2018). https://doi.org/10.17487/RFC8341
42. Bierman, A., Bjorklund, M., Watsen, K.: RESTCONF protocol. RFC 8040, RFC Editor (2017). https://doi.org/10.17487/RFC8040
43. Cisco: Management plane protection commands. Cisco Online Documentation, https://www.cisco.com/c/en/us/td/docs/iosxr/cisco8000/security/b-system-security-cr-cisco8000/management-plane-protection-commands.html. Accessed 23 Nov 2021
44. Cisco: Implementing management plane protection. Cisco Online Documentation. https://www.cisco.com/c/en/us/td/docs/routers/crs/software/crs_r4-0/security/configuration/guide/sc40crsbook_chapter7.html#con_999455_sysseccg_4. Accessed 23 Nov 2021

Chapter 10
Network Security and Privacy Architecture

Network security and privacy are one of the most critical considerations in computer networking. Encompassing all aspects and components of a network, they profoundly influence network performance and the behavior of network functions. Designing a security and privacy architecture that effectively safeguards the entire network while minimizing its impact on performance and behavior is a formidable challenge due to the fact that the threats originate both internally and externally. To tackle this challenge, a systematic approach to network security and privacy design is essential, especially by employing a top-down methodology and systems approach. This chapter aims to provide guidance towards achieving this goal.

Before delving into the various facets of security and privacy architecture, let us briefly clarify the concepts of network security and privacy. Although related, security and privacy are distinct concepts. Network security revolves around establishing rules and configurations to protect computer networks and data from unauthorized access, with its primary objectives governed by the principles of Confidentiality, Integrity, and Availability (CIA). On the other hand, privacy focuses on implementing measures to safeguard personal and other sensitive information. While security can be achieved independently without privacy, privacy cannot be achieved without ensuring security. Nonetheless, the term *security* is commonly used to encompass both security and privacy when no ambiguity arises.

10.1 Overall Recommendations

To develop a network security and privacy architecture, it is necessary to understand the process of security and privacy design. The process is briefly described below:

- Identify network assets, analyze security risks, and assess security requirements and trade-offs.

Y.-C. Tian and J. Gao, *Network Analysis and Architecture*, Signals and Communication Technology, https://doi.org/10.1007/978-981-99-5648-7_10

- Develop a security plan, define security policies, and establish procedures for applying the security policies.
- Choose appropriate security mechanisms and develop an implementation strategy for these mechanisms.
- Design an overall security and privacy architecture, ensuring the security of each important module, and implement the developed technical strategy and security procedures.
- Test the effectiveness of the security and privacy measures and regularly update the system to address any identified issues.
- Provide training to users, managers, and technical staff on network security and privacy.
- Develop a comprehensive plan to maintain network security and privacy, including:
  - Conducting regular and independent audits.
  - Monitoring audit logs and responding to any identified issues.
  - Monitoring the network and promptly responding to incidents.
  - Staying informed about new alerts, vulnerabilities, and risks.
  - Regularly reviewing and updating the security plan and policies as necessary.

This chapter will discuss some of the steps involved in the process of security and privacy design within the framework of security and privacy architecture.

The IETF RFC 2196 (September 1997) entitled *Site Security Handbook* [1] has covered various aspects of site security architecture, although new security and privacy mechanisms have emerged since its publication. The term *site* refers to any organization that owns computers and network-related resources. The handbook serves as a guideline for securely setting up computers and establishing procedures for sites with network connections. It provides a framework for developing security policies and procedures, and lists several issues and factors that sites must consider when designing their own policies. However, the handbook does not address the specific mechanisms that should be chosen or designed to meet security and privacy requirements. Therefore, how to select or design security and privacy mechanisms need to be considered separately.

Recently, Network Security Agency (NSA) has published a new Cybersecurity Technical Report entitled *Network Infrastructure Security Guidance* [2]. This report describes best practices in cybersecurity based on the extensive experience in responding to cybersecurity risks. The recommendations in the report include implementing perimeter and internal network defenses to enhance monitoring capabilities and access controls throughout the network. It is worth mentioning that Cybersecurity and Infrastructure Security Agency (CISA) also encourages to review and follow NSA's security guidance along with CISA's own recommendation of implementing network security through segmentation [3].

10.2 Architectural Considerations for Security

The design of network security and privacy component architecture is to fulfill the functional requirements of security and privacy. In this section, we will discuss the architectural considerations for security. These considerations help in establishing a robust and effective security framework.

10.2.1 Objectives

The security component architecture aims to achieve four main objectives, as discussed in the IETF RFC 2196 [1, pp. 11–14]:

(1) **Comprehensive Security Plan**: The first objective is to define a comprehensive security plan for each site. A security plan is a high-level framework that provides guidelines for the design of specific security policies and procedures. It outlines the overall approach and goals of the security measures to be implemented.
(2) **Separation of Services**: The second objective is to define the separation of services between external and internal provisioning, dedicated hosts, or different groups of users or hosts. By segregating services, better security management can be achieved. This separation helps control access and restrict unauthorized entities from accessing sensitive resources.
(3) **Access Control Philosophy**: The third objective is to decide the access control philosophy to be adopted when defining the security plan. Two common philosophies are *deny all* or *allow all*. The *deny all* philosophy maintains a white list of allowed access, meaning that access is denied by default unless explicitly permitted. By contrast, the *allow all* philosophy maintains a black list of blocked access, where access is allowed by default unless explicitly restricted. The chosen philosophy determines the default behavior of access control mechanisms.
(4) **Identification of Real Service Needs**: The fourth objective is to identify the real needs for services from the available network services. With a growing number of network services, it is important to assess and determine which services are essential for the network. Some services may be unnecessary or pose security risks, and they can be removed from the list of services under consideration. This helps streamline the network services and reduce potential vulnerabilities.

These four objectives provide a foundation for the development of a comprehensive security plan, which will be discussed in detail in the next section.

10.2.2 Network and Service Configuration for Security

In the security component architecture, it is important to consider network and service configuration for security protection from cybersecurity threats. The following aspects are essential for network security:

(1) Securing network infrastructure, such as hardware servers, routers, switches, hosts and computers, cables, and other devices. Network infrastructure also includes network management system (e.g., SNMP), services (e.g., DNS and WWW), and security (e.g., authentication and access control). Physical security and the avoidance of possible human errors should also be considered for securing network infrastructure.
(2) Securing the network against cybersecurity threats, such as DoS and DDoS attacks, which flood routers, firewalls, servers, and/or network links with extraneous traffic. Other cybersecurity threats include spoofing attacks, which cause the misrouting of traffic, e.g., to an intruder's host.
(3) Securing network services against cybersecurity attacks, e.g., on standard services (e.g., DNS, mail, WWW, FTP, and VPN) and user applications (e.g., cloud services).
(4) Securing the security protection systems, e.g., remote-access VPN. This is an often overlooked issue and may become the most obvious weakness in network security.

General and specific mechanisms will be introduced in the next few sections for the security of all these aspects.

10.2.3 NSA's Guide for Secure Network Architecture

In NSA's guidance for network infrastructure security recently released [2], a zero trust model is considered. Zero trust is a security model, a set of system design principles, and a coordinated cybersecurity and system management approach based on the understanding that cyber threats exist both inside and outside traditional network boundaries. Therefore, much of the NSA guidance can be applied at different boundaries of the network.

For a secure network, multiple layers of defense are recommended in network architecture design. NSA's recommendations include the following aspects [2, pp. 2–8]:

(1) Installing perimeter and internal defense devices,
(2) Grouping similar network systems,
(3) Removing backdoor connections,
(4) Implementing strict perimeter access controls,

(5) Implementing a Network Access Control Solution, and
(6) Limiting VPNs.

These recommendations will be briefly discussed in the following.

Installing Perimeter and Internal Defense Devices

To ensure network security, a multi-layer defense approach should be designed at the network's perimeter to provide protection against external threats and to monitor and restrict inbound and outbound traffic. Some examples of recommended measures are:

- Installing a border router.
- Deploying multi-layer next-generation firewalls throughout the network.
- Placing publicly accessible servers and systems in one or more DeMilitarized Zone (DMZ) subnets with appropriate access control.
- Implementing a network monitoring system such as an Intrusion Detection System (IDS).
- Deploying multiple dedicated remote log servers.
- Implementing redundant devices in core areas.

These measures contribute to strengthening the network's security posture and mitigating potential threats.

Grouping Similar Network Systems

Similar network systems within a network should be logically grouped together and then isolated either through subnetting or VLANs at the logical level, or physically through firewalls or filtering routers. This segmentation design helps enhance security by preventing adversarial lateral movement between different types of systems. For instance, it is advisable to separate workstations, servers, and printers from each other to mitigate the potential impact of a compromise in one system on the others. By implementing this segregation, the security protection against lateral movement and potential propagation of threats is improved.

Removing Backdoor Connections

A backdoor network connection exists between two or more network devices located in different network areas, which typically have distinct types of data and security requirements. With a backdoor connection, if one device is compromised, an adversary can exploit this connection to circumvent access restrictions and gain unauthorized access to other areas of the network.

Implementing Strict Perimeter Access Controls

Strict perimeter access controls aim to regulate inbound and outbound traffic and are achieved through the implementation of ACLs. For this purpose, the NSA recommends a deny-by-default and permit-by-exception approach for network connections.

Implementing a Network Access Control Solution

Network access control is responsible for monitoring authorized physical connections and preventing unauthorized physical connections on a network. It is strongly recommended because an adversary attempting to gain internal access to a network must either obtain access from within the network or find a means to infiltrate from the network's external perimeter.

Limiting VPNs

VPNs serve as a security mechanism for establishing encrypted communication between two endpoints. However, VPN gateways are susceptible to network scanning, brute force attempts, and zero-day vulnerabilities. To mitigate these risks, the NSA provides the following recommendations:

- Reserve the use of VPNs for situations where confidentiality and integrity cannot be adequately maintained through alternative means.
- Disable any unnecessary features on VPN gateways.
- Implement rigorous traffic filtering rules to enhance security.

Further details on VPN techniques will be covered later in this chapter.

10.3 Security and Privacy Plan

While security and privacy are always essential in networks, it is still necessary to understand what aspects of security and privacy should be protected and what mechanisms should be chosen to best protect security and privacy. This necessitates the development of a well-defined security and privacy plan, followed by a comprehensive analysis of requirements pertaining to different facets of security and privacy within the network.

10.3.1 Basic Approach

For the development of a security and privacy plan, the basic approach adopted by the IETF RFC 2196 [1] has been demonstrated over the years to be effective. This approach consists of the following steps [1, p. 6]:

(1) Identify the elements that require protection.
(2) Define the threats against which protection is needed.
(3) Evaluate the likelihood of the identified threats.
(4) Determine the appropriate mechanisms for security and privacy protection.
(5) Continuously review the process and make improvements whenever weaknesses are identified.

While RFC 2196 primarily focuses on step 4), it provides a brief discussion of the first three items. This chapter will provide more comprehensive discussions on all the first four steps.

10.3.2 Identifying What to Protect

While security is a crucial aspect of any network, it is important to have a clear understanding of the assets that require protection when designing a security and privacy architecture. Typical categories of assets include, but are not limited to, the following:

- Hardware: This includes all networking and networked devices such as personal computers, workstations, servers, access points, switches, and routers.
- System software: This refers to system-level software and software packages/systems, including operating systems, utilities, server programs, client agents, management systems, network services, and system services.
- Application software: This category covers various applications and user-developed software.
- Data: It encompasses databases, system and network logs, audit logs, and other data generated before or during execution, stored online, and archived offline.
- Documentation: This represents a special type of data that includes information about systems, networks, services, applications, and other relevant details.

Particularly, RFC 2196 mentions that the individuals who actually use the systems have been largely overlooked in risk analysis for security protection. Users, administrators, and maintainers of the systems should be adequately protected in the design of network security and privacy measures.

10.3.3 Clarifying Security Threats

After the assets that require protection are identified, a further *threat analysis* will assist in determining:

- The types of threats that pose a risk to those assets.
- The levels of risk associated with each identified threat.

Typical threats to general networks include, but are not limited to, the following:

(1) Unauthorized access to network resources and/or data.
(2) Unauthorized and/or unintended disclosure of information.
(3) Theft of data and/or other assets.
(4) Corruption of data, service, software, and/or hardware.
(5) Denial of Service (DoS) and Distributed DoS (DDoS) attacks.
(6) Viruses, worms, and spam.
(7) Physical damage to devices and/or systems.

Assessing the levels of risks associated with the identified threats helps in selecting and designing appropriate mechanisms to fulfill the requirements of security and privacy protection. Threat analysis involves a certain degree of subjectivity. However, certain assets, especially critical ones, require robust protection, and some threats are commonly associated with these assets, potentially posing severe risks. For example, in a banking system, mission-critical databases require special considerations for security and privacy protection. The risks of data theft and unauthorized/unintended disclosure of information are considerably high in terms of potential loss. In comparison, email services are not as critical as financial databases, even though they are also important assets that require protection.

10.3.4 Developing a Security and Privacy Plan

One of the initial steps in the design of security and privacy is the development of a *security plan* that considers both security and privacy. This plan serves as a high-level description of the actions required to meet security and privacy requirements. It provides a framework of broad guidelines that will inform the design of specific policies and procedures, ensuring consistency with the overall site security architecture. For instance, if there is a strong security restriction on external network access, all individual policies concerning various types of external access should adhere to similar restrictions. Thus, a security plan should be based on an analysis of network assets and risks while supporting the goals of the site.

In particular, a security plan should define a framework for various network services. RFC 2196 [1, pp. 10–11] outlines several key questions that need to be addressed, including:

- What network services will be provided?
- Which areas of the site will offer these services?

- Who will have access to the services?
- How will access be granted?
- Who will administer the access?

There is a plethora of network services, and their number continues to grow. While some services are essential to a site, others may not be truly useful. When developing a security plan, it is important to identify the genuine requirements for network services. Managing the security of internal network services within a site is comparatively easier than ensuring the security of external network services provided by third parties over the Internet. Therefore, it is advisable to identify unnecessary network services and eliminate them from the list of services to be provided.

Network services can possess different levels of access and, consequently, different trust models. Two fundamental and contrasting options for providing network services are the *delay all* and *allow all* approaches. The *deny all* option disables all services and selectively enables them on a case-by-case basis as needed. By contrast, the *allow all* option activates all services and permits all traffic flows by default. Each of these options can be applied to different parts of the site. For example, the *allow all* option might be used for internal traffic within the site, while the *deny all* option could be employed for all traffic outgoing from the site to the Internet.

A security plan should specify the various resources required for developing security policies and procedures and implementing them. These resources include time, personnel, and others. Special attention should be given to the individuals who must be involved in the development and implementation of the security plan. For example, will skilled security people be involved in the security administration? How will end users and their managers get involved in, and be trained on, security policies and procedures? In general, the individuals that must be involved should represent key stakeholders, management with budget and policy authority, technical staff who are network and security experts, and legal counsel with knowledge of the legal implications of different policy decisions.

It is worth mentioning that designing network security and privacy involves trade-offs. The following list illustrates some typical trade-offs:

- Providing security and privacy can be costly. Therefore, it is essential to evaluate the cost of implementing security measures against the potential losses resulting from an insecure network.
- Security and privacy mechanisms may impact network performance. For instance, data encryption and packet filtering can consume significant CPU, memory, and other computing resources.
- When encryption is employed, if all traffic passes through a single encryption device, that device becomes a single point of failure. Consequently, additional redundancy measures need to be implemented to ensure network availability.
- Offering more services may introduce additional risks. The risks associated with certain services may outweigh the benefits they provide.

These trade-offs should be carefully considered and aligned with the goals of the site.

10.4 Security Policies and Procedures

According to IETF RFC 2196 [1, pp. 7–8], a security policy is a formal statement that outlines the rules and guidelines to be followed by individuals who have authorized access to a site's technology and information assets. The primary objective of a security policy is to communicate the responsibilities of users, staff, and managers regarding the protection of technology and information assets.

On the other hand, security procedures are formal statements that provide detailed instructions on how to implement the security policies. These procedures define specific processes, such as configuration, login, audit, and maintenance processes, that are necessary for the effective implementation of the security policies. They serve as a practical guide for meeting the requirements specified in the security policy.

10.4.1 Characteristics of Security Policies

What characterizes a good security policy? A good security policy is practical, supported by administration procedures, accompanied by user training and guidelines, incorporates appropriate security tools, and defines the responsibilities of individuals involved. By encompassing these characteristics, the policy can effectively guide security practices and safeguard the site's assets.

Firstly, a good security policy should be practical and feasible to implement within the specific site or organization. This means that the policy should consider the site's resources, infrastructure, and operational capabilities, ensuring that it can be effectively enforced and maintained.

Secondly, a good security policy should be supported by appropriate administration procedures. These procedures outline the steps and protocols for implementing and managing the security policy, including aspects such as access controls, incident response, and security audits. Clear and well-defined procedures help ensure consistent enforcement and accountability.

Thirdly, user training and guidelines are crucial for a good security policy. Users need to be educated on their roles and responsibilities in maintaining security, as well as the specific measures and practices they should follow. Training programs, user awareness campaigns, and documentation can help promote a culture of security awareness and compliance among users.

Furthermore, a good security policy should align with industry best practices and use appropriate security tools and technologies. This includes hardware devices, software packages, and security services that aid in the enforcement and monitoring of the policy. These tools can provide functionalities such as access control, intrusion detection, encryption, and vulnerability management.

Moreover, a well-developed security policy should clearly define the areas of responsibility for individuals involved in the site's management and usage. This includes users, administrators, managers, and other stakeholders. By clearly outlining the roles and expectations of each party, the policy ensures that everyone understands their specific security obligations and contributes to the overall protection of the site's technology and information assets.

10.4.2 Components of Security Policies

A security and privacy policy typically encompasses several important components, which help establish guidelines and procedures for protecting assets and ensuring compliance with security requirements. The following components are commonly included:

- **Access Policy**: This component defines the access rights and privileges of users, operations staff, and managers. It outlines the rules and restrictions for accessing sensitive information and resources, aiming to prevent unauthorized access and protect assets from loss or unauthorized disclosure.
- **Privacy Policy**: The privacy policy sets forth the organization's stance on privacy and outlines the expectations users can have regarding the confidentiality of their information. It addresses concerns such as monitoring of email communications, logging of keystrokes, and access to users' files and private information. The policy clarifies the organization's approach to privacy protection and promotes transparency in data handling practices.
- **Authentication Policy**: This policy focuses on establishing trust and ensuring secure authentication mechanisms. It typically includes guidelines for creating strong passwords, enforcing password expiration and complexity requirements, and recommending the use of multi-factor authentication. Moreover, the authentication policy may provide instructions for secure remote authentication to ensure secure access from remote locations.
- **Accountability Policy**: The accountability policy defines the responsibilities of individuals involved in the organization's security and privacy practices. It outlines the roles and obligations of users, operations staff, and managers in maintaining security, specifies the need for an audit capability to track system activities, and provides guidelines for incident handling and response. This policy promotes a culture of accountability and helps ensure that security incidents are appropriately addressed.
- **Computer Technology Purchasing Guidelines**: These guidelines specify the security requirements and considerations that need to be taken into account when purchasing computer technology. They outline the necessary security features and standards that should be met by the acquired hardware and software, ensuring that the organization's security objectives are upheld.

While these components form the foundation of a security and privacy policy, additional components may be included if needed [1, pp. 9–10]. These could include sections on data classification, data retention and disposal, encryption practices, network security measures, and acceptable use policies. The goal is to develop a comprehensive policy that addresses all relevant aspects of security and privacy within the organization's context.

10.5 Security and Privacy Mechanisms

There are a few fundamental security and privacy mechanisms applicable to all network environments. A typical example is Authentication, Authorization, and Accounting (AAA). Also, specific security and privacy mechanisms, such as firewalls and encryption, can be applied to particular scenarios. In general, not all mechanisms are suitable for every network environment. Each mechanism must be carefully evaluated to determine its appropriateness for the specific network that requires protection. This section will discuss commonly used security and privacy mechanisms, including security awareness, physical security, AAA, encryption and decryption, packet filters, and firewalls. These mechanisms will be used in the modularized security design later on.

10.5.1 Security Awareness

Security awareness refers to users' understanding of the following two main aspects:

- The potential risks of violating security policies and procedures.
- The way of avoiding situations that might cause breaches.

Security awareness training should have the following elements:

- Training of users on different types of threats, such as viruses, spams, malware, phishing, whaling, and fraud.
- Training of users on IT, Internet, and privacy policies.
- Training of users on password policies and other authentication methods.
- Training of users on threat recognition and response.
- Training of technical staff, who administer and manage the IT system and network, on new threats and recent developments of security technologies.

10.5.2 Physical Security

Physical security aims to protect hardware assets from physical access, damage, and theft. These hardware assets include network devices like routers and switches, various servers and server farms, specialized devices, and security devices. In any case, core routers and servers, as well as firewalls and other security devices, must be physically secure. An effective approach to physical security is to keep key hardware devices behind locked doors and restrict physical access to authorized personnel only. Access to the locked rooms where hardware devices reside should be limited to individuals with proper authorization.

Physical security is the most basic form of security. However, due to its apparent simplicity and ease of implementation, it is often overlooked during the development of a security plan. Particularly, different hardware devices may require varying levels of physical security protection. For example, a network printer may be physically accessible to all members of a work group but should remain physically inaccessible to visitors and other users. Switches, routers, and wireless access points should not be physically accessible to regular users, but rather to authorized network administrators and engineers. Limited physical access should be granted to specific managers and network engineers as needed. Moreover, primary and backup storage may be located in separate areas or rooms. Primary and backup communication cables should be physically separated from each other. Redundant power supplies should be designed, particularly for critical servers and other devices.

10.5.3 Authentication, Authorization, and Accounting

AAA forms a framework that controls access to network resources, enforces policies, audits usage, and provides the information required for service billing. These processes work together to ensure effective network management and security.

Authentication

In security, the term *authentication* refers to the process of identifying which user is requesting network services. For user authentication, a user must be given a user ID as *identification*. Then, a password is set as an initial level of security. Other methods of user validation include access cards, face recognition, and fingerprints. To ensure a high level of security, multi-factor authentication can be employed. The factors used in authentication should be something the user knows, has, or is. For instance, two-factor authentication may involve a password followed by personal information questions. It is noted that user authentication confirms the user's possession of the correct ID, password, and other identity proofs, but it does not guarantee the user's true identity.

Furthermore, the term *authentication* is also used to describe the process of establishing the identity of non-user subjects within a network. These subjects can include networking devices such as routers, switches, servers, and hosts, as well as software entities like applications, routing processes, and software agents. Authentication for non-users is used in protocols such as IPsec [4], where IP Authentication Header (AH) is employed to provide integrity and data origin authentication.

Authorization

While authentication focuses on identifying users or non-user subjects for accessing network resources, authorization refers to the process of granting privileges, rights, and permissible actions to processes and ultimately users. It determines what a user or subject is allowed to do after successful authentication.

Authorization typically varies from one user to another. For instance, a network administrator should possess higher-level privileges compared to a regular user. However, listing the authorized privileges of each user for all network resources or objects is impractical. Therefore, simplified approaches are employed to grant privileges to users. Two common approaches are as follows:

- Assign each object, e.g., a Unix/Linux file, three classes of users: owner, group, and other (world). Owner permissions determine what actions (e.g., read, write, and execute) the owner of the object can perform on the object. Group permissions determine what actions a user, who is a member of the group that the object belongs to, can perform on the object. Other (world) permissions indicate what actions all other users can perform on the object.
- Attach an Access Control List (ACL) to an object so that only the users on the ACL are permitted to access the object. Implementing an ACL is straightforward but requires additional resources for storage, as there could be a large number of ACLs in a real system.

Accounting

Accounting or *auditing* is the process of monitoring and measuring the resources consumed by a user during network access. It serves various purposes such as authorization control, billing, trend analysis, resource utilization monitoring, capacity planning, and security assessment.

The resources and activities that are monitored can include:

- System time usage.
- Data transmitted and received by a user during a session.
- Authentication and authorization attempts made by a user.
- User attempts to modify their privileges.

It is important to note that passwords should not be collected during the auditing process. This practice helps prevent security breaches that could occur if audit logs were shared improperly. Audit logs are valuable sources of information for analyzing security incidents.

10.5.4 Firewalls

A firewall is an internetwork gateway that controls and restricts data communication traffic between the internal network and external networks. By doing so, it protects the system resources of the internal network from potential threats originating from external networks [5, p. 130]. A firewall is also commonly understood as a device or system that enforces security policies to regulate incoming and outgoing network traffic. Typically, a firewall is deployed to safeguard a smaller network, such as a LAN or host, from a larger network like the Internet. It is placed at strategic locations, typically at the point where the networks connect, e.g., at the boundary between multiple networks.

Firewalls can take the form of software, hardware, or a combination of both. They can also be provisioned through cloud services. The main types of firewalls include:

- Packet-filtering firewalls
- Stateful inspection firewalls
- Proxy firewalls
- Next-generation firewalls

The security policies enforced by a firewall come from various sources, including end users, application developers, host administrators, network administrators, and firewall vendors.

Software, Hardware, and Cloud Firewalls

A firewall can be implemented as either a hardware or software solution, and sometimes both. A *software firewall*, also known as a *Host Firewall*, is installed directly on a host device. Each device requiring protection must have its own software firewall installed. For example, a user's personal computer can have a software firewall installed, and routers often come embedded with firewall functions for packet inspection and filtering.

A *hardware firewall*, also referred to as an *appliance firewall*, is a dedicated security device placed at the boundary between internal and external networks. This type of firewall operates as a standalone hardware device with its own computing resources. It serves as a gateway for traffic entering and leaving an internal network.

Both software and hardware firewalls have their specific use cases. For smaller networks, it may be cost-effective to install software firewalls on individual network devices. However, in larger-scale networks, deploying hardware firewalls becomes more practical than configuring software firewalls on each device.

In addition, there is a concept of *cloud firewall*, which is not a distinct type of firewall technology but rather a service model provided by cloud service providers. Cloud firewalls, typically offered as Firewall-as-a-Service, operate on the Internet within the cloud infrastructure. Users often use cloud firewalls as proxy servers, and their configurations can vary depending on specific use cases. The key advantages of cloud firewalls are scalability and reliability. The physical infrastructure supporting cloud firewalls is managed by the cloud service provider, ensuring reliable firewall services. Consequently, cloud firewalls are particularly useful for safeguarding an organization's cloud-based services, such as IaaS and PaaS.

Sources of Firewall Rules

As a general requirement [6], "the introduction of a firewall and any associated tunneling or access negotiation facilities MUST NOT cause unintended failures of legitimate and standards-compliant usage that would work were the firewall not present." However, according to RFC 7288 [7, p. 3], many firewalls currently implemented do not adhere to this requirement and may behave in unintended ways. This raises questions about the intended purpose of firewalls and the types of traffic that are considered unwanted by specific entities. For example, who is meant to be protected from what? What traffic is unwanted by whom?

These questions raise important considerations when it comes to firewall implementation. It is essential to define the scope of protection and identify the specific threats or risks that the firewall aims to mitigate. Additionally, understanding which types of network traffic are deemed unwanted by specific entities is crucial for configuring effective firewall rules.

To address these fundamental questions and provide clarity, RFC 7288 presents a list of typical sources for creating *allow* versus *block* rules. These sources are outlined in Table 10.1.

Packet-Filtering Firewalls

A *packet-filtering firewall* selectively allows or prevents the passage of data packets to and from a network according to a security policy. The security policy should

Table 10.1 Common sources of firewall rules [7, p. 5]

Source	Consumer scenario		Enterprise scenario	
	Host firewall	Network firewall	Host firewall	Networkfirewall
End user	Sometimes as host admin	Sometimes as network admin		
App developer	Yes	Sometimes via protocols		
Host admin	Sometimes		Yes	
Network admin		Sometimes		Yes
Firewall vendor	Yes	Yes	Yes	Yes

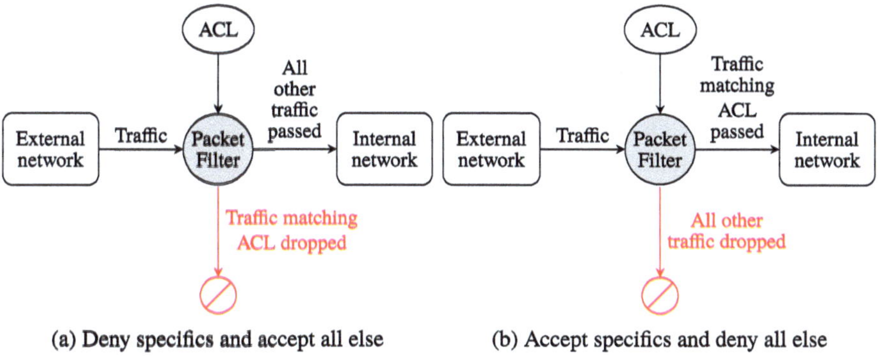

Fig. 10.1 The use of an ACL in two types of security policies

include well-defined rules about to whom, what, and where security is applied. It can be either of the following two types:

- *Deny specifics and accept all else* in an open network philosophy, or
- *Accept specifics and deny all else* in a closed network philosophy.

The first type that denies specifics and accepts all else looks like putting specifics onto a black list. It requires a good understanding of specific security threats, and is generally difficult to test and verify. By contrast, the second type that accepts specifics and denies all else sounds like maintaining a white list. It is relatively easy to test and verify because the length of the white list is finite. The list for allowing or blocking traffic is known as Access Control List (ACL). Figure 10.1 shows the use of an ACL as a black list and white list for the two types of security policies, respectively, in ACL-based packet-filtering firewalls.

A packet-filtering firewall can be loaded into a router. It can be used as a component of a more complex firewall system. In the case of a filtering router, when a data packet is received, the router examines the packet and makes a decision based on a predefined security policy. This decision determines whether the packet should be dropped or forwarded. The filtering rules of the firewall usually consider the values of control fields within the data packet, such as the protocol header, including the source and destination IP addresses and port numbers.

A packet filter has both advantages and disadvantages. Some of its disadvantages include:

- It does not inspect the payload of data packets, focusing only on control fields.
- It lacks the ability to filter at the application layer, limiting its scope to lower-level network protocols.
- It does not provide user authentication, meaning it cannot verify the identity of users.
- It is susceptible to IP spoofing, where malicious users forge their IP addresses.

Due to these limitations, it is recommended to complement a packet filter with other security mechanisms in a general network environment. However, packet filters offer several advantages, such as:

- Fast operation and high performance.
- Efficient utilization of system resources.
- Cost-effectiveness compared to other security mechanisms.

As a result, packet filters are well-suited for tasks such as protecting devices within an internal network or segregating traffic between different segments of an enterprise network.

Stateful Inspection Firewalls

A *stateful inspection firewall* keeps track of the state of a communication session by monitoring the TCP handshake, and thus denies or allows traffic more intelligently. When a connection is started, the stateful inspection builds a state table, such as source and destination IP addresses and port numbers, and stores the information of the connection session. During the communication in the session, it dynamically creates firewall rules to allow anticipated traffic and block other traffic. This helps prevent attacks that exploit existing connections.

Unlike stateless packet filtering that only inspects the packet header, a stateful inspection firewall inspects the actual data transmitted across multiple packets. This provides better security protection than packet filtering does although more system resources will be consumed. Therefore, a stateful inspection firewall can be considered in security design if an elevated level of security protection is required over a packet filtering firewall.

As a stateful inspection firewall checks the actual data across multiple packets, it may impact the performance of some applications. Also, it has no authentication support. Moreover, stateful inspection is known to be vulnerable to DDoS attacks. If stateful inspection is to be used, these disadvantages will need to be addressed in network security design.

Proxy Firewalls

A *proxy firewall* is the most advanced, and thus the least common, type of firewall that filters messages at the application layer. It acts as an intermediate device or system between internal and external systems communicating over the Internet. More specifically, when a client sends a request for service, e.g., to visit a website, the request message is intercepted by the proxy server. Then, the proxy masks the request message as its own and performs message forwarding, e.g., to the web server that the original client intends to visit, pretending to be the client. This hides the identification and geographic location information of the original client, protecting the client from potential attacks.

A proxy firewall has some disadvantages. It lacks compatibility with many network protocols. It may affect the performance of some applications. Moreover, it requires additional configuration to ensure overall encryption.

Nevertheless, a proxy firewall prevents clients from direct contact with other networks, thus providing a good level of security protection. It also hides the information of the clients, ensuring user anonymity and unlocking geographical location restrictions. A proxy firewall is particularly useful for web applications to secure the server from malicious users.

Next-Generation Firewalls

A *next-generation firewall* is a security device or system that integrates a number of security functions from other firewalls. It goes beyond header inspection and incorporates packet, stateful, and deep-packet inspections. Additionally, it adds application-level inspection, intrusion prevention, and information from outside the firewall. In simple terms, a next-generation firewall inspects the entire data transaction.

A next-generation firewall provides adequate protection against malware attacks, external threats, and intrusion. However, it is costly and may require additional and complicated configuration. Moreover, due to the flexibility of the next-generation firewall, there is no clear-cut definition of the functionalities it offers. Therefore, it is necessary to clarify the actual security requirements for the network and determine whether the functionalities offered by the next-generation firewall will meet those security requirements.

10.5.5 Intrusion Detection and Prevention

While firewalls are effective security tools, they are not sufficient on their own to protect an entire network from attacks. A robust security plan should also incorporate an Intrusion Detection System (IDS) and an Intrusion Prevention System (IPS) to handle suspicious traffic that passes through the firewall and enters the network. The NIST has developed a comprehensive guide to Intrusion Detection and Prevention Systems (IDPS) for cybersecurity, providing valuable insights and recommendations for implementing effective intrusion detection and prevention measures [8].

Definition of IDS

An IDS is a device or software that monitors and analyzes traffic, detects malicious events, and alerts an administrator if a threat is detected. Key functions of an IDS include [8, p. 2]:

- Record information related to observed events.

- Alert administrators for important observed events.
- Produce reports that summarize the monitored events and provide detailed information about particular events of interest.

In addition, an IDS also finds its applications in the following use cases: identify security policy problems, document the existing threat to an organization, and deter individuals from violating security policies. Usually, an IDS does not block any traffic even if a malicious event is detected.

Definition of IPS

An Intrusion Prevention System (IPS) performs similar functions to an IDS, but it goes a step further. When it detects something unusual, an IPS actively stops traffic until certain actions are taken, such as investigation by the administrator and a decision on whether to continue packet forwarding.

The similarities between an IDS and IPS are as follows:

- Monitor: Both IDS and IPS monitor traffic.
- Alert: Both IDS and IPS alert an administrator if a malicious event is detected.
- Learn: Both IDS and IPS can be designed with learning ability to understand patterns and emerging threats.
- Log: Both IDS and IPS keep logs, particularly about attacks and responses.

The differences between an IDS and IPS are as follows:

- Response: An IDS is passive, while an IPS is active in dealing with a malicious attack. If no manual action is taken after an IDS generates an alert for the administrator, the system remains under attack until the administrator takes action.
- Protection: An IDS relies on an administrator for providing protection after an attack is detected, whereas an IPS automatically prevents an attack, thus providing automatic system protection.
- False positives: If an IDS gives a false positive, it does not affect the system. However, a false positive from an IPS may block traffic, impacting network performance or even functionality.
- Security environment: An IPS has the capability to change the configuration of other security controls to disrupt an attack, whereas an IDS does not have such capability unless the administrator decides to do so.

Types of IDS and IPS

IDS can be categorized in terms of where intrusion detection takes place (network or host), resulting in five main types of IDS: *network-based IDS*, *host-based IDS*, *protocol-based IDS*, *application protocol-based IDS*, and *hybrid IDS*. These types of IDS are described below:

- *Network-based IDS*: This type of IDS is strategically placed within the network to monitor traffic from and to all devices on the network. It can be used to monitor an entire subnet, for example.
- *Host-based IDS*: Host-based IDS runs on individual hosts or devices on the network and examines traffic from and to those hosts or devices.
- *Protocol-based IDS*: Protocol-based IDS focuses on detecting intrusions between a device and a server, monitoring all traffic between the device and the server.
- *Application protocol-based IDS*: This type of IDS implements detection within a group of servers, monitoring all communication traffic among the servers.
- *Hybrid IDS*: Hybrid IDS combines some of the approaches mentioned above into a system designed for more complex detection scenarios.

Regardless of the type of IDS used, their fundamental functionality remains the same: intrusion detection without taking active actions for the detected events.

An IPS can be configured to act as an IDS. Therefore, it can also be network-based or host-based. However, because an IPS generally takes active actions when a malicious event is detected, it has some use cases different from those of an IDS. Overall, in terms of the type of events an IPS monitors and the ways in which the IPS is deployed, four main types of IPS exist: *network-based IPS*, *host-based IPS*, *wireless network IPS*, and *network behavior-based IPS*:

- A *network-based IPS* monitors network traffic and provides protection for particular network segments or network devices. It is typically deployed at the boundary between networks, e.g., around VPN servers and wireless networks, or right behind the border firewalls or routers.
- A *host-based IPS* monitors traffic from and to an individual host or device and takes active actions to protect the host or device if any suspicious activity is identified. It is most commonly deployed on critical hosts such as publicly accessible servers and servers that contain critical or sensitive information.
- A *wireless network IPS* monitors anything happening within a wireless network and defends against attacks launched from there. However, as the transport- and application-layer protocols that the wireless traffic is transferring are not unique to wireless networks, a wireless network IPS cannot, and actually is not designed to, detect and prevent suspicious activity in these layers. A wireless network IPS is commonly deployed within the range of an organization's wireless network.
- A *network behavior-based IPS* identifies and prevents threats that generate unusual traffic on the network, such as DDoS attacks, malware behavior, and policy violations. It is most often deployed to monitor traffic on an internal network. It can also be deployed to monitor traffic between networks within an organization, or between internal and external networks.

All these types of IPS monitor the network for suspicious activity or attacks in progress. When an anomaly is detected, active actions will be taken by the IPS.

Detection Methodologies

Many detection methodologies have been developed over the years. They can be classified into three main classes: signature-based, anomaly-based, and stateful protocol analysis, respectively. As indicated in the NIST SP 800-94 [8, pp. 2–3], most IDPS technologies use multiple detection methodologies, either alone or integrated, thus providing broader and more accurate detection of malicious events.

Signature-based detection identifies attacks by looking for specific patterns that are already known. The pattern of a known threat is referred to as a signature. Examples of signatures include a remote connection with a"root" username and an email with a subject of "free picture". These patterns are well understood to be malicious.

Anomaly-based detection aims to identify unknown attacks by comparing the definitions of what activity is considered normal against observed events. The normal behavior of activity is defined in *profiles*. For example, a profile of network traffic shows an average of 20% of network traffic going to the Internet during peak hours. A significant increase in Internet traffic, such as 80%, may indicate something wrong. In order to develop accurate profiles and increase the detection success rate, methods from artificial intelligence and machine learning are often employed.

Stateful protocol analysis, also known as deep-packet inspection, is a concept that has been employed in stateful inspection firewalls and next-generation firewalls. It identifies attacks by comparing predetermined profiles of generally accepted definitions of benign protocol activity for each protocol state against observed events. It is worth noting that both stateful protocol analysis and anomaly-based detection use the concept of profiles. However, they differ in the types of profiles used. Anomaly-based detection uses host- and network-specific profiles, while stateful protocol analysis uses vendor-developed universal profiles that characterize how particular protocols should and should not be used in specific scenarios.

Each of the three detection methodologies discussed above has its advantages and limitations. Signature-based detection is the simplest and easiest to implement and deploy. It is able to detect known threats but cannot detect new attacks for which no signatures are available. Anomaly-based detection has the capability of detecting new attacks, but it relies on well-developed profiles. Stateful protocol analysis performs deep-packet detection, providing more powerful detection capabilities. However, it relies on vendor-developed profiles, which may need to be updated periodically.

10.5.6 Encryption and Decryption

Encryption refers to the process of scrambling information to protect it from being read by anyone but the intended receiver. It encrypts the original representation of the information, known as *plaintext* or *clear data*, into an alternative form known as *ciphertext*, *ciphered data*, or *encrypted data*. Ideally, the ciphertext cannot be deciphered back to the original plaintext by unauthorized users, thus keeping the original information inaccessible to unintended users. When the ciphertext arrives at

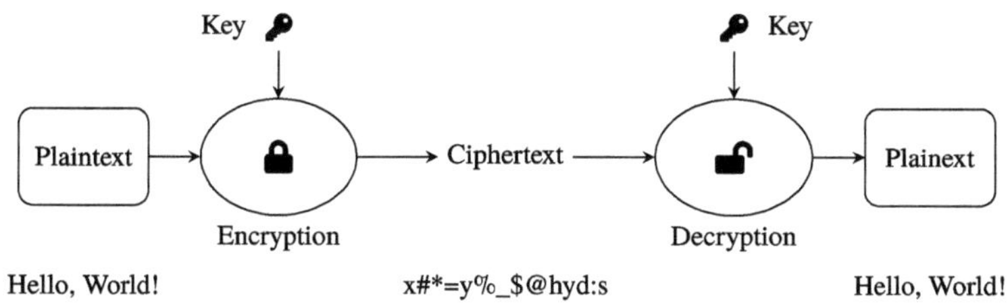

Fig. 10.2 Encryption and decryption

its intended destination, the receiver converts it back to plaintext and is then able to access the original information. The process of encryption/decryption is illustrated in Fig. 10.2.

Encryption is a useful security mechanism for data confidentiality. In addition, it can also be used to identify the sender of data. This means that encryption can provide both confidentiality and authentication features. While authentication and authorization should also protect data confidentiality and identify the sender of data, encryption provides a complementary mechanism for confidentiality and authentication.

In general, encryption consists of two essential parts:

- A cipher algorithm, which scrambles and unscrambles data, and
- An encryption key scheme (public or private key), which is used by a cipher algorithm to scramble and unscramble data.

Public Key and Private Key

In encryption, ciphers can be either symmetric or asymmetric. Symmetric ciphers utilize a single *private key* for both the encryption and decryption of data. This private key is a shared secret between the sender and receiver of the data. It represents a special case depicted in Fig. 10.2. Consequently, symmetric ciphers are also referred to as *secret key encryption.*

Asymmetric ciphers, also known as *public key encryption*, employ two distinct but logically related keys: a public key and a private key. In public key encryption, every secure node on a network publishes its public key, which is accessible to all other nodes. Any node can use the public key of another node to encrypt data intended for that specific node. However, the receiving node uses its private key to decrypt the received data. Since other nodes do not possess the private key of the receiving node, they are unable to decrypt the data. Consequently, data confidentiality is maintained. Figure 10.3 illustrates the utilization of public and private keys in network communication involving encryption and decryption.

As mentioned previously, encryption can also be used for authentication. Let us use the scenario depicted in Fig. 10.3 as an example. In this scenario, Host B receives encrypted data from Host A and successfully decrypts it. However, Host B cannot

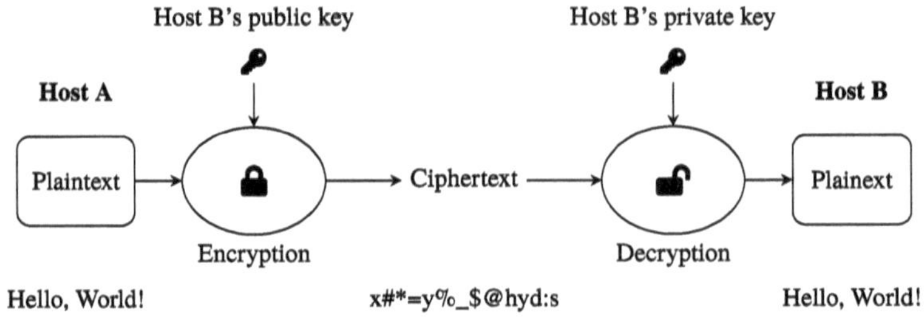

Fig. 10.3 Public and private keys are used in encryption and decryption

ascertain whether the data truly originated from Host A or not. To address this, Host A needs to encrypt the data with its own private key, resulting in what is known as *a digital signature* of Host A. By using Host A's public key, Host B can then decrypt the data and subsequently verify its origin as Host A. The digital signature serves as proof of authenticity and ensures that the data has not been tampered with during transmission. This process allows Host B to confirm that the data indeed came from Host A.

By employing digital signatures in this manner, the authentication aspect of encryption is strengthened, providing a means to verify the identity of the sender and the integrity of the transmitted data. For both authentication and data confidentiality, Host A will encrypt data twice:

(1) Use Host A's private key to generate a digital signature, and
(2) Use Host B's public key to encrypt the digital signature.

Upon receiving the encrypted data, Host B will need to decrypt the data twice:

(1) Use Host B's private key to recover Host A's digital signature, and
(2) Use Host A's public key to recover authentication and data information.

This process is demonstrated in Fig. 10.4

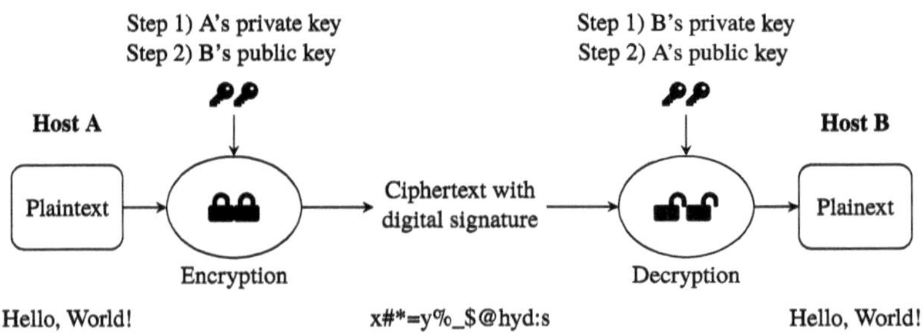

Fig. 10.4 Encryption with digital signature for authentication

Cipher Algorithms

A number of cipher algorithms have been developed. Three widely used ones are Advanced Encryption Standard (AES), Elliptical Curve Cryptography (ECC), and Rivest-Shamir-Adleman (RSA). They are briefly introduced below:

- **AES** is a symmetric encryption algorithm selected by the United States government to safeguard sensitive data. Established by the National Institute of Standards and Technology (NIST), AES is implemented in software and hardware globally. It has been adopted as a federal information processing standard [9] and an ISO/IEC Standard ISO/IEC 18033-3:2010(E) [10].
- **ECC**: ECC uses algebraic functions to generate security between key pairs. It offers faster computation and greater efficiency compared to other cipher algorithms while delivering a comparable level of security with shorter key lengths. As a result, ECC is particularly suitable for resource-constrained devices like IoT devices.
- **RSA**: RSA is currently the most widely used asymmetric encryption algorithm. In RSA, a user generates a public key from two large prime numbers, along with an auxiliary value, while keeping the prime numbers secret. Anybody can encrypt data using the public key, but only those with the secret prime numbers can decrypt it. RSA is based on the difficulty of factoring the product of two large prime numbers. While the principle of RSA is straightforward, breaking its security is computationally difficult if the key is sufficiently large. No published work has demonstrated a successful attack against RSA with a sufficiently large key.

These three cipher algorithms have proven their reliability and security in various applications and are widely adopted for encryption purposes.

Encryption Protocols and Infrastructure

Widely used encryption protocols and infrastructure include Diffie-Hellman key exchange, Internet Key Exchange (IKE), Transport Layer Security (TLS), and Public Key Infrastructure (PKI). These protocols are briefly discussed below:

- **Diffie-Hellman key exchange**, also known as *exponential key exchange*, is one of the first public-key protocols for secure exchange of cryptographic keys over a public communication channel. It uses numbers raised to specific powers to produce decryption keys based on components that are never directly transmitted, making the task of defeating it mathematically overwhelming. Before the Diffie-Hellman method is developed, secure encrypted communication between two nodes requires an exchange of keys by some secure physical means, such as handing over paper key lists by a trusted person. The Diffie-Hellman method has revolutionized the establishment of a secure communication link over an insecure channel between two parties that have no prior knowledge of each other. It is formally defined in the IETF RFC 2631 (June 1999) [11].

- **IKE** is the protocol used in the IPsec protocol suite [4] to establish a Security Association (SA). Its current version is Version 2 (IKEv2), which is specified in the IETF RFC 7296 (October 2014) [12]. IKEv2 performs the following main tasks:

  - Mutual authentication between two parties, and
  - The establishment of an IKE SA.

  The SA includes: 1) shared secret information, and 2) a set of cryptographic algorithms. All IKE communications are composed of request/response message pairs, know as exchanges. The first two exchanges are IKE_SA_INIT and IKE_AUTH, which establish an IKE SA. The subsequent exchanges are CREATE_CHILD_SA that creates a child SA, and INFORMATIONAL that deletes an SA, reports error conditions, or does other housekeeping jobs.
- **TLS**, which has evolved from the Secure Sockets Layer (SSL), is a security protocol with the goal of providing a secure channel between two communicating peers for CIA. It is widely used in email, VoIP, HTTPs, and other network services and applications. Its current version is Version 1.3 (TLS 1.3), which is defined in the IETF RFC 8446 (August 2018) [13]. TLS consists of two main components [13, p. 6]:

  - A handshake protocol, which authenticates the communicating peers, negotiates cryptographic modes and parameters, and establishes shared keying materials.
  - A record protocol, which uses the parameters established through the handshake protocol to protect traffic between the communicating parties.

  TLS runs on top of a reliable transport protocol, e.g., TCP, to provide encryption to higher layers. However, it is generally used by applications as if it were a transport-layer protocol, although applications that use TLS must actively control the handshake and authentication processes.
  Relevant to SSL, an SSL Library is a programming library that provides programming-level support for secure communication. Secure sockets APIs are available for developing secure communication systems with RSA-based authentication and RC4-based encryption.
- **PKI** is a security infrastructure that uses both public and private keys. It integrates a set of mechanisms, policies, procedures, hardware, and software tools into an infrastructure to:

  - Create, manage, distribute, use, store, and revoke digital certificates.
  - Manage public-key encryption.

  PKI employs one or more Certificate Authoritys (CAs) to associate public keys with their respective identities, such as individuals and organizations. To ensure a high level of security and privacy protection, PKI adopts a hierarchical structure with issuing authorities, registration authorities, authentication authorities, and local registration authorities. It facilitates secure data communication for various network services and applications that require a high level of security and privacy, such as e-commerce and Internet banking.

10.6 Securing Internet Connections

An important aspect of network security is to ensure the security of connections between the network and external networks, including the Internet. Internet connections are typically secured using multiple security mechanisms, such as physical security, packet filters, firewalls, and AAA. Some general considerations for securing Internet connections are outlined below:

- Deploy packet filters on border routers that are connected to external networks or the Internet.
- Install additional filters on the firewalls located behind the border routers to further enhance network security.
- Use IDS/IPS to monitor incoming and outgoing traffic to and from routers, subnets, servers, and other devices accessible from the Internet.
- Secure remote access to the enterprise network over the Internet through the use of AAA, encryption, and/or VPN tunneling.

This section will discuss how to secure Internet connections from three main perspectives: the DMZ, securing public servers, and securing mission-critical servers.

DeMilitarized Zone (DMZ)

A common practice for securing Internet connections is to design a DMZ, which serves as a small and isolated network located between the internal and external networks. The DMZ adds an extra layer of security to the internal network by exposing only specific devices to external access. The remainder of the internal network is located behind the firewall and is therefore inaccessible from external networks and the Internet.

Figure 10.5 illustrates a secure DMZ topology, which can be implemented with either a single firewall or dual firewalls. In both single-firewall and dual-firewall configurations, it is generally recommended to employ a dedicated firewall in conjunction with a router for larger-scale networks. This allows for the implementation of security features and policies on both the router and dedicated firewall, thereby maximizing security protection. However, for performance optimization, it is possible to choose not to run security functions on the router. This decision requires a careful analysis of the requirements for network performance and security.

In the single-firewall configuration depicted in Fig. 10.5, the network interfaces are organized as follows:

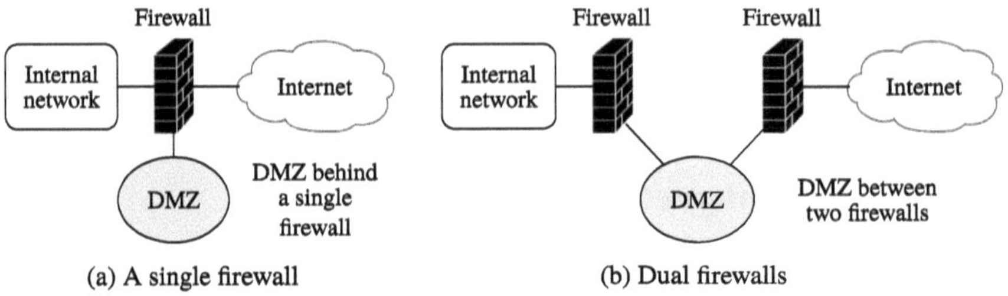

Fig. 10.5 DMZ secure topology

- An external interface connecting to the Internet.
- An internal interface connecting to the internal network.
- The DMZ interface connecting to a screened subnet, housing public servers accessible to users from the Internet.

The management of network traffic in this configuration follows these rules:

- Traffic originating from the internal network is allowed to flow towards both the DMZ and the external network.
- Traffic originating from devices within the DMZ is allowed to flow towards both the Internet and the internal network.
- Traffic originating from the Internet is only permitted to access the DMZ.

In the dual-firewall configuration illustrated in Fig. 10.5, a DMZ segment is positioned between two firewalls: an exterior (front-end) firewall and an interior (back-end) firewall. The exterior firewall is specifically configured to control access to and from the DMZ segment and the Internet. The interior firewall is configured to regulate access to and from the DMZ segment and the internal network. As a result, compared to the single-firewall topology, the dual-firewall topology offers a higher level of control over traffic traversing the firewalls, thereby enhancing security.

Securing Public Servers

In many organizations, there is a requirement to make certain public servers and resources accessible from the Internet while ensuring security protection and information privacy. These public servers and resources can include web servers, file servers, DNS servers, mail servers, and more. A common approach is to place these servers and resources within the DMZ. Figure 10.6 depicts the placement of public servers and resources within the DMZ, both in single-firewall and dual-firewall configurations, respectively.

A host located within the DMZ is commonly referred to as a *bastion host*. A bastion host is designed to support a restricted set of applications for external use, specifically from outside the DMZ. Consequently, the servers and resources situated

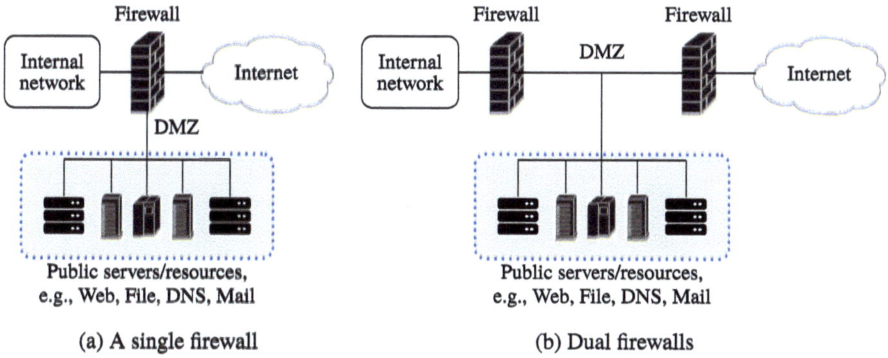

Fig. 10.6 Public servers and resources on a DMZ network

within the DMZ, as depicted in Fig. 10.6, should employ firewall software. They should be configured to offer only a limited range of services that are accessible to users external to the DMZ.

Securing Mission-Critical Servers

Mission-critical servers, such as those utilized in financial services and e-commerce, are vulnerable to cyberattacks just like other types of public servers. However, due to the highly sensitive and confidential information stored on these servers, a compromise can result in significant losses. Therefore, it is crucial to provide robust security protection for these servers.

The first aspect of enhanced security protection for critical servers is the prevention of DoS attacks. This can be accomplished through packet filtering and the implementation of a security policy that denies successive connection attempts within a short period of time.

The second aspect of enhanced security protection for critical servers involves segregating the sensitive and confidential components of the services from other components. These components should be placed within a separate network in the DMZ, equipped with its own firewall, and used as the back-end support. The publicly accessible components are used as the front-end interface to serve users external to the DMZ. This configuration is illustrated in Fig. 10.7. By adopting this approach, even if the public front-end interface is compromised, the back-end components remain unaffected, safeguarding them from potential damage originating from the front-end server.

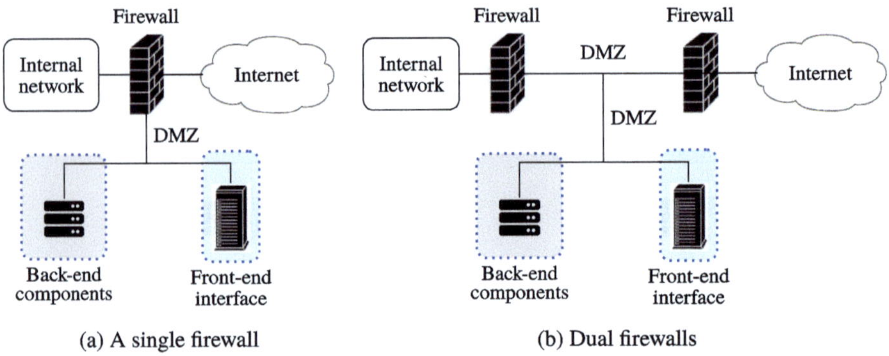

Fig. 10.7 Critical servers separated with front- and back-end components

10.7 Securing Network Services and Management

The provision of network services heavily depends on network devices, making the security of these devices and their management critical for ensuring the availability and integrity of network services. Key devices that require security measures include server farms, routers, and end-user hosts.

10.7.1 Securing End-User Hosts and Applications

To enhance the security of end-user hosts and the applications running on them, it is advisable to implement a comprehensive security policy that outlines specific guidelines and restrictions. Some key aspects that can be addressed in the security policy include:

- Application Execution: Specify which applications are permitted or prohibited from running on specific hosts. This helps prevent the execution of malicious or unauthorized software that may compromise the security of the system.
- Application Downloads and Installations: Define which hosts are restricted from downloading and installing certain applications. This can prevent the introduction of potentially harmful software or unapproved applications that may introduce vulnerabilities or cause system instability.
- User Login Restrictions: Determine which users are authorized to log in to specific hosts. By limiting access to authorized personnel only, the risk of unauthorized access or misuse of resources can be minimized.
- Firewall and Antivirus Requirements: Mandate the installation and regular updating of firewall and antivirus systems on designated hosts. This ensures that proper protection mechanisms are in place to detect and mitigate potential security threats.

For any organization, a general security policy should be established which provides overarching rules to guide or restrict end users for accessing and using network resources. This policy can address practices such as safeguarding login credentials, refraining from downloading irrelevant multimedia files, and refraining from disclosing private or sensitive information to unauthorized parties. This is related to security awareness, which has been discussed previously.

10.7.2 Securing Sever Farms

Server farms are an important part of an enterprise network. They are usually composed of storage servers, database servers, print servers, application servers, and other servers. Serving a large number of users and applications, they not only are essential for the provisioning of network and application services, but also store private, sensitive, and critical information. Therefore, they must be protected through various security mechanisms.

A challenge in securing server farms is balancing security with high performance requirements. While high performance is necessary to serve a large number of users and applications, security mechanisms must still be in place to mitigate the risk of unauthorized data access and compromise. An effective security measure is to deploy an IDS to monitor specific servers or specific subnets of servers.

It is generally recommended to avoid active connections between servers within the network design. This can be achieved through network filters that restrict network connections from one server to another. However, there are cases where active connections between servers are necessary, such as in active FTP. These requirements should be identified through requirements analysis and traffic flow analysis, and incorporated into the network filter design accordingly.

To prevent unauthorized access to servers and files, authentication and authorization features should be implemented. Security policies and procedures should be designed with specific rules for password management and usage. This includes specifying password formats (e.g., a minimum length of eight characters with a combination of upper-case and lower-case letters, digits, and special characters), password change frequency (e.g., every quarter), and when passwords should be used.

In addition to access controls, encryption mechanisms can be employed for server applications, access to server files, and other interactions with the servers. Encryption helps protect data stored on the servers. For example, files on disks can be encrypted specifically for an application, ensuring they can only be read by that specific application. This encryption complements any network communication encryption implemented to protect data packets during transmission over the network.

10.7.3 Hardening Routers and Switches

Routers and switches are the two most common devices on networks. As critical network devices, they make up the bulk of the network infrastructure. However, routers and switches are vulnerable to various cyberattacks. Therefore, they pose a significant risk because if they are compromised, there would be no path for data packets to flow. This highlights the importance of protecting routers, switches, and other network devices from attacks. This protection is known as *hardening*. Hardening devices, particularly routers and switches, are an essential requirement for securing network services.

In general, firewalls together with IPS help provide security protection to routers and switches. It is always recommended to protect routers with a firewall and ACLs. However, there are additional steps to take for hardening routers and switches within a network. The Network Infrastructure Security Guidance recently released by the NSA [2] can be used as a foundation for securing network devices including routers and switches.

Some basic aspects for securing routers and switches include:

- Physically securing routers and devices.
- Protecting routers and switches with passwords for both the login mode and privileged mode. The login mode grants access to the devices, while the privileged mode allows configuration changes.
- Ensuring the security of routers and switches on console and AUX ports in addition to SSH access.
- Setting the correct time and date, typically achieved through NTP-based clock synchronization.
- Recording log data locally, but preferably on a centralized server such as a Syslog server.
- Backing up the configuration to a central server.
- Implementing other security measures such as stateful firewalls or ACLs, encryption of sensitive traffic, and more.

More broadly, there are three main functions or planes within network devices that need to be protected: management, control, and data:

- The management plane handles various traffic flowing to the router or switch, enabling the functioning of applications and protocols that manage the devices. Typical applications and protocols involved include SSH, SNMP, RMON, FTP, and HTTP/HTTPS.
- The control plane processes traffic that contributes to the functionality of the network infrastructure. It involves applications and protocols between network devices, such as various layer-2 protocols (e.g., STP), layer-3 protocols (e.g., routing protocols), and application protocols (e.g., storage protocols).
- The data plane is responsible for the actual data forwarding through a network device. This encompasses layer-2 forwarding, layer-3 forwarding, and data forwarding for storage.

In addition to the security mechanisms and procedures discussed above, good practices for securing routers also include:

- Securing the control plane.
- Controlling privilege levels.
- Ensuring routing update authentication.

For hardening switches, it is a good practice to implement DHCP snooping and port security, along with activating security guards such as Bridge Protocol Data Unit (BPDU) guard, root guard, and loop guard. BPDU guard is an enhancement to STP designed to enhance switch security.

10.7.4 Securing Wireless Networks

Nowadays, wireless networks have become ubiquitous to provide various network services. They are integrated to enterprise networks for easy access from mobile and wireless devices. Wireless networks are particularly useful for rapid network deployment in locations where no wired infrastructure is available, such as disaster management and emergency services. However, due to their reliance on airwaves, wireless networks are more vulnerable to security threats compared to wired networks. Therefore, securing wireless networks has become increasingly important.

There are a number of security mechanisms designed specifically for the security of wireless networks. These mechanisms are developed with regard to Wireless LAN (WLAN) design, access point management, user access and authentication, and data protection.

In WLAN design, it is generally recommended to isolate the wireless network from the rest of the LAN. This enables to create a wireless DMZ or perimeter network that is separated from the wired networks. The implementation of a firewall between the wireless network and wired networks adds an extra layer of security protection. It requires authentication for wireless network users to access network services provided by the wired networks.

For the management of access points, security measures include:

- Change the default administrative password,
- Change the default Service Set IDentifier (SSID),
- Turn off SSID broadcasting,
- Turn off access points when not in use,
- Control the wireless signal, e.g., through adjusting signal strength and direction, and
- Use a different frequency, for example, use 5 GHz (802.11a) instead of 2.4 GHz (802.11b/g).

For the management of user access, authentication and authorization could be implemented. MAC filtering could also be used, implying that a white list of hosts can be set up which are allowed to connect to the wireless network based on their MAC addresses.

For the protection of data transmitted over the wireless network, use strong encryption in data transmission. Encryption is considered as the top security measure in wireless networking. It is worth mentioning that most access points support the Wired Equivalent Privacy (WEP) protocol. But WEP is not enabled by default. Make sure it is enabled and also set the WEP authentication method for *shared key* rather than *open system*. The *open system* mode does not provide encryption at all. As WEP has its weakness, use Wi-Fi Protected Access (WPA) protocol if available and feasible instead of WEP.

10.7.5 Securing Network Management

Network management systems host extremely sensitive data with regard to the network, network services, network applications, and security configurations. Therefore, their security protection is important and should be designed carefully.

Some general guidelines for securing network management are described below:

- Protect network management systems from unauthenticated and unauthorized access. Particularly, design the network management systems to prevent impersonation of administrators. This can be achieved through strong authentication mechanisms, such as a multi-factor authentication measure, which requires more than one proof of identity to gain access.
- When SNMP is used for network management, SNMPv3 should be used, which provides authentication for various SNMP operations.
- In the presence of a separate network over which network management traffic flows, such as in out-of-band network management, this separate network for network management should be designed with proper security protection.
- To maximize security protection, place network management systems in their own DMZ behind a firewall.

10.8 Remote Access and VPNs

Remote access is a network technology that allows an organization's enterprise network to be accessed remotely by its employees over a third-party's network or the Internet. It also allows other authorized users, such as customers and business partners, to gain off-premise access to the extranet of the network. Remote access is also deployed for an individual user to access the ISP's network resources.

Remote access is implemented through traditional dial-in, point-to-point connections, and/or VPN connections. Its security is critical for the enterprise edge and service provider's network. In general, the security of remote access is achieved through what is called Authentication, Authorization, Accounting, and Allocation (AAAA). AAAA represents the popularly known AAA plus the Allocation of configuration information. It is normally supported by a Network Access Server (NAS) or Subscriber Management System (SMS). For VPNs, the NSA and CISA have recently released a cybersecurity information sheet that provides general guidelines on selecting and hardening standards-based remote access solutions [14].

Point-to-point connectivity to a remote network over the Internet is typically provided through advanced encryption and tunneling, as we mentioned previously. Tunneling uses encapsulation to embed data packets of one protocol into another protocol. The data packets of the original protocol are treated as payload and can be encrypted before they are encapsulated into the new protocol. The encryption process is optional and can be omitted for applications where security and privacy are not a concern.

VPNs are an effective tool for secure remote access to the resources of an organization's enterprise network from any off-premise location. The creation of a VPN tunnel requires support for three types of protocols:

- A carrier protocol, which is a network transport protocol supported by the transit internetwork. For example, Point-to-Point Protocol (PPP) is a carrier protocol used in IP-based transit networks [15, 16].
- An encapsulation protocol, which is a protocol to encapsulate the data packets to be transported through the tunnel. Layer-2 L2TP [17, 18] and Layer-3 IPsec [4] are examples of encapsulation protocols.
- A passenger protocol, which is a protocol used by the tunnel-connected networks for the transport of the data packets encapsulated using an encapsulation protocol. IP and NetBEUI are examples of passenger protocols.

Our further discussions on VPNs will focus on layer-2 and layer-3 encapsulation mechanisms.

10.8.1 Layer-2 and Layer-3 Tunneling Protocols

Tunneling can be implemented at Layer 2 and Layer 3. Layer-2 tunneling means encapsulation of data packets at the data link layer. Layer-3 tunneling encapsulates data packets at the network layer.

Layer-2 Tunneling

A widely used layer-2 tunneling protocol is Layer 2 Tunneling Protocol (L2TP). L2TP has its origins of two old tunneling protocols for point-to-point communications: Cisco's Layer 2 Forwarding (L2F) and Microsoft's Point-to-Point Tunneling Protocol (PPTP). L2F is specified in the IETF RFC 2341 (May 1998) [19]. PPTP is defined in the IETF RFC RFC 2637 (July 1999) [20]. These two RFCs are now classified into the "Historic" and "Informational" categories, respectively. Thus, the focus of our discussions here on layer-2 tunneling will be on L2TP.

L2TP is originally specified in the IETF RFC 2661 (August 1999) [17]. The latest version of L2TP is version 3, known as L2TPv3, which is defined in the IETF RFC 3931 (March 2005) [18]. The base L2TP protocol defined in this RFC consists of two main parts:

- A control protocol for dynamic creation, maintenance, and termination of L2TP sessions, and
- An L2TP data encapsulation mechanism to multiplex and demultiplex L2 data streams between two L2TP nodes over an IP network.

Overall, L2TPv3 is a lightweight and robust layer-2 tunneling technique.

L2TP operates between two L2TP nodes, which could be L2TP Access Concentrator (LAC) or L2TP Network Server (LNS). Topologically, there are three predominant tunneling reference models for L2TP: LAC-LAC, LAC-LNS, and LNS-LNS, as shown in Fig. 10.8. They are briefly described in the following:

- In the LAC-LAC model, each side of the LAC-LAC connection may initiate an L2TP session, and each LAC forwards L2 traffic to the peer LAC.
- In the LAC-LNS model, the establishment of an L2TP session between the LAC and LNS may be driven by either the LAC as an incoming call or the LNS as an outgoing call. The LAC receives L2 traffic from the remote system and forwards

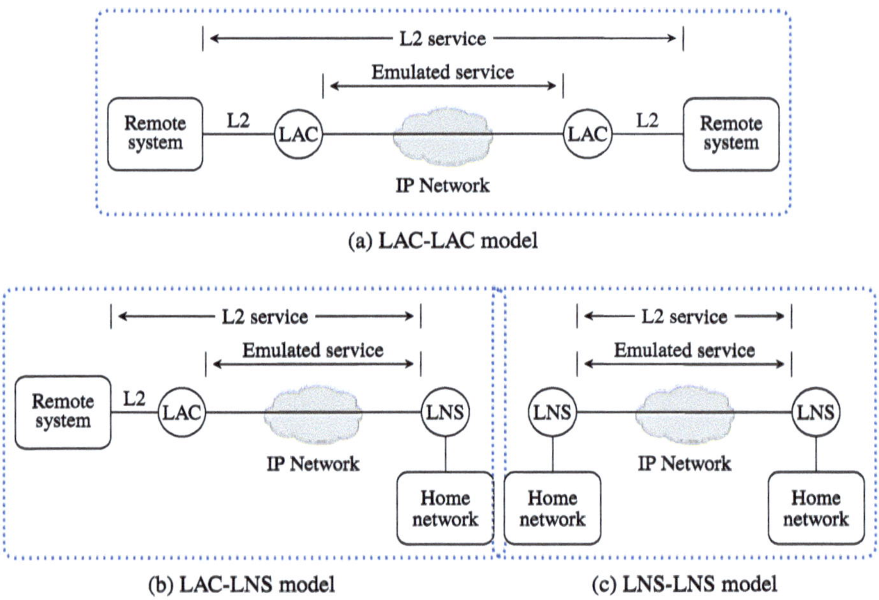

Fig. 10.8 L2TP topological reference models [18, pp. 8–9]

the traffic to the LNS of the other side. The LNS logically terminates the L2 circuit locally and routes the traffic to the home network.
- In the LNS-LNS model, the establishment of an L2TP session is typically driven by a user-level, traffic-generated, or signaled event. In the earlier version L2TPv2, an LNS acting as part of a software package on a host is referred to as an *LAC Client* [17, p. 8]. User-driven tunneling is known as *voluntary tunneling*.

L2TP has two types of messages: control messages and data messages. Control messages are used to establish, maintain, and terminate L2TP control connections and sessions. Data messages encapsulate the L2 traffic that is being carried over the L2TP session. An L2TP session must be established before L2TP can start to forward frames. An L2TP control connection can handle multiple L2TP sessions. Multiple L2TP control connections may exist between two L2TP nodes.

Layer-3 Tunneling

A popular layer-3 tunneling protocol is IPsec [4], which operates directly on top of the IP protocol. It has been previously discussed as a mandatory inclusion in IPv6. Overall, IPsec is available for use in both IPv4 and IPv6 networks for secure VPN applications. The IPsec connections involve the following main steps:

- Key exchange: IPsec sets up a key exchange between the connected devices so that each device is able to decrypt messages from the other devices.
- Packet encapsulation and decapsulation: On the sending device of a VPN tunnel, IPsec adds headers and trailers to the data packets to be transported over the VPN tunnel. Two types of such headers are Authentication Header (AH) header [21] and Encapsulating Security Payload (ESP) header [22]. On the receiving device of the tunnel, IPsec conducts decapsulation to get the original data.
- Authentication: IPsec provides authentication for each data packet through its AH header [21]. This ensures that the received data packets are from a trusted source.
- Encryption and decryption: To keep data sent over IPsec secure and private, IPsec encrypts the payload within each packet in the IPsec transport mode. It additionally encrypts the original IP header of each packet in the IPsec tunnel mode. The IPsec encryption is defined in its ESP header and trailer [22]. On the receiving side of the communication, IPsec decrypts the packets.
- Transmission: Encrypted IPsec packets are sent to their destination along the established VPN tunnel. They are treated as normal IP traffic over the networks on which the VPN tunnel is established. Most often, IPsec traffic is transported by using UDP as its transport protocol (rather than TCP). This will allow IPsec packets to get through firewalls.

Comparisons Between Layer-2 and Layer-3 Tunneling

Some comparisons between layer-2 and layer-3 tunneling are tabulated in Table 10.2.

Table 10.2 Comparisons between layer-2 and layer-3 tunneling

Feature	Layer-2 tunneling	Layer-3 tunneling
Encapsulation	Layer-2 encapsulation	Layer-3 encapsulation
Use cases	Predominantly for remote-access VPNs	Site-to-site VPNs, and remote-access VPNs
Protocol support	IP networks, and non-IP networks (e.g., NetBEUI)	Only IP-based networks
Assignment of IP addresses	Dynamic assignment of IP addresses VPN clients by the VPN server at the time of tunnel establishment	IP addresses are assigned prior to tunnel establishment
Authentication	User-based authentication via authentication protocols; hardware-based authentication mechanisms, e.g., smart cards	Device-based authentication methods, e.g., digital certificate
Encryption	Various encryption mechanisms	Similar encryption mechanisms to those from layer-2 tunneling protocols. In addition, IPsec protocol suite provides strong encryption

10.8.2 Site-to-Site, Remote-Access, and Client-to-Site VPNs

Applications of VPNs for remote access can be classified into three main categories: site-to-site VPNs, remote-access VPNs, and client-to-site VPNs. Site-to-site VPNs aim to connect geographically dispersed networks, extending the traditional WAN with additional security. In comparison, remote-access and client-to-site VPNs provide secure access to the enterprise network and its applications for remote users.

Site-to-Site VPNs

Site-to-site VPNs connect multiple networks located at geographically disparate sites, cities, states, or countries. For example, they are used to connect enterprise headquarters securely to its branch offices at different locations. They are also used to expand a data center across a group of service branches or centers. Therefore, site-to-site access looks like network-to-network access. Dedicated equipment is required to establish and maintain a site-to-site VPN connection. The VPN link may run over a dissimilar intermediate network. For example, two IPv6 networks are connected via a VPN over an IPv4 network.

As site-to-site VPNs look like a private WAN, the design of the topology of a site-to-site network should consider the same requirements as those for a private WAN. The requirements include high reliability and availability with automatic failover, performance, scalability, and security. To fulfill these requirements, a full- or partial-mesh VPN topology can be designed for the core layer of the network. For large-scale enterprise networks, a hierarchical VPN topology is also an option, though it is more complicated to design, operate, manage, and maintain.

Remote-Access VPNs

A remote-access VPN connects the host of a remote user securely to a corporate network. It is similar to a host-to-network connection in a LAN, but the VPN connection is established over a third-party network or the Internet. This allows the employees of an organization to work off-premise (e.g., from home or on the go) with secure connections to the corporate network, and enables the customers and business partners of the organization to securely access authorized resources from the corporate network.

Typically, a remote-access VPN connection is initiated by a remote user. The host of the remote user must have a VPN client installed. By using protocols such as IPsec or L2TP, the VPN client requests a remote-access VPN connection to the corporate network from the VPN server. Once the VPN connection is established, data packets can be exchanged over the secure VPN connection.

There are many VPN clients available on the market. Some are freeware, and others are licensed. In our organization, we use Cisco's AnyConnect Secure Mobility Client for remote-access VPN connections to our corporate network. Our VPN server, remote.qut.edu.au, is a high-speed VPN server that allows a large number of VPN connections from around the world.

Client-to-Site VPNs

Similar to remote access VPNs, a client-to-Site VPN also provides individual users with secure remote access to specific network resources of a network over a third party's network or the Internet. In this sense, the terms of client-to-site and remote VPNs are often used interchangeably when there is no confusion. However, client-to-site VPNs are distinct from remote access VPNs in many aspects. Some of the distinctions are summarized in Table 10.3. Typically, a client-to-site VPN provides secure access to a specific network application or service, which can be hosted by a third-party provider.

10.8.3 Choosing and Hardening VPNs

For security protection, NSA has recommended that VPNs be used only when the confidentiality and integrity of the traffic cannot be maintained through other methods [2, p. 5]. From the NSA's guidance, when VPNs are used, disable all unnecessary features on the VPN gateways and implement strict traffic filtering rules.

Table 10.3 Comparisons between client-to-site and remote access VPNs

Feature	Client-to-site VPN	Remote access VPN
Scope	Typically provides access to a network specific application or service, which can be hosted by a third-party provider. It focuses on enabling secure access to a specific resource rather than the entire network.	Offers secure access to a private network, allowing users to access various resources within that network, such as files, applications, or internal systems.
Deployment	Typically involves installing a client software/application on the user's device, which establishes a secure connection with the service provider's network.	Often utilizes VPN server software or hardware located within the organization's network. Users connect to the network using VPN client software or built-in VPN support on their devices.
Management	Management and control are primarily handled by the third-party provider hosting the application or service. They are responsible for maintaining and securing the infrastructure.	The organization's IT department manages and controls the VPN infrastructure, including server configuration, user access controls, security policies, and network resources.
Scalability	Offers flexibility in accessing specific applications or services hosted externally. It can easily accommodate varying numbers of users, making it suitable for organizations with a dispersed user base.	Designed for scalability within an organization's network infrastructure, accommodating multiple users and providing access to various resources within the network.
Security	Security measures and policies are typically defined and implemented by the third-party provider. It is crucial to review their security practices to ensure data privacy and protection.	Security measures can be customized and controlled by the organization's IT department, ensuring adherence to their security policies and compliance requirements.

In general, it is recommended that the guidelines jointly released by NSA and CISA [14] in September 2021 be followed when choosing and hardening remote-access VPN solutions. More specifically,

- Select standards-based VPNs from reputable vendors.
- Harden the VPN against compromise by reducing the VPN server's attack surface. This can be achieved from the following three aspects:
  - Configure strong encryption and authentication,
  - Run only strictly necessary features, and
  - Protect and monitor access to and from the VPN.

10.9 Summary

Security and privacy are paramount concerns in computer networks, necessitating a meticulous requirements analysis and systematic architectural design. To ensure robust security, it is recommended to adhere to the cybersecurity recommendations and guidance provided by reputable organizations such as the IETF, NSA, and CISA when crafting the architecture of security and privacy components.

The design process of security and privacy component architecture starts with the development of a security plan. This plan helps understand what assets need to be protected, what threats and risks may exist, what security requirements are, and what trade-offs may be needed and achieved. Building upon the framework established by the security plan, specific security policies and procedures can be developed and implemented. To enforce these policies and procedures effectively, appropriate security mechanisms are carefully evaluated and selected for various network areas, functions, and services.

There are various security mechanisms, which encompass physical awareness, physical security, authentication, authorization, and access control. Among the commonly employed security mechanisms are AAA, firewalls, IDS/IPS, and encryption. Each of these mechanisms incorporates a range of techniques tailored to fulfill its designated functions.

Within the security and privacy component architecture, it is essential to safeguard the network infrastructure such as various hardware servers and Internet connections, protect the network against cybersecurity threats such as DoS and spoofing, secure network services such as DNS and WWW, and particularly protect the security protection systems. This chapter has comprehensively addressed all these aspects from the architectural planning perspective. This should be integrated into the overall network architecture to ensure a robust and resilient network environment.

References

1. Fraser, B.: Site security handbook. RFC 2196, RFC Editor (1997). https://doi.org/10.17487/RFC2196
2. (NSA), N.: Network infrastructure security guidance. Cybersecurity Technical Report PP-22-0266, Version 1.0, NSA (2022)
3. Cybersecurity, (CISA), I.S.A.: Layering network security through segmentation. CISA Online Documentation: https://www.cisa.gov/sites/default/files/publications/layering-network-security-segmentation_infographic_508_0.pdf (2022). Accessed 2 Oct. 2022
4. Kent, S., Seo, K.: Security architecture for the Internet protocol. RFC 4301, RFC Editor (2005). https://doi.org/10.17487/RFC4301
5. Shirey, R.: Internet security glossary, version 2. RFC 4949, RFC Editor (2007). FYI 36, https://doi.org/10.17487/RFC4949
6. Thaler, D.: Behavior of and requirements for Internet firewalls. RFC 2979, RFC Editor (2000). https://doi.org/10.17487/RFC2979
7. Thaler, D.: Reflections on host firewalls. RFC 7288, RFC Editor (2014). https://doi.org/10.17487/RFC7288
8. Scarfone, K., Mell, P.: Guide to intrusion detection and prevention systems (IDPS). NIST SP 800-94, National Institute of Standards and Technology (2007). DOI: 6028/NIST.SP.800-94
9. NIST: Advanced encryption standard (AES). Federal Information Processing Standards Publication 197 (2001). https://doi.org/10.6028/NIST.FIPS.197
10. ISO/IEC: Information technology - security techniques - encryption algorithms - part 3: Block ciphers. ISO/IEC 18033-3:2010(E) (2010)
11. Rescorla, E.: Diffie-Hellman key agreement method. RFC 2631, RFC Editor (1999). https://doi.org/10.17487/RFC2631
12. Kaufman, C., Hoffman, P., Nir, Y., Eronen, P., Kivinen, T.: Internet key exchange protocol version 2 (IKEv2). RFC 7296, RFC Editor (2007). STD 79, https://doi.org/10.17487/RFC7296
13. Rescorla, E.: The transport layer security (TLS) protocol version 1.3. RFC 8446, RFC Editor (2018). https://doi.org/10.17487/RFC8446
14. (NSA), N., (CISA), C.: Selecting and hardening remote access VPN solutions. Cybersecurity Information Sheet U/OO/186992-21/PP-21-1362, Ver. 1.0, NSA and CISA (2021)
15. Simpson, W.: The point-to-point protocol (PPP). RFC 1661, RFC Editor (1994). STD 51, https://doi.org/10.17487/RFC1661
16. Varada, S., Haskins, D., Allen, E.: IP version 6 over PPP. RFC 5072, RFC Editor (2007). https://doi.org/10.17487/RFC5072
17. Townsley, W., Valencia, A., Rubens, A., Pall, G., Zorn, G., Palter, B.: Layer two tunneling protocol L2TP. RFC 2661, RFC Editor (1999). https://doi.org/10.17487/RFC2661
18. Lau, J., Townsley, M., Goyret, I.: Layer two tunneling protocol - version 3 (L2TPv3). RFC 3931, RFC Editor (2005). https://doi.org/10.17487/RFC3931
19. Valencia, A., Littlewood, M., Kolar, T.: Cisco layer two forwarding (protocol) “L2F". RFC 2341, RFC Editor (1998). https://doi.org/10.17487/RFC2341
20. Hamzeh, K., Pall, G., Verthein, W., Taarud, J., Little, W., Zorn, G.: Point-to-point tunneling protocol (PPTP). RFC 2637, RFC Editor (1999). https://doi.org/10.17487/RFC2637
21. Kent, S.: IP authentication header. RFC 4302, RFC Editor (2005). https://doi.org/10.17487/RFC4302
22. Kent, S.: IP encapsulating security payload (ESP). RFC 4303, RFC Editor (2005). https://doi.org/10.17487/RFC4303

Part III
Network Infrastructure

This part comprises three chapters:

- Chapter 11: Data Centers.
- Chapter 12: Virtualization and Cloud.
- Chapter 13: Building TCP/IP Socket Applications.

This part covers architecture for significant network infrastructure entities including data centers, virtualization and cloud, and sockets. For data centers, the focus is on various standards for design and operations. Virtualization of computing and network resources is the fundamental technique that supports cloud computing. The discussions of cloud computing will cover the aspects of its characteristics, deployment models, service models, implementation framework, and security. To build practical TCP/IP client-server communication systems, the concepts and tools of sockets will be discussed in detail with many practical examples.

Chapter 11
Data Centers

A data center is a facility that centralizes the shared IT operations and equipment of an organization for the storage and processing of data, information, and applications. Traditionally, data centers were dedicated physical spaces within a building or group of buildings, meticulously controlled and managed on premises.

However, the advent of the public cloud has brought about significant changes in the physical infrastructure of data centers. Modern data centers have evolved considerably in a relatively short period. The shift from traditional on-premises physical servers to virtualized environments has transformed the IT service infrastructure, enabling the delivery of services across multiple physical and virtual facilities. While some regulatory requirements still necessitate on-premises infrastructure without Internet connectivity, the overall landscape of data centers has shifted towards virtualized facilities capable of supporting applications and workloads across diverse multi-cloud environments. Notably, even public clouds themselves are composed of interconnected data centers. For example, Google maintains a network of data centers worldwide to provide a wide range of data and cloud services.

In the context of data centers, data and IT services can be interconnected or provisioned across various locations, including multiple data centers, edge servers, public and private clouds. Therefore, a data center communicates across these multiple sites, both within an organization's own premises and in the cloud. When applications are hosted in the cloud, they leverage the resources provided by data centers belonging to the cloud service provider.

Data centers are fundamental infrastructure nationally and globally for information, communication, networking, and computing services. They form the backbone of modern cloud computing and cloud services. This chapter will delve into various aspects of data centers, with particular emphasis on national and international standards pertaining to data centers.

Y.-C. Tian and J. Gao, *Network Analysis and Architecture*, Signals and Communication Technology, https://doi.org/10.1007/978-981-99-5648-7_11

11.1 Data Centers Around the World

More and more data centers are being built in response to the significant rise of data generation and use across a wide range of industries. According to a brief report by Brian Daigle [3], new data of 1.2 trillion gigabytes was created globally in 2010, representing a 50% increase from the previous year. It was estimated in 2010 that 35 trillion gigabytes of data would be created annually by 2020, but this level was actually reached in 2018. In 2020, the data created amounted to about 59 trillion gigabytes, indicating an amount over 68% higher than the estimated level. It is now estimated that the creation of new data will reach 175 trillion gigabytes. With such a huge amount of data being created annually, the data storage and processing market is expected to reach US$90 billion by 2025. In relation to these generated data, the global revenue of the data market, the business analytics sector, the banking sector, manufacturing industries, and professional services are expected to be several times higher than that from data storage and processing.

The exponential increase in data generation demands more data servers and data centers. The actual number of data centers around the world is believed to be in the millions. According to statistics from CloudScene, as of March 29, 2020, the total number of data centers globally exceeds 8,384 [1]. Spread across 110 countries, these data centers are counted based on publicly available information and cloud service provisioning to some extent. Among these 8,384 data centers, the United States has the highest number with 2,762 (33%). This is followed by Germany with 488 (5.8%), UK with 459 (5.5%), and China with 447 (5.3%). In addition to these top four countries with the most data centers, other countries with a significant number of data centers include the Netherlands (approximately 290), Australia (270+), Canada (approximately 270), France (approximately 250), Japan (approximately 210), and Russia (approximately 150). The top 10 countries with the most data centers are tabulated in Table 11.1 [2]. They are graphically shown in Fig. 11.1.

The statistics presented in the report by Brian Daigle [3] demonstrate the widespread use of data and data centers throughout the United States economy. Each state in the United States has at least one data center, with certain states like California known for housing numerous data-intensive firms such as Google and Twitter. Similarly, countries like Germany, UK, and China are also experiencing an

Table 11.1 Top 10 countries with the most data centers [2]

Rank	Country	Rank	Country
1	USA	6	Australia
2	Germany	7	Canada
3	UK	8	France
4	China	9	Japan
5	Netherlands	10	Russia

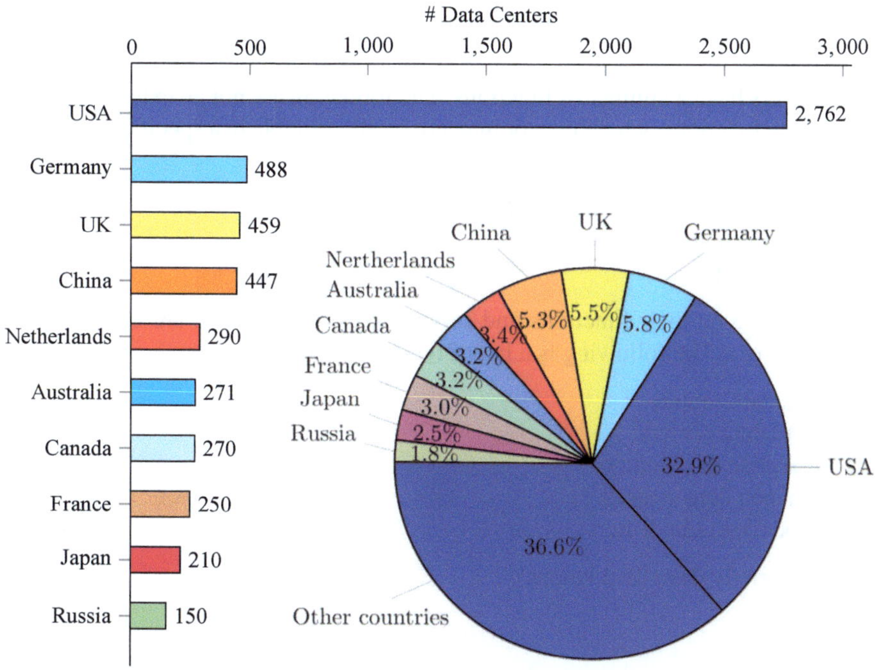

Fig. 11.1 Illustration of top 10 countries with the most data centers [2]

increasing demand for the widespread use of data and data centers, making significant contributions to their national and global economies.

A valuable source of general information regarding data centers is Data Center Magazine [2]. This magazine provides various top 10 statistics, including the top 10 countries with the highest number of data centers, top 10 hyper-scale data center operators, top 10 data center brands, and top 10 characteristics of data centers projected for 2025. Our interest here is the top 10 data center brands, which from top 1 to top 10 are listed as follows:

(1) Equinix (from the USA).
(2) Lumen (from the USA).
(3) Digital Realty/Interxion (from the USA).
(4) NTT Communications (from Japan).
(5) AWS (from the USA).
(6) Google (from the USA).
(7) Switch (from the USA).
(8) China Telecom (from China).
(9) Cyxtera (from the USA).
(10) BT (from UK).

11.2 Functions and Categories of Data Centers

Data centers have a number of functions. The following are not an exhaustive list of data center functions, which may or may not be all available in a specific data center:

- Traditional enterprise data and IT services.
- On-demand enterprise data and IT services.
- High performance computing.
- Internet service.

Depending on their functions and service roles, data centers are classified into different types. The following is a list of typical data center types:

(1) Edge data centers.
(2) Cloud data centers.
(3) Enterprise data centers.
(4) Managed data centers.
(5) Colocation data centers.

They are briefly described below.

11.2.1 Edge Data Centers

Edge data centers are relatively small facilities that are strategically positioned near the population or devices they serve. In comparison to cloud data centers, edge data centers possess less computing and network resources. Nevertheless, they offer an affordable solution by providing connectivity to end users or devices. In close proximity to the end users or devices, these edge data centers enable the delivery of data, network services, and other related functionalities with minimal latency. Edge data centers are expected to support the growing demand for Internet of Things (IoT) networks and services in a more efficient manner.

11.2.2 Cloud Data Centers

Cloud data centers are typically large-scale facilities that are owned and operated by cloud service providers. These data centers serve as the backbone of cloud computing infrastructure. While cloud service providers are responsible for managing and maintaining these data centers, they may also collaborate with third-party managed service providers for assistance.

Cloud service providers offer a range of services to end users from their data centers through various service models. These models include Infrastructure as a Service (IaaS) where infrastructure resources are provided, Platform as a Service

(PaaS) where platforms and development tools are offered, and Software as a Service (SaaS) where complete software applications are made available.

Some cloud service providers also offer customized cloud services, allowing clients to have dedicated access to their own cloud environment, referred to as private clouds. On the other hand, public clouds provide resources and services to clients over the Internet. Well-known public cloud service providers include Amazon Web Services (AWS), Microsoft Azure, Google Cloud Platform, and IBM Cloud.

One of the key advantages of using cloud services from a cloud service provider is the flexibility and cost-efficiency it offers. Clients have the flexibility to request and use only the resources they need, resulting in cost savings. Also, the management and maintenance tasks related to the underlying data center infrastructure are handled by the cloud service providers, relieving clients of those responsibilities and further reducing their operational costs.

11.2.3 Enterprise Data Centers

Enterprise data centers are highly private facilities designed exclusively for a single organization. They can be located either on the organization's premises or off-site. For instance, a university may establish its own enterprise data center on or off campus to cater primarily to its students, staff, and external visitors.

The scales and capacities of enterprise data centers can vary from one organization to another. However, the distinguishing factor of enterprise data centers lies primarily in their ownership and target usage, rather than their size and available resources. For example, data centers belonging to two universities may have similar functions and requirements, regardless of their differences in size and available resources.

Typically, an enterprise data center consists of several sub-data centers, each serving a distinct purpose. These sub-data centers can be categorized into three groups: Internet data centers, extranet data centers, and intranet data centers:

- An **Internet data center** provides services over the Internet, such as hosting web applications.
- An **extranet data center** supports business-to-business transactions within the organization's data network. Extranet services are generally accessible through private Wide Area Network (WAN) connections or secure Virtual Private Network (VPN) links.
- An **intranet data center** hosts and maintains application and data services within the data center itself. Intranet services are provisioned for various business functions, including financial management, teaching and learning, research, and asset management.

Depending on the services they provide, an enterprise data center may have one or more of the following hardware platforms: Internet server farms, extranet server farms, and intranet server farms. These platforms are tailored to meet the specific requirements of each data center category within the enterprise data center ecosystem.

11.2.4 Managed Data Centers

Managed data centers follow a logical facility model where a third-party service provider assumes full or partial responsibility for various aspects of data deployment, management, monitoring, and technical details. The level of involvement and control granted to businesses can vary based on the agreement with the service provider. In some cases, businesses may retain certain administration privileges, access to back-end data, and other rights, while in other instances, they may entrust complete control to the service provider.

For example, IBM offers a comprehensive range of managed data center services to clients, catering to medium- to large-scale enterprises. These services include security services, network services, and managed mobility, among others. By leveraging the expertise of the service provider, businesses can offload the complexities of data center management and focus on their core operations.

11.2.5 Colocation Data Centers

Colocation data centers are humongous facilities used by organizations that lack sufficient space and resources to maintain their own data centers. These organizations opt to rent rack space and other essential resources from colocation data centers to accommodate their servers and other devices. This service model is particularly beneficial for businesses operating across multiple geographic locations, cities, or even countries, allowing them to distribute their hardware devices among various colocation data centers.

It is worth mentioning that there is a difference between a colocation data center and a colocation rack. A colocation data center refers to a company that rents out the entire facility to another organization, providing comprehensive data center services. In comparison, a colocation rack involves a company that leases rack space within a data center to multiple organizations, allowing each organization to house their equipment separately within the shared facility.

11.3 Standards for Building Data Centers

Designing a data center is a complex endeavor that combines both technical expertise and creative thinking. It is not merely a large-scale IT project, but also an art that draws on past successful projects and experiences in network planning. This chapter primarily focuses on the IT project perspective of building data centers, encompassing compliance with existing standards and adherence to best practices.

While there are numerous standards indirectly relevant to data centers, such as those pertaining to structured cabling systems, this chapter will not delve into them unless explicitly mentioned. Instead, the emphasis will be placed on standards specifically developed for data centers. These standards may be international, national, or industry-specific. The following provides brief discussions on some widely used data center standards: ANSI/TIA-942, Telcodia's GR-3160, and EN 50600/ISO/IEC 22237.

11.3.1 ANSI/TIA-942

The American National Standards Institute (ANSI) and the Telecommunications Industry Association (TIA) have collaborated to develop a comprehensive standard called *Telecommunications Infrastructure Standard for Data Centers*, known as ANSI/TIA-942 [4]. This standard specifies the minimum requirements for telecommunications infrastructure elements in private- and public-domain data centers and computer rooms. Therefore, it provides guidelines and specifications for the design, implementation, and operation of data center infrastructures.

The most recent version of ANSI/TIA-942 is ANSI/TIA-942-B, released in 2017, which supersedes the previous version ANSI/TIA-942-A. ANSI/TIA-942-B incorporates updates and refinements based on industry advancements, technological developments, and lessons learned from previous implementations.

More specifically, ANSI/TIA-942 covers four main aspects:

(1) Site space and layout
(2) Cabling infrastructure.
(3) Tiered reliability.
(4) Environmental considerations.

In addition to these four main aspects, ANSI/TIA-942 also covers other significant issues encountered in data center design and operation. These include:

- Network architecture.
- Physical security measures.
- Fire safety protocols.
- Energy and power management strategies.
- The incorporation of electrical, mechanical and other technological features in data centers.

The topological architecture proposed in ANSI/TIA-942 is intended to be universal, suitable for data centers of any size.

While ANSI/TIA-942 is primarily focused on the United States, its concepts and principles are universally applicable for data center design. However, it is important to note that some specifications within the standard may not align with the requirements of other countries, such as European nations. Therefore, different standards have been developed specifically to cater to the building and operation of data centers in various regions.

11.3.2 Telcodia's GR-3160

Telcodia's GR-3160 is a standard that specifies *Generic Requirements for Telecommunications Data Center Equipment and Spaces* [5]. It is developed with input from 19 participating companies, including AT&T, Brocade, DELL, Intel, Juniper Networks, Telcordia, and Verizon, listed in alphabetical order. The current Issue 2 of GR-3160-CORE, published in 2013, replaces the earlier Issue 1 of GR-3160.

The GR-3160 standard focuses more on critical and high-value areas, relying on existing popular standards when possible and encouraging the use of well-designed and robust commercial off-the-shelf equipment. Briefly speaking, GR-3160 describes the minimum spatial and environmental requirements for data center equipment and spaces, particularly those operated by Telecommunications Carriers. According to the standard, the equipment housed in a Telecommunications Data Center may be used for various purposes, e.g.,:

(1) Support and manage a Telecommunications Carrier's own telecommunications network,
(2) Provide data-center-based services and applications directly to the Telecommunications Carrier's customers,
(3) Provide hosted services and applications for a third party to provide services to their customers, and
(4) Provide a combination of these and similar data center services and applications.

By using GR-3160-CORE, data center reliability can be systematically tested and evaluated. The main tests include:

- Acoustic noise,
- Airflow (temperature and humidity),
- Fast transient impulse immunity test (EFT),
- Fire performance,
- Energy efficiency,
- Electrostatic discharge (ESD),
- Electromagnetic Pulse (EMP) testing,
- Electromagnetic radiation and anti-interference,
- Handling shock (earthquake and office vibration resistance),
- Illumination, and
- Surface Temperature.

GR-3160 represents the current industry practice, particularly concerning operating temperatures and equipment airflow. It also allows for determining the environmental equipment requirements based on the characteristics and service features of a data center. Thus, the generic criteria specified in GR-3160 "are intended to help avoid equipment damage and malfunction caused by environmental impacts such as temperature, humidity, and vibrations; and to minimize fire ignitions and fire spread, as well as provide for improved space planning and thermal management" [5].

Similar to ANSI/TIA-942, GR-3160 is also considered a national standard, with a focus on requirements in the United States. Nevertheless, the concepts and principles from GR-3160 are applicable to the design and operation of data centers in other countries.

11.3.3 EN 50600 and ISO/IEC 22237

Since 2014, the European standardization organization CENELEC, i.e., the European Committee for Electrotechnical Standardization, has developed the EN 50600 series of data center standards [6]. The EN 50600 is the first European-wide, transnational standard on data centers. With considerations of the current best industry practices, it provides comprehensive specifications for the planning, construction, and operation of data centers in a holistic approach. Especially, the EN 50600 standard clarifies the requirements of construction, power supply, air conditioning, cabling, and security systems of data centers. It also defines criteria for the operation of data centers. The standard offers flexibility in the design of a data center and also allows a modular system design. Part of the EN 50600 standard is republished by ISO as an ISO/IEC standard ISO/IEC 22237 [7].

EN 50600 Structure

The EN 50600 standard is structured in four main sections, each of which is further divided into a series of documents within the large grouping. The overall structure of EN 50600 is depicted in Fig. 11.2, where the relationships of the EN 50600 standard and two ISO/IEC standards (ISO/IEC 22237 and ISO/IEC 30134) are also illustrated.

As shown in Fig. 11.2, the four sections of the EN 50600 standard are:

- EN 50600-1: General concepts for design and specification,
- EN 50600-2: Design,
- EN 50600-3: Operations and management, and
- EN 50600-4: Key performance indicatorss (KPIs).

A final section, known as CLC-TR 50600-99, is a subcategory devoted to technical reports:

- CLC/TR 50600-99-1: Energy management – Recommended Practices
- CLC/TR 50600-99-2: Environmental sustainability – Recommended Practices
- CLC/TR 50600-99-3: Guidance to the application of EN 50600 series.

EN 50600-2

For data center design specified in EN 50600-2, specifications are described for data centers from various aspects including:

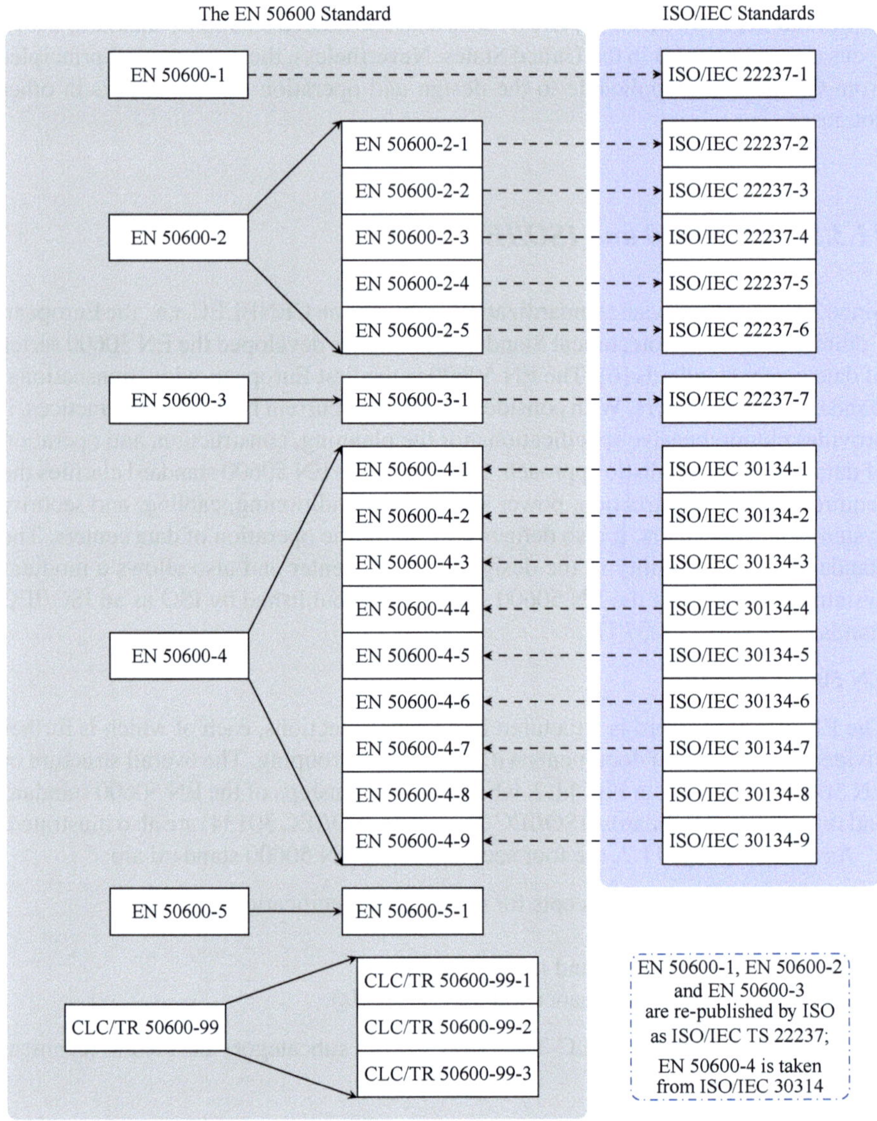

Fig. 11.2 EN 50600 Structure

- EN 50600-2-1: Building construction,
- EN 50600-2-2: Power distribution,
- EN 50600-2-3: Environmental control,
- EN 50600-2-4: Telecommunications cabling infrastructure, and
- EN 50600-2-5: Security systems.

EN 50600-4

EN 50600-4, which deals with KPIs, directly references the standardized KPIs in Section 4 of the ISO/IEC 30314 standard [8]. This section is divided into nine parts:

- EN 50600-4-1: Overview and general requirements (ISO/IEC 30134-1)
- EN 50600-4-2: Power Usage Effectiveness (PUE) (ISO/IEC 30134-2)
- EN 50600-4-3: Renewable Energy Factor (REF) (ISO/IEC 30134-3)
- EN 50600-4-4: IT Equipment Energy Efficiency for Servers (ITEEsv) (ISO/IEC 30134-4)
- EN 50600-4-5: IT Equipment Energy Utilization for Servers (ITEUsv) (ISO/IEC 30134-5)
- EN 50600-4-6: Energy Reuse Factor (ERF) (ISO/IEC 30314-6)
- EN 50600-4-7: Cooling Effectiveness Ratio (CER) (ISO/IEC 30314-7)
- EN 50600-4-8: Carbon Usage Effectiveness (CUE) (ISO/IEC 30314-8)
- EN 50600-4-9: Water Usage Effectiveness (WUE) (ISO/IEC 30314-9)

EN 50600 Design Certification

There are various organizations that offer EN 50600 certification. Through EN 50600 Design certification, a data center can receive an EN 50600 rating based on the EN 50600 standard. For instance, based on the specifications outlined in EN 50600-2-1 through EN 50600-2-4, a data center can receive an Availability Class rating ranging from 1 to 4, as well as an overall rating level between 1 and 4 for the entire data center. Additionally, according to the criteria specified in EN 50600-2-5, a data center will receive a Protection Class rating between 1 and 4. The topic of tied reliability for data centers will be discussed later.

ISO/IEC 22237 [7] **and ISO/IEC 30314** [8]

The EN 50600 standard, while transnational, is not a global standard for data centers. In response to the demand for a globally consistent data center standard, ISO/IEC JTC1 SC39 WG3 has adopted the European EN 50600 standard as the basis for a new international standard, ISO/IEC TS (Technical Specification) 22237. It is interesting to note that the ISO/IEC TS 22237 documents are exact copies of the EN 50600 documents, with the same content and title.

After refinement by all participating national standards bodies, including ISO, the refined documents are formally published as an ISO/IEC standard, specifically ISO/IEC 22237 instead of a TS version. The current version of the ISO/IEC 22237 standard was published in 2021 and is denoted by the standard number followed by the year of publication, for example, ISO/IEC 22237-1:2021. The relationships between the EN 50600 standard and ISO/IEC 22237 are illustrated in Fig. 11.2.

Many professionals in the data center sector may be unaware that globally standardized KPIs have already been defined and published by ISO under the standard ISO/IEC 30314. This standard specifies KPIs from various perspectives and encompasses nine documents numbered from ISO/IEC 30134-1 through ISO/IEC 30134-9. It is worth highlighting that Part 4 of the EN 50600 standard, EN 50600-4, is directly derived from the ISO/IEC 30134 standard. As illustrated in Fig. 11.2, the nine documents of the ISO/IEC 30134 standard are republished within the EN 50600 standard as EN 50600-4-1 through EN 50600-4-9, respectively.

It is important to distinguish between standards that support data center design and those that are applicable to improving data center operations. The ISO/IEC 30134 standard, and therefore the EN 50600-4 standard, focus on enhancing data center operations rather than data center design. By contrast, the ISO/IEC 22237 standard, and hence the EN 50600-2 standard, place more emphasis on data center design rather than data center operation. However, the operational impacts of a data center design should be considered in the design phase. Once a design is implemented, comprehensive tests and evaluations are necessary to confirm whether the data center design meets the specifications, operational requirements, and relevant business objectives.

11.4 Tiered Reliability of Data Centers

Data centers are classified into multiple classes, levels, or tiers in various standards, each providing a different perspective. The classification of data centers, along with their corresponding details, is summarized in Table 11.2. Detailed discussions of these classifications will be give below in the reminder of this section.

Table 11.2 Four classes/levels/tiers of data centers

TIA-942 levels from the physical perspective [4]	
Level 1	**Basic** site infrastructure
Level 2	**Redundant capacity component** site infrastructure
Level 3	**Concurrently maintainable** site infrastructure
Level 4	**Fault-tolerant** site infrastructure
EN50600 and ISO/IEC P22237 classes from the availability perspective [6, 7]	
Class 1	Single path solution: **low availability**
Class 2	Single path with redundancy solution: **medium availability**
Class 3	Multiple paths with concurrent repair/operate solution: **high availability**
Class 4	Multiple paths with fault tolerance: **very high availability**
Uptime Institute tiers from the capacity perspective [9]	
Tier I	**Basic Capacity**
Tier II	**Redundant Capacity**
Tier III	**Concurrently Maintainable** ensuring **Any** component cab be taken out of service without affecting production
Tier IV	**Fault Tolerant** allowing any production capacity to be insulated from **Any** type of failure

11.4.1 ANSI/TIA-942 Data Center Levels

The ANSI/TIA-942 standard defines four rating levels of data centers from the perspective of physical infrastructure [4]. These rating levels are commonly referred to as tiers in various documents. Therefore, the terms "levels" and "tiers" are used without differentiation in this chapter unless otherwise indicated explicitly.

Level 1/Tier 1: Basic Site Infrastructure

A Level-1 data center has single capacity components and a single, non-redundant distribution path serving the computer equipment. It possesses limited protection against physical events. Specific requirements include:

- Single non-redundant power and cooling distribution path serving the IT equipment,
- Non-redundant capacity component,
- Annual downtime of 28.8 h,
- Shutting down completely for preventive maintenance, and
- Basic site infrastructure guaranteeing 99.671% availability.

Level 2/Tier 2: Redundant Capacity Component Site Infrastructure

A Level-2 data center has redundant capacity components and a single, non-redundant distribution path serving the computer equipment. It possesses improved protection against physical events. Specific requirements include:

- Fulfilling all Tier-1 requirements,
- Single path for power and cooling distribution with redundant components,
- Raised floor, Uninterruptible Power Supply (UPS), and backup generator,
- Annual downtime of 22.0 h,
- Maintenance of power path and other parts of the infrastructure will require a processing shutdown, and
- Redundant site infrastructure capacity component guaranteeing 99.741% availability

Level 3/Tier 3: Concurrently Maintainable Site Infrastructure

A Level-3 data center is characterized by having redundant capacity components and multiple independent distribution paths that serve the computer equipment. Typically, only one distribution path is active at any given time, while the other serves as a backup. The site is designed to be concurrently maintainable, allowing planned removal, replacement, or servicing of all capacity components, including those within the distribution path, without disrupting the ICT capabilities for end users. Furthermore, Level-3 data centers provide protection against most physical events.

Specific requirements for a Level-3 data center include:

- Fulfilling all Tier-1 and Tier-2 requirements.
- Having multiple power and cooling distribution paths, with only one path active at a time.

- Requiring all IT equipment to be dual-powered and fully compatible with the site's architectural topology.
- Incorporating redundant components to ensure high availability.
- Utilizing a raised floor and sufficient capacity and distribution to handle the load on one path while performing maintenance on the other.
- Aim for an annual downtime of 1.6 h.
- Targeting an availability of 99.982%.

Level 4/Tier 4: Fault Tolerance Site Infrastructure

A Level-4 data center has redundant capacity components and multiple independent distribution paths that serve the computer equipment, with all paths being active simultaneously. It allows for concurrent maintainability, meaning that maintenance activities can be performed while the data center remains operational, and a single fault anywhere in the installation does not result in downtime. Level-4 data centers provide protection against almost all physical events.

Specific requirements for a Level-4 data center include:

- Fulfilling all Tier-1, Tier-2, and Tier-3 requirements.
- Having multiple power and cooling distribution paths.
- Requiring all IT equipment, as well as chillers and heating, ventilating, and air-conditioning systems, to be independently dual-powered.
- Incorporating redundant components for increased reliability.
- Aiming for an annual downtime of 26.3 min.
- Implementing a fault-tolerant site infrastructure with power storage and distribution facilities to ensure a guaranteed availability of 99.995%.

11.4.2 EN50600 and ISO/IEC 22237 Data Center Classes

The EN50600 and ISO/IEC 22237 classify data centers into four classes from the service availability perspective [6, 7]. The four classes are specified according to the criteria across three categories: Power supply, environment control, and telecommunications cabling. They are tabulated in Table 11.3.

11.4.3 Uptime Institute Data Center Tiers

Data centers are categorized into four tiers in the Uptime Institute standard [9]. Each tier has specific requirements that determine the reliability and availability of the services provided. A brief summary of the tier requirements from the Uptime Institute standard is listed in Table 11.4. The four tiers of data centers and their requirements are described below in terms of fundamental requirements, performance tests, and operational impacts.

Table 11.3 Availability classes of data centers from EN50600/ISO/IEC 22237 (Class 1: single path; Classes 2 through 4: multi-path)

Class	EN50600-2-2 Power supply	EN50600-2-3 Environmental	EN50600-2-4 Cabling
1	No redundancy	Not specified	Direct connections
2	System redundancy	No redundant components	Fixed infrastructure
3	System redundancy	Component redundancy	Fixed infrastructure
4	Fault tolerance	System redundancy allowing maintenance during operation	Fixed infrastructure with diverse pathways

Availability classes: 1 (low), 2 (medium), 3 (high), 4 (very high)

Table 11.4 Summary of tier requirements [9]

<table>
<tr><th>Requirements</th><th>Tier I</th><th>Tier II</th><th>Tier III</th><th>Tier IV</th></tr>
<tr><td>Min capacity components (N: necessary capacity)</td><td>N</td><td colspan="2">N + 1</td><td>N after any failure</td></tr>
<tr><td>Distribution paths</td><td colspan="2">1</td><td>1 active + 1 alternate</td><td>2 simultaneously active</td></tr>
<tr><td>Critical distribution</td><td colspan="2">1</td><td colspan="2">2 simultaneously active</td></tr>
<tr><td>Concurrently maintainable</td><td colspan="2">No</td><td>Yes</td><td rowspan="4">Yes</td></tr>
<tr><td>Fault tolerance</td><td colspan="3" rowspan="3">No</td></tr>
<tr><td>Compartmentalization</td></tr>
<tr><td>Continuous cooling</td></tr>
</table>

Fundamental Requirements

Tier I: Basic capacity. A Tier I data center provides dedicated site infrastructure to support information technology beyond an office setting. Tier I infrastructure includes a dedicated space for IT systems; an UPS to filter power spikes, sags, and momentary outages; dedicated cooling equipment that will not be shut down at the end of normal office hours; an engine generator to protect IT functions from extended power outages; and 12 h of on-site fuel storage for on-site power production, e.g., engine generator or fuel cell.

Tier II: Redundant capacity. Tier II facilities include redundant critical power and cooling components to provide select maintenance opportunities and an increased margin of safety against IT process disruptions that would result from site infrastructure equipment failures. The redundant components include power and cooling equipment such as UPS modules, chillers or pumps, and engine generators. It is also a requirement for a Tier II data center to have an engine generator and 12 h of on-site fuel storage.

Tier III: Concurrently maintainable. A Tier III data center requires no shutdowns for equipment replacement and maintenance. A redundant delivery path for power and cooling is added to the redundant critical components of Tier II facilities so that each and every component needed to support the IT processing environment can be shut down and maintained without impacting IT operations. All IT equipment must be dual-powered and installed properly to be compatible with the topology of the site's architecture. As for Tier I and Tier II data centers, an engine generator and 12 h of on-site storage are required for a Tier III data center.

Tier IV: Fault tolerance. Building on Tier III, Tier IV site infrastructure adds the concept of fault tolerance to the site infrastructure topology. Fault tolerance means that when individual equipment failures or distribution path interruptions occur, the effects of the events are stopped short of the IT operations. This requires:

- Multiple, independent, and physically isolated systems,
- Multiple, independent, diverse, and active distribution paths simultaneously serving the critical environment,
- Dual-powering of all IT equipment,
- Complementary systems and distribution paths being physically isolated from one another, and
- Continuous cooling.

Moreover, as in other tiers of data centers, it is an essential requirement to have an engine generator and 12 h of on-site fuel storage in a Tier IV data center.

Power supply. The Uptime Institute standard for data centers has specific requirements for engine-generator systems. While local power utility is an economic choice, on-site power generation systems are considered the primary power source for data centers. They must automatically start and assume load upon loss of utility. Also, all critical equipment not backed up by UPS power must automatically restart after power restoration. For a Tier III or Tier IV data center, its engine-generator system, along with its power paths and other supporting elements, should meet the concurrently maintainable and/or fault-tolerant requirements while they are carrying the site on engine-generator power.

Performance Tests

The performance confirmation tests for the four tiers of data centers discussed above are tabulated in Table 11.5 with detailed explanations.

Operational Impacts

From the basic requirements and performance confirmation tests described above for different tiers of data centers, the reliability of data centers increases from Tier I to Tier IV.

Both Tier I and Tier II data centers are vulnerable to disruptions caused by both planned and unplanned activities. In particular, human operational errors can lead to data center disruptions.

Tier III data centers are designed to withstand planned activities. Maintenance tasks can be carried out using redundant capacity components and distribution paths without impacting IT operations. However, Tier III data centers are still susceptible to disruptions resulting from unplanned events.

Table 11.5 Performance confirmation tests for data centers [9]

Tier I	
(a)	Sufficient capacity to meet the needs of the sites
(b)	Planned work will require most or all of the site infrastructure systems to be shut down, affecting critical environment, systems, and end-users
Tier II	
(a)	Redundant capacity components can be removed from service on a planned basis without causing any of the critical environment to be shut down
(b)	Removing distribution paths from service for maintenance or other activities requires the critical environment to be shut down
(c)	Sufficient permanently installed capacity to meet the needs of the site when redundant components are removed from service
Tier III	
(a)	Each capacity component or element in the distribution paths can be removed from service on a planned basis without impacting any part of the critical environment
(b)	Sufficient permanently installed capacity to meet the needs of the site when redundant and distribution components are removed from service
Tier IV	
(a)	A single failure of any capacity system, capacity component, or distribution element will not impact the critical environment
(b)	Autonomous response of the infrastructure control system to a failure while sustaining the critical environment
(c)	All tests for Tier III
(d)	Capability of detecting, isolating, and containing any potential fault while maintaining the required capacity to the critical load

By contrast, a Tier IV data center is resilient to both single unplanned events and planned activities. Like Tier III, planned maintenance can be performed without affecting IT operations by using redundant components and distribution paths. However, certain operations such as triggering the fire alarm, activating fire suppression systems, or initiating the Emergency Power Off (EPO) feature may cause disruptions in a Tier IV data center.

11.4.4 The Choice of a Data Center Tier

When choosing a data center tier, several key considerations need to be taken into account to ensure the balance between meeting service needs and maintaining cost-effectiveness. These considerations include:

- Uptime needs and the potential impact of downtime.
- The mission-criticality of hosted servers and/or data.
- Available budget.
- security requirements.

- Relevant legal obligations.
- Customer expectations.

Typically, the primary considerations when deciding the data center tier are uptime needs and costs.

According to the operational impacts of different data center tiers described previously, the following guidelines can be used:

- **Tier 1** data centers are best suited for small businesses and start-ups that are looking for the most affordable hosting option. Small firms without complex IT requirements are able to tolerate frequent downtime occurrences.
- **Tier 2** data centers are a good choice for small businesses that expect a cost-effective and more reliable option than Tier 1. Small to medium-sized firms typically use Tier 2 facilities, often for hosting data backups and non-mission-critical databases as well as other IT services.
- **Tier 3** data centers are an ideal choice for large companies with IT operations that require enhanced safety, reliability, and availability. Businesses that host extensive data sets and other IT services are the main users of Tier 3 facilities.
- **Tier 4** data centers are the choice of enterprises without major budget constraints. Government organizations, commercial data center service providers, and large enterprises with mission-critical services and intense customer or business demands are typical candidates for Tier 4 data centers.

11.5 Site Space, Cabling, and Environments

In addition to tiered reliability, three of the main parts of the TIA-942 standard are site apace, cabling, and environmental considerations. These three aspects are important in data center design to ensure the functionality, efficiency, and sustainability of data centers.

11.5.1 Site Space and Layout

Space design and allocation for a data center should take into consideration changing environments and future growth, including the possibility of relocation. Reserve some empty space to accommodate future racks or cabinets, allowing for scalability and adaptability. The surrounding space of the data center should be considered to facilitate future growth and allow for easy expansion.

A significant portion of the TIA-942 standard focuses on facility specifications, such as the space and layout of data centers. According to the specifications outlined in TIA-942, there are distinct functional areas within a data center. These key areas, as identified in TIA-942, include:

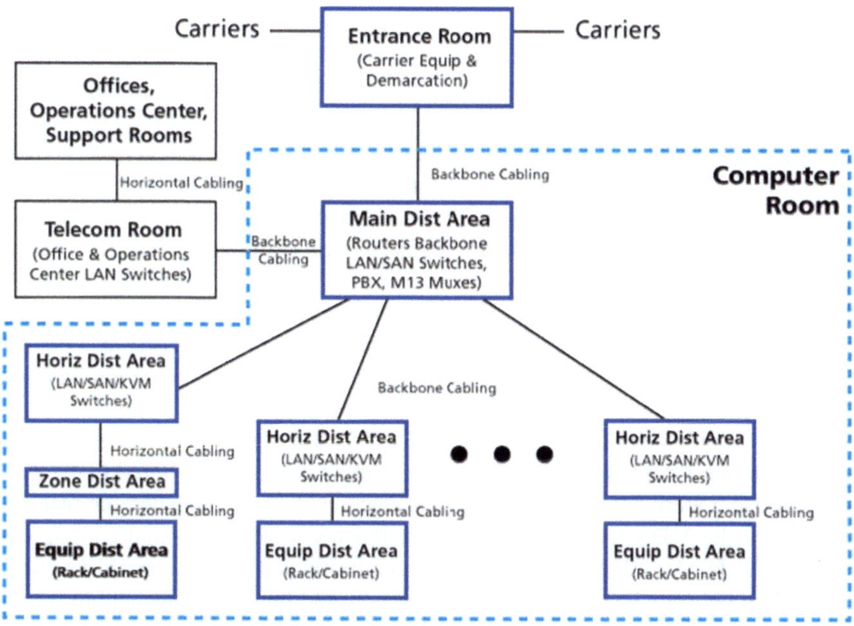

Fig. 11.3 Key functional areas in a TIA-942 compliant data center [4]

(1) Entrance rooms.
(2) Main Distribution Area (MDA).
(3) Horizontal Distribution Area (HDA).
(4) Equipment Distribution Area (EDA).
(5) Zone Distribution Area (ZDA).
(6) Backbone and horizontal cabling.

These key areas in a data center are shown in Fig. 11.3. They are described below in more detail.

One or More Entrance Rooms. The entrance room can be located either inside or outside the computer room. The TIA-942 standard recommends locating the entrance room outside the computer room for enhanced security. Large data centers may require multiple entrance rooms to accommodate the cabling systems efficiently.

Main Distribution Area (MDA). The MDA is a central space housing core routers and switches for LAN and SAN infrastructure. It also contains the main cross-connects. TIA-942 specifies that at least one MDA is necessary, and separate racks for fiber, UTP, and coaxial cable are required in the MDA.

One or More Horizontal Distribution Areas (HDA). The HDA functions as the distribution point for horizontal cabling. It accommodates cross-connects and active equipment to distribute cables to the equipment distribution area. Similar to the MDA, separate racks for fiber, UTP, and coaxial cable are mandated in the HDA. The placement of switches and patch panels is recommended to minimize patch cord lengths and facilitate cable management. The number of HDAs depends on the total number of connections in the data center, and the HDA is typically limited to 2,000 connections.

Equipment Distribution Area (EDA). The EDA is where equipment racks and cabinets are located. Patch panels are commonly used to terminate horizontal cables in the EDA. The TIA-942 standard specifies that racks and cabinets should be installed in an alternating pattern to create hot and cold aisles for effective heat dissipation from electronics.

Zone Distribution Area (ZDA). The ZDA is an optional interconnection point in the horizontal cabling between the HDA and EDA. It can be used as a consolidation point for reconfiguration flexibility or house freestanding equipment that cannot accommodate patch panels, such as mainframes and servers. TIA-942 allows for only one ZDA within a horizontal cabling run, with a maximum of 288 connections. Cross-connects and active equipment are not permitted in the ZDA.

Backbone and Horizontal Cabling. Backbone cabling establishes connections between the MDA, HDAs, and Entrance Rooms, while horizontal cabling connects the HDAs, ZDA, and EDA. Optional redundant backbone cabling can be installed between HDAs to ensure redundancy and fault tolerance within the network infrastructure.

11.5.2 Cabling Infrastructure

Regarding cabling infrastructure, the TIA-942 standard builds upon the existing TIA-568 and TIA-569 standards. It specifies a generic, permanent telecommunications cabling system. More specifically, it provides specifications for recognized cabling media, including:

- Standard single-mode fiber.
- 62.5 and 50 μm multimode fibre (recommended).
- 75 ohm coaxial cable (recommended for E-1, E-3, and T-3 circuits).
- 4-pair Category 6 UTP and ScTP cabling (Cat 6 recommended).

The TIA-942 standard recommends the use of 50 μm multimode fiber for backbone cabling. This fiber type can support higher network speeds over longer distances with cost-effectiveness compared to single-mode fiber.

For horizontal cabling, it is recommended to use the highest capacity media available to minimize the need for future re-cabling. It is worth mentioning that while Category 6 is the highest recognized horizontal cabling media in the TIA-942 standard, 10 Gigabit Ethernet over UTP solutions have already been verified and are available.

In the TIA-942 standard, backbone cabling and horizontal cabling are typically limited to 300 m and 100 m, respectively. However, for small data centers, it is possible to combine the HDA with the MDA, allowing for horizontal fiber cabling to be extended to 300 m.

To ensure better cabling management, a data center should be designed with separate racks and pathways for each type of cabling media. Power and communication cables must be placed in separate pathways or separated by a physical barrier. Adequate space should be provided within and between racks and cabinets, as well as in pathways, to facilitate proper cable installation and maintenance.

The TIA-942 standard also extends the TIA-606-A Administrator Standard for labeling purposes. It specifies the requirements for labeling racks, cabinets, patch panels, patch cords, and cables within the data center. Proper labeling aids in the identification, organization, and troubleshooting of the cabling infrastructure.

11.5.3 Environmental Considerations

Environmental considerations in data center design encompass a range of factors, including electrical and mechanical specifications, architectural considerations, fire suppression, operating temperatures, and humidity control. Some of these considerations are closely linked to the tiered specifications discussed earlier.

Power Supply

Determining the power requirements is primarily based on the desired reliability tier. It is essential to have one or two power feeds from the utility, supplemented by UPS systems, multiple circuits, and on-site engine generators. When planning for power, it is essential to account not only for the existing devices but also for future growth and redundancy of the data center. Moreover, all supporting equipment such as UPS systems, on-site generators, and cooling systems must be taken into consideration.

Cooling

Cooling contributes significantly to the overall energy cost of operating a data center. Therefore, special attention is given to cooling systems and their efficiency in various standards like TIA-942. A well-designed airflow helps reduce the heat generated by densely packed equipment. Energy-efficient equipment and raised-floor systems are recommended for more flexible cooling. Also, arranging racks and cabinets in an alternating pattern creates hot and cold aisles, as depicted in Fig. 11.4.

11.6 Considerations of Data Center Locations

Data centers are located everywhere in the world, as discussed at the beginning of this chapter. Before constructing a data center, selecting a good location is challenging. It entails striking a balance among various priorities to achieve the best trade-off for the particular data center design. Some important considerations are discussed below to aid in the selection of data center locations [10].

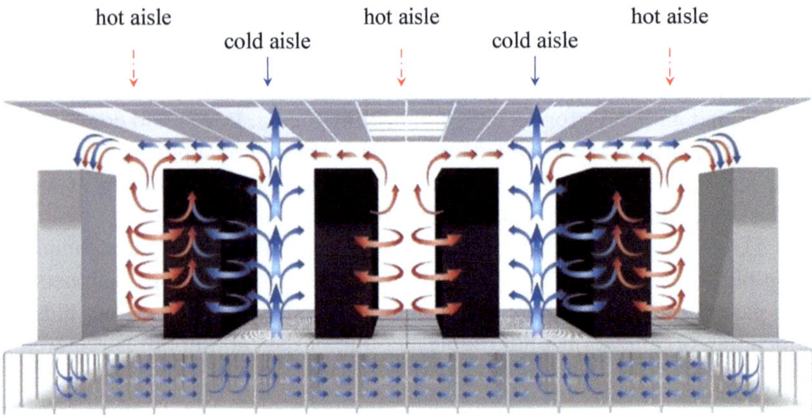

Fig. 11.4 Hot and cold airflows in data center cooling

11.6.1 Safety and Security of Physical Locations

The first consideration in choosing data center locations is ensuring safety and security. Downtime is a critical performance metric for data centers, as it results in substantial costs such as revenue loss, decreased productivity, recovery expenses, and other associated costs. The average cost of data center outages can reach millions of dollars. Therefore, minimizing potential downtime is of utmost importance when selecting a suitable data center location.

To achieve this, data center locations must be carefully chosen to avoid areas prone to natural disasters, such as earthquakes, tsunamis, hurricanes, tornadoes, floods, fires, and landslides. Geographic data from various sources can be used to identify regions with a high risk of such disasters. Exclude such areas from consideration for data center construction.

Although natural disasters cannot be completely avoided, there are geographical regions that offer greater safety compared to others. Some locations are even considered completely safe from all types of natural disasters. Notably, IBM's search for the site of its largest data center concludes that Kelowna in Canada is one of the safest places in North America, based on an assessment of environmental safety.

11.6.2 Power Supply

Energy consumption is a significant factor in data center operations, accounting for approximately 70% of the total operating costs. Therefore, the price of electricity at a data center location greatly impacts its operating expenses. Consequently, data center locations with lower-cost power are preferred choices.

For instance, the IBM Kelowna data center in British Columbia, Canada, benefits from hydroelectric power sourced from a dam on the Columbia River, costing as low as 2 cents per kilowatt-hour. It is also interesting to note that through the Grand Coulee Dam, which is downriver from Kelowna, the Columbia River is powering a number of large data centers in Eastern Washington, such as Microsoft, Yahoo, and Dell data centers.

Naturally, the price of power should be evaluated alongside the reliability and proximity of the power supply. Proximity to the power supply reduces the risk of transmission issues. Unreliable power supply poses the potential for power loss, necessitating the implementation of well-designed backup power generators or risking significant downtime. The latter would require a higher capital investment for engine generator systems or lead to increased operational costs due to the resulting downtime.

Certain locations may have a higher likelihood of floods, hail, or thunderstorms. While such natural disasters are unlikely to cause damage to a data center, they can result in power outages. This factor should be considered when assessing the reliability of the power supply.

11.6.3 Environment Temperatures

A significant portion of the energy consumption in data center operation results from cooling. Therefore, choosing locations with lower environmental temperatures will help reduce the operation cost of data centers significantly.

However, keeping data centers cool is not easy. Different firms have tried to solve this problem in different ways. For example, Microsoft has been exploring the possibility of underwater data center facilities. Google uses its AI expertise to reduce costs. Facebook has a simple solution: move to a cold country. Facebook's Luleå Data Center is located in the deep forests of the northernmost part of Sweden. The small town of Luleå is less than 70 miles south of the Arctic Circle. It is typically pretty cold, with daily mean temperatures being lower than 8 °C from October through May.

The Facebook Luleå data center is Facebook's first investment outside of the United States. Some of the reasons for locating the data center in Luleå are the natural cooling due to the climate, cheap electricity, reliable electrical networks, and clean energy. The data center is the largest in Europe, with an area of 84,000 m^2, comparable to 11 football fields. A picture of the data center from Facebook is shown in Fig. 11.5.

The solution of moving to a cold country is easy to implement. Therefore, a large number of data centers have been build in Europe, particularly Northern Europe, by various firms.

It is worth noting that there are also some potential issues when locating data centers offshore, or using data center and IT services from overseas counties. Recently, the data protection authorities of Ireland has decided to rule on the legality of meth-

Fig. 11.5 Facebook's Luleå data center (*Source*: https://m.facebook.com/LuleaDataCenter, accessed on 10 Apr. 2022)

ods used by platforms like Facebook and Google to transfer European data to the United States. This has actually been a long debate between European countries and Giant IT companies from the United States.

Not long time ago, Facebook suddenly stopped providing social network services to Australia without a notice. While the services were restored shortly, this did threaten the use of overseas IT services, indicating that replying on overseas IT services was not always reliable. Therefore, localized IT services are important, particularly for national critical systems and applications.

11.6.4 Access to Communication Networks

There is no doubt that access to high-speed and reliable communication networks is crucial for data centers. To ensure the reliability and availability of data center services, it is generally recommended to have at least two such networks accessible. One network can serve as the primary connection, while the other acts as a backup. Alternatively, both networks can be used simultaneously for load balancing and improved service performance. This setup allows for seamless switching to the backup network in the event of a service outage, thus minimizing disruption for end users.

It is worth mentioning that two network service providers may use the same physical infrastructure to provide their network connections. In such cases, one provider owns the infrastructure and leases it to the other provider, or both providers rent the network infrastructure from a third party. By connecting to the networks of these two providers, data centers gain access to a single physical network. As an example, in Australia, TPG Telecom offers mobile and network services using the network infrastructure of Vodafone Hutchison Australia, one of the country's largest telecommunications companies. Vodafone also provides similar services to end users. Interestingly, the recent merger between TPG and Vodafone has been approved by the Federal Court, further shaping the telecommunications landscape in Australia.

11.6.5 Proximity to End Users and Skilled Labor

The proximity to end users is an important consideration when latency and other critical QoS requirements are crucial. By reducing the distance between the data center and end users, latency performance can be improved, resulting in reduced response times. While latency may not be a significant issue for the majority of end users who use best-effort network services, it becomes more critical for a small number of users who require low-latency services. Meeting different latency requirements for various users typically needs to be assessed on a case-by-case basis.

Similarly, the proximity to skilled labor is a factor to consider for supporting the construction and operation of data centers. Although labor costs are relatively low compared to the energy consumption costs, remote operation alone may not be practically feasible for data centers. Having a qualified labor force readily available is beneficial for building and operating data centers, even though much of the work can be conducted remotely.

The IBM Kelowna data center provides a good example of balancing safety from natural disasters, power costs, and proximity to skilled labor:

- Kelowna is located in an area safe from natural disasters, as discussed previously.
- The availability of hydroelectric power from a dam on the Columbia River in Kelowna enables power costs as low as 2 cents per kilowatt-hour.
- In addition to local skilled labor, qualified individuals from the nearby city of Vancouver, British Columbia, are also accessible. Kelowna is just a four-hour drive from Vancouver.

11.6.6 Other Considerations

There are other factors to consider when selecting a location for a data center. Real estate is one of these factors. Choosing a location with affordable real estate prices can significantly reduce the capital investment required for building a data center. For example, the IBM Kelowna data center, which occupies approximately $85,000$ square feet (nearly 8,000 m^2), costs around US$100 million. If the same data center were to be built in a high-cost area like New York, the expenses would be significantly higher.

Building a data center entails substantial costs. For those who find it financially challenging, colocation can be a viable option. Colocation involves renting data center space and equipment from another company, with the service provider managing the servers and storage. This approach is particularly beneficial for small to medium service providers and startups in the data center industry. It allows them to rent the necessary resources, pay only for what they use, and avoid the upfront costs associated with building and maintaining a dedicated data center facility.

11.7 Network Architecture

A data center is the home to various IT services, including Internet/Extranet/Intranet services, network services, computing services, data services, and many others for enterprise businesses. Alongside its physical infrastructure, such as the site and layout, the network infrastructure of a data center holds significant importance in the overall IT architecture. It serves as the conduit through which all network traffic flows from its source to its destination. Consequently, designing the network architecture for a data center necessitates careful consideration of its impact on performance, resiliency, scalability, security, and future expandability.

11.7.1 Three-layer Data Center Architecture

Among all these considerations, the flexibility of quick deployment and support for new services and applications is a crucial factor to be taken into account during the initial design phases. This requirement aligns with the design principles of other large-scale computer networks. To meet these needs, a suitable choice is a layered logical architecture, which offers not only flexibility but also benefits in terms of security, scalability, and future expandability for data centers.

Cisco, a prominent networking company, recommends a three-layer architecture for data centers, similar to their Core/Distribution/Access architecture for general networks. This three-layer architecture, comprising the core layer, aggregation layer, and access layer, forms the foundation of Cisco's data center design [11]. The architecture is depicted in Fig. 11.6. It provides a framework from which various variations can be derived, depending on specific requirements and objectives. The layered approach enables efficient management and control of data center operations, facilitating the seamless integration of new services and applications.

The *core layer* provides a highly reliable and high-speed packet switching backplane for managing incoming and outgoing traffic flows within the data center. It also establishes connectivity to the aggregation layer. To ensure maximum reliability, the core layer is typically designed with Tier 3 or Tier 4 reliability standards, thus minimizing the risk of a single point of failure. High-speed packet switching is achieved through the utilization of 10 Gigabit Ethernet interfaces. In addition, the core layer employs an interior routing protocol such as OSPF or EIGRP for efficient routing of data.

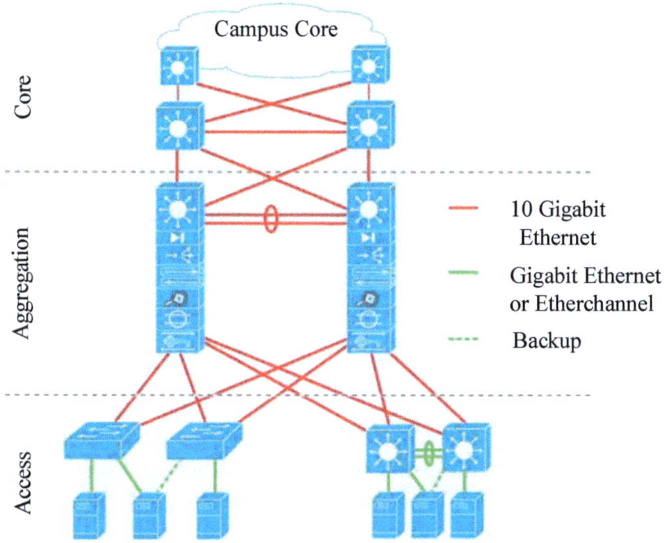

Fig. 11.6 Cisco's three-layer data center architecture [11]

The *aggregation layer* fulfills several essential functions, such as service module integration, spanning tree processing, default gateway redundancy, and defining layer-2 domains for server-to-server traffic. Service modules within the aggregation layer facilitate various tasks such as content switching, intrusion detection, firewall services, SSL offloading, network analysis, and more.

The *access layer* primarily focuses on providing physical connectivity between servers and the network infrastructure. Depending on the specific requirements, different types of servers may be present within the data center, such as blade servers with integrated switches, blade servers with pass-through cabling, cluster servers, and mainframes with IBM's Open System Adapters (OSA). The access layer network architecture incorporates modular switches that support both Layer-2 and Layer-3 topologies, enabling the fulfillment of server broadcast domain requirements and other administrative needs.

11.7.2 Data Center Design Models

In the design of data centers, two commonly adopted frameworks or models are the **multi-tier model** and the **server cluster model**. Each of these models is suitable for different application scenarios and environments. Both models are extensively documented in Cisco's online resources [12, 13].

The *multi-tier model* is the prevailing design in enterprise data centers. It follows a layered approach, with distinct tiers dedicated to web services, applications, and databases that support commerce, enterprise resource planning (ERP), and customer relationship management (CRM) systems. The multi-tier model relies on network infrastructure that incorporates security measures and application optimization services.

The *server cluster model* is commonly associated with high-performance computing (HPC), parallel computing, high-throughput computing, and grid computing. It finds extensive usage in the scientific community, including universities and research institutions. However, the server cluster model is also applicable in enterprise data centers serving financial, manufacturing, and other industries. In this model, customized application architectures are typically employed to meet specific application requirements and optimize performance.

Multi-tier Model

The multi-tier data center model is characterized by the presence of dominant web applications organized in a multi-tier hierarchical structure. Most web-based applications in data centers are built as multi-tier applications. Typically, the multi-tier model consists of three tiers:

- The *web-server tier*,
- The *application tier*, and
- The *database tier*.

In the multi-tier data center model, software can be executed either as separate processes on the same physical machine or on different physical machines with communication over the network. Designing multi-tier server farms with processes distributed across physical machines provides enhanced resiliency and security. Improved resiliency is achieved by allowing a physical server to be taken out of service without affecting the system functions provided by other servers within the same application tier. Security is enhanced because compromising a web server does not grant access to the application or database servers. While web and application components can be deployed on the same physical machine, database servers are typically kept on separate machines.

The requirements of resiliency and security drive the need for segregation between the tiers in the multi-tier model. This segregation can be achieved through physical or logical means. Physical segregation involves deploying separate infrastructures for each tier, operating within both the aggregation layer and access layer of the data center. This is accomplished using aggregation and access switches or Virtual Local Area Networks (VLANs). Logical segregation uses VLANs to separate server farms. Figure 11.7 illustrates a diagram depicting the segregation between the tiers.

Server Cluster Model

The server cluster model focuses on deploying clusters of physical servers to achieve increased computational power, specifically targeting HPC applications. This model is commonly used in environments where a large number of CPUs need to be integrated into a unified high-performance system. From the perspective of end users, the CPUs, GPUs, storage, and other computing resources are transparent, appearing

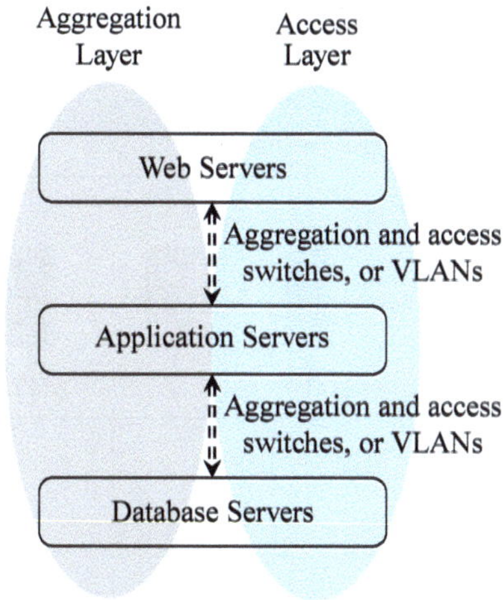

Fig. 11.7 Segregation between tiers in the multi-tier model

to come from a single system. This simplifies and streamlines the management of computing tasks for end users.

Historically, the server cluster model has been associated with higher education and research institutions that have demanding high-performance computing requirements, such as universities and research institutes. However, it has also found applications in various sectors including meteorology, seismology, defense, and now expanding to enterprise data centers, such as financial and banking systems, government departments and agencies, medical organizations, and manufacturing.

Logically, a server cluster consists of several components, including a front-end public interface, one or more master nodes, a back-end private interface, back-end compute nodes, back-end storage, and a back-end high-speed fabric. The high-speed fabric serves as the primary medium for communication between master nodes and compute nodes, as well as inter-computer node communication. Figure 11.8 provides a logical view of a server cluster.

Physically, the front-end interface terminals are connected to the master nodes through local switches. The master nodes, compute nodes, file systems, and storage area networks are located in the access layer. The compute nodes are organized into one or more clusters, and the storage servers can be placed in separate clusters. Each of these compute and storage clusters is connected to network switches. The switches of all these clusters are further interconnected through a high-speed fabric in the core layer. Figure 11.9 illustrates a physical view of a server cluster data center.

11.7.3 HPC Cluster Types

With regard to server clusters, there is a particular interest in HPC clusters, as the HPC capacity of a country more or less indicates its competitiveness in information technology and associated manufacturing industries. As of June 2023, the top 10

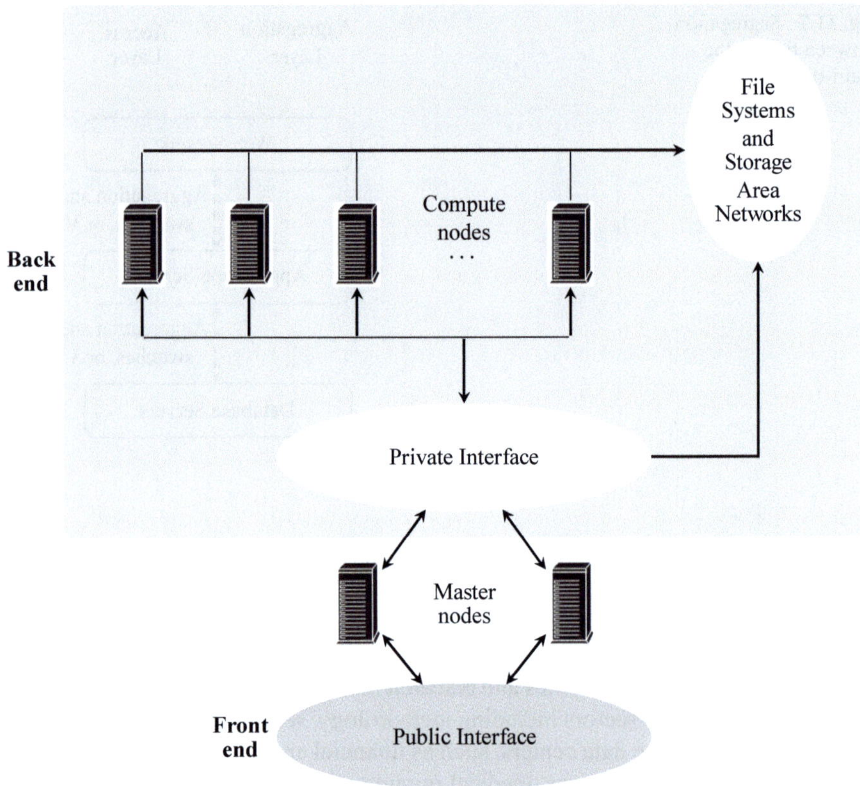

Fig. 11.8 Logical view of a server cluster

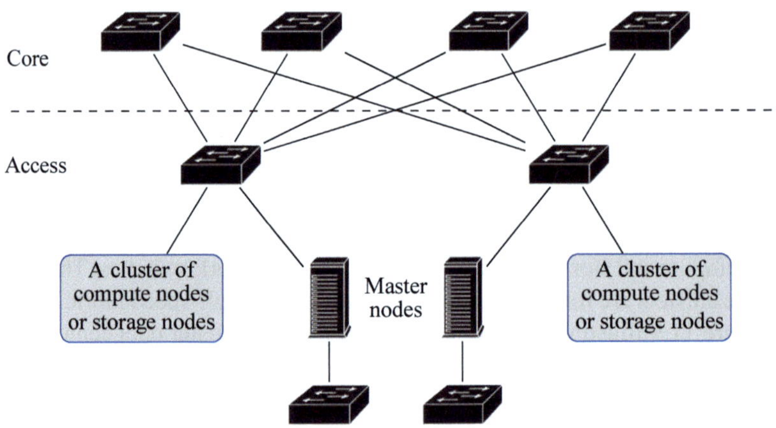

Fig. 11.9 Physical view of a server cluster data center

Table 11.6 Top 10 supercomputers as of June 2023 [14]

Rank	Supercomputer	Country	Rmax (Pflops/s)
1	Frontier	USA	1, 194.00
2	Supercomputer Fugaku	Japan	442.01
3	LUMI	Finland	309.10
4	Leonardo	Italy	238.70
5	Summit	USA	148.60
6	Sierra	USA	94.64
7	Sunway TaihuLight	China	93.01
8	Perlmutter	USA	70.87
9	Selene	USA	63.46
10	Tianhe-2A	China	61.44

supercomputers in the world are listed in Table 11.6. However, this list is changing over time.

While these top-ranked supercomputers are obviously not affordable for general organizations, less powerful yet cost-effective high-performance computers have found wide applications in enterprise data centers. These computers typically rely on Fast Ethernet and Gigabit Ethernet for interconnecting their cores and nodes, as opposed to the InfiniBand interconnection used in many of the top 10 supercomputers listed in Table 11.6. InfiniBand and 10 Gigabit Ethernet are increasingly employed for interconnection in general high-performance computers to reduce latency and improve bandwidth.

High-performance clusters are categorized into three main types in Cisco's data center architecture overview [11]:

- Type 1: Parallel messaging passing (also known as tightly coupled).
- Type 2: Distributed I/O processing (e.g., search engines).
- Type 3: Parallel file processing (also known as loosely coupled).

For each of these three types of HPC clusters, there is a master node that coordinates various computing tasks among all the nodes.

Tightly coupled Type-1 HPC clusters are traditionally used in universities and other research-intensive institutions for intensive and high-performance computing. They run applications on all nodes simultaneously in parallel.

Type-2 and Type-3 HPC clusters are being increasingly adopted in enterprise data centers to support industries such as finance, banking, and manufacturing. In the presence of multiple master nodes, Type-2 HPC clusters balance the requests from clients across the multiple master nodes and then distribute the computing tasks to the compute nodes for parallel processing. Loosely coupled Type-3 HPC clusters split and distribute the source data across the compute nodes for parallel processing. The processed results from each compute node are rejoined to obtain the final results.

11.7.4 Data Center Network Virtualization

Virtualization is a broad topic in the management of computing resources. However, this section specifically focuses on its impact on the architecture of data centers.

In modern data centers, virtualization has been used to provide virtualized computing environments. Server virtualization and network virtualization are two common use cases of virtualization in data centers. In traditional data center networks, virtual servers and Virtual Machines (VMs) are typically assigned IP addresses based on their physical location. This approach works well for physical servers that remain stationary and are not frequently moved to different locations. However, VMs may need to be migrated from one data center to another. Unfortunately, traditional data centers face technical limitations in terms of supporting VM mobility. These limitations have become evident with the increasing demands of cloud applications and multi-tenant networks [15].

Data Center Overlay Networks

To address aforementioned issue, the IETF RFC 7364 [15] published in October 2014 has introduced the concept of data center Network Virtualization over Layer 3 (NVO3) networks. NVO3 networks are overlay-based virtual networks that separate the communication among multi-tenant systems from the physical infrastructure systems within a data center. These virtual networks over layer 3 provide network virtualization services to a set of tenant systems [16], effectively hiding the tenancy separation from the underlying physical infrastructure. Also, NVO3 networks limit the scope of packets sent on the virtual network. This enables correct traffic forwarding without the need for the underlying transport network to be aware of tenancy separation.

As summarized in the IETF RFC 8151 [17, p. 2], NVO3 networks allow a physical network architecture to perform the following functions:

- Carry multiple NVO3 virtual networks and isolate the traffic of different NVO3 virtual networks.
- Provide independent address spaces (such as MAC and IP) within each NVO3 virtual network.
- Support flexible VM placement and workload placement without requiring changes in address or physical infrastructure network configuration.

The IETF RFC 8151 [17, p. 2] has also discussed some NVO3 use cases that can be deployed in various data centers to serve different data center applications.

In the overlay approach, each virtual network instance is implemented as an overlay. The original packet is encapsulated by the first-hop network device, known as a Network Virtualization Edge (NVE). The encapsulation process identifies the remote NVE that will perform the decapsulation of the tunneled packet before delivering the original packet to its destination. The encapsulated packet is then tunneled to the identified remote egress NVE. The rest of the network simply performs packet forwarding based on the encapsulation header.

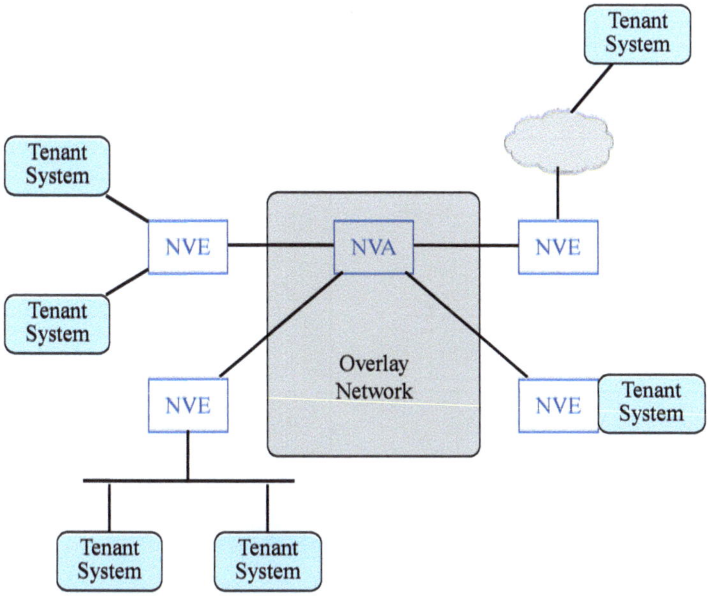

Fig. 11.10 Generic reference model for data center network virtualization overlays [16, p. 9]

Overlays in NVO3 networks follow the "map-and-encap" architecture as described in the IETF RFC 7364 [15]. The packet processing and forwarding involve three steps:

(1) Upon receiving a packet to be forwarded, the ingress overlay NVE maps the destination address (L2 or L3) of the packet to the corresponding address of the egress NVE responsible for decapsulation.
(2) Then, the ingress NVE encapsulates the packet within an overlay header.
(3) After that, the encapsulated packet is forwarded by the underlay (i.e., the IP network) based entirely on its outer address. When received by the egress overlay NVE, the packet is decapsulated and delivered to its destination.

This encapsulation and forwarding mechanism enables efficient communication within NVO3 networks and facilitates the seamless VM migration across data centers.

NVO3 Reference Model

A generic reference model for data center network virtualization overlays is recommended in the IETF RFC 7365 [16]. It is illustrated in Fig. 11.10. In this reference model, NVEs may exchange information directly with each other via a control-plane protocol to obtain reachability information. NVEs may also communicate with an external Network Virtualization Authority (VNA) to obtain reachability and forwarding information. For redundancy, multiple NVAs can be used, and they may be organized in clusters.

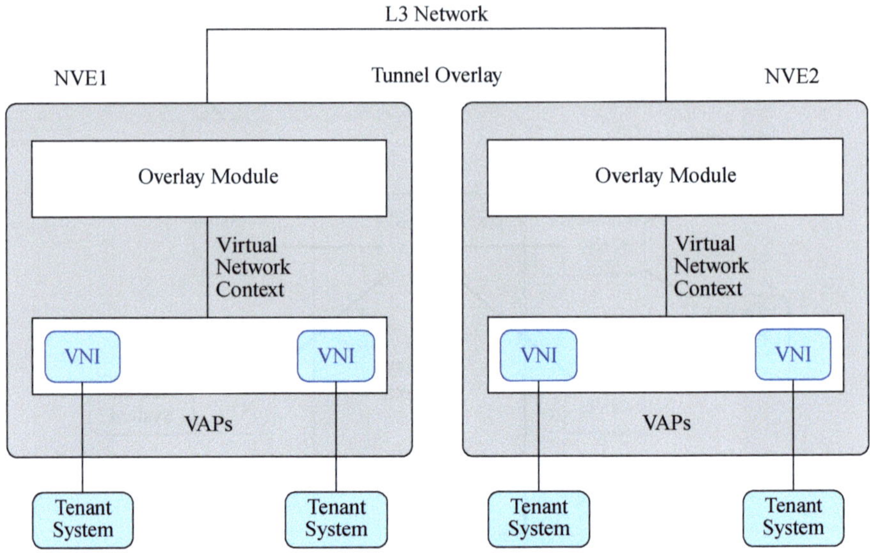

Fig. 11.11 Generic NVE reference model [16, p. 11]

The reference model shown in Fig. 11.10 also indicates that a tenant system can be attached to an NVE in different ways, such as

- Locally, by being co-located in the same end device.
- Remotely, via a point-to-point connection or a switched network.

Local attachment allows for easy determination of the state of the tenant system without protocol assistance. However, the state of the tenant system or the NVE needs to be exchanged. The information exchange can be performed directly or through a management entity using a control-plane protocol, or directly using a data-plane protocol.

Generic NVE Reference Model

As part of data center network virtualization overlays, the NVE has been defined in the IETF RFC 7365 [16] with a general reference model depicted in Fig. 11.11. In this reference model, an NVE provides different types of virtualized network services to multiple tenant systems. It is capable of offering both Layer 2 (L2) and Layer 3 (L3) services for a tenant. An L2 NVE provides Ethernet LAN-like services, while an L3 NVE provides IP-like services. Overall, NVO3 services are overlay services over an IP underlay.

The generic reference model illustrated in Fig. 11.11 is composed of several functional components, including:

- Virtual Access Points (VAPs).
- Virtual Network Instances (VNIs),
- Overlay modules and Virtual Network Context.

- Tunnel overlays and encapsulation options.
- Control-plane components.

Among these components, a VAP is an interface through which a tenant system is connected to a VNI. It can be a physical or virtual port.

A VNI is a virtual network instance on an NVE. It defines a forwarding context that includes reachability information and policies. One or more VNIs can be instantiated on the NVE.

The overlay module on an NVE provides tunneling functions, including tunnel initiation/termination and encapsulation/decapsulation of frames. In the case of simultaneous overlay of multiple tenant services over the same underlay Layer 3 network topology, mechanisms are required to identify each tenant service. Therefore, in the data plane, when sending a tenant packet, each NVE must encode the virtual network context for the destination NVE, in addition to the information about L3 tunneling. In a multi-tenant scenario, tunneling aggregates traffic from/to different VNIs. The virtual network context identifier is used for tenant identification and traffic demultiplexing.

For tunnel overlays and encapsulation options, Multi-Protocol Label Switching (MPLS) tunneling and various IP tunnel options such as Layer 2 Tunneling Protocol (L2TP) and IPSec can be used. Stateless or stateful tunneling can be applied. After adding the virtual network context identifier to the frame, an L3 encapsulation is performed to transport the frame to the destination NVE.

The control-plane components on an NVE deal with distributed or centralized control and management, auto-provisioning and service discovery, address advertisement and tunnel mapping, and overlay tunneling. These aspects have been discussed in detail in the IETF RFC 7365 [16, pp. 14–16].

A Specific VNO3 Architecture

The VNO3 framework discussed in RFC 7365 [16] covers various approaches within the general design space for data center VNO3 overlays. RFC 8014 [18] further develops a specific architecture for data center VNO3, introducing new concepts such as network virtualization domain and network virtualization region. It provides comprehensive discussions on key aspects of an overlay system.

In this specific VNO3 architecture, a service model is discussed in detail, which offers L2 Ethernet or L3 IP services to tenant systems. It also considers the provision of a combined L2 and L3 service.

Regarding NVEs in the VNO3 architecture, the concept of a Tenant System Interface (TSI) is introduced, which logically connects to the NVE via a VAP. Thus, tenant systems connect to NVEs through a TSI. The NVE reference model of this specific NVO3 architecture is depicted in Fig. 11.12. It is similar to the generic NVE reference model shown in Fig. 11.11, but has a notable addition of TSIs. It is worth mentioning that conceptually, the NVE is a single entity that implements the NVO3 functionality. However, in practice, there are scenarios where multiple entities are used.

Furthermore, for building and maintaining a mapping table in an NVE, it is realized that the learning method as used in 802.1.Q bridges can easily flood traffic. Therefore,

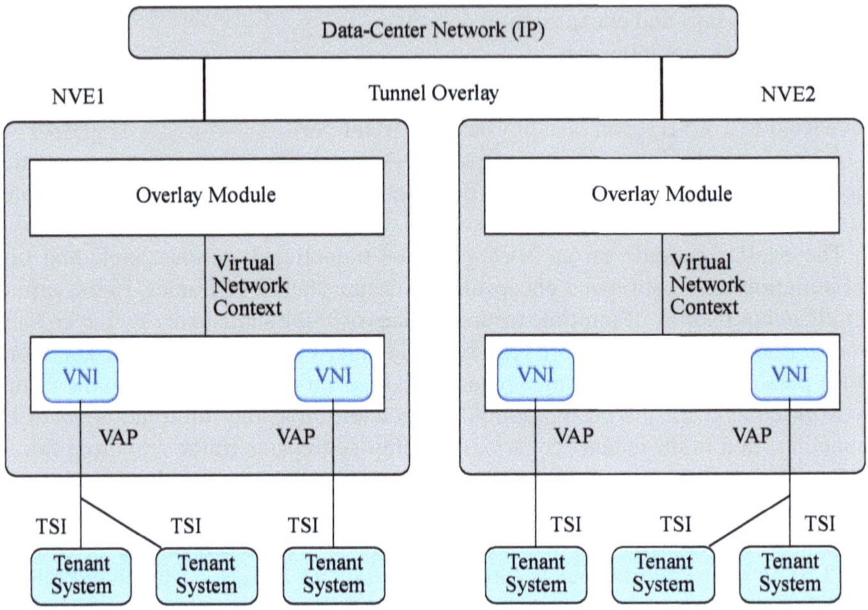

Fig. 11.12 NVE reference model from the IETF RFC 8014 [18, p. 9]

the architecture recommends using an NVA. NVEs interact with an NVA to obtain address and forwarding information required for traffic forwarding.

Moreover, VM orchestration systems are discussed in the architecture for the management of server virtualization across a set of servers. While VM orchestration and network virtualization are separate topics, they are closely related. This is because when a VM is created, it is attached to a new or existing virtual network. Before a VM is deleted, it should be detached from the virtual network it belongs to. The management of the virtual network also requires the management of VMs. Therefore, NVEs need to interact with the hypervisor that manages the VM and its physical host. For this purpose, VM orchestration systems run a protocol with an agent on the hypervisor, carrying information about which virtual network a VM belongs to. When an orchestrator instantiates a VM on a hypervisor, the hypervisor communicates with the NVE to connect the VM to the appropriate virtual network. It is worth mentioning that VM orchestration systems may not have full information about the addresses used by a VM. This implies that a VM may use additional MAC or IP addresses that the orchestration systems are unaware of.

11.8 Data Center Security

Security is one of the major concerns in data centers. Data center security encompasses policies, precautions, and practices to keep the data center secure from threats, attacks, unauthorized access, and resource manipulation. Denial of Service (DoS) attacks, theft of confidential data, and data loss are common security issues in data center environments.

Several areas must be considered to ensure data center security. The following three types of security are essential for data centers:

- Physical security: This type of security focuses on securing the physical site of the data center. Measures such as access control systems, surveillance cameras, alarms, and security personnel are implemented to prevent unauthorized access and tampering.
- Network security: Network security is responsible for maintaining the security of the data center's networks. It involves the implementation of firewalls, intrusion detection and prevention systems, network segmentation, encryption protocols, and secure authentication mechanisms. These measures protect against unauthorized access, network attacks, data interception, and manipulation.
- Virtual security: Virtual security ensures the security of virtualization services and software within the data center. Since a wide range of data center services are provisioned in virtualized environments, it is crucial to implement security measures at the virtualization layer. This includes hypervisor security, virtual machine isolation, access control policies, vulnerability management, and secure configurations and patch management for virtualization software and virtual machines.

In addition to these security aspects, social engineering security is always important for maintaining data center security. It involves educating users about good security practices, particularly the importance of not disclosing information to unauthorized individuals. Negligent insiders and malicious attacks are the main causes of data breaches, with 39% of organizations attributing data breach incidents to negligent insiders [19].

11.8.1 Physical Security

Physical security is to secure the physical site of the data center. This includes the building and surveillance of the data center. Some important techniques are listed below for the physical security of data centers:

(1) **Building in the correct location.** The data center itself should be physically isolated away from any other commercial places and away from potential hazards, e.g., areas prone to fires and flooding. It should also have its own road that is at least 30 meters away from the main road.

(2) **Buffer zone.** Keep at least a 30-meter buffer zone around the site. Use fencing or walls to restrict access to the site. Possibly use the environment to hide the data center, e.g., hills or slopes.
(3) **Walling.** This is a basic requirement to ensure that any physical attacks would be extremely difficult. Use thick concrete walls for enhanced physical security.
(4) **Avoiding windows.** Windows are one of the main ideal entry points for burglars/vigilantes due to their ease of breakage. If any windows are required, there should be at least some indoor restrictions, such as security doors.
(5) **Vehicle entry points.** There should be sufficient control access to the parking lot and the loading area with guards. Use gates at entry points with security officers to control physical access to the site.
(6) **Exit-only fire doors.** This is a simple way to ensure no one can enter the facility using fire escape doors (fire doors can require no authentication). It can be designed to trigger the alarms if one opens a fire door.
(7) **Two-factor authentication.** As one gets closer to the core of the data center, two-factor authentication should be installed. These can include biometric identification (e.g., fingerprint) and key cards.
(8) **Surveillance.** Cameras should be installed around the entire perimeter of the data center. Specific places that should be monitored include exits and entrances.
(9) **Security layers.** As one gets closer to the core of the data center, authentication needs to ramp up. Three layers of authentication can be implemented before one reaches the main core.
(10) **Bomb detection.** If the data center has valuable data or provides critical services, it is imperative to have constant checks for any possible explosives.

11.8.2 Network Security

The network security infrastructure for a data center includes various security methods and tools that are employed to enforce security policies. These security methods and tools encompass packet filtering, firewalls, and Intrusion Detection Systems (IDSs).

Packet Filtering

Packet filtering, such as Access Control Lists (ACLs), is used at multiple points within the data center to control network access based on specific criteria. Standard ACLs primarily filter packets based on source IP addresses. Extended ACLs offer more flexibility and granularity by making filtering decisions based on additional factors such as source and destination IP addresses, layer-4 protocols (e.g., TCP or UDP) and ports, layer-3 protocols (e.g., ICMP message type and code), and type of service.

Firewall

Firewalls are advanced filtering devices or software that segregate LAN segments of the data center, providing security protection for the data center networks. By assigning separate security levels to each segmented LAN, firewalls establish a security perimeter and enable control over network traffic between segments. The packet filtering technique mentioned earlier is a specific type of firewall.

Intrusion Detection

The use of IDSs helps enhance network security within the data center. IDSs aim to identify intruders and detect suspicious network activities in real-time. They trigger alerts when potential intrusions or malicious activities are detected. From these alerts, security responses are initiated to enforce security policies. IDSs employ two key intrusion detection techniques: signature-based and anomaly-based intrusion detection. Signature-based intrusion detection compares network traffic and log data against existing attack patterns to identify potential threats. In comparison, anomaly-based intrusion detection focuses on detecting unknown attacks by identifying deviations from normal network behavior. A combination of both signature-based and anomaly-based intrusion detection techniques forms a hybrid intrusion detection approach.

11.9 Summary

Data centers are critical infrastructure both nationally and globally, serving as the backbone of cloud computing and cloud services. They offer a wide range of data and IT services, including traditional enterprise services, on-demand services, HPC services, and Internet services. Depending on their functions and roles, data centers are classified into various types such as edge data centers, cloud data centers, enterprise data centers, managed data centers, and colocation data centers.

The design and operation of a data center should comply with various standards. Many national, transnational, and international standards are indirectly relevant, and thus applicable, to data center design and operation. A few others are specifically developed for data centers. Prominent standards in this domain include ANSI/TIA-942, Teccodia's GR-360, EN 50600 and ISO/IEC 22237. Particularly, data centers are classified into four reliability tiers, indicating the level of site infrastructure: Tier 1 (basic), Tier 2 (redundant capacity), Tier 3 (concurrently maintainable), and Tier 4 (fault-tolerant).

Site space, cabling, and environment considerations are among the aspects specified in various standards for data centers. For site space, key functional areas are defined for data centers, such as MDA, HDA, EDA, and ZDA. For cabling infrastructure, detailed specifications are developed to provide high-speed and cost-effective cabling solutions in data center environments. For environmental considerations, a special attention is paid to cooling.

Choosing where to build a data center is motivated by the requirement to reduce the capital investment, downtime, and other operational costs. Factors critical for the choosing data center locations include the safety and security of physical locations, power supply, environment temperatures, access to communication networks, and proximity to end users and skilled labor.

Data centers are typically designed with a three-layer architecture consisting of the core layer, aggregation layer, and access layer. In the design of data centers, two types of design frameworks or models are commonly adopted, which are multi-tier model and server client model, each suitable for different application environments. To decouple the communication among multi-tenant systems from data center physical infrastructure systems, data center NVO3 overlay networks are developed to provide L2 and L3 network services.

Finally, security is one of the major concerns in data centers. Data center security requires well-designed physical security protection. This has been specified in data center standards, particularly on tired data center security. Moreover, as part of data center security, cybersecurity is also important in data center networks. It can be enforced through packet-filtering techniques, ACLs, firewalls, IDSs, and other security measures.

References

1. Scene, C.: Top markets. https://discover.cloudscene.com/search/markets, accessed 29 Mar. 2022
2. Magazine, D.C.: https://datacentremagazine.com/, accessed 29 Mar. 2022
3. Daigle, B.: Data centers around the world: A quick look. Technical report, United States International Trade Commission (2021). https://www.usitc.gov/publications/332/executive_briefings/ebot_data_centers_around_the_world.pdf, accessed 29 Mar. 2022
4. (TIA), T.I.A.: Telecommunications infrastructure standard for data centers TIA-942-B. ANSI/TIA 942-B, TIA (2017)
5. Telcordia: GR-3160 generic requirements for telecommunications data center equipment and spaces. GR 3160 Issue 2, Telcordia (2013)
6. CENELEC: Information technology - data centre facilities and infrastructures. EN 50600 Series, European Committee for Electrotechnical Standardization (CENELEC) (2019)
7. ISO: ISO/IEC 22237 information technology - data ceter facilities and infrastructures. ISO/IEC 22237 Series, ISO (2018-2022)
8. ISO: ISO/IEC 30314 information technology – data centres – key performance indicators. ISO/IEC 30314 Series, ISO (2017)
9. Uptime Institute: Data center site infrastructure tier standard: Topology. Technical report, Uptime Institute (2018). https://uptimeinstitute.com/publications/asset/tier-standard-topology, accessed on 8 Apr. 2022
10. Bittner, J.: Data center location considerations (principles and resources). Posted in: Data Center at exITtechnologies.com (2021). https://www.exittechnologies.com/blog/data-center/where-locate-data-center, accessed 10 Apr. 2022
11. Cisco: Data center architecture overview (2018). Cosco Online Document, https://www.cisco.com/c/en/us/td/docs/solutions/Enterprise/Data_Center/DC_Infra2_5/DCInfra_1.html, accessed 12 Apr. 2022

12. Cisco: Data center multi-tier model design (2018). Cosco Online Document. https://www.cisco.com/c/en/us/td/docs/solutions/Enterprise/Data_Center/DC_Infra2_5/DCInfra_2.html#wpxref31529, accessed 12 Apr. 2022
13. Cisco: Server cluster designs with ethernet (2018). Cosco Online Document. https://www.cisco.com/c/en/us/td/docs/solutions/Enterprise/Data_Center/DC_Infra2_5/DCInfra_3.html#wpxref64494, accessed 12 Apr. 2022
14. Top500: Top 10 supercomputers (2023). https://top500.org/lists/top500/2023/06/, accessed 22 Jun. 2023
15. Narten, T., Gray, E., Black, D., Fang, L., Kreeger, L., Napierala, M.: Overlays for network virtualization. RFC 7364, RFC Editor (2014). https://doi.org/10.17487/RFC7364
16. Lasserre, M., Balus, F., Morin, T., Bitar, N., Rekhter, Y.: Framework for data center (DC) network virtualization. RFC 7365, RFC Editor (2014). https://doi.org/10.17487/RFC7365
17. Yong, L., Dunbar, L., Toy, M., Isaac, A., Manral, V.: Use cases for data center network virtualization overlay networks. RFC 8151, RFC Editor (2017). https://doi.org/10.17487/RFC8151
18. Black, D., Hudson, J., Kreeger, L., Lasserre, M., Narten, T.: An architecture for data-center network virtualization over layer 3 (NVO3). RFC 8014, RFC Editor (2016). https://doi.org/10.17487/RFC8014
19. Anixter: The four layers of data center physical security for a comprehensive and integrated approach (2012). Anixter White Paper. https://www.anixter.com/content/dam/Anixter/White%20Papers/12F0010X00-Four-Layers-Data-Center-Security-WP-EN-US.pdf, accessed 12 Apr. 2022

Chapter 12
Virtualization and Cloud

Computer networks were initially designed to connect physical network devices through wired or wireless connections. With advancements in network technologies including hardware and software, computer networks have undergone significant evolution. This evolution has driven the research, development, and implementation of flexible, programmable, and configurable networks in various applications and services. Consequently, virtual networking has emerged and gained popularity, particularly in data centers and large-scale networks. Virtualization and virtual networking not only decouple computing environments from physical infrastructure but also enable the pooling and sharing of physical resources in virtualized environments. This has paved the way for cloud computing, a service that provides on-demand access to shared computing resources (such as infrastructure, platforms, software, and data) over the Internet. Cloud computing has become ubiquitous worldwide, serving both enterprise and individual customers.

This chapter will begin with an exploration of the fundamental concepts of virtual networking. Then, it will discuss various aspects of Network Function Virtualization (NFV) and cloud computing. The content covered in this chapter will facilitate a solid understanding of virtual networking architecture and cloud-based network services.

12.1 Virtualization and Virtual Networking

Virtualization is a technology for the management of general computing and network resources. It transforms physical resources into logically manageable virtual resources. This is achieved by creating a virtual version of a server, storage, network functions, or other resources on the hardware platforms under management. The virtual system created on the hardware infrastructure is abstracted away from the true underlying hardware and software. It simulates hardware functionality in software, thus enabling much-improved flexibility and efficiency of resource man-

Y.-C. Tian and J. Gao, *Network Analysis and Architecture*, Signals and Communication Technology, https://doi.org/10.1007/978-981-99-5648-7_12

agement. With virtualization, a single Physical Machine (PM) is able to host multiple virtual systems, such as Virtual Machines (VMs), each of which can serve a different purpose.

The concept of virtualization appeared in the 1960s as a method of logically dividing the system resources of mainframe computers for different users and/or applications. An example is the IBM Control Program/Cambridge Monitor System (CP/CMS), which is a time-sharing operating system of the late 1960s and early 1970s. The system provides each user with a logically standalone computer. Therefore, multiple users can work with their own logical computers abstracted from the same physical mainframe computer.

Due to the popularity of personal computers, the wide adoption of virtualization happened in the early 2000s. As many companies had physical servers and single-vendor IT stacks, running legacy applications on different hardware became extremely difficult. This also led to the problem of low utilization of expensive hardware because one server normally ran only one vendor-specific application. This has driven the rapid development and deployment of virtualization in the 21st century.

Before introducing different types of virtualization, let us discuss how virtualization works through a key use case of virtualization technology: server virtualization.

12.1.1 How Virtualization Works

In server virtualization, which is a key use case of virtualization, a virtual instance of a computer system, known as a VM, is created, operated, and managed in a layer abstracted from the actual PM. Most commonly, a PM hosts multiple VMs, implying that multiple VMs are running on the same PM simultaneously. Each VM is installed with an OS, which is referred to as the guest OS. It hosts various computing tasks or applications. To these tasks and applications, it appears as if they were on their own dedicated machine with their OS, libraries, and other resources.

Server virtualization uses a virtualization layer to abstract and emulate the underlying hardware. The emulated hardware resources typically include compute, memory, storage, I/O, and network. The virtualization layer is managed by software known as a hypervisor.

Figure 12.1 shows a comparison between traditional and virtual computing models for three applications. In the traditional computing architecture, each of the three applications is running on a different PM. Thus, the three PMs are all active but show a low level of resource utilization. Through virtualization, the three applications are packed into three VMs, each with its own OS. As resource requirements from these three VMs are within the capacity of PM1, all these three VMs are placed into PM1. As a result, the other two PMs, PM2 and PM3, become idle and can be powered off. VM placement to PMs has a significant impact on the energy consumption and efficiency in data centers [1].

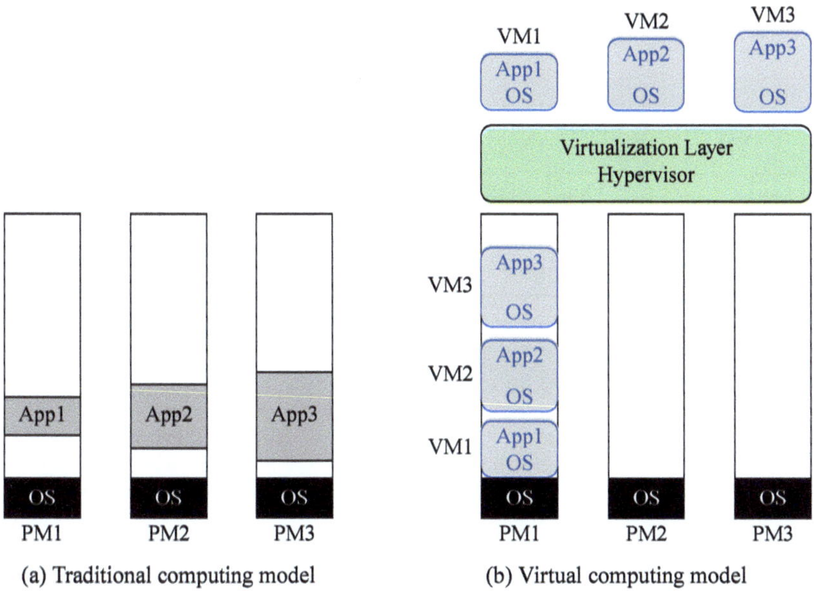

Fig. 12.1 Traditional versus virtual computing models for three applications. In the virtual computing model, depending on the type of hypervisor, each PM may or may not need an OS

In Fig. 12.1, the virtualization layer takes (i.e., virtualizes) the physical resources from the PMs, logically places them into a resource pool, and separates them as required so that they can be used in the virtual environment. Effectively, all VMs on a PM share the PM's hardware resources, such as CPU, memory, storage, I/O, and networks. This means that the sum of the resources allocated to these VMs on the PM is bounded by the capacity of the PM. To users, a VM appears essentially the same as a physical computer running the same software and applications.

The hypervisor in the virtualization layer is a software program to create and manage VMs and allocate virtualized resources to VMs. There are generally two types of hypervisors: Type 1 and Type 2:

- A **Type 1 Hypervisor** is installed on a computer before any OS is installed. Therefore, it is also referred to as a "bare-metal" hypervisor. A Type 1 hypervisor acts like a lightweight OS and runs directly on the host's hardware. Actually, a Type 1 hypervisor is itself a minimal operating system typically built on Linux. Examples of Type 1 hypervisors include Kernel-based Virtual Machine (KVM), Xen, and Microsoft Hyper-V. Hyper-V is now available with a free license.
- A **Type 2 Hypervisor** runs as a software layer on top of an OS on a computer. Therefore, it is also known as a "hosted hypervisor". An example of a Type 2 hypervisor is Oracle's VirtualBox, with which one or more VMs can be created and managed on top of the OS of a PM.

Figure 12.2 illustrates the difference between Type 1 and Type 2 hypervisors.

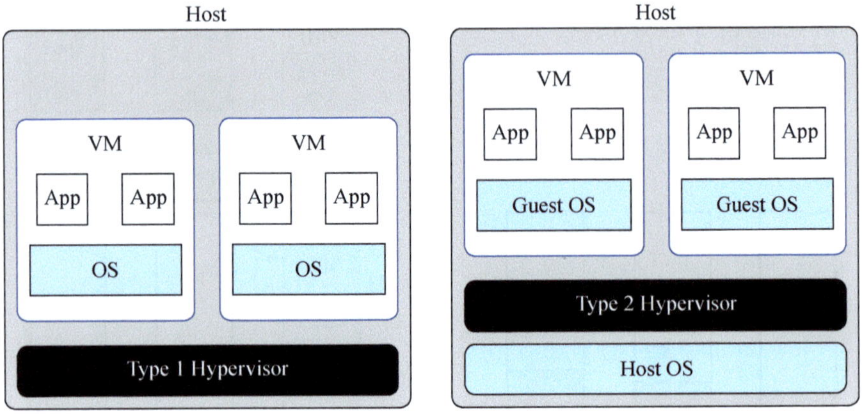

(a) Type 1 hypervisor (b) Type 2 hypervisor

Fig. 12.2 Comparison between Type 1 (bare-metal) and Type 2 (hosted) hypervisors

It is worth mentioning that KVM is native to Linux. It has elements of both Type 1 and Type 2 hypervisors. When KVM is installed, it converts the existing Linux OS into a Type 1 hypervisor. However, the original Linux OS is still accessible and can be used to run applications, for example, to host VMs with support from a hosted hypervisor. The main difference between KVM and Xen is that KVM is a virtualization module in the Linux kernel that allows the kernel to function as a hypervisor, while Xen is a Type 1 hypervisor that provides services to allow multiple computer operating systems to execute on the same computer hardware concurrently.

When a virtual environment is running, a user or application may request additional resources. Upon receiving this request, the hypervisor will relay this request to the physical system and cache changes. This process happens at close to native speed, particularly if the request is sent through a kernel-based hypervisor such as KVM. This means that the size of a VM, measured in the amount of allocated resources, can grow (scale up) or shrink (scale down) as needed. However, this is not an easy resource management task. For example, the work reported in [2] has investigated memory management across VMs on a consolidated server. Our recent work in [3] makes it possible to manage memory resources across PMs in data centers. For a Map/Reduce scenario, theory developed in [4] focuses on elastic resource management with automated scaling-up and scaling-down of resources.

12.1.2 Types of Virtualization

There are different types of virtualization in IT systems and services. In addition to server virtualization that is discussed above, other types of virtualization include application virtualization, data virtualization, desktop virtualization, memory virtualization, network virtualization, OS-level virtualization, and storage virtualization. They are briefly discussed below in alphabetical order.

Application Virtualization

Application virtualization abstracts the application layer away from the OS. It allows users to run applications in an encapsulated form from a separate computer than the one on which the applications are originally installed. In application virtualization, the virtualization layer between the applications and OS replaces part of the runtime environment that is normally provided by the OS. Particularly, it intercepts all disk operations of virtualized applications and redirects them to a single virtualized location, often a single file, rather than many files throughout the system in the traditional way of running applications. However, this process of I/O redirection is transparent to the virtualized applications.

Data Virtualization

Data virtualization virtualizes data and data management in an abstract layer. It makes the data and data management independent of the underlying database systems, structures, and storage. It also masks other details of the data and data management, such as performance and physical location. In the database scenario, database virtualization can be regarded as a special case of data virtualization.

Desktop Virtualization

Desktop virtualization is typically used in conjunction with application virtualization. It is an abstraction of a physical desktop environment and its related application software. Desktop virtualization virtualizes a workstation load rather than a physical server. It enables a user to access the desktop remotely, typically using a thin client such as a tablet. The back-end workstation essentially runs on a physical server in a data center, while the functioning of the workstation remains transparent to the user. Moreover, the physical location of the workstation is also transparent to the user.

Memory Virtualization

Memory virtualization is an abstraction of memory resources in a virtualization layer. It virtualizes the memory resources from a physical server or a networked system. In the case of a networked system, memory virtualization aggregates the memory resources from the system into a single memory resource pool. Then, virtual memory is allocated to VMs along with virtualized CPU, storage, and other resources. From the perspective of a user or application, the virtual memory appears as if it were dedicated physical memory.

Network Virtualization

Broadly speaking, network virtualization includes network resource virtualization, network function virtualization, and network protocol virtualization. Therefore, it abstracts physical network resources, network functions, and network protocols away from their physical or implementation layer. We will discuss network virtualization in detail later.

It is worth mentioning that a virtual private network, such as a Virtual Local Area Network (VLAN), is slightly different in concept from the network virtualization mentioned above. A VLAN is a logical LAN that uses protocols to replace physical media, whether wired or wireless, in a physical network with an abstract layer.

OS-level Virtualization: Containerization

OS-level virtualization, also known as containerization, is an OS feature that allows the creation and operation of multiple isolated user-space instances. These instances, known as *containers*, appear to users and applications as real computers with CPU, memory, storage, hard disk, I/O, and other resources. However, the resources allocated to containers are virtual resources assigned from the virtualized pool of resources on the physical server.

One or more containers can run within a VM, sharing the same OS. These containers are user-space instances of the OS. They can include their own software, applications, libraries, and configuration files. While containers are isolated from each other, they can communicate with one another. Figure 12.3 illustrates a comparison between virtualization with VMs and containers.

A popular implementation of containers is Docker. Docker is a set of PaaS products that deliver various platforms in virtual containers. Docker containers are lightweight, allowing a physical server or even a VM to host multiple containers. This significantly improve the utilization of computing resources.

Server Virtualization

Server virtualization has been previously introduced when discussing how virtualization works. In essence, it is the masking of the physical server resources, such as CPU, memory, storage, and operating system, from server users. This is typically achieved through the creation, management, and operation of VMs. Server virtualization is facilitated by a software hypervisor in the virtualization layer.

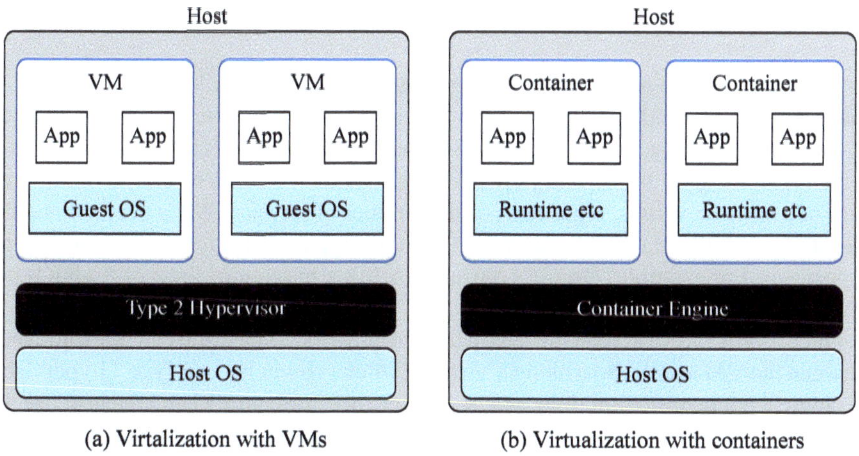

Fig. 12.3 Virtualization with VMs and containers

It is worth noting that virtualized resources, such as VMs, can themselves be further virtualized. This means that a VM can host other VMs. This is known as *nested virtualization*, in which one or more VMs can run within another VM.

Storage Virtualization

Storage virtualization is the abstraction and pooling of networked physical storage devices into a virtualized single storage device. This virtualized storage is centrally managed and allocated to VMs as required. Storage virtualization is commonly implemented in storage area networks. From the perspective of users and applications, the storage allocated from the pooled storage resources appears as a single storage unit.

12.1.3 Internetworking of VMs

If a VM does not require communication with any other virtual or physical devices, it can function as a standalone VM without the need for networking with other devices. However, in cases where network connectivity with other virtual or physical devices is required, establishing a network connection becomes necessary for the VM.

Virtual Devices for Networking VMs

To connect a VM to a network, whether virtual or physical, a Virtual NIC (vNIC)) needs to be configured on the VM. The vNIC is created within the VM and is assigned its own MAC address and IP address. Similar to a physical NIC, the vNIC operates at the data link layer to provide network access to the VM. A VM can be configured with one or more vNICs, regardless of the number of physical NICs available on the host. However, the maximum number of vNICs is typically limited by the underlying hypervisor. For example, Oracle's VirtualBox has a maximum limit of 8 vNICs per VM.

Once a vNIC is created on the VM, a logical network connection can be established between the VM and the host using a Virtual Switch (vSwitch). In a Type 1 hypervisor, this logical connection functions as a *switch* because both the VM and the host use the vSwitch for network communications. In a Type 2 hypervisor, this logical connection is referred to as a *bridge* as it bridges two networks: the network of VMs, and the network of the host and external physical network. In this case, the VM relies on the vSwitch, while the host maintains its direct connection to the external network through the physical NIC. Figure 12.4 demonstrates the difference between a switch and a bridge for VM networks.

As a logical switch, the vSwitch operates similarly to a physical switch, functioning at the data link layer to transmit data frames between nodes. Depending on the network communication requirements, the vSwitch may or may not be connected to the physical NIC of the host. This connectivity is illustrated in Fig. 12.5.

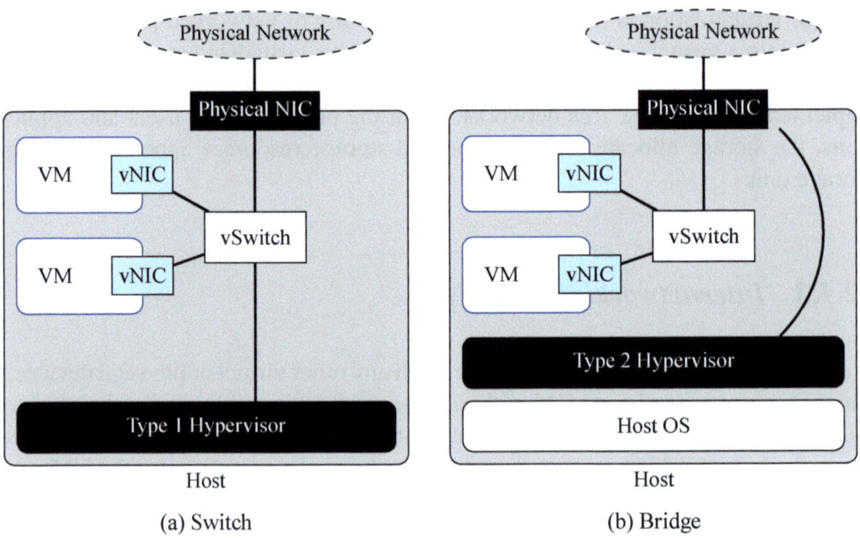

Fig. 12.4 Switch and bridge of VM connection

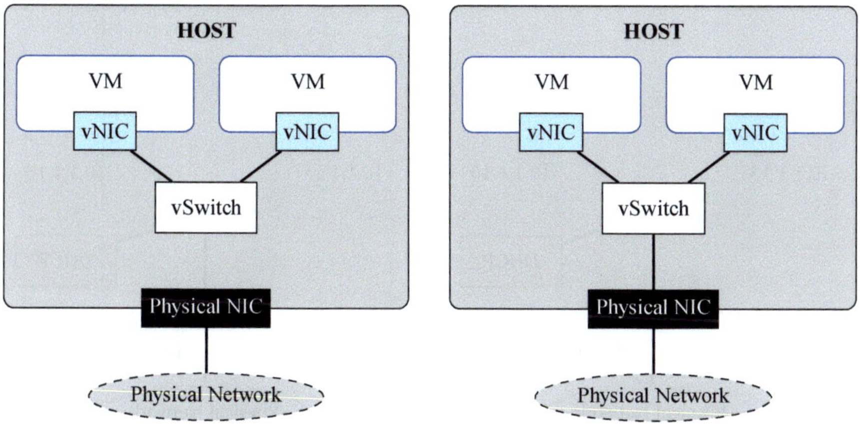

(a) No connection to the Host NIC (b) Connection to the Host NIC

Fig. 12.5 Internetworking of VMs

Communication Requirements of VMs

Basically, there are two scenarios for a VM to communicate with other devices:

(1) A VM can communicate with other VMs residing within the same host.
(2) A VM can communicate with devices external to the host on which the VM is located.

The combination of both (1) and (2) will allow a VM to communicate with other VMs within the host as well as external devices outside the host.

To cater to these requirements, the vSwitch of a VM can be configured in different connection modes. These connection modes determine how the VM interacts with the network. The three common connection modes for VMs are: host-only mode, NAT mode, and bridge mode. These connection modes are graphically shown in Fig. 12.6. They will be discussed below in more detail.

It is worth mentioning that in Oracle's VirtualBox, the connection mode of the vSwitch on the host can be configured through network settings in the hypervisor. For example, Fig. 12.7 illustrates how to set the network connection mode of a VM in Oracle's VirtualBox.

Host-Only Mode of VM Networking

In *host-only mode*, as shown in Fig. 12.6a, a VM on a host is able to communicate with the host and other VMs within the host. But it cannot communicate with any devices external to the host. This implies that the vSwitch does not receive frames from the physical NIC of the host or send frames to it. In order for each VM on the host to obtain an IP address, the host must provide DHCP service through the virtualization

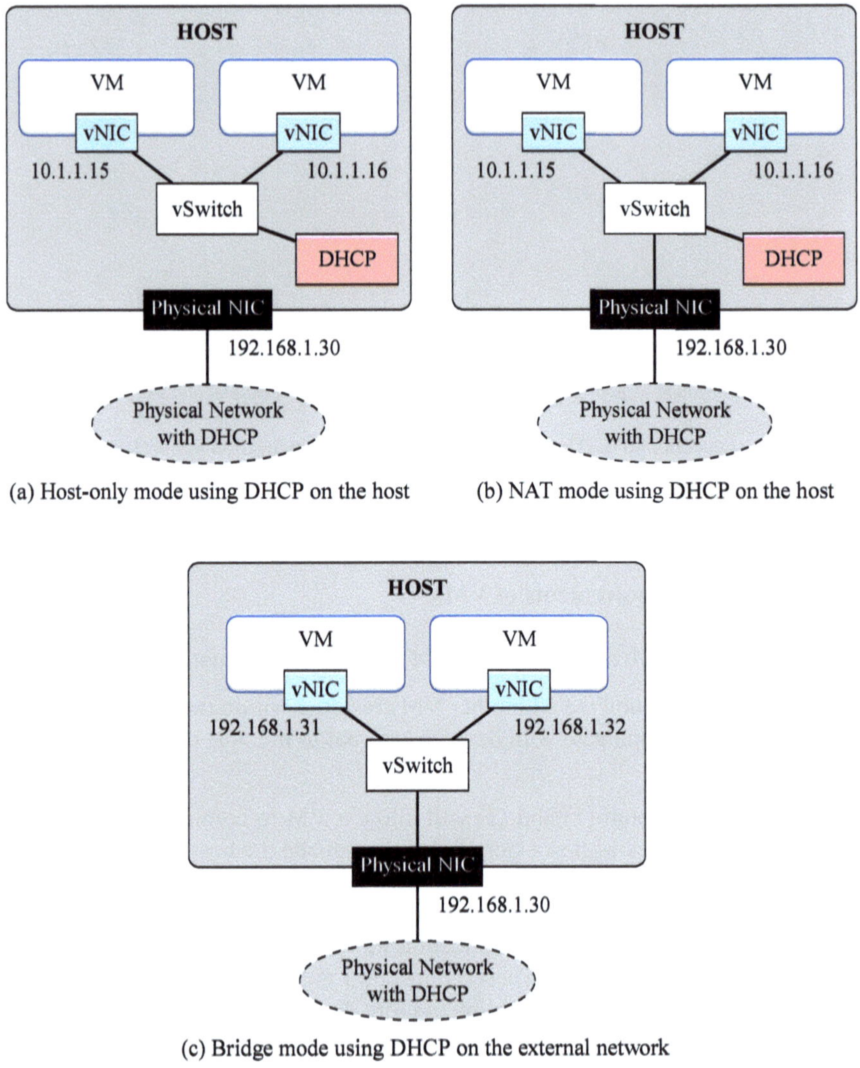

Fig. 12.6 Three modes of vSwitch configuration: host-only mode, NAT mode, and bridge mode

software. In Fig. 12.6a, the DHCP on the host assigns private IP addresses 10.1.1.15 and 10.1.1.16 to the two VMs, respectively.

NAT Mode of VM Networking

In *NAT mode*, as depicted in Fig. 12.6b, the VMs on the host get IP addresses from the DHCP service provided by the host. They are interconnected through the vSwitch

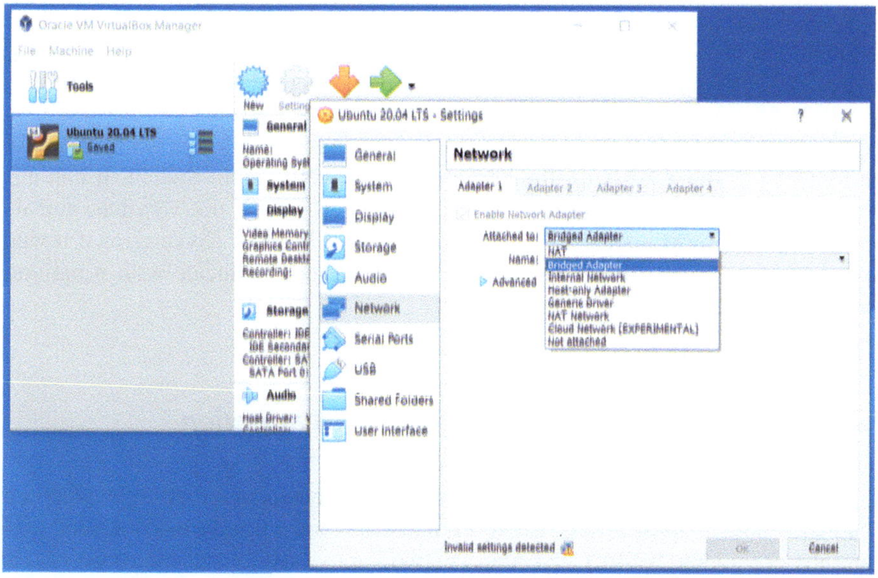

Fig. 12.7 Network settings in the Oracle VirtualBox

to form a private network. In the scenario shown in Fig. 12.6b, the two VMs obtain private IP addresses 10.1.1.15 and 10.1.1.16, respectively, from the host's DHCP service. Therefore, the VMs within the host can communicate with each other through their vNICs and the vSwitch.

In NAT mode, the private network of VMs on the host is invisible to the external physical network. The host acts as a NAT device for the connectivity of the internal virtual VM network and the external physical network. This means that a VM within the host can communicate with other devices on the physical network through the NAT service provided by the host.

As part of the physical network, the physical NIC of the host obtains an IP address lease from a DHCP server installed within the physical network. In the scenario shown in Fig. 12.6b, the physical NIC of the host obtains an IP address 192.168.1.30 from the DHCP server. Comparing this IP address with the IP addresses of the two VMs, we conclude that the physical network and the virtual network of VMs are two different networks.

Bridge Mode of VM Networking

In *bridge mode*, as illustrated in Fig. 12.6c, the physical NIC of the host is bridged with the vSwitch that interconnects the VMs on the host. This means that the virtual interface and the physical interface are bridged. As a result, all VMs and physical network nodes belong to the same network. The DHCP service on the physical network assigns IP addresses to the vNICs (of the VMs) and the physical NIC (of the host). In the scenario shown in Fig. 12.6c, the physical NIC is assigned an IP address 192.168.1.30, while the two vNICs are assigned IP addresses 192.168.1.31 and 192.168.1.32, respectively.

In bridge mode, all VMs are visible to the physical network. Each VM can communicate with other VMs on the host through its vNIC and vSwitch. It can also communicate with the physical network through its vNIC, the vSwitch, and the physical NIC. A VM appears to other nodes, either virtual or physical, as if it were just another node on the network. Other nodes that communicate with it may not realize that it is actually a virtual node.

12.1.4 Advantages and Limitations of Virtualization

Virtualization has many advantages that have contributed to its rapid development and wide range of applications. However, it also has some limitations that need to be carefully considered before converting a network into a virtualized environment. The advantages and limitations of virtualization are briefly discussed in the following.

Main Advantages of Virtualization

The main advantages of virtualization, though not an exhaustive list, are summarized below:

(1) **Increased resource utilization**: Virtualization allows for the pooling of physical resources, such as servers, storage, and networking devices, enabling more efficient utilization. Multiple VMs can be hosted on a single physical server, leading to higher resource utilization and reduced hardware costs.
(2) **Reduced costs and energy consumption**: By consolidating multiple VMs on fewer physical servers, organizations can save on hardware costs, power consumption, and cooling requirements. Virtualization enables better resource allocation and eliminates the need for dedicated servers for each application or service, resulting in cost and energy savings.
(3) **Enhanced fault isolation**: Virtualization provides isolation between VMs, ensuring that faults or issues in one VM do not impact others. This isolation improves system reliability and makes it easier to test new software versions or configurations without affecting the stability of other VMs.
(4) **Quicker backups and recovery**: Virtualization simplifies the backup and recovery process by allowing for the creation of VM snapshots or images. These snapshots can be quickly backed up and restored, enabling faster recovery times in case of system failures or data loss.
(5) **Eliminated vendor dependency**: Virtualization abstracts the underlying hardware, making the VMs independent of the specific hardware vendor. This flex-

ibility allows organizations to migrate VMs between different physical hosts or cloud environments without being tied to a particular vendor's hardware or infrastructure.

(6) **Easier migration to the cloud**: Virtualization serves as a stepping stone for organizations looking to adopt cloud-based infrastructures. VMs can be easily deployed in both on-premises and cloud environments, facilitating the migration of workloads to the cloud and providing flexibility in managing resources.

Limitations of Virtualization

Main limitations of virtualization come from the following aspects:

(1) **Complexity of implementation**: Converting a network into a virtualized environment can be complex and requires a good understanding of the existing computing environment and the requirements for the conversion. It may involve significant changes in software and hardware configurations.

(2) **Capital investment**: Virtualization may require an initial investment in new software and hardware resources. The cost of virtualization software, virtualization management tools, and hardware upgrades should be carefully assessed to determine the short-term and long-term costs.

(3) **License costs**: Some commercial virtualization management software may have high license costs. For example, VMWare license fees are charged in terms of the number of cores or nodes under management. Organizations need to consider these costs when budgeting for virtualization implementation.

(4) **Learning curve and IT support**: Virtualization may introduce a learning curve for IT support staff, requiring them to acquire new skills and knowledge. The transition to a virtualized environment may also involve interruptions to IT services during the migration process. Organizations may need to invest in training or consider outsourcing IT services to mitigate these challenges.

(5) **Performance considerations**: In certain cases, the performance of applications running in a virtualized environment may be lower compared to a dedicated physical environment. The virtualization layer introduces overhead, and critical applications or services should be evaluated to ensure their performance meets the required standards in a virtualized environment.

12.2 Network Functions Virtualization (NFV)

In traditional networking, there are a large number of hardware appliances, many of which are proprietary. The deployment of new network services often requires the installation of additional proprietary hardware appliances. This increases not only capital investment but also energy consumption for powering these new devices. Moreover, the operation and management of the resulting network become more complex due to the variety of hardware appliances involved. These challenges have driven the development of software-defined NFV.

Building upon the concept of virtualization discussed earlier, where a single PM can host multiple VMs with their own network connections, operating systems, and applications, NFV extends virtualization to network devices. This means that network devices, such as firewalls and routers, can be implemented as virtualized software-based instances running on a PM. For example, after installing a VM on a PM, we can also install a virtual firewall and a virtual router on the same PM to provide security protection and network connectivity for the VM. These virtualized network devices eliminate the need for dedicated proprietary hardware, and enable the consolidation of network functions on shared infrastructure.

While NFV is a software-based approach, it is not dependent on Software Defined Networking (SDN) for its implementation and functionality. The relationship between NFV and SDN is described in the NFV white paper [5, pp. 5–6]. NFV and SDN are highly complementary concepts, but they can also exist independently of each other. NFV can be implemented without SDN support, and SDN can be deployed without relying on NFV. However, incorporating the concepts and solutions of NFV and SDN can provide additional benefits and value. The separation of the control plane and data plane in SDN can enhance the performance and flexibility of NFV deployments. Moreover, running SDN software requires underlying infrastructure support, which can be provided in traditional networks or facilitated by NFV.

12.2.1 NFV Objectives and Standards

Since 2012, the European Telecommunications Standards Institute (ETSI) NFV Industry Specification Group (ISG) has made continuous efforts to develop the concept and technology of virtualization to consolidate many network equipment types into industry standard high-volume servers, switches, and storage. The efforts have led to software-defined network functions, which can run a variety of industry-standard server hardware and can be loaded into existing equipment anywhere in the network as required, without the need for installing new hardware equipment.

Among many objectives of NFV, the high-level objectives of NFV include [5, 6]:

- Rapid service innovation through software-based deployment and operationalization of network functions and end-to-end services.
- Improvement of operational efficiencies through common automation and operating procedures.
- Power usage reduction via consolidating/migrating workload and powering off idle hardware.

Table 12.1 ETSI GS NFV specifications [7]

No.	Specification and latest version	Year	Content
1	ETSI GS NFV 001 V1.1.1	Oct. 2013	Use cases [6]
2	ETSI GS NFV 002 V1.2.1	Dec. 2014	Architecture framework [8]
3	ETSI GS NFV 003 V1.4.1	Aug. 2018	Terminology [9]
4	ETSI GS NFV 004 V1.1.1	Oct. 2013	Virtualization requirements [10]
5	ETSI GS NFV 005 V1.2.1	Nov. 2019	Proofs of concept, framework [11]
6	ETSI GS NFV 006 V2.1.1	Jan. 2021	Management and orchestration, architectural framework [12]

- Standardized and open interfaces between network functions and their management entities, leading to the decoupling of network elements, which can then be provided by different service providers.
- Increased flexibility in the assignment of Virtualized Network Functions (VNFs) to hardware equipment.
- Improved capital efficiencies in comparison with dedicated hardware implementations and deployment.

These objectives also directly translate into benefits of NFV.

All specifications and documentation from the ETSI NFV ISG are available from this link: www.etsi.org/deliver/etsi_gs/NFV/ (Accessed on 3 Jun 2022) [7]. Particularly, the terminology for main concepts in NFV is defined in the ETSI GS NFV 003 V1.4.1 (Aug. 2018), which evolves from previous versions v.1.1.1 (Oct. 2013), v.1.2.1 (Dec. 2014), and v.1.3.1 (Jan. 2018). The ESTI NFV ISG specifications [7] are summarized in Table 12.1.

Some of the ETSI NFV ISG specifications form part of the IETF RFC 8568, which is published in April 2019 [13].

12.2.2 NFV Use Cases

NFV has a broad applicability to various network functions, encompassing both data-plane packet processing and control-plane functions in mobile and fixed networks. It plays a crucial role in transforming and virtualizing these network functions. The NFV white paper has provided an extensive list of NFV use cases, highlighting the diverse applications of NFV in different domains. Moreover, the ETSI GS NFV 001 specification elaborates on nine specific NFV use cases, providing detailed descriptions and insights into their implementation and benefits. These use cases are listed below:

(1) Network Functions Virtualization Infrastructure as a Service (NFVIaaS),
(2) Virtual Network Function (VNF) as a Service (VNFaaS),
(3) Virtual Network Platform as a Service (VNPaaS),
(4) VNF Forwarding Graphs,
(5) Virtualization of Mobile Core Networks and IP Multimedia Subsystem (IMS),
(6) Virtualization of Mobile base station,
(7) Virtualization of the Home Environment,
(8) Virtualization of Content Delivery Network (vCDN), and
(9) Fixed Access NFV

For a better understanding of NFV, let us discuss these NFV use cases briefly in the following.

NFVIaaS

Infrastructure as a Service (IaaS), Platform as a Service (PaaS), and Software as a Service (SaaS) are the three basic service models of cloud computing [14]. In the context of cloud infrastructure, IaaS characterizes the capacity of a cloud provider to offer computing resources such as processing power, storage, and networking to cloud consumers. With the implementation of NFV, networking becomes an integral part of the infrastructure and is referred to as Network Functions Virtualization Infrastructure (NFVI). In certain cases, a service provider may need to run VNF instances within an NFVI, which can be provided by the same or a different cloud provider. When the NFVI is offered as a service to others, including different departments within a single service provider, it is referred to as the NFV service model NFVIaaS. NFVIaaS enables the provision of value-added commercial services that directly support and accelerate the deployment of NFV infrastructure.

In NFVIaaS, the resources to be pooled are the compute, storage, and physical network resources. They are interpreted differently in NFV and cloud computing. In the NFV model, these resources are considered as the compute, hypervisor, and network domains of the NFVI. In the cloud computing model, they are considered as elements that support IaaS and Network as a Service (NaaS). It is worth mentioning that NaaS indicating network connectivity as a service is not a formally defined term in cloud computing standards. An example of using NFVIaaS is illustrated in Fig. 12.8 for NFVIaaS multi-tenant support [6]. In this example, NFVIaaS supports both cloud computing applications and VNF instances from different administrative domains.

In the use case of NFVIaaS, VNFs would coexist with non-VNFs. Also, VNFs from multiple service providers may coexist within the same NFVI. In this case, an appropriate isolation should be provided between the resources allocated from the NFVI to different service providers. It is a general requirement that the operation of the VNF instances of a service provider should not be affected by VNF failures or resource demand from other service providers.

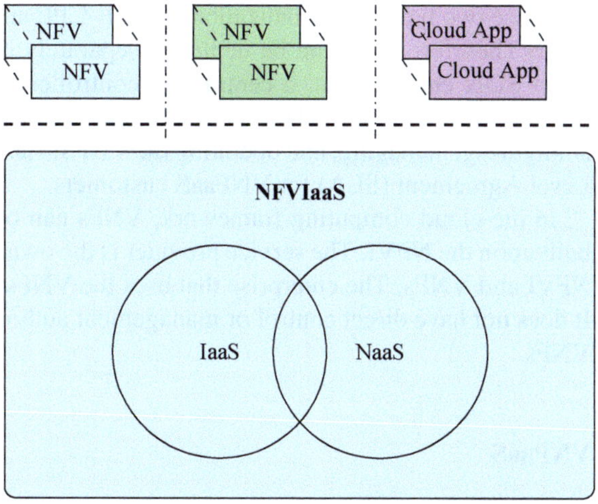

Fig. 12.8 NFVIaaS multi-tenant support of both cloud computing apps and NFVs from different administration domains [6, p. 12]

VNFaaS

In traditional networking, the network infrastructure of a service provider typically comprises a Provider Edge (PE) router positioned at the edge of the core network. The PE router interfaces with the Customer Premises Equipment (CPE). Two distinct business models exist in this context, allowing either the service provider or the enterprise to assume ownership and operational responsibility for the CPE.

The virtualization of enterprise networks encompasses two distinct types, as outlined in [6, p. 16]:

- Virtualization of the CPE (vE-CPE) in the cloud of the service provider to replace the hardware CPE.
- Virtualization of the PE (vPE) on which the virtual network service functions and core-facing PE functions can be executed in the cloud of the service provider.

In the vE-CPE model, a virtualized CPE is implemented in the service provider's cloud environment, replacing the physical hardware CPE. This virtualized solution offers a variety of services, including router functionality with QoS management and security features like stateful firewall, intrusion detection, and prevention. The vE-CPE can be strategically placed at different locations within the backbone network to serve geographically dispersed branches and offices of an enterprise.

In the vPE model, a virtualized PE is employed within the service provider's cloud infrastructure to execute both the virtual network service functions and core-facing PE functions. Compared to vE-CPE, the implementation of vPE is more intricate since the same vPE serves multiple customers concurrently. It can be achieved by integrating the vPE functions into a single VM to replicate the functionality of a single hardware PE or a subset of its capabilities, thereby enhancing scalability and performance. The vPE functions can be split across the core set of functions and virtual network service functions, with or without CPE functions.

These two types of virtualization, i.e., vE-CPE and vPE, are independent of each other. Therefore, they can be deployed separately based on specific requirements. In an SDN environment, a centralized controller can be implemented to manage both vE-CPE and vPE. The service provider assumes responsibility for deploying, configuring, managing, and operating the VNF instances to ensure the desired Server Level Agreement (SLA) for VNFaaS customers.

In the cloud computing framework, VNFs can be considered as a type of SaaS built upon the NFVI. The service provider is the owner, manager, and operator of the NFVI and VNFs. The enterprise that uses the VNFs is the consumer of the service. It does not have direct control or management authority over either the NFVI or the VNFs.

VNPaaS

It is understood from the preceding discussions that VNFs can be delivered as a type of SaaS. They can also be provided as a type of PaaS on top of the NFVI to consumers. In modern networks, an enterprise network can be hosted on the infrastructure of a service provider. The service provider may offer the enterprise a suite of NFVI and applications as a platform, enabling the enterprise to deploy their network applications. This gives rise to the concept of VNPaaS.

With VNPaaS, the enterprise can deploy various network services and applications on the VNPaaS platform. Examples of such services and applications include firewall services, DHCP, DNS, proxy services, email services, FTP services, and many others. These services and applications could be under certain control and management of the enterprise through some interfaces. However, if all of them are fully controlled and managed by the service provider, they are provisioned to the consumer as part of VNFaaS.

In comparison to VNFaaS, which primarily provides individual VNFs, VNPaaS offers a more comprehensive service, typically in a virtual network rather than one or more VNFs. Specifically, VNFaaS configures a predefined set of VNF instances that are made available by the service provider, whereas VNPaaS empowers the consumer to introduce their own VNF instances. As a result, VNPaaS and VNFaaS are fundamentally distinct from each other.

An illustrative example of VNPaaS is depicted in Fig. 12.9, which shows a scenario involving three-party enterprises that share the infrastructure of a service provider. The hosting service provider owns the underlying infrastructure and offers its resources to third-party enterprises. Enterprise A uses two VNF instances, with one of them connected to a VNF instance within the service provider's network. Enterprise B employs a standalone VNF instance hosted on the service provider's infrastructure. Operator C uses three VNF instances, which are connected to one of the service provider's VNF instances as well as the operator's home network.

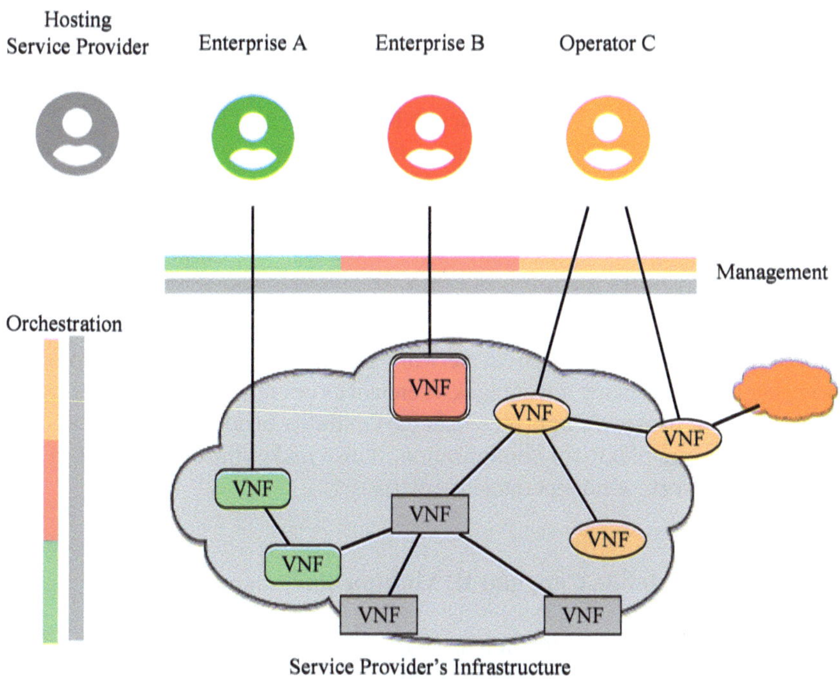

Fig. 12.9 Example of three-party enterprises sharing a service provider's infrastructure [6, p. 22]

VNF Forwarding Graphs

A network function forwarding graph characterizes the sequence of network functions that packets traverse. A VNF forwarding graph is a software-based network function forwarding graph. It is an analogue of connecting existing hardware appliances via physical cables. The VNF forwarding graph provides the logical connectivity between virtual appliances, i.e.,VNFs.

The VNF forwarding graph consists of a number of logical components including [6, pp. 24–25]:

- Physical network function, which is not virtualized but is a part of the overall service.
- Physical network logical interface, which is the boundary between a VNF forwarding graph and physical network function.
- Packet flow.
- NFV network infrastructure, which provides connectivity services between the VNFs that implement the forwarding graph links between VNF nodes in hardware and software.
- Physical network association, which describes the relationship between the NFV network infrastructure and a physical network port on a physical network function

known by management and orchestration at the boundaries between the VNFs and physical elements.

- Physical network port, which is a physical port on a physical function, a physical network router or switch, or a physical NIC.
- Network forwarding path, which describes the sequence of hardware/software switching ports and operations in the NFV network infrastructure as configured by management and orchestration.
- Virtual machine environment, which characterizes the compute, storage, and network environment for a specific set of VNF software elements as configured by management and orchestration.

In comparison with the physical network function forwarding graph, the VNF forwarding graph has many advantages. It enhances operational efficiency, improves resilience against backup failure, increases the flexibility of deployment and upgrades, and reduces the complexity of configuration. It also makes the deployment of VNFs easier in the network of the operator or enterprise.

Virtualization of Mobile Core and IP Multimedia Subsystems

Mobile networks, including 4G, 5G, ad-hoc vehicular networks, and the Internet of Things (IoT), are rapidly evolving and gaining popularity worldwide. While proprietary hardware appliances and software protocols/packages have traditionally been used in these networks, their integration and interoperability can be complex, leading to performance limitations. NFV for mobile networks aims to address these issues by replacing certain hardware appliances with VNFs.

The adoption of NFV in the mobile core and IP Multimedia Subsystem (IMS) brings several benefits. It allows enterprises to reduce the total cost of ownership by eliminating the need to purchase, install, and manage hardware appliances. The flexibility in allocating network resources and functions improves network usage efficiency. Additionally, NFV enables higher service availability, resiliency, elastic capacity, and configurable network topology for customers.

The design of network functions for mobile and converged networks follows the standards defined by the 3rd Generation Partnership Project (3GPP). These standards define the network architecture and specifications for various network functions. However, due to the rapid evolution of mobile networks, 3GPP standards are continuously evolving, and challenges such as 5G inter-slice mobility management drive ongoing research and development efforts [15]. Comprehensive surveys, like the one provided in a recent report [16], shed light on network slicing management for Industrial IoT.

Within the Evolved Packet Core (EPC), which is a core network architecture for cellular networks, network functions such as the Mobility Management Entity (MME), Packet Gateway (PGW), Serving Gateway (SGW), among others, can be virtualized using NFV, as described in the ETSI GS NFV 001 [6, p. 30]. Figure 12.10 illustrates an example of EPC virtualization.

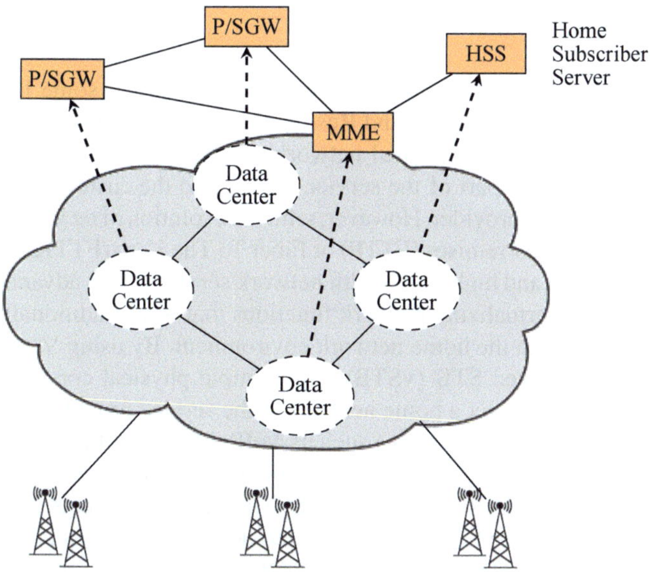

Fig. 12.10 An example of EPC virtualization [6, p. 29]

It is worth noting that in some cases, only a portion of the mobile core network is virtualized, resulting in the coexistence of virtualized and non-virtualized components. For virtualized components such as MME, SGW, PGW, and others, a virtualized Home Subscriber Server (HSS) is required. However, a non-virtualized HSS is required for non-virtualized MME, PSW, SWG, and other components.

Virtualization of Mobile Base Station

Similar to the virtualization of the mobile core and IMS, a mobile base station can also be virtualized. This concept is thoroughly discussed in ETSI GS NFV 001, which focuses on leveraging virtualization technology to implement certain radio access network nodes on standard servers, storage devices, and switches [6, pp. 33–36]. By virtualizing the mobile base station, it becomes possible to consolidate multiple radio access network nodes and platforms from different vendors into a single physical base station. This consolidation simplifies system design and deployment while enhancing resource utilization through resource pooling and sharing within the virtualized environment.

The virtualization of the mobile base station targets various cellular networks, including 4G, 5G, WiMAX, and other mobile networks. Just like the virtualization of other network functions, both virtualized and non-virtualized network functions will coexist in the context of the mobile base station.

Virtualization of the Home Environment

To maintain connectivity of home networks to the Internet, some hardware CPE devices, e.g., Set Top Box (STB) and Residential Gateway (RGW), are required on the customer premises in traditional networking. These CPE devices are typically provided at a cost or as part of the service package to the customer, indicating the presence of the service provider. However, with the evolution of technology, the availability of Fiber To The Premises (FTTP) or Fiber To The Kerb (FTTK) has increased, enabling high-speed and high-bandwidth network services. This advancement opens up possibilities for virtualizing network functions that were traditionally performed by hardware devices in the home network environment. By using Virtualized RGW (vRGW) and Virtualized STB (vSTB), only simple physical connectivity devices are required to interconnect a home network to the service provider's core network, facilitating various network services such as VoIP, IPTV, and Internet.

Examples of virtualizing CPE network functions in the home network environment include vSTB and vRGW. Just like the virtualization of network functions in other use cases, virtualized and non-virtualized network functions in the home environment may coexist, such as non-virtualized RGW and vRGW, and non-virtualized STB and vSTB.

Figure 12.11 presents a simplified logical diagram depicting a scenario where virtualized and non-virtualized network functions coexist in the home environment. In the Home A network, both the RGW and STB are virtualized. In Home B, only the RGW is virtualized while the STB remains non-virtualized. Home C does not have any virtualization.

For Home A, where both the STB and RGW are virtualized, the virtualized vSTB and vRGW are physically connected to the Broadband Network Gateway (BNG). Communication from the home to vSTB and vRGW occurs through private IP addresses. By contrast, communication beyond the vRGW towards the BNG and other network services takes place using public IP addresses. The vRGW connects to the Internet via the BNG for Internet services, while for IPTV and VoIP, it connects to the respective servers through the routing device.

In Home B, the STB remains installed on the premises, while the RGW is virtualized. Communication from the home to the vRGW, via the STB, occurs via private IP addresses. Communication beyond the vRGW towards the BNG and other network services is performed using public IP addresses. The vRGW is directly connected to certain network services such as IPTV and VoIP. Moreover, it is connected to the BNG for Internet services. In the home network, the vRGW performs NAT functions.

For Home C, where no NFV is implemented, both the STB and RGW are installed on the premise, representing the traditional home networking setup. In this scenario, all services are received by the RGW, converted to private IP addresses, and delivered within the home network. The RGW is connected to the BNG for Internet services, while VoIP and IPTV services bypass the BNG, as depicted in Fig. 12.11.

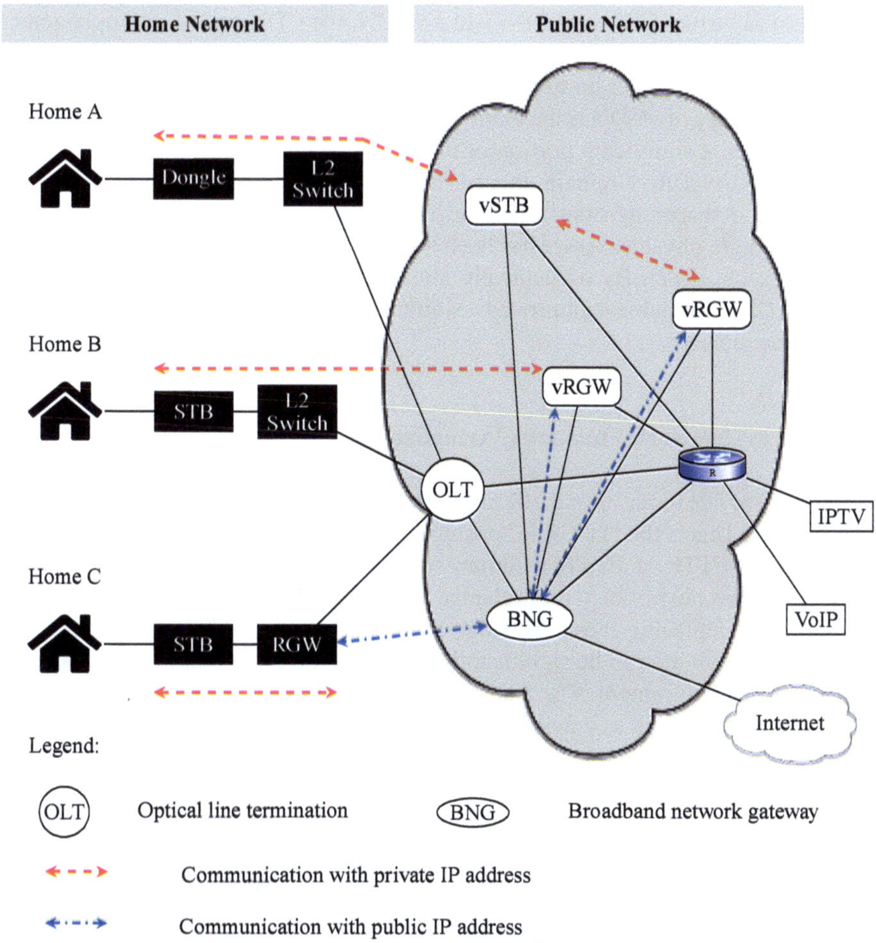

Fig. 12.11 An example of EPC virtualization, in which both STB and RGW are virtualized for Home A, RGW is virtualized for Home B, and no NFV for Home C

vCDN

A CDN is a distributed network of interconnected servers and data centers designed to enhance the delivery speed of content, such as web pages and files, by bringing the content closer to the end user. The CDS stores copies of content temporarily, in terms of some criteria [17], in caches located in various nodes across different geographical locations. This allows the CDN to deliver the content from the nearest or most optimal node when a user requests it.

In traditional CDN deployments, the caching nodes within the CDN are dedicated physical appliances or software running on dedicated hardware devices. However, with the emergence of virtualization, it is now possible to virtualize CDN nodes, referred to as virtual CDNs (vCDNs). In a vCDN, the CDN nodes are implemented as virtual appliances running on the infrastructure owned by an operator, similar to virtualizing a physical server to provide specific services.

One advantage of vCDN is the ability to utilize resources more efficiently. During off-peak hours, a significant portion of the allocated resources for vCDN nodes on the physical server may remain unused. In such cases, these unused resources can be pooled and shared by other services. The sharing of resources among multiple VMs on a single physical server has been a topic of recent investigation, as outlined in studies such as [2]. By dynamically allocating and sharing resources based on demand, vCDN can achieve improved resource utilization and flexibility in managing network resources.

Fixed Access Network Functions Virtualization

In network service requests and offerings, the main costs and bottlenecks often occur in the access. This is the "last mile" problem in network communication. To address this problem, FTTP or FTTK solutions are increasingly being deployed in wireline fixed access networks, with exchange equipment installed in close proximity to end-users. By replacing these exchange-based devices with VNF components, the data transmission rate can be significantly increased, often several times higher than with traditional equipment. The adoption of VNFs in fixed access networks brings additional benefits such as enhanced service-delivery flexibility and reduced energy consumption. These advantages are strong motivators for the deployment of NFV in fixed access networks.

The target network functions for virtualization in fixed access networks encompass various Layer-2 functionalities and control planes, including Optical Line Termination (OLT), Optical Network Terminal (ONT), Optical Network Unit (ONU), and others.

It is worth noting that although the virtualization of Layer-1 functionalities, such as signal processing and error correction, is possible, it is not currently within the scope of study within the ETSI Industry Specification Group (ISG). However, virtualization techniques for Layer-1 functions can be observed in areas like software-defined radio for wireless communications, where the functionalities are implemented in software.

12.2.3 NFV Architecture Framework

NFV implements network functions as software-based entities that run on the NFVI. From the architectural perspective, NFV is composed of three main components: VNFs, NFVI, and NFV Management and Orchestration:

- The VNFs are implemented in software running on the NFVI.

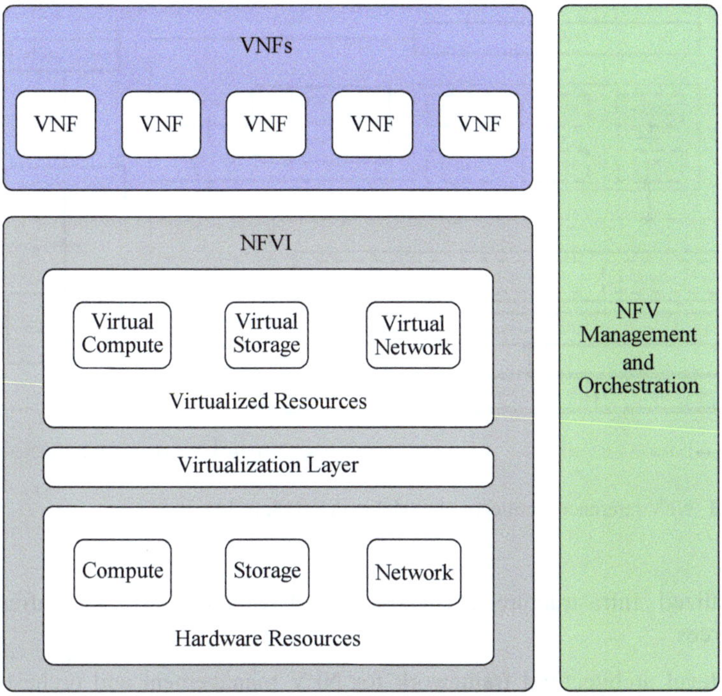

Fig. 12.12 High-level NFV architectural framework [8, p. 10]

- The NFVI supports the execution of VNFs. It provides physical resources, such as compute, storage, and network, at the hardware layer. These physical resources are pooled and virtualized through a virtualization layer, creating virtual compute, virtual storage, and virtual network environments.
- The component of NFV management and orchestration deals with the orchestration and management of VNFs, as well as the physical and software resources that support infrastructure virtualization.

A logical diagram of the high-level NFV architectural framework is depicted in Fig. 12.12. It shows the interaction and interdependencies between the VNFs, NFVI, and NFV Management and Orchestration components in an NFV environment.

In the implementation of the high-level NFV architecture, additional building blocks are required for management and operations. For example, the Element Management (EM) module is essential for directly managing VNFs. On top of the EM modules, Operations Support Systems and Business Support Systems (OSS/BSS) are also needed. For NFV management and orchestration, the following components are required:

- An NFV Orchestrator for NFV orchestration,
- VNF Managers for the management of VNFs and EM components, and

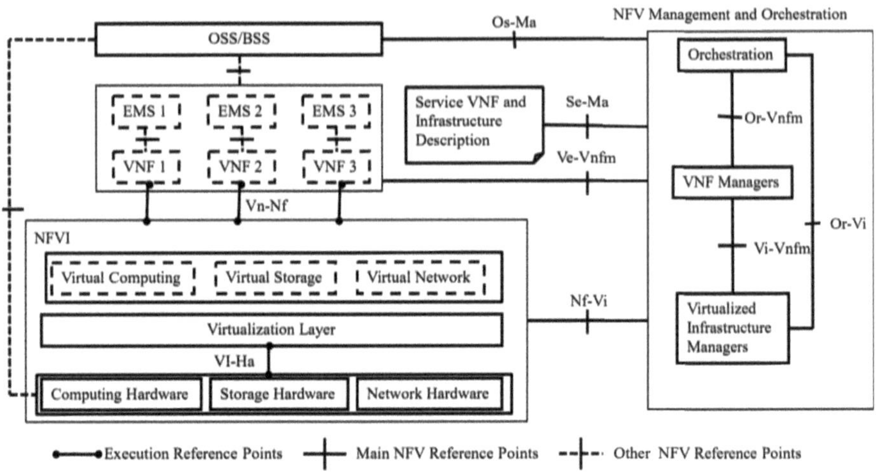

Fig. 12.13 NFV reference architectural model in detail [8, p. 14]

- Virtualized Infrastructure Managers for the management of infrastructure resources.

A high-level architectural framework for NFV management and orchestration is developed in the ETSI GS NFV 006 [12]. Combining all these components, a detailed NFV reference architectural model is shown in Fig. 12.13.

12.2.4 NFV Framework Requirements

NFV implements software-based network functions on high-volume physical servers, switches, and other hardware components. It facilitates the consolidation of network equipment, resulting in cost savings and flexible network operations. To ensure the design and deployment of VNFs with interoperability, a common set of requirements must be met. This has driven the development of the ETSI GS NFV 004, which outlines the NFV virtualization requirements [10].

High-level NFV requirements specified in the ETSI GS NFV 004 are classified into 11 categories. They are listed below, each with a brief description:

(1) General: Ability to partially or fully virtualize network functions needed to create, deploy, and operate the services to be provided.
(2) Portability: Decoupling from underlying infrastructure and promoting interoperability among multi-vendor implementations.
(3) Performance: Ability to monitor performance and configure any VNF for optimal performance.
(4) Elasticity: Scalability to meet Service Level Agreements (SLAs) and mobility to migrate to other servers.

(5) Resiliency: Ability to recover and recreate VNFs after failures, with specified metrics such as packet loss rate, call drops, and time to recover.
(6) Security: Role-based authorization and authentication for secure access.
(7) Service Continuity: Seamless or non-seamless continuity of services after failures or migration.
(8) Service Assurance: Time-stamping and forwarding copies of packets for fault detection and troubleshooting.
(9) Operation and management: Providing comprehensive management and orchestration functionality for efficient operation.
(10) Energy Efficiency: Enabling the ability to put a subset of VNFs in a power-conserving sleep state to optimize energy consumption.
(11) Transition: Ensuring coexistence with legacy systems during the transition to NFV.

For the deployment of VFNs, service providers may use NFV infrastructure operated by other operators. This enables the providers to more efficiently serve their customers located at different geographical areas.

12.2.5 Cloud-native Network Functions (CNFs)

While NFV has many benefits, it also has limitations in its traditional implementations. Examples of these limitations include slow upgrades and restarts of VNFs, CLI interfaces, difficulties in installing hypervisors like OpenStack, and limited elasticity and scaling capabilities.

To address these issues, VNF providers such as Cisco have pioneered the development of next-generation VNFs by migrating their VNFs to cloud-native environments. These VNFs are container-based and known as Cloud-native Network Functions (CNFs). When deployed in telecommunications premises, CNFs form a private cloud and can effectively use the same principles as public clouds.

CNFs offer several advantages over traditional VNFs. Some of their notable advantages include a micro-services architecture, built-in micro-service discovery mechanisms, resilient services, dynamic elasticity and auto-scaling, and continuous deployment and automation principles. By leveraging service discovery and orchestration, a network based on CNFs becomes more resilient to infrastructure resource failures. Moreover, by using containers, the overhead associated with traditional virtualization with a guest operating system can be eliminated. This leads to improved infrastructure resource efficiency.

12.2.6 Open-Source Network Virtualization and Orchestration

For the development of network virtualization and orchestration, the open-source community is active with many interesting and encouraging initiatives. The IETF RFC 8568 has listed some open-source initiatives in this area [13, pp. 15–16]. They are briefly described below:

OpenStack: One of the most popular open-source cloud computing platforms. It is managed through a dashboard or APIs. It controls large pools of compute, storage and networking resources virtualized from physical resources. Website: https://www.openstack.org/ (accessed on 10 June 2022).

Kubernetes: Also known as k8s, originally developed by Google and currently maintained by the Native Cloud Computing Foundation. It is an open-source container orchestration system for automating deployment, scaling, and management of container-based applications. It can schedule and run application containers on clusters of PMs or VMs. A particular feature of Kubernetes is its capability of scaling-up or scaling-down on the fly. Website: https://kubernetes.io/ (accessed on 10 June 2022).

OpenDayLight: A modular, extensible, and scalable multi-protocol controller infrastructure for SDN deployments on heterogeneous multi-vendor networks. It is funded by the Linux Foundation. Website: https://www.opendaylight.org/ (accessed on 10 June 2022).

ONOS: The Open Network Operating System (ONOS), which is a Linux-based SDN operating system designed with scalability, high performance, and high-availability for service providers. It provides the control plane for an SDN. Website: https://opennetworking.org/onos/ (accessed on 10 June 2022).

OPNFV: The Open Platform for NFV (OPNFV), which is a collaborative project under the Linux Foundation to accelerate the transformation of enterprise and service provider networks. It provides a common NFVI, continuous integration with upstream projects, stand-alone testing tool sets, and a compliance and verification program for industry-wide testing and integration. The OPNFV goals include accelerating time to market for NFV solutions, easing operational burdens, and ensuring the platform meets the industry's needs. Website: https://www.opnfv.org/ (accessed on 10 June 2022).

OSM: The Open Source Mano (OSM), which is an ESTI-hosted open-source Management and Orchestration (MANO) stack aligned with ETSI NFV Information Models. It is a community-led project with the aim to deliver a production-quality MANO stack that meets operators' requirements for commercial NFV deployments. Website: https://osm.etsi.org/ (accessed on 10 June 2022).

OpenBaton: An extensible and customizable NFV MANO-compliant framework, which is capable of orchestrating network services across heterogeneous NFV Infrastructures. While OpenStack is the major supported Virtual Infrastructure Manager, OpenBaton provides a driver mechanism for supporting additional types of Virtual Infrastructure Manager. Website: https://openbaton.github.io/ (accessed on 10 June 2022).

ONAP: The Open Network Automation Platform (ONAP), which is a comprehensive open-source software platform for orchestration, management, and automation of network and edge computing services for network operators, cloud providers, and enterprises. Its real-time, policy-driven orchestration and automation of physical and virtual network functions enable rapid automation of new services and complete lifecycle management critical for 5G and next-generation networks. Website: https://www.onap.org/ (accessed on 10 June 2022).

SONA: The Simplified Overlay Network Architecture (SONA), which is an optimized tenant network virtualization service for cloud-based data centers. SONA is regarded as an SDN-based NFV solution for cloud data centers. Basically, it is a set of ONOS applications which provides an almost full SDN network control in OpenStack for virtual tenant network provisioning. Therefore, it is listed as a use case of ONOS in the ONOS wiki https://wiki.onosproject.org/display/ONOS/SONA%3A+DC+Network+Virtualization (accessed on 10 June 2022).

Tungsten Fabric: A scalable and multi-cloud SDN platform re-branded in 2018 from OpenContrail. The name of Contail is retained in the commercial version of Tungsten Fabric. Website: https://tungsten.io/ (accessed on 10 June 2022).

12.2.7 NFV Challenges

As in cloud computing, which migrates computing functions and storage from local and dedicated physical resources to remote virtual functions accessible to customers through the Internet, NFV also migrates network functions from local and dedicated physical resources to a virtualized pool of network resources. The virtualized pool of network resources can then be shared by customers locally and remotely. However, in addition to the virtualization of computing and storage that cloud computing deals with, the virtualization of network itself that cloud computing does not address is also essential in NFV. Therefore, NFV is generally considered to be more complex than cloud computing.

In spite of significant progress in research, development, and practice of NFV over the years, there are still many challenges that need to be addressed. The IETF RFC 8568 (April 2019) [13] has summarized 10 such challenges, which are listed below:

(1) QoS guarantee in a virtualized network environment is more difficult compared to traditional dedicated networks. The diversity of performance metrics further complicates QoS assurance. Instead of developing specific technical implementations of NFV, the ESTI has aimed to develop a few sets of NFV requirements through a series of ESTI GS NFV 001–006 specifications. The ESTI assumes that meeting these requirements in the design and implementation of NFV and its MANO will provide some sort of QoS guarantee. However, guaranteeing QoS remains challenging in NFV.

(2) Performance improvement in terms of energy efficiency and link usage is a complex problem, especially in large-scale networks. Computationally-efficient algorithms incorporating heuristics and meta-heuristics are needed for energy-aware VNF placement, migration, and scaling. This is still an active field of research and development.
(3) Cross-domain orchestration and services across multiple administrations or multi-domain single administrations pose challenges for seamless integration and management.
(4) 5G and network slicing present specific challenges, particularly for virtual network operators and IoT networks. Inter-slice mobility management in 5G is an active area of research. A recent survey paper on inter-slice mobility management in 5G has addressed some issues relevant to 5G network slicing [15].
(5) Service composition with respect to the optimization of Network Function Forwarding Graph and Service Function Path is generally a scheduling problem, which is difficult in large-scale systems particularly in a dynamic virtual network environment. Many of the issues related to this challenge have not been well addressed.
(6) End-user device virtualization is an area that requires further investigation in research and development.
(7) Security and privacy are understandably important everywhere in both physical and virtual networks. Since virtual networks are more dynamic than physical networks and may also span more than one network infrastructure, protecting their security and privacy them becomes more challenging. Formal standards need to be developed for security and privacy in NFV.
(8) Separation of control concerns is an ongoing challenge that requires attention.
(9) Testing methodologies, including testing of new functionality and the softwarization of tests, need to be developed.
(10) Function placement is a complex task in dynamic virtual network environments.

The challenges identified above have garnered significant interest in research and development in recent years. Various initiatives have been undertaken to address these challenges, some of which are discussed in the IETF RFC 8568 [13, p. 33]. For example, the re-architecting of functions has emerged as a potential solution to address issues related to QoS guarantee, performance improvement, network slicing, security, end-user device virtualization, and separation of control concerns. By redesigning and restructuring network functions, these initiatives aim to align them with the requirements of NFV.

Furthermore, new management frameworks have been proposed to tackle challenges arising from multiple domains, service decomposition, and end-user device virtualization. These frameworks provide enhanced management and orchestration capabilities for NFV, enabling more efficient and effective handling of these issues.

In addition to the aforementioned challenges, there are other open problems in the realm of NFV. One such challenge is virtual network consolidation and migration, which poses difficulties in achieving improved resource allocation, energy efficiency, and overall performance in dynamic virtual network environments. A recent research article published in IEEE Transactions on Cloud Computing has investigated multi-objective virtual network migration based on reinforcement learning [18].

These ongoing initiatives and research endeavors demonstrate the commitment to overcoming the obstacles faced in NFV and driving advancements in the field. By addressing these challenges, the NFV community strives to fully realize the potential of NFV, enabling scalability, flexibility, optimized resource allocation, and improved service delivery in network environments.

12.3 Cloud Computing and Its Characteristics

Cloud computing is being increasingly used in a variety of systems and applications, ranging from thin mobile platforms to powerful workstations. It is provisioned via various cloud services over networks. Let us start our discussion of cloud computing with its basic concepts and characteristics.

12.3.1 Cloud Computing and Cloud Services

Cloud computing is a networking model that delivers various computing services over the Internet. In the Special Publication SP 800-145 of the National Institute of Standards and Technology (NIST) [14], cloud computing is formally defined as "*a model for enabling ubiquitous, convenient, on-demand network access to a shared pool of configurable computing resources (e.g., networks, servers, storage, applications, and services) that can be rapidly provisioned and released with minimal management effort or service provider interaction.*" Since the publication of the NIST SP 800-145 in 2010, the cloud computing technologies have experienced significant evolution. Nevertheless, this simple definition of cloud computing has been widely accepted worldwide.

In addition to this simple definition of cloud computing, a few supporting concepts are also introduced for cloud computing in the NIST SP 800-145. They include five characteristics, four deployment models, and three service models for the provisioning of cloud services. These supporting concepts are further developed later in other NIST's publications, e.g., NIST SP 500-292 [19], SP 500-293 [20], and SP 500-322 [21].

The terminology of *cloud service* has a specific meaning. In the NIST definition, a *cloud service* refers to one or more capabilities, which are offered in the cloud computing model and exhibit the essential characteristics specified in NIST SP 800-145. Some organizations and vendors label their service offerings as "cloud services," even though these offerings do not actually support the essential characteristics of a cloud service [21, p. 1].

Computing services available from the cloud include servers, storage, databases, networking, software, and applications. In the context of cloud computing, the term

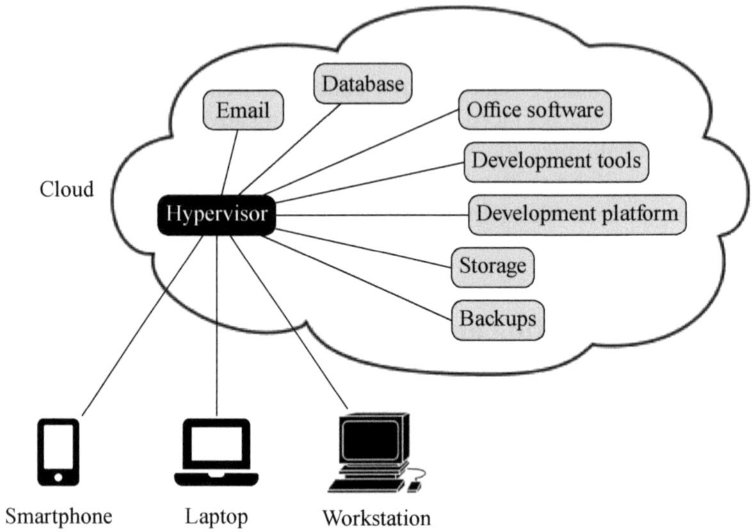

Fig. 12.14 A simple scenario of cloud computing accessible from anywhere and any platforms in an on-demand and pay-as-you-go manner

application may refer to a cloud-enabled SaaS, a web or mobile app, or a software system running on a VM. Therefore, when describing a cloud application, it is preferable to clarify its type if necessary to avoid any confusion [21, p. 3].

Let us consider a simple scenario of cloud computing. An organization outsources its IT services to a cloud service provider. It consists of many departments, all of which require some common cloud services such as email and word processing. Some departments require specific cloud services that other departments do not need, such as software development tools and platforms, specific databases, storage, and data backup. Also, the staff of the organization may access the cloud services from different types of devices, ranging from thin equipment (e.g., smartphones) to powerful workstations. The cloud service provider provides the organization with all the required cloud services, accessible from any location, with the on-demand, elastic, and pay-as-you-go features. Figure 12.14 shows a high-level logical diagram of this simple scenario of cloud computing.

12.3.2 Essential Characteristics of Cloud Computing

From the NIST definition stated above, cloud computing is described with five essential characteristics: on-demand self-service, broad network access, resource pooling, rapid elasticity, and measured service. These characteristics have been summarized in NIST SP 800-145 [14]. They are described in more detail in NIST SP 500-322 [21].

On-Demand Self-Service

This means that cloud services are available to users at any time upon their request. To the users, the provisioning of the user interface for on-demand request of cloud services is fully automated. This is the front end of cloud computing. However, the provisioning of the cloud infrastructure, i.e., the back end of cloud computing, is transparent to the users. It can be fully automated or may involve manual labor within the cloud service provider's environment. The benefit of on-demand self-service is the access to cloud services as needed.

Broad Network Access

This indicates the capability of accessing cloud services from a wide range of locations, networks, devices, and software environments using standard network protocols. Physical locations can be in different cities or countries. Networks can include the Internet, 5G, broadband, or other data communication networks. Devices range from thin to thick devices such as smartphones, laptops, desktops, and workstations. Software environments can include Android, iOS, macOS, Linux, Windows, and others. The standard protocols for network access imply network protocols such as HTTP/HTTPS, TCP/IP, TCP, UDP, and other Internet protocols. The obvious benefit of broad network access is the ability to access cloud services anytime, from anywhere, and from any machine within policy and security constraints.

Resource Pooling

This implies that all cloud resources to be provisioned to users are pooled or consolidated, and the pooled cloud resources are shared among users. For example, a cloud provider may host many websites for a large number of customers on just a few physical servers. Resource pooling and sharing are managed by cloud service providers and are transparent to cloud service users. Resource pooling and sharing enable multiple users to pay for resources running on the same hardware, leading to reduced costs for users.

Rapid Elasticity

The characteristic of cloud computing is described in NIST SP 800-145 as elastic provisioning and releasing to scale rapidly outward and inward commensurate with demand. This means that when more resources are required, resource scaling-up is performed, and when reduced resources are needed, resource scaling-down is conducted to save costs. Resource scaling-up and scaling-down are automated and near-real-time. They may or may not be fully automated from the cloud service provider's perspective. To cloud service users, the cloud resources appear to be unlimited and can be requested in any quantity whenever needed. In practice, cloud providers generally offer a limited number of options with different resource configurations for cloud service users to choose from. The benefit of rapid elasticity is the ability to quickly grow and reduce computing capability and associated costs dynamically whenever needed.

Measured Service

Cloud services offered by a cloud provider, including applications, storage, bandwidth, processing power, etc., are metered with enough detail for automatic control and optimization of resources to support the requirements of cloud service users. The usage of cloud resources can be monitored, controlled, and reported, providing transparency for both the provider and consumer of the utilized service. Metering is typically used for charging the use of cloud services but can also be used for non-charging purposes such as showback and chargeback. In a private cloud, metering may be used to show how much resources are consumed by specific applications, network domains, or groups of users.

12.4 Deployment Models of Cloud Computing

Depending on how cloud resources are managed and accessed, cloud services are provisioned in different deployment models. In NIST SP 800-145 [14] and SP 500-322 [21], four deployment models have been specified for delivering cloud computing services. They are public cloud, private cloud, community cloud, and hybrid cloud. These deployment models are developed from the NITS Cloud Computing Technology Roadmap, which is known as NIST SP 500-293 [20].

12.4.1 Public Cloud

In the public cloud deployment model, the cloud infrastructure is provisioned for open use by the general public. This implies that cloud services are provided over public transmission media such as the Internet. Public cloud infrastructure exists on the premises of a cloud provider, which could be a business, a university, or a government department. The cloud provider may own, manage, and operate the cloud infrastructure. A logical view of the public deployment model is shown in Fig. 12.15.

Public clouds are available almost everywhere as long as there is Internet connectivity. Typical examples of public cloud include Amazon's AWS and Microsoft's Azure, which serve the general public.

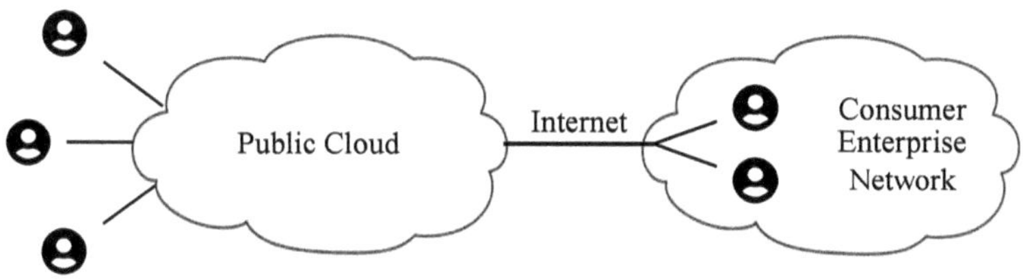

Fig. 12.15 A logical view of public cloud [21, p. 15]

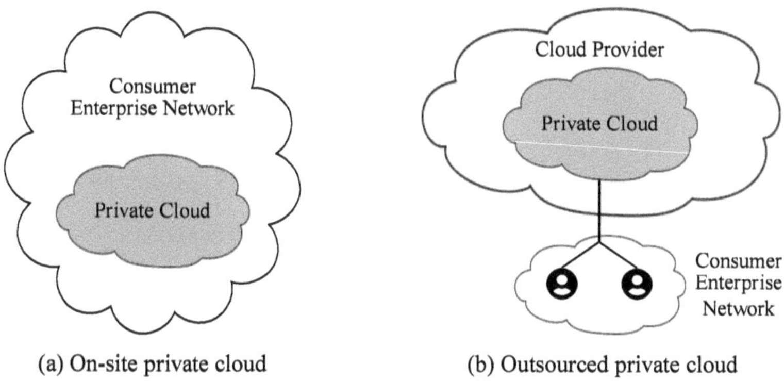

Fig. 12.16 Logical diagrams of private cloud [21, p. 13]

12.4.2 *Private Cloud*

The deployment model for a private cloud is characterized by the provisioning of the cloud infrastructure for exclusive use by a single organization comprising multiple cloud service customers. Cloud services are hosted either internally on the organization's servers in its own data center or externally in the virtualized environment of a third-party's cloud infrastructure. If hosted internally, the private cloud infrastructure is owned, managed, and operated by the organization for which the cloud services are provisioned. Logical diagrams of on-site private cloud and outsourced private cloud are illustrated in Fig. 12.16.

Private clouds are widely used in a variety of organizations. As an example, an organization's intranet could be part of the private cloud of the organization.

12.4.3 *Community Cloud*

In a community cloud, the cloud infrastructure is provisioned for exclusive use by a specific community of cloud service customers from multiple organizations with shared concerns or common interests such as security requirements, perfor-

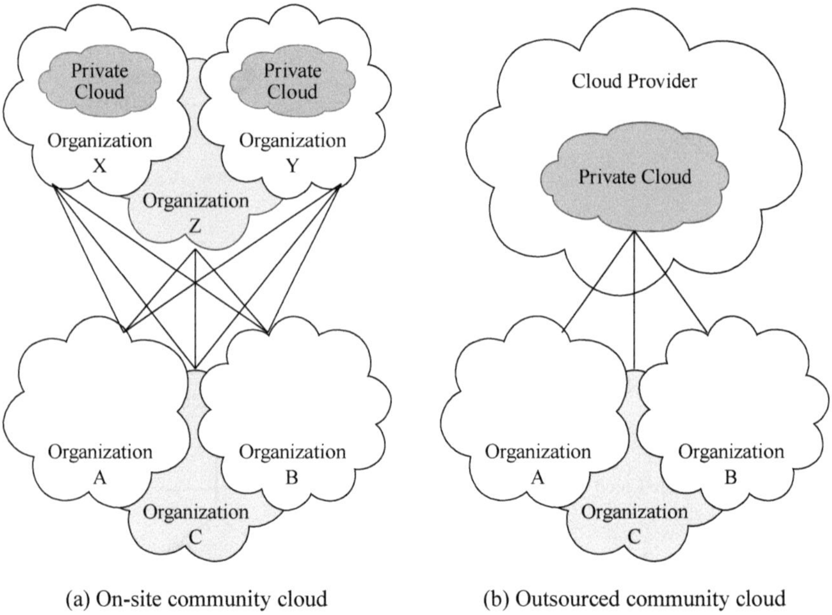

Fig. 12.17 Logical diagrams of community cloud [21, p. 14]

mance requirements, management requirements, and access to the same datasets. It is owned, managed, and operated by one or more organizations involved in the community. Community cloud services could be hosted internally on the servers of the involved organizations or externally by a third-party provider. Logical views of on-site community cloud and outsourced community cloud are shown in Fig. 12.17.

A community cloud is a popular deployment cloud model. A typical example of a community cloud is a cloud eHealth system that shares the medical records of patients across multiple hospitals.

12.4.4 Hybrid Cloud

In a hybrid cloud, the cloud infrastructure is a composition of two or more distinct cloud infrastructures that are private, community, and/or public. These distinct cloud infrastructures remain unique entities but are bound together by standardized or proprietary technology that enables data and application portability, such as cloud bursting for load balancing between clouds. A logical diagram of a hybrid cloud is illustrated in Fig. 12.18.

In the real world, many companies have their own private cloud for intranet services and simultaneously use email services from a public cloud. This is a typical example of hybrid cloud services.

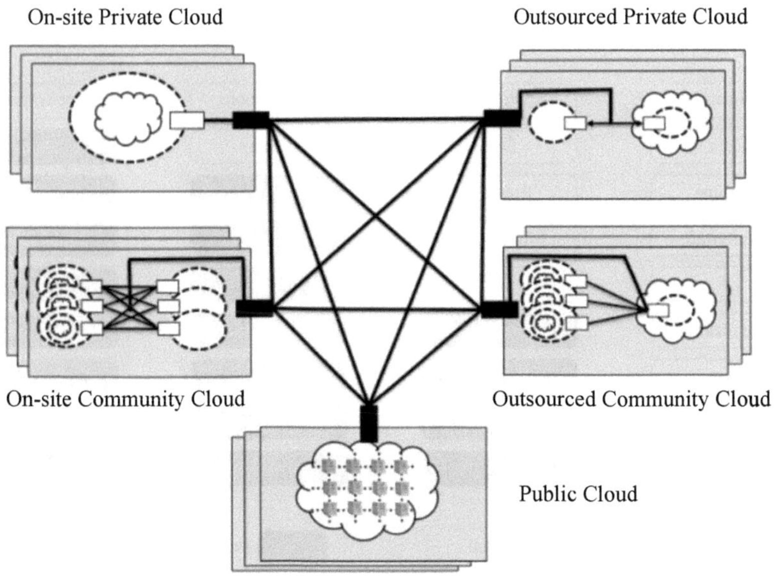

Fig. 12.18 A logical diagram of hybrid cloud consisting of multiple distinct cloud infrastructures [21, p. 15]

12.5 Service Models of Cloud Computing

Cloud computing involves both the cloud service provider and cloud service customers. The provisioning of cloud services is the responsibility of the cloud service provider. Using the cloud services provisioned by the cloud provider, the customers also have their own responsibility to manage their own resources shared out from the cloud. Therefore, there is a dividing line between the responsibilities of the cloud provider and customers. Depending on how to share the responsibilities in the management of the cloud computing environment, this dividing line can move up from infrastructure to platform and to software, or move down from software to platform and to infrastructure. Consequently, there are three service models in cloud computing: IaaS, PaaS, and SaaS.

These three service models of cloud computing are graphically depicted in Fig. 12.19, in which the dividing line for responsibility sharing is clearly shown. For comparison, traditional on-premise provisioning of computing services is also shown in the figure. It is a computing service model that fully relies on local resources and management without any support from any part of a cloud.

It is worth mentioning that the term "as a Service" is a shortened phrase of "as a Cloud Service". It describes a (cloud) computing capability that supports all five essential characteristics of cloud computing [21, p. 3], which have been discussed

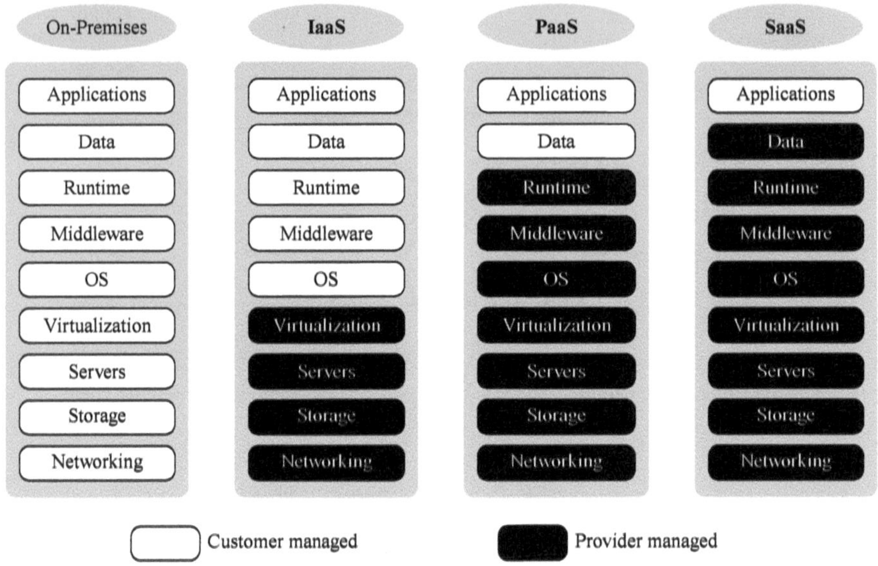

Fig. 12.19 Service models of cloud computing

previously. Simply speaking, it means that the resources available to cloud service customers reside remotely on a different server or network from the one that actually uses the resources.

12.5.1 IaaS (Infrastructure as a Service)

In the IaaS service model, the services provisioned by a cloud provider to cloud service customers are infrastructure. Infrastructure services include processing (CPU), storage, networks, and other fundamental computing resources. Network infrastructure services could be virtual servers, DNS services, and firewalls. All these hardware infrastructure resources in IaaS are owned and managed by the cloud provider. They are typically software-defined.

IaaS requires the cloud service customers to deploy and run any software of their choice, including operating systems, software development platforms, and applications. Therefore, it is the responsibility of the cloud service customers to configure and manage their operating systems, software development platforms, applications, data storage and backup, and network settings. While the customers do not manage or control the underlying infrastructure, they have control of their installed operating systems and deployed applications. They also have control of the storage they are using and possibly limited control of some networking components such as firewalls.

IaaS is provisioned virtually by a cloud provider to cloud service customers. Typically, IaaS servers are virtualized, implying that they run VMs on the cloud provider's PMs. A simple user interface is generally available to the customers to access the IaaS resources. It is web-based.

As an example of IaaS, customers may use a cloud provider's servers for computing tasks, data storage, and network services such as DHCP, DNS, email, FTP, and website hosting. In the real world, many cloud providers offer various IaaS services. For instance, AWS EC2 (Elastic Compute Cloud) allows customers to create and run their own VMs on AWS PMs in the cloud. The VMs can be configured (and requested) to have the required computing power, memory and storage space, and deployment services. They can be installed with the operating systems and applications of the customers' choice.

An interesting use case of IaaS is the *Virtual Desktop Infrastructure* (VDI). VDI deploys operating systems and applications using a virtualized desktop through IaaS. Then, IaaS customers access the virtual desktop remotely from a range of platforms, including thin clients (with limited hardware resources) and thick desktop computers (with old versions of operating systems and applications). It is becoming popular as a way for companies to deliver desktop operating systems and applications. This is an alternative way to local software installation on every machine.

12.5.2 PaaS (Platform as a Service)

In the context of PaaS, a platform is a software development environment that consists of an operating system, programming languages, software development tools, and runtime libraries. This platform is provided by the cloud provider and runs on the hardware infrastructure of the cloud infrastructure.

PaaS allows cloud service customers to deploy requested platforms onto the cloud infrastructure for the development and deployment of cloud-enabled applications. Often, multiple platforms with different environments, such as different operating systems, may be required, and the cloud provider can fulfill these requests.

Applications developed and deployed in PaaS environments are specifically designed as cloud-enabled applications, such as web and mobile applications. They inherently support broad network access, which is one of the essential characteristics of cloud computing. This distinguishes PaaS applications from general VM and desktop applications that may run on a VM.

PaaS shares similarities with IaaS in providing hardware infrastructure like servers and storage. In both IaaS and PaaS, the cloud service customers need to install the operating system and software development tools for software development in a cloud environment. However, PaaS differs significantly from IaaS. In IaaS, the customers are responsible for purchasing, installing, and managing their own operating systems and software development tools, whereas in PaaS, the cloud provider takes care of providing, installing, and managing the operating systems and software development tools.

In the PaaS service model, the applications developed by the customer are owned by the customer rather than the cloud provider. These applications can be deployed to third parties, such as the cloud service customer's users, using the cloud provider's servers.

PaaS can be provisioned through VMs or other cloud environments. For example, *containers* can be used as lightweight versions of servers to host PaaS, providing only the necessary resources to run an application. *Serverless* compute can also be used to host PaaS, allowing cloud service customers to execute their code directly in the cloud without the need to manage a server environment.

Examples of PaaS include SAP Cloud Platform, Microsoft Azure for Windows, AWS Lambda, Google App Engine, and IBM Cloud Foundry, among others. SAP Cloud Platform is an open business platform that assists PaaS developers in developing and deploying applications more easily. Microsoft Azure provides a development and deployment environment based on the PaaS concept, supporting a wide range of tools, languages, and frameworks. It also offers IaaS and SaaS services. As part of the Amazon Cloud, AWS Lambda is suitable for the development and deployment of various applications. Google App Engine, which is part of the Google Cloud ecosystem, is a highly scalable serverless PaaS for rapid application development, although some users have expressed concerns about limited language support, development tools, application compatibility, and vendor lock-in. IBM Cloud provides an open-source version of its PaaS. It has been shown to be a scalable environment for a wide range of applications.

12.5.3 SaaS (Software as a Service)

In cloud computing, SaaS allows cloud service customers to utilize the cloud service provider's applications running on a cloud infrastructure. These applications are either owned or purchased by the cloud provider. Customers do not acquire licenses for the applications themselves. Instead, they pay for the usage of these applications that are hosted on the cloud provider's network.

SaaS applications can be accessed from anywhere as long as the cloud service customers have an Internet connection. They are designed to be accessible through various client devices, either through a thin client interface like a web browser or a program interface. This enables mobile users and telecommuters to have the same access to applications they use in the office without the need to install the applications on their mobile phones, laptops, or home computers. Many SaaS applications cannot be locally installed on these devices. For example, installing the full version of Microsoft Office on a mobile phone is not feasible. With the SaaS service model, the applications run on remote servers and can be accessed through mobile phones and other devices without local installation.

SaaS is typically provided as a subscription service, either for individuals or companies. Local installation and maintenance of the applications are no longer necessary. This is particularly advantageous as it allows for faster software upgrades compared to traditional local deployments. In some cases, hardware devices may not even support software upgrades, such as upgrading Windows 10 applications to the Windows 11 environment. However, local network and Internet access must be maintained for accessing SaaS applications.

Both SaaS and PaaS involve running applications on the cloud provider's infrastructure. The key distinction lies in the ownership of the applications. In PaaS, cloud service customers create or acquire the applications and then deploy them onto the cloud provider's network or a third-party platform. However, in the SaaS service model, the cloud provider owns the applications and runs them on their own infrastructure.

There are numerous examples of SaaS, including widely used applications such as Gmail and Google Docs, which are available to many users free of charge. A popular SaaS example is Microsoft Office 365, which many companies use to access applications hosted in the Microsoft cloud, such as email, calendar, word processing, PowerPoint, and spreadsheets. Figure 12.20 provides a screenshot of the web-based Office 365 interface, displaying all the available applications accessible to the author at the time of writing this section.

12.5.4 The Use of Cloud Service Models

The choice of a specific cloud service model from SaaS, PaaS, and IaaS by cloud service customers determines where to place the dividing line between the responsibilities of the customers and cloud provider (Fig. 12.19). Customers need to understand what this choice means with respect to their privilege and responsibility in the management of the provider's resources and infrastructure. As shown in Fig. 12.21a, moving down towards IaaS gives customers more control of the resources and infrastructure. Moving up towards SaaS implies that less control of the resources and infrastructure will be available to the customers.

The level of control over the provider's resources and infrastructure corresponds to the customer's understanding of the underlying infrastructure and its access. End users typically opt for SaaS as it requires minimal knowledge of cloud computing. PaaS is commonly used by application developers who need familiarity with the chosen cloud platform but do not require deep knowledge of the underlying infrastructure. IaaS is often used by network architects who possess a comprehensive understanding of the cloud infrastructure. Figure 12.21b visually represents the distribution of users across the three service models, with a larger user base for SaaS and a smaller user base for IaaS.

The NIST SP specifications, such as NIST SP 800-145 [14], NIST SP 500-292 [19], NIST SP 500-293 [20], and NIST SP 500-322 [21], provide clear definitions of the three cloud service models (IaaS, PaaS, and SaaS) to encompass all types of cloud services. These models serve as high-level categorizations of cloud services. However, cloud service providers often use various "aaS" terms for marketing purposes. Many of such marketing "aaS" terms are listed in the NIST SP 500-322

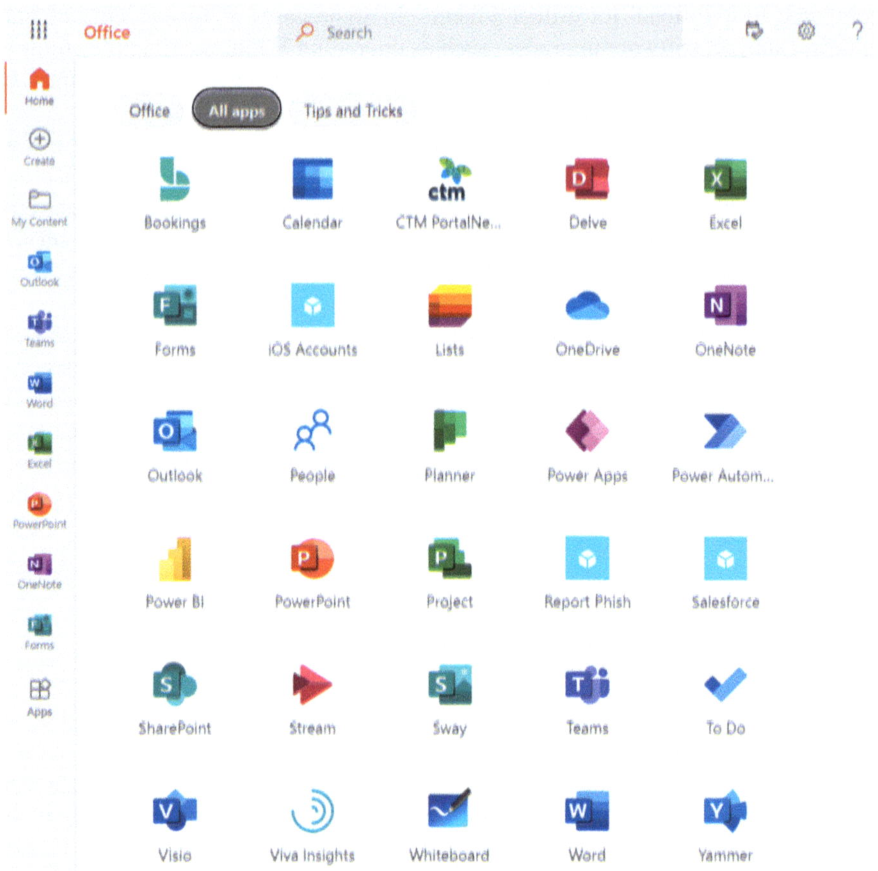

Fig. 12.20 A screenshot of web-based Office 365 interface showing all applications available to the author from cloud Office 365

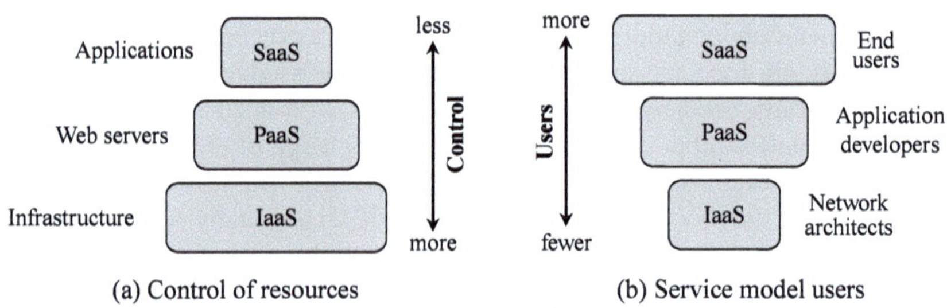

Fig. 12.21 The use of cloud service models from the perspective of cloud service customers

[21, p. 20]. Some of these terms are seen frequently, such as Anything as a Service (XaaS), Business process as a Service, Database as a Service, Email as a Service, Security as a Service, and Website as a Service. While these marketing terms facilitate informal communication about specialized cloud services, they do not replace the three high-level service models (IaaS, PaaS, and SaaS). The three models remain fundamental for understanding and discussing cloud computing services.

12.6 Cloud Computing Reference Architecture

The NIST SP 500-292 has recommended a cloud computing reference architecture [19]. This reference architecture focuses on the requirements of "what cloud services" provide, rather than providing a 'how to design' solution. With this philosophy in mind, the main intention of the architecture is to facilitate understanding of the operational intricacies in cloud computing. Therefore, the reference architecture is a generic high-level conceptual model as an effective tool for describing, discussing, and developing a system-specific architecture in a common framework of reference. It does not intend to represent the system architecture of a specific cloud computing system.

12.6.1 Conceptual Reference Model

The overall conceptual reference model of cloud computing consists of five main actors: cloud consumer (which is frequently referred to as cloud service customer), cloud provider, cloud carrier, cloud auditor, and cloud broker. These five actors and their relationships are shown in Fig. 12.22 [19]. Their definitions are tabulated in Table 12.2 [19]. On top of the cloud carrier, the other four actors (i.e., cloud consumer, cloud provider, cloud broker, and cloud auditor) have direct interactions for collecting information, information exchanges, or service provisioning. This is graphically shown in the logical diagram of Fig. 12.23.

Before discussing the details of the five actors in the reference architecture, let us take a look at where IaaS, PaaS, and SaaS are located in the reference architecture depicted in Fig. 12.22. From the cloud provider's perspective, IaaS, PaaS, and SaaS are made available in the service layer, which is part of the service orchestration stack residing in the Cloud Provider component of the reference architecture specified in the NIST SP 800-292 [19]. It is interesting to note that IaaS, PaaS, and SaaS are represented as "L-shaped" horizontal and vertical bars rather than a simple "three-layer cake" stack. This is because PaaS and SaaS are not necessarily built on top of IaaS and SaaS, respectively, although cloud services can have dependencies on each other within the stack. This means that, depending on the architecture of each layer shown in Fig. 12.22, it is possible to implement cloud services that interact directly with the resource abstraction and control layer.

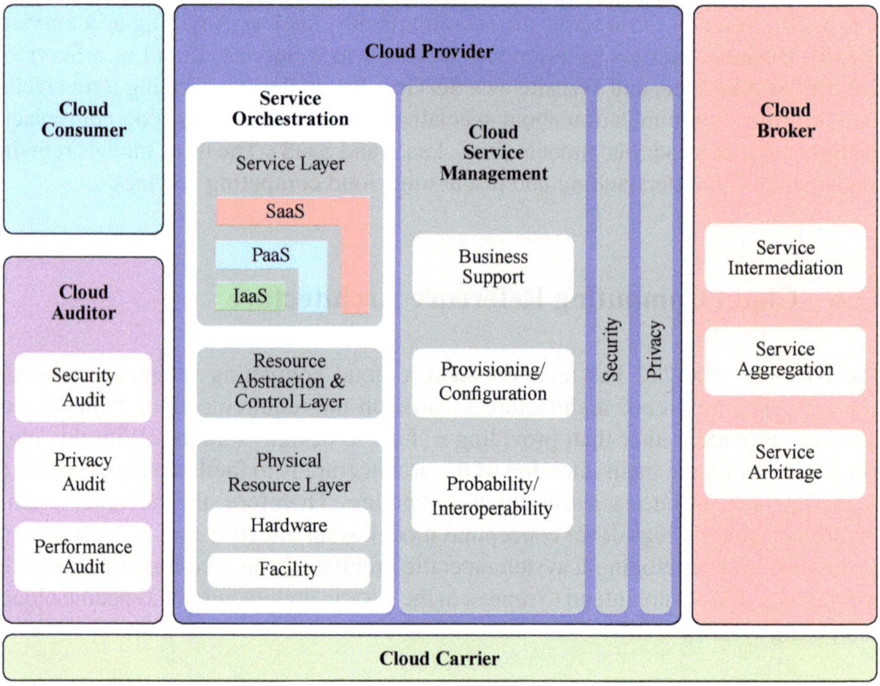

Fig. 12.22 Cloud computing reference architecture [19, p. 3]

Table 12.2 Actors in cloud computing [19, p. 4]

Actor	Description
Cloud consumer	A person or organization that maintains a business relationship with, and uses cloud services from, cloud providers
Cloud provider	A person, organization, or entity that is responsible for making cloud services available to interested parties
Cloud auditor	A party that can conduct independent assessment of cloud services, information system operations, performance and security of the cloud implementation
Cloud broker	An entity that manages the use, performance, and delivery of cloud services, and negotiates between cloud providers and cloud consumers
Cloud carrier	An intermediary that provides connectivity and transport of cloud services from cloud providers to cloud consumers

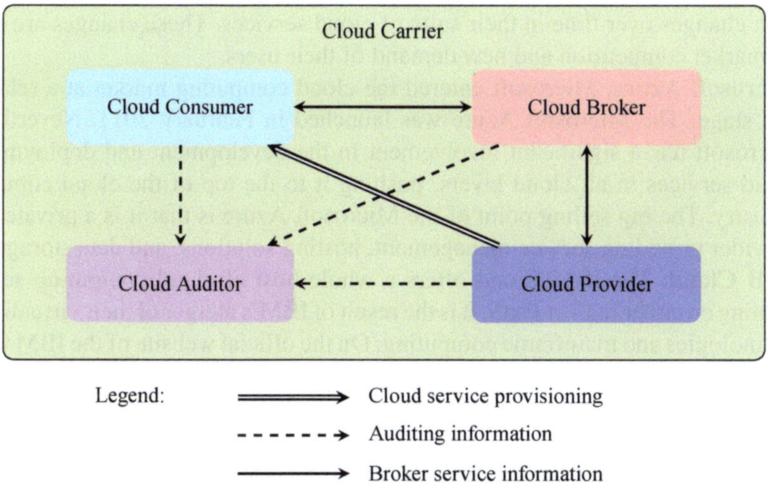

Fig. 12.23 Interactions among the five actors in cloud computing [19, p. 4]

12.6.2 Cloud Provider

In order to make cloud services available to cloud consumers, a cloud provider needs to manage the computing infrastructure, run cloud software that provides services, and ensure security and privacy. Therefore, the main activities of a cloud provider include service deployment, service orchestration, cloud service management, security, and privacy, as shown in the middle part of Fig. 12.22.

For service deployment, a cloud provider provides cloud services to cloud consumers via three types of service models: IaaS, PaaS, and SaaS. This function sits in the service layer within the Service Orchestration as discussed previously. The service orchestration is composed of three main layers: a service layer, a resource abstraction and control layer, and a physical resource layer.

The cloud service management manages all cloud services. Typically, it includes business support, service provisioning and configuration, and interoperability.

A Google search of "top cloud providers 2022" returns a list of big cloud providers in the world. On the list are Amazon Web Service (AWS), Google Cloud Platform (GCP), Microsoft Azure, IBM Cloud, and Oracle:

- **Amazon AWS**. In cloud services, Amazon is currently the world leader in cloud services. It will remain as the leader for at least the foreseeable future because no other cloud providers can really compete with it. Amazon is the first choice for many companies, either big or small, to consider using some cloud services or even converting their data centers to cloud computing.
- **Google GCP**. The Google GCP offered by Google includes a range of cloud services, such as storage, application development, and computing. All these cloud services run on Google's hardware infrastructure. The Google GCP is still evolving

with changes over time in their suite of cloud services. These changes are driven by market competition and new demand of their users.

- **Microsoft Azure**. Microsoft entered the cloud computing market at a relatively late stage. The Microsoft Azure was launched in February 2011. Nevertheless, Microsoft has a significant involvement in the development and deployment of cloud services in all cloud layers, pushing it to the top of the cloud computing industry. The big selling point of the Microsoft Azure is that it is a private cloud provider providing service management, hosting solutions, and data storage.
- **IMB Cloud**. The IBM Cloud offers a whole host of cloud computing services running on either IaaS or PaaS. It is the result of IBM's merger of their virtualization technologies and mainframe computing. On the official website of the IBM Cloud, it is claimed that IBM Cloud offers the most open and secure public cloud for businesses, a next-generation hybrid multi-cloud platform, and advanced data and AI capabilities.
- **Oracle**. Oracle offers private cloud computing services through its global network of data centers. The services encompass a range of components such as network, storage, services, applications, and servers. These services enable cloud consumers to deploy, build, extend, and integrate applications over the cloud. Oracle cloud services can be requested on-demand by consumers via the Internet

12.6.3 Cloud Consumer

Cloud consumers are the primary stakeholders in a cloud computing environment. They request cloud services from a cloud provider, use these services, and pay for what they use. Cloud services are provided to cloud customers via IaaS, PaaS, or SaaS models, typically through a web-based interface.

The required QoS for cloud services can be explicitly defined in SLAs, which are later fulfilled by the cloud provider. SLAs can cover various aspects beyond QoS, such as performance commitments, limitations, and obligations that cloud consumers are required to accept.

12.6.4 Cloud Broker

Since the evolution and integration of cloud services can be too complex for a cloud consumer to manage, the involvement of a cloud broker can help simplify the management of cloud services. Cloud consumers may choose to request cloud services through a cloud broker instead of directly from a cloud provider. The cloud broker assists in managing the use, performance, and delivery of cloud services, as well as negotiates relationships between the cloud provider and cloud consumers.

The services provided by a cloud broker are generally classified into three categories: service intermediation, service aggregation, and service arbitrage, as depicted on the right side of Fig. 12.22:

- Service Intermediation: A cloud broker offers cloud services with performance enhancements and/or added value compared to direct access from a cloud provider.
- Service Aggregation: A cloud broker aggregates multiple services into one or more new cloud services, facilitating data integration and ensuring secure data transfer between the cloud consumer and multiple cloud providers.
- Service Arbitrage: Similar to service aggregation, but with the flexibility to select cloud services from multiple agencies.

A Google search for "top cloud brokers" provides a list of popular cloud brokers. Some examples from the search results for 2022 include AWS Service Broker, IBM Multicloud Management Services, Cloudmore, Jamcracker Cloud Services Brokerage, and Boomi.

Using a cloud broker offers several benefits and features, including but not limited to:

- Using integrated or aggregated cloud infrastructure.
- Managing cloud user identity and addressing risks.
- Enhancing manageability of cloud services through user-friendly interfaces.
- Identifying cost-effective and high-performance cloud service providers.

12.6.5 Cloud Auditor

A cloud auditor is an independent party separate from the cloud provider and cloud consumers. The primary role of a cloud auditor is to conduct an unbiased evaluation of cloud service controls and assess their compliance with established standards through the examination of objective evidence. This evaluation encompasses various aspects, such as security controls, privacy impact, performance, and more.

Auditing is particularly important for government departments and agencies, as well as other organizations that have specific and critical requirements regarding security, privacy, and performance. For example, in the context of security auditing, a cloud auditor evaluates to what extent the security controls of the information system are appropriately implemented and deployed, functioning as intended, adhering to security standards and regulations, and achieving the expected outcomes in alignment with the system's security requirements. Such audits ensure the protection of system Confidentiality, Integrity, and Availability (CIA).

12.6.6 Cloud Carrier

The delivery of cloud services from cloud providers to cloud consumers, whether directly or through cloud brokers, heavily relies on network connectivity. This crucial role of providing connectivity and transporting cloud services between cloud providers and cloud consumers is fulfilled by a cloud carrier through network infrastructure, telecommunication systems, and other access devices.

To ensure the consistent and desired QoS for cloud consumers, a cloud provider may establish SLAs with a cloud carrier. These SLAs outline the specific requirements and performance expectations. In order to meet these SLAs, a cloud carrier may need to establish dedicated and secure connections, enabling the reliable and efficient delivery of cloud services offered by cloud providers to cloud consumers.

12.7 Cloud Security

Cloud security is a comprehensive set of technologies, protocols, policies, controls, services, and best practices designed to safeguard cloud computing environments, cloud data, cloud applications, and cloud infrastructure from potential threats. To enhance the security of cloud environments, it is crucial to have a clear understanding of what needs to be protected, what aspects require management, and how security measures can be implemented effectively.

12.7.1 Cloud Security Responsibilities

Cloud security entails shared responsibilities between the cloud service provider and cloud customers, each having their respective areas of responsibility:

- Certain aspects of cloud security are the sole responsibility of the cloud service provider, such as safeguarding the cloud infrastructure, managing physical servers and networks, and handling personal information and personally identifiable information. According to the NIST SP 800-292 [19, p. 17], the cloud service provider should also protect the assured, proper, and consistent collection, processing, communication, use and disposition of personal information and personally identifiable information in the cloud. This is a cloud privacy issue.
- Conversely, other security aspects fall under the purview of the cloud customers. These aspects include user management, access privileges, VPN configuration, protection of cloud accounts and services, and safeguarding of cloud data assets.
- There are also security responsibilities that are shared between the cloud service provider and cloud customers, as outlined in the NIST SP 800-292 [19, p. 16]. Examples of such shared responsibilities include cloud email services and cloud-based Internet banking.

Overall, regardless of the specific responsibilities, cloud security is designed to protect physical networks, data storage, data servers, virtualization frameworks and environments, operating systems, platforms, software and applications, as well as end-user hardware.

12.7.2 Cloud Security Challenges

Cloud security is challenging due to a number of factors. The following aspects highlight some of these factors:

- A noticeable challenge is the increased attack surface with the increased deployment of cloud computing services.
- The lack of visibility and tracking in a cloud environment is also a big challenge for cloud computing. This is different from the computing environments offered via traditional networks.
- Increased complexity in the management and operation of the computing environments makes cloud security difficult. While to end users, cloud computing appears with simple interfaces, typically web pages, the required back-end support for automation, orchestration, and other management tasks are complicated. For example, cloud security needs to be managed seamlessly across public cloud, private cloud, and on-premise deployments.
- Cloud service models and cloud deployment models have significant impacts on cloud security [19, p. 16]. The three cloud service models, i.e., IaaS, PaaS, and SaaS, create different attacking surfaces for adversaries. The variations of cloud deployment models, i.e., private cloud, public cloud, community cloud, and hybrid cloud, also have important security implications.

12.7.3 Cloud Security Methods

General security methods can be applied to cloud computing environments. For example, encryption can be used to reduce the inherent risks of cloud computing. Also, choosing an appropriate method to connect a network to cloud resources will also improve the security protection of cloud computing. A VPN or even a dedicated private connection to cloud resources will enhance cloud computing security.

Cloud service providers, such as Amazon's AWS, Microsoft's Azure, and Google's GCP, offer many cloud-native security features. This is good but not enough for enterprise cloud computing because third-party service providers, such as ISPs or cloud brokers, are also involved in the cloud computing environment. Therefore, additional security measures are also essential for the security of cloud computing. Cloud-native and third-party security could be integrated to provide enterprise-grade cloud security protection from various threats.

There are security methods that are more specific to, or fine-tuned for, cloud computing. Some of them are summarized in the following:

- Granular and policy-based Identity and access management (IAM). IAM aims to create digital identities for all users so that they can be actively monitored and restricted whenever necessary during their access activities. Therefore, it enables enterprises to deploy policy-driven enforcement protocols for all users attempting to access both on-premises and cloud-based services. For cloud security enforcement, always grant access in terms of groups and roles rather than individuals. Also, grant the minimal required access privileges to assets and services that are essential for a group or role.
- Stateless versus stateful security. Stateless security does not maintain an awareness of how traffic streams, e.g., incoming and outgoing traffic, are related to each other. For example, the Access Control List (ACL) security used in cloud and other network applications is a stateless security method. By contrast, stateful security inspects details inside data packets, assesses the characteristics and behavior of data packets, and examines the channels of communication. Therefore, it maintains an awareness of the communication behavior. If a data packet is revealed to be suspicious, it can be filtered out. This feature of stateful security helps improve the security in a cloud environment.
- Examination of default security configuration. Each tool from a cloud service provider or third party comes with default configuration including settings for security. It is important to examine if the default configuration fulfills the security requirements. If not, modify or override the default configuration.
- Zero-trust cloud network security controls. For enhanced reliability and security, deploy critical resources and applications in logically isolated sections of the cloud network infrastructure. Meanwhile, use subnets for granular micro-segmentation management of data, applications, and services. Whenever needed, use dedicated WAN links in hybrid cloud. Moreover, it will be more secure to use static routing to virtual devices and virtual networks on the cloud.
- Safeguarding of all applications. Web application firewalls are available to safeguard web-based applications such as cloud applications, which are typically web-based. A web application firewall helps protect web applications by filtering and monitoring HTTP traffic between a web application and the Internet, thus protecting the web application from attacks. Deploying a web application firewall in cloud computing will enable granular inspection and control of traffic to and from web application servers. It will also enable automatic update of security rules in response to traffic behavior changes.
- Multiple levels of firewalls. Multi-firewalling in the cloud refers to the common practice of deploying and configuring multiple firewalls at different points to create a layered defense against unauthorized access and network-based attacks. Multiple levels of firewalls may include perimeter firewalls at the edge of the cloud infrastructure, network firewalls within the internal network of the cloud environment, host-based firewalls installed directly on individual VMs or cloud instances, web

application firewalls for the protection of web applications hosted in the cloud, and container firewalls.

12.8 Summary

Virtualization is an approach that pools physical resources into a virtualized environment for sharing among multiple users and applications. It achieves this by abstracting physical resources from the underlying hardware and software. Due to its advantages in many aspects, virtualization has found widespread applications in network infrastructure, particularly in data centers and large-scale networks. It has become a fundamental technology in modern cloud computing. Today, various types of virtualization are deployed, such as virtualization of application, data, desktop, memory, network, OS, server, and storage. This has motivated the development and deployment of applications and services in virtualized environments.

Among these virtualization types, NFV provides virtualized network functions by abstracting network devices, protocols, functions, and services from their underlying physical entities. NFV encompasses use cases such as NFVIaaS, VNFaaS, VNPaaS, VNF forwarding graphs, mobile core network virtualization, and mobile base station virtualization, among others. An NFV architecture framework has been recommended in the ESTI GS NFV specifications.

Cloud computing, which extensively use virtualization, is increasingly adopted to offer a broad range of cloud services over networks. With resource pooling and rapid elasticity, it aims to provide on-demand and measured self-service with broad network access. Cloud computing can be deployed in various models, including public cloud, private cloud, community cloud, and hybrid cloud.

There are three fundamental service models in cloud computing: IaaS, PaaS, and SaaS. These models constitute the service layer of the service orchestration component in the NIST-recommended cloud computing reference architecture. The reference architecture is composed of five actors: provider, consumer, broker, carrier, and auditor. Within the cloud provider, there are components such as service orchestration, service management, security, and privacy. The service orchestration component comprises a service layer (IaaS, PaaS, and SaaS), a resource abstraction and control layer, and a physical resource layer.

Cloud security poses challenges due to the involvement of not only the cloud provider and consumer but also potentially one or more third parties. Centralized control is not feasible to manage resources, applications, services, and access from all involved parties. While some aspects of cloud security fall under the full responsibility of either the cloud consumer or the cloud provider, other aspects are shared responsibilities of both and potentially third parties. General security methods like encryption and VPN connections are applicable in cloud computing environments. Additional security solutions tailored to cloud computing should also be implemented for effective cloud security enforcement. Examples include granular and policy-based IAM, zero-trust network security controls, and web application firewalls.

References

1. Ding, Z., Tian, Y.C., Tang, M., Li, Y., Wang, Y.G., Zhou, C.: Profile-guided three-phase virtual resource management for energy efficiency of data centers. IEEE Trans. Industr. Electron. **67**(3), 2460–2468 (2020)
2. Zhang, W.Z., Xie, H.C., Hsu, C.H.: Automatic memory control of multiple virtual machines on a consolidated server. IEEE Trans. Cloud Comput. **5**(1), 2–14 (2017)
3. Ge, Y., Tian, Y.C., Yu, Z.G., Zhang, W.: Memory sharing for handling memory overload on physical machines in cloud data centers. J. Cloud Comput. **12**(1), Article No. 27, 1–12 (2023)
4. Xu, X., Tang, M., Tian, Y.C.: Theoretical results of QoS-guaranteed resource scaling for cloud-based MapReduce. IEEE Trans. Cloud Comput. **6**(3), 879–889 (2018)
5. ETSI: Network functions virtualisation: an introduction, benefits, enablers, challenges & call for action. NFV Introductory White Paper Issue 1, ETSI (2012). https://portal.etsi.org/NFV/NFV_White_Paper.pdff, Accessed 3 June 2022
6. ETSI: Network functions virtualisation (NFV); use cases. ETSI GS NFV 001 V1.1.1, ETSI NFV ISG (2013). www.etsi.org/deliver/etsi_gs/NFV/001_099/001/01.01.01_60/gs_nfv001v010101p.pdf, Accessed 3 June 2022
7. ETSI: NFV specifications and documentation. Online Documentation (2013–2021). www.etsi.org/deliver/etsi_gs/NFV/001_099/, Accessed 3 June 2022
8. ETSI: Network functions virtualisation (NFV); architectural framework. ETSI GS NFV 002 V1.2.1, ETSI NFV ISG (2014). https://www.etsi.org/deliver/etsi_gs/NFV/001_099/002/01.02.01_60/gs_nfv002v010201p.pdf, Accessed 3 June 2022
9. ETSI: Network functions virtualisation (NFV); terminology for main concepts in NFV. ETSI GS NFV 003 V1.4.1, ETSI NFV ISG (2018). https://www.etsi.org/deliver/etsi_gs/NFV/001_099/003/01.04.01_60/gs_nfv003v010401p.pdf, Accessed 3 June 2022
10. ETSI: Network functions virtualisation (NFV); virtualisation requirements. ETSI GS NFV 004 V1.1.1, ETSI NFV ISG (2013). https://www.etsi.org/deliver/etsi_gs/NFV/001_099/004/01.01.01_60/gs_nfv004v010101p.pdf, Accessed 3 June 2022
11. ETSI: Network functions virtualisation (NFV); proofs of concept; framework. ETSI GS NFV 005 V1.2.1, ETSI NFV ISG (2019). https://www.etsi.org/deliver/etsi_gs/NFV/001_099/005/01.02.01_60/gs_nfv005v010201p.pdf, Accessed 3 June 2022
12. ETSI: Network functions virtualisation (NFV) release 2; management and orchestration; architectural framework specification. ETSI GS NFV 006 V2.1.1, ETSI NFV ISG (2021). https://www.etsi.org/deliver/etsi_gs/NFV/001_099/006/02.01.01_60/gs_nfv006v020101p.pdf, Accessed 3 June 2022
13. Bernardos, C.J., Rahman, A., Zuniga, J.C., Contreras, L.M., Aranda, P., Lynch, P.: Network virtualization research challenges. RFC 8568, RFC Editor (2019). https://doi.org/10.17487/RFC8568
14. Mell, P., Grance, T.: The NIST definition of cloud computing. NIST SP 800-145, National Institute of Standards and Technology (2011). https://doi.org/10.6028/NIST.SP.800-145
15. Sajjad, M.M., Bernardos, C.J., Jayalath, D., Tian, Y.C.: Inter-slice mobility management in 5G: motivations, standard principles, challenges, and research directions. IEEE Commun. Stand. Mag. **6**(1), 93–100 (2022)
16. Wu, Y., Dai, H.N., Wang, H., Xiong, Z., Guo, S.: A survey of intelligent network slicing management for industrial IoT: integrated approaches for smart transportation, smart energy, and smart factory. IEEE Commun. Surv. Tutor. **24**(2), 1175–1211 (2022)
17. Fang, Q., Zeitouni, K., Xiong, N., Wu, Q., Camtepe, S., Tian, Y.C.: Nash equilibrium based semantic cache in mobile sensor grid database systems. IEEE Trans. Syst. Man Cybern.: Syst. **47**(9), 2550–2561 (2017)

18. Wang, D., Zhang, W., Han, X., Lin, J., Tian, Y.C.: A multi-objective virtual network migration algorithm based on reinforcement learning. IEEE Trans. Cloud Comput. **11**(2), 2039–2056 (2023)
19. Liu, F., Tong, J., Mao, J., Bohn, R., Messina, J., Badger, L., Leaf, D.: NIST cloud computing reference architecture. NIST SP 500-292, National Institute of Standards and Technology (2011). https://doi.org/10.6028/NIST.SP.500-292
20. Liu, F., Tong, J., Mao, J., Bohn, R., Messina, J., Badger, L., Leaf, D.: NIST cloud computing technology roadmap, volumes i and ii. NIST SP 500-293, National Institute of Standards and Technology (2014). https://doi.org/10.6028/NIST.SP.500-293
21. Simmon, E.: Evaluation of cloud computing services based on NIST SP 800-145. NIST SP 500-322, National Institute of Standards and Technology (2018). https://doi.org/10.6028/NIST.SP.500-322

Chapter 13
Building TCP/IP Socket Applications

In the networked world, various general-purpose application-layer systems are available, such as web browsers, email clients, and social network apps. These systems are already integrated with existing network protocols to provide their functionality. For example, web browsers use the HTTP application protocol for processing hypertext. They also use TCP as the underlying transport protocol. Since these off-the-shelf application systems already embed the necessary network protocols and communications, no further development is required before they are used.

However, many networked application systems have unique requirements and specifications that differ significantly from one system to another. There is no one-size-fits-all network design for such systems. For example, the network design for vehicle control systems differs from that for aircraft control systems, and IoT networking has its own distinct characteristics. Communication in vehicular networks, as a type of ad-hoc network, is different from that in networks with a fixed network topology like wired office networks. Therefore, developing practical networked systems necessitates customized design and implementation of networks and communications.

While understanding the theory of computer networks is one aspect, developing TCP/IP network software systems for real applications is another challenge. As famously stated by Linus Torvalds, "Talk is cheap. Show me the code." This chapter aims to guide readers in building actual TCP/IP communication software systems for applications. It will also provide code examples.

There are many books and references with comprehensive discussions of TCP/IP programming. A relatively simple guide for socket programming is *Beej's Guide to Network Programming* (Brian "Beej Jorgensen" Hall, version 3.0.20, March 11, 2016) [1]. However, this guide, which starts with detailed discussions of various APIs, may be challenging to follow for individuals with limited programming experience. Taking a top-down approach, this chapter introduces TCP/IP socket programming from the application-oriented perspective. It is specifically written for those with limited programming experience, particularly in C programming language. For this purpose, the chapter begins by analyzing system requirements, proceeds to system

Y.-C. Tian and J. Gao, *Network Analysis and Architecture*, Signals and Communication Technology, https://doi.org/10.1007/978-981-99-5648-7_13

design, and finally implements the design. During the implementation phase, socket APIs will be introduced when they are used.

All Linux code examples presented in this chapter have been successfully compiled and tested using GNU's `gcc` compiler on Linux and Mac OS systems. For Windows users, a separate section is presented to focus on Windows socket programming because there are differences in socket data structures and APIs. The compiler used in Windows is also different from the `gcc` compiler in Linux and Mac OS. In this chapter, two terminal-window compilers are introduced for Windows users: the gcc compiler ported to Windows, and Microsoft's Visual Studio command line compiler (cl). All Windows code examples presented in this chapter have been compiled and tested successfully on Windows 10. It is worth mentioning that Integrated Development Environments (IDEs) are available to build a system from C code. But they are heavy and complicated, and thus are omitted here in this chapter.

13.1 Why Socket Programming

Connectivity and scalability are two of the most important issues in computer networking. A network system will not function without network connectivity. Scalability ensures that a network works on not only a small scale, e.g., a work group, but also a large scale, e.g., over the Internet. To maintain good connectivity and scalability of a network, it is important in computer networking to follow well-supported international standards for network architecture and protocols.

From the network architecture perspective, the design and deployment of computer networks and networked application systems follow ISO's seven-layer architecture. These seven layers are implemented in four layers in the TCP/IP suite of protocols. The upper three layers (Layers 5 to 7) of ISO's seven-layer architecture are integrated into a single layer in the TCP/IP model, known as Application Layer. The bottom two layers (e.g., Layers 1 and 2) of the ISO's seven-layer architecture) are also combined into a single layer in the TCP/IP architecture, known as Network Access Layer. The mapping from ISO's seven-layer architecture to the TCP/IP four-layer architecture is shown in Fig. 13.1. Nevertheless, when discussing network layers, we normally assume the ISO's seven-layer architecture unless otherwise specified explicitly.

As for network protocols, there are many well-supported protocols at each of the seven layers. Some of these protocols are illustrated in Fig. 13.1. For the purpose of this chapter on socket programming, many other protocols are omitted in this figure. For TCP/IP network communications, obviously we will use TCP or UDP protocols as the underlying transport protocol. IP is also essential and will be directly used. Media access control, such as Ethernet for wired networks or Wi-Fi for wireless networks, is also critical in all TCP/IP communication systems. Other protocols, such as ICMP and ARP, do not appear explicitly in the programming of a network communication system, but network communications rely on these protocols to function.

In a physical computer system, the functions of Layers 1 and 2 are implemented in hardware, typically a Network Interface Card (NIC). Layers 3, Layer 4, and standard

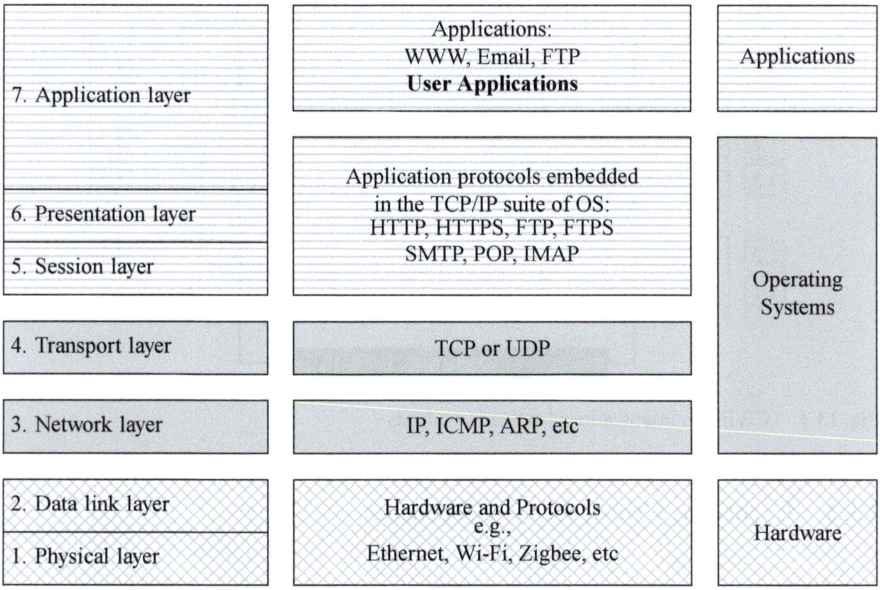

Fig. 13.1 Layered architecture of computer networks

application protocols of Layer 7 are implemented in operating systems. Application systems and user programs are located in Layer 7, i.e., the Application layer. Where each layer is implemented in a computer system is clearly depicted in Fig. 13.1.

As many network protocols are already implemented, as indicated in Fig. 13.1, it does not make sense, and actually it is very hard if not impossible, to program these protocols again in our user programs for TCP/IP network communications. We do not need to reinvent the wheel. Therefore, in our network system development, we should reuse the functions already available from the operating systems and NIC. This will largely simplify our system design, development, and operations. Now, the question is how to make use of these available functions. The answer is socket programming.

What is a socket? Socket is formally defined in the IETF RFC 147 [2], which is an update of the IETF RFC 129. In physical networking, a physical network socket on the wall enables us to plug in a network cable to connect to a network. Analogous to the function of a wall network socket, a socket in network socket programming is a software interface that enables us to plug in our user programs to connect to the network protocols implemented in the operating systems and NIC. This is conceptually shown in Fig. 13.2 for TCP/IP network communications between two hosts interconnected over a network. With the software socket as part of the operating system, when developing a user network communication system, we do not need to program anything on TCP, UDP, IP, and other protocols that are available in the operating system but simply connect our user program to the socket. Therefore, the only thing we need to understand is how to connect to the socket. This requires an understanding of socket APIs, which will be discussed in the next few sections.

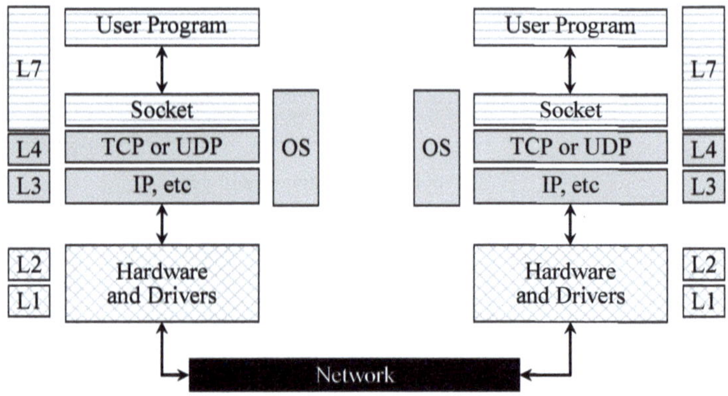

Fig. 13.2 TCP/IP communications between two hosts

13.2 Example Client-Server Systems

Consider an example client-server network system as shown in Fig. 13.3. In this client-server system, there is a server that manages all client-server communications. All clients, i.e., hosts in the figure, communicate with the server for data transmission and other tasks.

Three client-server communication scenarios are illustrated in Fig. 13.3: communications within a LAN, across LANs, and over the Internet. The three hosts, acting as clients in the client-server systems, are located in three different LANs. Host 1 reaches the server through Router 1. Host 2 and the server are within the same LAN. Host 3 communicates with the server over the Internet. For all these three scenarios,

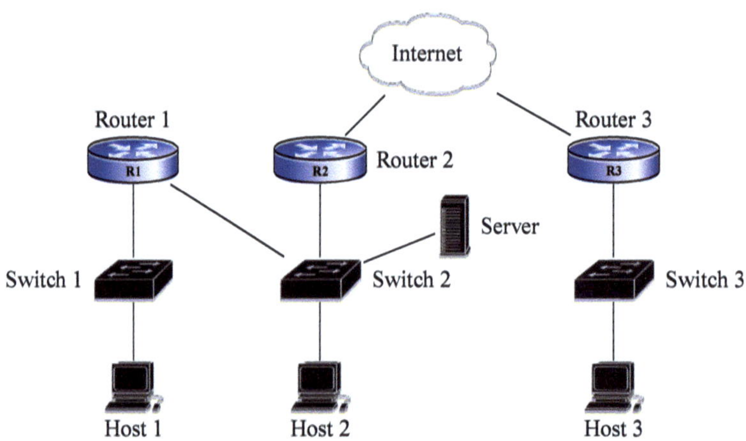

Fig. 13.3 A client-server network system with communicating hosts as clients

socket programming for TCP/IP communications is the same without any difference. Therefore, the discussions below apply to all these three communication scenarios.

For a server-client communication system, a server program needs to be developed and executed on the server. Similarly, a client program is required to be developed and run on the client. As the server and client have different functions and roles, the server program and client program are different. However, they will communicate with each other for client-server communications. Therefore, there are also some similarities between the server and client programs.

The communication requirements and logical flows of the server and client in a client-server system are illustrated in Fig. 13.4. The server (Fig. 13.4a) starts by opening a socket as a communication channel. It then associates the opened socket with an IP address (as the identification of the local machine) and a port number (as the identification of the communication process). After that, the server specifies the number of pending client connections allowed. Once this step is completed, it waits in a loop for requests from clients for connection and decides whether to accept or reject the requests. After establishing a connection with a client, the server enters a loop to send data to, or receive data from, the client. Once all communications are completed, the server closes the communication socket and exits from the program. It is worth noting that the server program may also proceed to the Close block before establishing any connections with clients, as indicated by the dashed arrow in Fig. 13.4a.

As for the client in a client-server system (Fig. 13.4b), similar to the server, it starts by opening a communication socket. It then requests a connection to the server. After the request is accepted by the server, the communication connection is established. Consequently, the client enters a loop to send data to, or receive data from, the server. Similar to the server, after all communications are completed, the client closes the communication socket and terminates the client program. It's worth pointing out that the steps of binding the socket and setting the number of connections allowed are unique operations on the server and do not exist on the client.

Apart from network communication tasks, a client-server system may involve other application tasks, such as data processing, data visualization, data-driven control, and alarms. However, as the focus of this chapter is on TCP/IP network communications, these application tasks are not shown in Fig. 13.4. Data-driven application tasks that need to execute continuously or periodically based on received data are typically integrated with the Send/Receive data block in both the server and client. Simple examples will be provided later to demonstrate such a system design.

13.3 Socket APIs for TCP/IP Communications

Converting the logical diagrams of Fig. 13.4 into C programming APIs, we have the logic flowchart in Fig. 13.5 to illustrate the operations and APIs of the server and client in a client-server network system. It is seen from this figure that by using just a

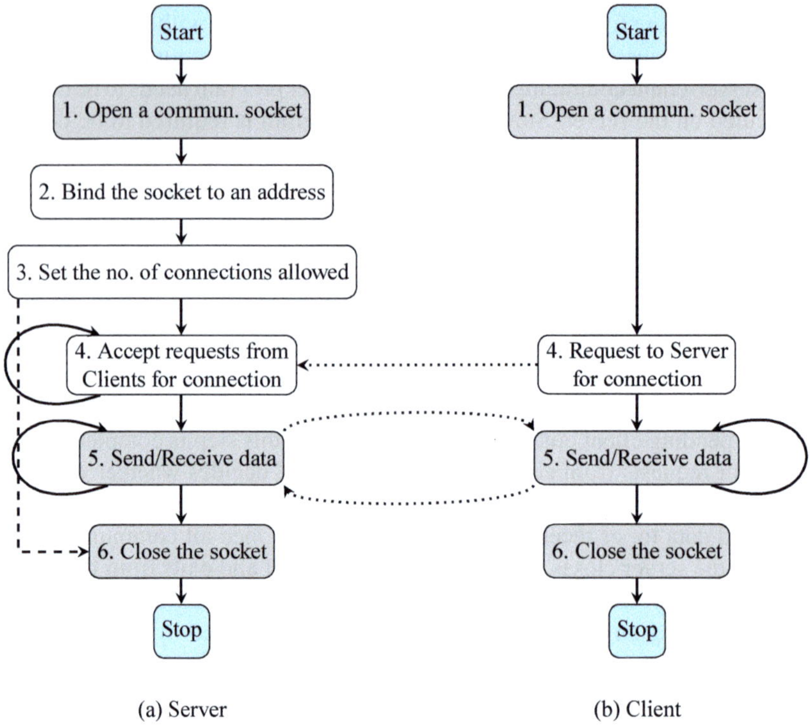

Fig. 13.4 The communication requirements and logical flows of the server and client in client-server network systems. The dotted arrows represent information flows between the server and client. Other application tasks that need to execute forever or periodically are normally integrated with the Block of Send/Receive data in both the server and client

few socket APIs, we are able to program a TCP/IP network communication system. The server and client share four common APIs, which are:

`socket()` for openning socket,
`send()/recv()` for sending/receiving data, and
`close()` for closing socket

The server has three unique APIs for server operations, namely,

`bind()` for binding socket,
`listen()` for setting the number of pending client connections allowed, and
`accept()` for accepting requests from clients for connections

The client has a unique API:
`connect()` for requesting a connection to the server.

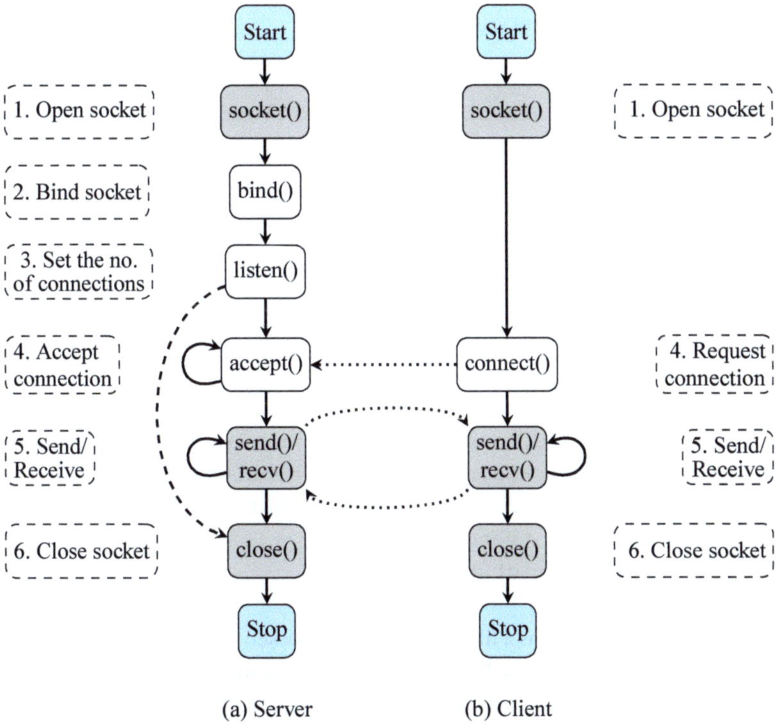

Fig. 13.5 The operations and APIs of the server and client in client-server network systems. The dotted arrows indicate logical information flows between the server and client. The bind() and listen() operations on the server are not needed on the client

All socket APIs shown in Fig. 13.5 are tabulated in Table 13.1. Avoiding lengthy discussions of some APIs and their required arguments, we list these APIs from a programmer's view. For example, in `socket()`, we directly specify the third argument as `0` because currently this is the default value commonly used for this argument.

13.3.1 Socket()—Open a Socket for Communications

The prototype of socket() is demonstrated in Table 13.2 together with its arguments and return value. A typical function call to `socket()` is:

Table 13.1 Programmer's view of socket APIs

```
int socket(int sockdomain,int socktype,0);
int bind(int sockfd,struct sockaddr *servaddr,int addrlen);
int listen(int sockfd,int backlog);
int connect(int sockfd,struct sockaddr *servaddr,int addrlen);
int accept(int sockfd,struct sockaddr *servaddr
     socklen_t *addrlen);
int send(int sockfd,const void *msg,int len, 0);
int recv(int sockfd,void *buf,int len,0);
int close(int sockfd);
```

```
int setsockopt(int sockfd,int level,int option_name,
     const void *option_value,size_t option_len);
```

```
int sendto(int sockfd,const void *msg,int length,int flags,
     const struct sockaddr *dest_addr,socklen_t destlen);
int recvfrom(int sockfd,void *restrict buf,int len,int flags,
     struct sockaddr *restrict addr,
     socklen_t *restrict addrlen);
```

Return value:

On success,	`socket()` returns a file descriptor of the socket; `accept()` returns a descriptor of the accepted socket; `send()/recv()` and `sendto()/recvfrom()` return the number of bytes sent/received; all other functions return `0`.
On error,	all functions return a value of `-1` and an error code is stored in the global variable `errno`.

```
int sockfd;
sockfd = socket(AF_INET,SOCK_STREAM,0); /* need error check */
```

This opens a socket in the Internet domain and configures the socket to use stream-oriented TCP protocol for communications. The function call returns a socket descriptor `sockfd` if successful, or `-1` otherwise. An error check is necessary in the function call to ensure the reliability of the network communication system. If an error is detected, issue a function call `exit(EXIT_FAILURE)` to exit the program.

After a socket is opened, some options associated with the socket could be manipulated through `setsockopt()` and `getsockopt()`. For `setsockopt()`, its prototype is:

```
#include <sys/socket.h>

int setsockopt(int sockfd,int level,int option_name,
const void *option_value,size_t option_len);
```

Its backward compatible version is:

Table 13.2 The prototype of `socket()`

int socket(int domain,int socktype,int protocol);		
Arguments	int sockdomain	AF_INET for IPv4
		AF_INET6 for IPv6
	int socktype	SOCK_STREAM for TCP
		SOCK_DGRAM for UDP
	int protocol	0 as the default value
Return	A socket descriptor on success, or -1 on error	

```
int setsockopt(int sockfd,int level,int option_name,
char *option_value,int option_len);
```

Upon successful completion, the function returns the value `0`. Otherwise the value `-1` is returned and the global variable `errno` is set to indicate the error.

Options can exist at multiple protocol levels, but they are always present at the uppermost "socket" level. To manipulate options at the socket level, the level is specified as SOL_SOCKET. When it comes to options, we normally need to specify that the rules used in validating addresses supplied by a `bind()` function should allow the reuse of local addresses. We should also allow more than one process to receive UDP datagrams destined for the same port. To achieve this, the `bind()` system call, which binds a process to that port, must be preceded by one or more `setsockopt()` system calls that specify these options. The option names for address reuse and port reuse are `OP_REUSEADDR` and `OP_REUSEPORT`, respectively. Thus, typical calls to `setsockopt()` are:

```
int optv = 1;
setsockopt(sockfd,SOL_SOCKET,SO_REUSEADDR,&optv,sizeof(optv));
setsockopt(sockfd,SOL_SOCKET,SO_REUSEPORT,&optv,sizeof(optv));
/* need error check in each of the above two statements */
```

As usual, an error check needs to be done for each of these function calls.

13.3.2 Bind()—Bind the Socket to an Address

The `bind()` function is used to binds the opened socket to the address (of the local machine). Its prototype is listed in Table 13.3, where the arguments and return value of `bind()` are also summarized. Before calling `bind()`, the address that the socket will be associated with needs to be formatted correctly. The data structure for server address is defined as follows in the header file `netinet/in.h` for IPv4:

```
struct sockaddr_in {
  short            sin_family;   /* e.g. AF_INET */
  unsigned short   sin_port;     /* e.g. htons(8886) */
  struct in_addr   sin_addr;     /* struct in_addr below */
```

Table 13.3 The prototype of bind()

int bind(int sockfd,struct sockaddr *servaddr, int addrlen);		
Arguments	int sockfd	returned from socket()
	struct sockaddr *	cast from struct sockaddr_in
	int addrlen	sizeof(servaddr)
Return	0 on success, or -1 on error	

```
    char              sin_zero[8];  /* zero this if you like */
  };
  struct in_addr {
    unsigned long s_addr;   /* load with inet_aton() */
  };
```

In this data structure, we set sin_family to AF_INET. The sin_port is assigned a port number through a conversion by a function call htons(), for example, htons(8886). To bind the socket to all available interfaces of the local machine, we set sin_addr.s_addr = INADDR_ANY. If we only bind the socket to the localhost, we set sin_addr.s_addr = htonl(INADDR_LOOP-BACK), or sin_addr.s_addr = inet_addr("127.0.0.1"). In summary, we set the values of the address struct members as follows:

```
sin_family =  AF_INET;
sin_port = htons(8886); /* 8886 as example */
sin_addr.s_addr = INADDR_ANY; /* all interfaces*/
/*********** or
sin_addr.s_addr = htonl(INADDR_LOOPBACK); //localhost
**** or ****
sin_addr.s_addr = inet_addr("127.0.0.1"); //localhost
************/
```

A typical call to function bind() looks like the following:

```
struct sockaddr_in servaddr;
servaddr.sin_family = AF_INET;
servaddr.sin_addr.s_addr = INADDR_ANY;
servaddr.sin_port = htons(8886); /* 8886 as example */

bind(sockfd,(struct sockaddr *)&servaddr,sizeof(servaddr));
/* need error check in the above statement */
```

This function call associates the opened socket sockfd with all available interfaces (addresses) of the local machine for TCP/IP communications. It returns 0 on success or -1 on error. Once again, an error check should be performed. If an error occurrence is detected, exit the program by calling exit(EXIT_FAILURE).

Table 13.4 The prototype of `listen()`

`int listen(int sockfd,int backlog);`		
Arguments	`int sockfd`	Bound socket
	`int backlog`	The number of pending connections allowed
Return	0 on success, or -1 on error	

13.3.3 Listen()—Set the Number of Pending Connections

After a socket is bound, the server begins listening to the incoming connection requests from clients. These requests are queued up for the server to accept. Therefore, we need to control the number of pending connections allowed in the incoming queue. This is performed through the `listen()` function. The function has the prototype shown in Table 13.4. A typical function call `listen()` looks like the following:

```
int sockfd;
/*Open, configure, and bind a socket before calling listen()*/
listen(sockfd,3); /*need error check */
```

13.3.4 Connect() from Client and Accept() from Server

The function `connect()` is used on a client to request connection to a server, while the function `accept()` is performed on a server to accept a connection request from a client. The prototypes of these functions are tabulated in Table 13.5.

After a socket is opened, a client can request a connection to a server. For this to happen, the server must be actively listening for incoming connection requests. In order for the client to connect to the server, it needs to know the IP address of the server and the port number that the server is using for the communication process. A function call `connect()` from a client typically looks like this:

Table 13.5 The prototypes of `connect()` and `accept()`

`int connect(int sockfd,struct sockaddr *servaddr,int addrlen);`		
`int accept(int sockfd,struct sockaddr *servaddr,socklen_t *addrlen);`		
	`socklen_t *addrlen);`	
Arguments	`int sock`	Socket file descriptor
	`struct sockaddr *servaddr`	Pointer to server address structure
	`int addrlen`	The length of server address
	`socklen_t *addrlen`	Pointer to the length of address
Return	0 on success, or -1 on error	

```
int sockfd;
struct sockaddr_in servaddr;

servaddr.sin_family = AF_INET;
servaddr.sin_addr.s_addr = inet_addr("127.0.0.1"); /*testing*/
servaddr.sin_port = htons(8886);          /* 8886 as example */

sockfd = socket(); /* need error check in socket()/connect() */
connect(sockfd, (struct sockaddr *)&servaddr, sizeof(servaddr));
```

Upon receiving an incoming connection request from a client, the server accepts the request by using the `accept()` function call. If the `accept()` operation is successful, a connection is established with the client. At this point, both the server and the client are ready to communicate with each other by sending and/or receiving data. A typical `accept()` operation is given below:

```
int sockfd, acptdsock;
struct sockaddr_in servaddr;

servaddr.sin_family = AF_INET;
servaddr.sin_addr.s_addr = INADDR_ANY;
servaddr.sin_port = htons(8886);          /* 8886 as example */

acptdsock = accept(sockfd,(struct sockaddr *)&servaddr,
                   (socklen_t*)&addrlen); /*need error check*/
```

13.3.5 send()/recv()-Send/Receive Data

Sending and receiving data are performed using `send()` and `recv()` function calls, respectively. The prototypes of these two functions are listed in Table 13.6. The `int sockfd` argument in both `send()` and `recv()` specifies the socket that has been established between two hosts through the `connect()` and `accept()` functions.

The `send()` function sends the message indicated by the `const void *msg` argument, up to the length specified by the `int len` argument. However, the actual length of the message pointed to by *msg may be longer or shorter.

Table 13.6 The prototypes of `send()` and `recv()`

`int send(int sockfd,const void *msg, int len,0);`		
`int recv(int sockfd,void *buf, int len, 0);`		
Arguments	`int sockfd`	Connected socket
	`const void *mgs`	Message to be sent
	`void *buf`	Buffer to store message
	`int len`	Length of message
Return	The number of bytes sent/received on success, or `-1` on error	

On the other hand, the `recv()` function stores the received message in the `void *buf` argument, up to the length specified by the `int len` argument. Similar to `send()`, the length of the received message may be larger or smaller. This approach helps avoid buffer overflow issues, which can be risky in practical system applications.

For both the `send()` and `recv()` functions, the number of bytes sent or received is returned upon successful completion. In case of an error, a value of `-1` is returned and the global integer variable `errno` is set to record the error code.

Typical function calls to `send()` and `recv()` are shown below:

```
int sockfd, acptdsock;
char buf[1024] = {0}, *hello = "Hello from server";
sockfd = socket();                          /* need error check */
/*insert bind() and listen() here ...*/

acptdsock = accept(sockfd,(struct sockaddr *)&servaddr,
                   (socklen_t*)&addrlen); /* need error check */
send(acptdsock,hello,strlen(hello),0);  /* need error check */
recv(acptdsock,buf,1023,0);             /* need error check */
```

For these function calls demonstrated above, error checking should be performed for system reliability in the development of practical applications.

It is worth mentioning that in our programming practice, we have observed that instead of using the `send()` and `recv()` functions, the `write()` and `read()` functions can also be used to respectively send (write) messages to, and receive (read) messages from, the socket using the socket file descriptor. As we understand, the `write()` and `read()` functions are typically used for writing to, and reading from, files, respectively. Since an opened socket is treated as a file with a file descriptor, it is natural to use the `write()` and `read()` operations on the socket (file) by using the `write()` and `read()` functions. The prototypes of the `write()` and `read()` functions are as follows:

```
int write (int fd, const void *msg, int len);
int read(int fd,void *buf, int len);
```

In these prototypes, the argument `int fd` indicates a file descriptor, which corresponds to the socket file descriptor when dealing with a socket. It is seen that the `write()` and `read()` functions have one fewer argument compared to `send()` and `recv()`. This implies that `send()` and `recv()` provide more control over the communication process than `write()` and `read()`. But in normal applications, the additional control provided by setting a flag in the last argument is not essential. In such cases, the flag is simply set to 0. Similar to the `send()` and `recv()` functions, the `write()` and `read()` functions also return the number of bytes sent or received upon successful completion, or return `-1` on error. In case of an error, the global integer variable `errno` is set to record the specific error code.

Table 13.7 The prototype of `close()`

`int close(int sockfd);`		
Argument	`int sockfd`	Socket file descriptor
Return	An integer	`0` on success, or `-1` on error

Fig. 13.6 A simple client-server network system

13.3.6 close()—Close Socket

The `close()` function simply deactivates and deletes a socket descriptor. Upon successful completion, a value of `0` is returned. Otherwise, a value of `-1` is returned and the global integer variable `errno` is set to indicate the error code. The prototype of the `close()` function is shown in Table 13.7.

13.4 Example Server-Client Programs

Having gained an understanding of client-server network applications and the knowledge of socket APIs for TCP/IP networks from the previous sections, we are now ready to design, develop, and implement a simple client-server network communication system. This section will present a minimum working program for a server and a minimum working program for a client. These programs will be extended later to incorporate additional functionality.

13.4.1 System Specifications and Requirements

A simple client-server network system is illustrated in Fig. 13.6. It consists of a single server and a single client, which communicate with each other over a TCP/IP network. The network can be one of the three scenarios discussed previously: a LAN, across a router, or over the Internet. As we have learned, socket programming remains the same for all these scenarios, so we do not specify the interconnections of the TCP/IP network in Fig. 13.6.

The communications between the server and client in a client-server network system follow the high-level logic flow depicted in Fig. 13.4. For socket programming in C, the low-level flowchart shown in Fig. 13.5 is followed. In order for a client-server network system to be truly useful, additional tasks are typically performed concurrently or independently alongside the communications. Let us outline a few simple tasks in addition to the TCP/IP communications:

(1) The server and client periodically perform the following operations:

  (1a) The server sends a hello message to the client, which can be used for handshaking or requesting data from the client.
  (1b) Upon receiving a hello message, the client sends an ACK message to the server. The ACK message can be embedded with data if requested by the server.
  (1c) Both the server and client display information about the communications.

(2) When the server no longer requires data from the client, it notifies the client of the termination of communications by sending a single character command, such as $0 \times 1b$, which represents the ASCII code for the ESC key on a keyboard. The server then closes its socket and terminates the server program.
(3) When the client receives a character command, such as $0 \times 1b$, from the server, it closes its socket and terminates the client program.

13.4.2 System Design

Where to implement these tasks in addition to TCP/IP communications? They are mostly embedded into the Send/Receive loop depicted in Fig. 13.4. In C socket programming, they can be incorporated into the `send()/recv()` loop illustrated in Fig. 13.5.

The timelines of these tasks and the TCP/IP communications are shown in Fig. 13.7. However, it is important to note that for different system requirements and specifications, the timelines of the TCP/IP communications and other tasks may need to be designed differently to meet those specific requirements.

13.4.3 System Implementation

With the system specifications and design discussed in the previous section, a client-server network communication system has been implemented using socket programming in C. Below is the server program:

```
/* server.c for a TCP server */

#include <unistd.h>
```

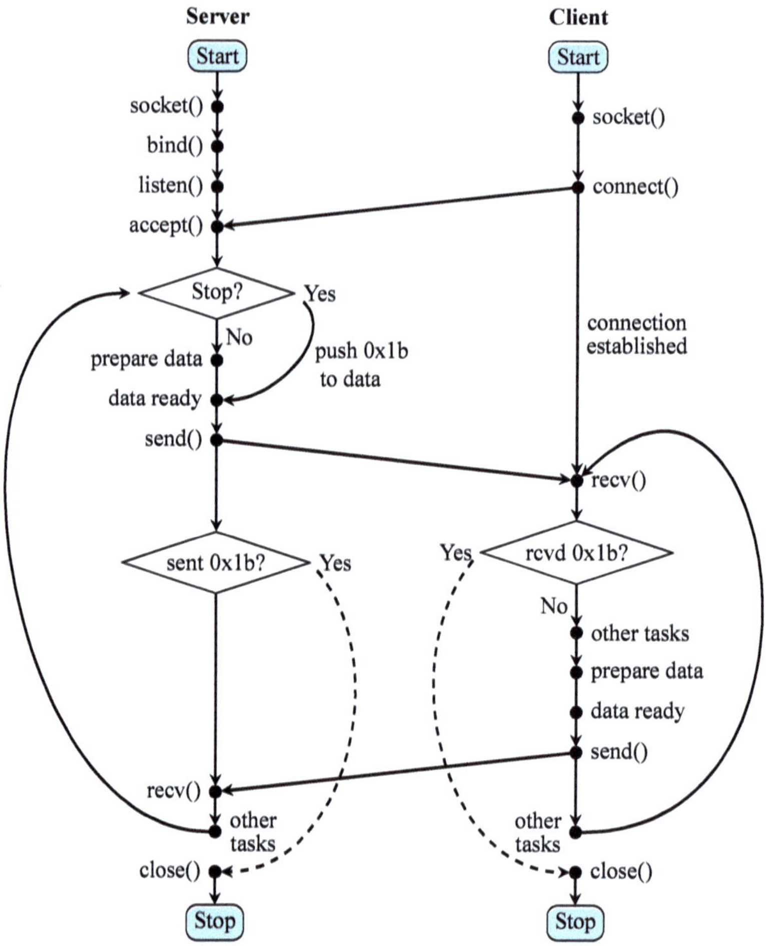

Fig. 13.7 Timelines of TCP/IP communications and other tasks

```
#include <stdio.h>
#include <stdlib.h>
#include <string.h>
#include <sys/socket.h>
#include <netinet/in.h>
#define PORT 8886     /* port number */
#define BUFLEN 1024   /* buffer length */
#define PERIOD 1      /* in seconds */
#define LOOPLIMIT 3   /* loop testing send()/recv() */
#define QUITKEY 0x1b /* ASCII code of ESC */

int main(int argc, char const *argv[]){
  int sockfd, acptdsock, optv=1,i=0;
  struct sockaddr_in servaddr;
  int addrlen = sizeof(servaddr);
  char buffer[BUFLEN] = {0};
```

```
    char *hello = "Hello from server";
    char cmd = QUITKEY; /* character ESC */

    if ((sockfd = socket(AF_INET,SOCK_STREAM,0)) == -1){
      perror("socket failed");
      exit(EXIT_FAILURE);
    }

    if (setsockopt(sockfd,SOL_SOCKET,SO_REUSEADDR,
                   &optv,sizeof(optv))){
      perror("setsockopt SO_REUSEADDR failed");
      exit(EXIT_FAILURE);
    }
    if (setsockopt(sockfd,SOL_SOCKET,SO_REUSEPORT,
                   &optv,sizeof(optv))){
      perror("setsockopt SO_REUSEPORT failed");
      exit(EXIT_FAILURE);
    }

    servaddr.sin_family = AF_INET;
    servaddr.sin_addr.s_addr = INADDR_ANY;
    servaddr.sin_port = htons(PORT);

    if (bind(sockfd,(struct sockaddr *)&servaddr,
             sizeof(servaddr)) == -1){
      perror("bind failed");
      close(sockfd);
      exit(EXIT_FAILURE);
    }

    if (listen(sockfd,3) == -1){
      perror("listen failed");
      exit(EXIT_FAILURE);
    }

    printf("\nServer waiting for connection request........\n");
    if ((acptdsock=accept(sockfd,(struct sockaddr *)&servaddr,
                          (socklen_t*)&addrlen)) == -1){
      perror("accept() tried but not succeded, keep trying...");
      exit(EXIT_FAILURE);
    }

    while (1) { /* loop for send()/recv() */
      if ((send(acptdsock, hello, strlen(hello) , 0 )) == -1){
        perror("send() failed ");
        close(sockfd);
        exit(EXIT_FAILURE);
      }
      printf("%2d Server sent:     %s\n",i,hello);

      if ((recv(acptdsock,buffer,BUFLEN-1,0)) == -1)
        perror("recv() failed ");
      buffer[BUFLEN-1]=0x00;  /* force ending with '\0' */
      printf("   Server received: %s\n",buffer);

      if ((++i) == LOOPLIMIT) /* LPLIMIT reached */
        break;

      sleep(PERIOD);/* unsigned int sleep(unsigned int seconds)*/
    }                /* end of while loop */
```

```
    if ((write(acptdsock, &cmd, 1)) == -1){ /* write() works */
      perror("send() failed ");
      close(sockfd);
      exit(EXIT_FAILURE);
    }

    if (close(sockfd)== -1){
      perror("close socket failed ");
      exit(EXIT_FAILURE);
    }
    printf(".......Server retuned with success!\n\n");
    return 0; /* with success */
}
```

As for the client, the IP address and port number of the server need to be specified in the client program. To make system testing easier, the loopback IP address of the local host, i.e., 127.0.0.1, has been used in the example code to be presented below. This allows to test both the server and client programs on the same machine. If the server and client programs are executed on two different hosts, the loopback IP address in this code should be changed to the IP address of the host on which the server program is executed.

A complete C program for the client is presented in the following:

```
        /* client.c for a TCP client */

#include <unistd.h>
#include <stdio.h>
#include <stdlib.h>
#include <string.h>
#include <sys/socket.h>
#include <arpa/inet.h>
#define SERVIPADDRESS "127.0.0.1" /* loopback for testing */
#define SERVPORT 8886             /* port number */
#define BUFLEN 1024               /* buffer length */
#define QUITKEY 0x1b              /* ASCII code of ESC */

int main(int argc, char const *argv[]){
      int sockfd = 0, i=0;
      struct sockaddr_in serv_addr;
      char buffer[BUFLEN] = {0};
      char *ackmsg = "ACK from client";

      if ((sockfd = socket(AF_INET,SOCK_STREAM,0)) == -1) {
        perror("Socket creation error \n");
        exit(EXIT_FAILURE);
      }

      serv_addr.sin_family = AF_INET;
      serv_addr.sin_addr.s_addr=inet_addr(SERVIPADDRESS);
      serv_addr.sin_port = htons(SERVPORT);

      printf("\nClient conneting to Server.......\n");
      if (connect(sockfd,(struct sockaddr *)&serv_addr,
                  sizeof(serv_addr)) == -1){
        perror("Server started? Connection failed \n");
        exit(EXIT_FAILURE);
      }
```

```
        printf(".......Connection established.\n");

        while (1) { /* loop for send()/recv() */
          if ((recv(sockfd,buffer,BUFLEN-1,0)) == -1)
            perror("recv() failed ");
          buffer[BUFLEN-1] = 0x00; /* force ending with '\0' */
          if (buffer[0] == QUITKEY)     /* prepare termination */
            break;
          printf("%2d Client received: %s\n",i++,buffer );

          if ((send(sockfd,ackmsg,strlen(ackmsg),0)) == -1){
            perror("send failed ");
            close(sockfd);
            exit(EXIT_FAILURE);
          }
          printf("    Client sent:      %s\n", ackmsg);
        } /* end of while loop */

        if (close(sockfd) == -1){
          perror("close socket failed ");
          exit(EXIT_FAILURE);
        }
        printf(".......Client retuned with success!\n\n");
        return 0; /* with success */
}
```

In Linux or Mac OS, open a terminal window and compile the server program by using gcc compiler. Then, run the resulting executable. The server starts to run, and listens to the network for connection requests from clients.

Open another terminal window and compile the client program. After that, execute the compiled file. The client will open a socket, requests to the server for connection. Then, the client and server will keep communicating with each other to perform the specified functions.

When the server is ready to terminate the socket communications, it sends a single character command 0x1b, i.e., ESC, to the client. Then, the server and client will close their respective sockets and terminate the system execution.

The processes of the compilation and execution of the server and client programs are depicted in Fig. 13.8, in which the dollar sign $ in the command lines is the Mac OS prompt. Note that in the current program design, in order for the client to connect to the server correctly, the server should be executed before the client starts to run.

13.5 IPv6 Sockets

Conceptually, socket programming remains the same for both IPv4 and IPv6 networks. However, due to the differences in address formats between IPv6 and IPv4, there are variations in the data structures and socket APIs used for these two versions. Additionally, certain changes are required for address conversion and socket options to support IPv6 in socket APIs.

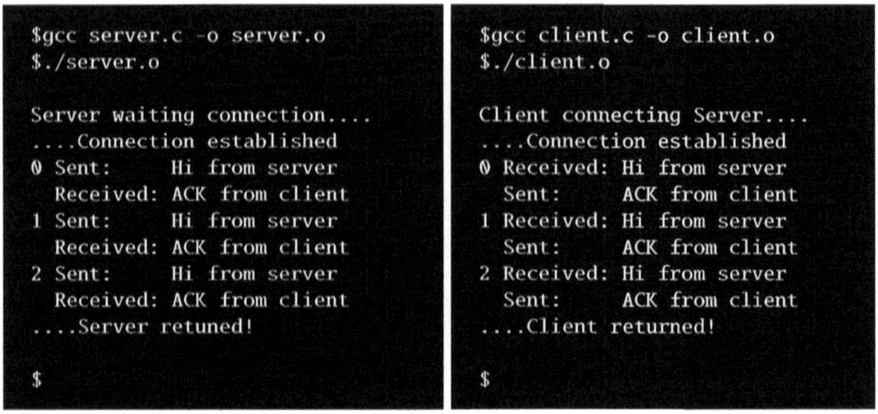

(a) Server (b) Client

Fig. 13.8 Screenshots of server and client operations on Mac OS

The basic socket interface extensions for IPv6 are documented in the IETF RFC 3493 [3], which is the successor to RFC 2133 and RFC 2553. These extensions outline the necessary modifications to support IPv6 in socket programming.

For more advanced features of the socket APIs for IPv6, such as raw sockets and header configuration, the IETF RFC 3542 provides a comprehensive summary [4]. Initially published as RFC 2292, this document delves into the intricacies of using these advanced features in IPv6 socket programming.

13.5.1 *Changes in IPv6 Sockets from IPv4*

The data structure of the server address discussed in Sect. 13.3.2 for IPv4 now becomes the following format for IPv6:

```
struct sockaddr_in6 {
  sa_family_t     sin6_family;        /* AF_INET6 */
  in_port_t       sin6_port;          /* e.g., htons(8886) */
  uint32_t        sin6_flowinfo;
  struct in6_addr sin6_addr; /* IPv6 address struct below */
  uint32_t        sin6_scope_id;
};
struct in6_addr {
  uint8_t s6_addr[16];            /* load with inet_pton() */
};
```

In this data structure, we set `sin6_family = AF_INET6`. The IP address of the server can be set as `sin6_addr = in6addr_any` for all available interfaces. If the server's socket is bound to the localhost, this can be explicitly set through `sin6_addr = in6addr_loopback`. As the loopback IP address in

IPv6 is `::1/128`, setting the loopback IP address in IPv6 can also be done through `inet_pton(AF_INET6, "::1", &sin6_addr.in6_addr)`. Note that `inet_addr()` is for IPv4 only and cannot be used to set an IPv6 address.

With the new data structure for IPv6 socket programming, all API functions that use an IP address should use this new data structure. For example, the following function calls are used to create a socket for TCP and UDP communications, respectively:

```
socket(AF_INET6, SOCK_STREAM, 0); /* for TCP */
socket(AF_INET6, SOCK_DGRAM, 0);  /* for UDP */
```

Once an application has created a socket of `AF_INET6` type, it must use the `sockaddr_in6` address structure when passing addresses to the system. The functions that the application uses to pass addresses into the system include `bind()`, `connect()`, and `send()`. The system will also use the `sockaddr_in6` address structure to return addresses to applications that are using `AF_INET6` sockets. The functions that return an address from the system to an application include `accept()`, `recvfrom()`, `recvmsg()`, `getpeername()`, and `getsockname()`.

For IPv4 socket programming, two functions, `inet_addr()` and `inet_ntoa()`, are defined to convert an IPv4 address between binary and text forms. IPv6 applications need similar functions. The following two functions are defined in `arpa/inet.h` to convert both IPv6 and IPv4 addresses:

```
int inet_pton(int addr_family, const char *src, void *dst);
const char *inet_ntop(int addr_family, const void *src,
                      char *dst, socklen_t size);
```

13.5.2 Example IPv6 Socket Programs

As an example, here is an IPv6 version of the server program implemented in the previous section:

```
/* server6.c for TCP in IPv6 */

#include <unistd.h>
#include <stdio.h>
#include <stdlib.h>
#include <string.h>
#include <sys/socket.h>
#include <netinet/in.h>
#include <arpa/inet.h>
#define PORT 8886     /* port number */
#define BUFLEN 1024   /* buffer length */
#define PERIOD 1      /* in seconds */
#define LOOPLIMIT 3   /* loop testing send()/recv() */
#define QUITKEY 0x1b /* ASCII code of ESC */
```

```
int main(int argc, char const *argv[]){
  int sockfd, acptdsock, optv=1,i=0;
  struct sockaddr_in6 servaddr;
  int addrlen = sizeof(servaddr);
  char buffer[BUFLEN] = {0};
  char *hello = "Hi from server";
  char cmd = QUITKEY; /* character ESC */

  if ((sockfd = socket(AF_INET6,SOCK_STREAM,0)) == -1){
    perror("socket failed");
    exit(EXIT_FAILURE);
  }

  if (setsockopt(sockfd,SOL_SOCKET,SO_REUSEADDR,
        &optv,sizeof(optv))){
    perror("setsockopt SO_REUSEADDR failed");
    exit(EXIT_FAILURE);
  }
  if (setsockopt(sockfd,SOL_SOCKET,SO_REUSEPORT,
          &optv,sizeof(optv))){
    perror("setsockopt SO_REUSEPORT failed");
    exit(EXIT_FAILURE);
  }

  servaddr.sin6_family = AF_INET6;
  servaddr.sin6_addr = in6addr_any;
  servaddr.sin6_port = htons(PORT);
  / *servaddr.sin6_flowinfo = 0;  */
  / *servaddr.sin6_scope_id = 0; */

  if (bind(sockfd,(struct sockaddr *)&servaddr,
        sizeof(servaddr)) == -1){
    perror("bind failed");
    close(sockfd);
    exit(EXIT_FAILURE);
  }

  if (listen(sockfd,3) == -1){
    perror("listen failed");
    exit(EXIT_FAILURE);
  }

  printf("\nServer waiting connection....\n");
  if ((acptdsock=accept(sockfd,(struct sockaddr *)&servaddr,
        (socklen_t*)&addrlen)) == -1){
    perror("accept() tried but not succeded, keep trying...");
    exit(EXIT_FAILURE);
  }

  while (1) { /* loop for send()/recv() */
    if ((send(acptdsock, hello, strlen(hello) , 0 )) == -1){
      perror("send() failed ");
      close(sockfd);
      exit(EXIT_FAILURE);
    }
    printf("%2d Sent:     %s\n",i,hello);

    if ((recv(acptdsock,buffer,BUFLEN-1,0)) == -1)
      perror("recv() failed ");
    buffer[BUFLEN-1]=0x00;  /* force ending with '\0' */
```

```
    printf("   Received: %s\n",buffer);

    if ((++i) == LOOPLIMIT) /* LPLIMIT reached */
      break;

    sleep(PERIOD);/* unsigned int sleep(unsigned int seconds)*/
  }               /* end of while loop */

  if ((write(acptdsock, &cmd, 1)) == -1){ /* write() works */
    perror("send() failed ");
    close(sockfd);
    exit(EXIT_FAILURE);
  }

  if (close(sockfd)== -1){
    perror("close socket failed ");
    exit(EXIT_FAILURE);
  }
  printf("....Server retuned!\n\n");
  return 0; /* with success */
}
```

For communicating with the IPv6 server program, here is an IPv6 implementation of the client program designed in the previous section:

```
/* client6.c for a TCP client in IPv6*/

#include <unistd.h>
#include <stdio.h>
#include <stdlib.h>
#include <string.h>
#include <sys/socket.h>
#include <arpa/inet.h>
#define SERVIPADDRESS "::1"  /* loopback for testing */
#define SERVPORT 8886        /* port number */
#define BUFLEN 1024          /* buffer length */
#define QUITKEY 0x1b         /* ASCII code of ESC */

int main(int argc, char const *argv[]){
  int sockfd = 0, i=0;
  struct sockaddr_in6 serv_addr;
  char buffer[BUFLEN] = {0};
  char *ackmsg = "ACK from client";

  if ((sockfd = socket(AF_INET6,SOCK_STREAM,0)) == -1) {
    perror("Socket creation error \n");
    exit(EXIT_FAILURE);
  }

  serv_addr.sin6_family = AF_INET6;
  inet_pton(AF_INET6,SERVIPADDRESS,&serv_addr.sin6_addr);
  serv_addr.sin6_port = htons(SERVPORT);

  printf("\nClient connecting Server....\n");
  if (connect(sockfd,(struct sockaddr *)&serv_addr,
              sizeof(serv_addr)) == -1){
```

```
      perror("Server started? Connection failed \n");
      exit(EXIT_FAILURE);
    }
    printf("....Connection established\n");

    while (1) { /* loop for send()/recv() */
      if ((recv(sockfd,buffer,BUFLEN-1,0)) == -1)
        perror("recv() failed ");
      buffer[BUFLEN-1] = 0x00; /* force ending with '\0' */
      if (buffer[0] == QUITKEY)     /* prepare termination */
        break;
      printf("%2d Received: %s\n",i++,buffer );

      if ((send(sockfd,ackmsg,strlen(ackmsg),0)) == -1){
        perror("send failed ");
        close(sockfd);
        exit(EXIT_FAILURE);
      }
      printf("   Sent:      %s\n", ackmsg);
    } /* end of while loop */

    if (close(sockfd) == -1){
      perror("close socket failed ");
      exit(EXIT_FAILURE);
      }
    printf("....Client retuned!\n\n");
    return 0; /* with success */
}
```

In Linux and Mac OS, use `gcc` compiler to compile the IPv6 server and client programs. The commands for Mac OS are as follows:

```
$gcc server6.c -o server6.o
$gcc client6.c -o client6.o
```

After compiling, execute the executable files `server6.o` and `client6.o` to test the IPv6 client-server network communications. Since the loopback IP address has been used for the server for easy testing on the same machine, run `server6.o` and `client6.o` in two separate terminal windows on the same machine. The information displayed in these two terminal windows will be similar to that shown in Fig. 13.8.

To test the server and client programs on two different physical machines, replace the server IP address in the client program with the actual IP address of the server.

In comparison with the IPv4 server and client programs presented in the previous section, the only changes in the IPv6 server and client programs are the address format and data structures of related variables and function calls. The names of the API functions defined in IPv4 remain the same in IPv6 socket programming. Therefore, transitioning from IPv4 to IPv6 for socket programming is relatively straightforward.

13.6 Keyboard Input Processing

Keyboard input is often used to control program execution. For example, input a command from the keyboard to terminate a client-server network communication system. It is important to ensure that keyboard processing does not disrupt normal tasks such as network communications. The previously discussed server and client programs do not have any keyboard input processing tasks. This section will show how to add keyboard processing to these example client and server programs.

Let us consider a simple keyboard input processing scenario. While a client-server network communication system is running, input characters from the keyboard on the server side. Use the ESC key (ASCII code `0x1b`) as a command to instruct the server to terminate the execution of the client-server system. Therefore, the server will scan its keyboard input and ignore all input characters except the ESC key. When the server reads the ESC key, it will notify the client of the system termination, close the socket, and terminate the server program. Upon receiving the notification of system termination from the server, the client will close the socket and stop the execution of the client program.

While this scenario is simple, it sufficiently demonstrates a general guideline for dealing with keyboard input. Standard functions in C for keyboard input wait for keyboard input or keep reading characters from the keyboard until a newline character is read. They will hang indefinitely if there is no keyboard input or if a newline character is not read. Therefore, a method needs to be designed and implemented to check if a key is pressed: if yes, read it and continue; if no, continue without delay. Meanwhile, once the ESC key is read, the system should be terminated after notifying the client of the system termination.

There are different ways to handle such a keyboard input scenario. For example, multithreading is an effective but complicated method. It is also possible to use multiple processes to handle multiple tasks. In this section, we will design and implement two methods: a simple one and a multithreading one.

13.6.1 Simple Keyboard Input Processing

Let us begin with a simple method, although its efficiency needs careful consideration and design. The basic idea is to design a function that can detect whether a key is pressed. If a key is pressed, read it; otherwise leave there immediately. Borrowing the name of the function `kbhit()` from Windows, this section implements a Linux version of `kbhit()` as shown below:

```
#include <stdio.h>
#include <stdlib.h>
#include <unistd.h>
#include <termios.h>
#include <sys/ioctl.h>
```

```
#include <stdbool.h>

int kbhit(void) {
      static bool initflag = false;
      static const int STDIN = 0;

      if (!initflag) {
            struct termios term;
            tcgetattr(STDIN,&term);
            term.c_lflag &= ~ICANON;
            tcsetattr(STDIN,TCSANOW,&term);
            setbuf(stdin,NULL);
            initflag = true;
      }
      int nbbytes;
      ioctl(STDIN,FIONREAD,&nbbytes);
      return nbbytes;
}
```

Place this function right above the `main()` function of the server program. Note that the library header files included in this code snippet should be merged with the existing ones in the server program.

With the `kbhit()` function developed above, two slight modifications to the IPv4 and IPv6 server programs will meet the aforementioned requirements:

(1) Introduce a boolean variable as a flag to indicate whether or not it is ready to terminate the system. Initialized it to `false`.
(2) Add a code segment to check keyboard input. If the ESC key is read from keyboard, set the boolean flag to `true` and terminate the client-server system.

In this example, the boolean variable is named `stop`:

```
bool stop = false;
```

Place it in the variable declaration segment of the `main()` function of the server program.

The code segment for keyboard input processing is implemented as follows:

```
while ((kbhit()) && (!stop)) {
  cmd = getchar();
  fflush(stdout);
  if (cmd == QUITKEY)
    stop = true;
}
if (stop)
  break;
```

Insert this segment of code right above the statement of the `sleep()` function call in the server program.

No other changes are needed to the server program. Also, no any changes are required to the client program either. Compile both server and client programs, and execute them for periodic client-server TCP/IP communications. The execution can be terminated by issuing an ESC command from the keyboard of the server.

13.6.2 Multithreading Keyboard Input Processing

Multithreading is an advanced level of programming. It allows for the creation of multiple threads within a process. These threads can execute independently and concurrently, sharing process resources. When the logic flows of a process become too complicated in a single-threaded program, using multiple concurrent threads can simplify the design. The overall computing efficiency of multithreading systems depends on how much the multiple threads interfere with each other. While a comprehensive discussion of multithreading programming is beyond the scope of this book, this section will show how to employ the multithreading technique to handle keyboard input processing in socket programming.

Our basic idea is to create two threads: one for the periodic socket communications and the other for sporadic keyboard input processing. The `main()` function simply create these two threads and let them run concurrently. When an ESC key is read from the keyboard in the keyboard thread, a Boolean stop flag will be passed to the socket thread. The socket thread will use this flag to perform the following tasks: (1) notify the client of system termination, (2) terminate the server, and (3) exit from the socket thread. Once the stop flag is passed to the socket thread, the keyboard thread will also exit. As a result, the entire client-server system will return with success.

The stop flag is a simple flag that is only altered in two cases: (1) when a keyboard input ESC is captured in the keyboard thread, and (2) when the loop limit is reached in the socket thread. No other locations change the value of the stop flag. Therefore, the stop flag can be declared as a global variable without much risk. The loop limit in the example programs is for demonstration and testing purposes only. Otherwise, when you have to use Ctrl-C or Ctrl-Z to terminate your program, some system settings may not be restored correctly or some system resources may not be reclaimed. In real application systems, such a loop limit does not exist, especially in embedded systems that run continuously and forever until power supply is lost or specific conditions provided by end users are met.

Multithreading functions and data structures are defined in the header file `pthread.h`. Include this header file in the server program. A thread is declared as a `pthread_t` type and is created in the `main()` function using the `pthread_create()` function call. After its creation, a thread can be joined using `pthread_join()` or detached using `pthread_detach()`, depending on how the results from the thread are used. When `pthread_join()` is called from another thread, which is usually the thread that created it, the calling thread will be blocked until the joined thread terminates and returns a value. In comparison, `pthread_detach()` can be called from the thread itself or another thread. It does not block the calling thread, indicating that the calling thread does not require the return value of the detached thread or has no ability to wait for the detached thread to finish. In our scenario, the socket thread and the keyboard thread should not mutually block each other. Therefore, `pthread_detach()` is called in the `main()` function to detach both threads. After both threads are detached, `pthread_exit(NULL)` is called to keep these two threads running until they terminate.

The complete multithreading code for the server with keyboard input processing is given below:

```
/* server_kbd_mt.c, multithreading TCP server with kbd input */

#include <unistd.h>
#include <stdio.h>
#include <stdlib.h>
#include <string.h>
#include <termios.h>
#include <stdbool.h>
#include <pthread.h>
#include <netinet/in.h>
#include <sys/ioctl.h>
#include <sys/socket.h>
#define PORT 8886       /* port number */
#define BUFLEN 1024     /* buffer length */
#define PERIOD 1        /* in seconds */
#define LOOPLIMIT 10    /* loop testing send()/recv() */
#define QUITKEY 0x1b    /* ASCII code of ESC */

/**** bool stop: a global flag for system termination ****/
bool stop = false;/* change via kbd ESC or looplimit reached */

/******** kbhit() utility ********/
int kbhit(void) {
      static bool initflag = false;
      static const int STDIN = 0;

      if (!initflag) {
         struct termios term;
            tcgetattr(STDIN,&term);
            term.c_lflag &= ~ICANON;
            tcsetattr(STDIN,TCSANOW,&term);
            setbuf(stdin,NULL);
        initflag = true;
      }
      int nbbytes;
      ioctl(STDIN,FIONREAD,&nbbytes);
      return nbbytes;
}

/******** keyboard input processing thread ********/
void *run_thread_kbd(void *vargp){
      char cmd = QUITKEY;     /* character ESC */

      printf("thread_kbd started...\n");
```

```
	while (1){
		fflush(stdout);
		while ((kbhit()) && (!stop)) {
			cmd = getchar();
			fflush(stdout);
			if (cmd == QUITKEY){
			    printf("\n");/* for better terminal looking */
			    stop = true;
			}
		}
	    if (stop)
	      break;
	    sleep(1);
	}
	printf("...thread_kbd returned\n");
	return NULL; /* thread ends */
}

/******** periodic socket commun. thread ********/
void *run_thread_sock(void *vargp){
	int sockfd, acptdsock, optv=1,i=0;
	struct sockaddr_in servaddr;
	int addrlen = sizeof(servaddr);
	char buffer[BUFLEN] = {0};
	char *hello = "Hi from server";
	char cmd = QUITKEY;   /* character ESC */

	printf("thread_socket started...\n");
	if ((sockfd = socket(AF_INET,SOCK_STREAM,0)) == -1){
		perror("socket failed");
		exit(EXIT_FAILURE);
	}

	if (setsockopt(sockfd,SOL_SOCKET,SO_REUSEADDR,
			&optv,sizeof(optv))){
		perror("setsockopt SO_REUSEADDR failed");
		exit(EXIT_FAILURE);
	}
	if (setsockopt(sockfd,SOL_SOCKET,SO_REUSEPORT,
			&optv,sizeof(optv))){
		perror("setsockopt SO_REUSEPORT failed");
		exit(EXIT_FAILURE);
	}

	servaddr.sin_family = AF_INET;
	servaddr.sin_addr.s_addr = INADDR_ANY;
	servaddr.sin_port = htons(PORT);

	if (bind(sockfd,(struct sockaddr *)&servaddr,
		sizeof(servaddr)) == -1){
		perror("bind failed");
		close(sockfd);
		exit(EXIT_FAILURE);
	}

	if (listen(sockfd,3) == -1){
		perror("listen failed");
		exit(EXIT_FAILURE);
	}
```

```
	printf("\nServer waiting connection....\n");
	if ((acptdsock=accept(sockfd,(struct sockaddr *)&servaddr,
				(socklen_t*)&addrlen)) == -1){
		perror("accept() tried but not succeded, keep trying...");
		exit(EXIT_FAILURE);
	}

	while (1) { /* loop for send()/recv() */
		if ((send(acptdsock, hello, strlen(hello) , 0 )) == -1){
			perror("send() failed ");
			close(sockfd);
			exit(EXIT_FAILURE);
		}
		printf("%2d Sent:     %s\n",i,hello);

		if ((recv(acptdsock,buffer,BUFLEN-1,0)) == -1)
		  perror("recv() failed ");
		buffer[BUFLEN-1]=0x00;  /* force ending with '\0' */
		printf("   Received: %s\n",buffer);

		if ((++i) == LOOPLIMIT)          /* LOOPLIMIT reached  */
		  stop = true; /* make sure i within the integer limit */

		if (stop)       /* stop may also be changed in thread_kbd */
		  break;

		sleep(PERIOD);/* unsigned int sleep(unsigned int seconds) */
	}               /* end of while loop */

	if ((write(acptdsock, &cmd, 1)) == -1){ /* write() works */
		perror("send() failed ");
		close(sockfd);
		exit(EXIT_FAILURE);
	}

	if (close(sockfd)== -1){
		perror("close socket failed ");
		exit(EXIT_FAILURE);
	}
	printf("....Server retuned!\n\n");
	return NULL; /* thread ends */
}

/******** main() function ********/
int main(int argc, char const *argv[]){
	char cmd = QUITKEY;    /* character ESC */
	pthread_t thread_sock,thread_kbd;

	if (!pthread_create(&thread_sock,NULL,run_thread_sock,NULL))
	  pthread_detach(thread_sock);
	if (!pthread_create(&thread_kbd,NULL,run_thread_kbd,NULL))
	  pthread_detach(thread_kbd);

	pthread_exit(NULL);
	return 0;
}
```

13.7 Socket Programming in Windows

Socket in Windows is known as Winsock. Comprehensive discussions on Winsock can be found in Microsoft's documentation entitled Windows Socket 2 [5]. This documentation includes a section with the heading of Porting Socket Applications to Winsock [6], which provides valuable information to assist Linux/Unix users to port their Linux socket programs to Windows socket programs.

13.7.1 Comparisons Between Winsock and Linux Sockets

Before delving into socket programming in Windows, it is worth indicating clearly that the logical flows and main components of socket programming remain the same for both Linux and Windows, as illustrated previously in Figs. 13.4, 13.5, and 13.7. However, there are differences between Linux and Windows in terms of data structures, initialization, and clean-up of sockets. Table 13.8 provides a summary of Winsock components, highlighting the differences and similarities compared to Linux sockets.

13.7.2 Example Code for Winsock Server and Client

Now, let us design and implement the Windows versions of the socket server and client programs shown in Sect. 13.6.1. A server program in Windows is provided below:

Table 13.8 Summary of sockets components in Windows

Item	Syntax	Example
Header file	`winsock2.h`	`#include <winsock2.h>`
Library	ws2_32.lib	`#pragma comment(lib,"ws2_32.lib")`
Socket type	`SOCKET`	`SOCKET sock, new_sock;`
Data type	`WSADATA`	`WSADATA wsa;`
Macros	`SOCKET_ERROR, INVALID_SOCKET`	
Get error	`WSAGetLastError()`	
Initialization	`WSAStartup()`	`WSAStartup(MAKEWORD(2,2),&wsa);`
Create socket	`socket()`	`socket(AF_INET,SOCK_STREAM,0);`
close socket*	`closesocket()`	`closesocket(sock);`
	`WSACleanup()`	`WSACleanup();`
	*Both functions must be called together to close socket	
Other APIs	`bind(), listen(), connect(), accept(), send(), sendto(), recv(), recvfrom()`, etc., are the same as, or similar to, those in Linux	

```
/**** serverWin.c, Winsock server in Windows ****/

#define _WINSOCK_DEPRECATED_NO_WARNINGS
#define _CRT_NONSTDC_NO_WARNINGS
#define _CRT_SECURE_NO_WARNINGS

#include<stdio.h>
#include<stdlib.h>
#include <windows.h>    /* winsock2.h included in windows.h */
#include <stdbool.h>

#define SERVER_IP_ADDR "127.0.0.1" /*loopback for testing */
#define PORT   8886  /* port number */
#define BUFLEN 1024  /* buffer length */
#define PERIOD 1000  /* in milliseconds */
#define LOOPLIMIT 8  /* loop testing send()/recv() */
#define QUITKEY 0x1b /* ASCII code of ESC */

#pragma comment(lib,"ws2_32.lib")

int main(void){
      WSADATA wsa_data;              /* type defined in winsock2.h */
      SOCKET sockfd, acptdsock;      /* type defined in winsock2.h */
      struct sockaddr_in servaddr;   /* struct in winsock2.h */
      int addrlen=sizeof(servaddr),recvStatus, i=0;
      char buffer[BUFLEN] = {0};
      char *hello = "Hi from server";
      char cmd = QUITKEY; /* character ESC */
      bool stop = false;  /* bool tyle in stdbool.h. stop running */

      printf("======== TCP Server ========\n");
      /* Step 1: startup winsocket - this is for Windows only */
      /* in pair with WSACleanup() */
      if(WSAStartup(MAKEWORD(2,2), &wsa_data) != 0){
            printf("WSAStartup failed: %d\n",WSAGetLastError());
            exit(SOCKET_ERROR);
      }

      /* Step 2: Create socket and check it is successful */
      /* in pair with closesocket()    */
      if ((sockfd = socket(AF_INET,SOCK_STREAM,0))==SOCKET_ERROR){
            printf("socket() failed: %d\n", WSAGetLastError());
            exit(INVALID_SOCKET);
      }

      /* Step 3: Bind to local TCP Server*/
      servaddr.sin_family = AF_INET;
      servaddr.sin_addr.s_addr = inet_addr(SERVER_IP_ADDR);
      servaddr.sin_port = htons(PORT);

      if(bind(sockfd,(struct sockaddr *)&servaddr,
              sizeof(servaddr)) == SOCKET_ERROR){
            printf("bind() failed: %d\n", WSAGetLastError());
            exit(SOCKET_ERROR);
      }

      /* Step 4: Listen */
      if ((listen(sockfd,8)) == SOCKET_ERROR){
            printf("listen() failed: %d\n", WSAGetLastError());
```

```
        exit(SOCKET_ERROR);
    }

    /* Step 5: Accept */
    printf("\nServer awaiting connection....\n");
    if ((acptdsock = accept(sockfd,(struct sockaddr *)&servaddr,
                             &addrlen))==INVALID_SOCKET){
        printf("accept() failed: %d\n", WSAGetLastError());
        exit(SOCKET_ERROR);
    }
    printf("....Connection established\n");

    /* Step 6: Send/Receive Data in loop */
  while (1){
        if ((send(acptdsock,hello,strlen(hello),0))==SOCKET_ERROR){
            printf("send() failed: \n",WSAGetLastError());
            exit(SOCKET_ERROR);
        }
        printf("%2d Sent:     %s\n",i,hello);

        recvStatus = recv(acptdsock,buffer,BUFLEN-1,0);
        if(recvStatus == 0)
          break;
        if (recvStatus == SOCKET_ERROR){
            printf("recv() failed: %d\n",WSAGetLastError());
            break;
        }
        buffer[recvStatus] = 0x00; /* force ending with '\0' */
        printf("    Received: %s\n",buffer);

      if ((++i) == LOOPLIMIT)      /* LOOPLIMIT reached */
        break;  /* make sure i within integrer limit */

      while ((kbhit()) && (!stop)){
          cmd = getch();
          fflush(stdout);
          if (cmd == QUITKEY){
              stop = true;
              printf("Terminating by kbd cmd 0x%x...\n",cmd);
              break;
          }
      }
      if (stop)
        break;

      Sleep(PERIOD); /* PERIOD in milliseconds */
  } */ end while loop */

    if ((send(acptdsock,&cmd,1,0)) == SOCKET_ERROR){
        printf("send() failed: \n",WSAGetLastError());
        closesocket(sockfd);
        WSACleanup();
        exit(SOCKET_ERROR);
    }

    /* Step 7: Close socket, in pair with socket() */
    Sleep(2000); /* Allow client without WSAECONNREST err*/
    if ((closesocket(sockfd))== SOCKET_ERROR){
        printf("closesocket() failed: %d\n",WSAGetLastError());
        exit(SOCKET_ERROR);
```

```
        }

        /* Step 8: Clean up winsocket - this is for Windows only! */
        /* in pair with WSAStartup() */
        WSACleanup();
        printf("....Server returned!\n\n");

        return 0;
}
```

Communicating with the Winsock server given above, the following is a complete client program in Windows for TCP/IP network communications. The client passively receives messages from the server, and upon receiving a message, it sends an ACK message back to the server. When an ESC character (ASCII code `0x1b`) is received, the client closes its socket and terminates the client program.

```
/**** clientWin.c, Winsock client in Windows ****/

#define _WINSOCK_DEPRECATED_NO_WARNINGS
#define _CRT_NONSTDC_NO_WARNINGS
#define _CRT_SECURE_NO_WARNINGS

#include <stdio.h>
#include <stdlib.h>
#include <windows.h> /* winsock2.h included in windows.h */
#include <stdbool.h>

#define SERVER_IP_ADDR "127.0.0.1"  /* loopback for testing */
#define SERVPORT  8886 /* server port number */
#define BUFLEN    1024 /* buffer length */
#define QUITKEY   0x1b /* ASCII code of ESC */

#pragma comment(lib,"ws2_32.lib")

int main(void){
        WSADATA wsa_data;              /* defined in winsock2.h */
        SOCKET  sockfd;                /* defined in winsock2.h */
        struct  sockaddr_in serv_addr; /* defined in winsock2.h */
        struct  hostent  *host = NULL; /* defined in winsock2.h */
        char buffer[BUFLEN] = {0};
        char *ackmsg = "ACK from client";
        int sendStatus, i=0;
        bool stop = false;

        printf("======== TCP Client ========\n");

        /* Step 1: startup winsocket - this is for Windows only */
        if(WSAStartup(MAKEWORD(2,2), &wsa_data) != 0){
              printf("WSAStartup failed: %d\n",WSAGetLastError());
              exit(SOCKET_ERROR);
        }

        /* Step 2: Create socket and check it is successful */
        if ((sockfd = socket(AF_INET,SOCK_STREAM,0))==SOCKET_ERROR){
              printf("socket() failed: %d\n", WSAGetLastError());
              exit(INVALID_SOCKET);
        }

        /* Step 3: Connect to the TCP Server */
```

```
    serv_addr.sin_family = AF_INET;
    serv_addr.sin_addr.s_addr = inet_addr(SERVER_IP_ADDR);
    serv_addr.sin_port = htons(SERVPORT);

    printf("\nConnecting to Server....\n");
    if ((connect(sockfd,(struct sockaddr *)&serv_addr,
             sizeof(serv_addr)))==SOCKET_ERROR){
        printf("connect() failed: %d\n", WSAGetLastError());
        exit(SOCKET_ERROR);
    }
    printf("....Connection established\n");

    /* Step 4: Send and receive data in loop */
    while (1){ /* loop for send()/recv() */
        /* receive message */
        if ((recv(sockfd,buffer,BUFLEN-1,0)) == SOCKET_ERROR){
            printf("recv() failed: %d\n", WSAGetLastError());
            break;
        }
        buffer[BUFLEN-1] = 0x00;  /* force ending with '\0' */
        if (buffer[0] == QUITKEY) // prepare termination
          break;
        printf("%2d Received: %s\n",i++, buffer);

        /* send message */
        sendStatus = send(sockfd, ackmsg, strlen(ackmsg), 0);
        if (sendStatus == 0)
          break;  /* nothing has been sent */
        if (sendStatus == SOCKET_ERROR){
            printf("send() failed: %d\n", WSAGetLastError());
            break;
        }
        printf("    Sent:     %s\n", ackmsg);
    }

    /* Step 5: Close socket, in pair with socket() */
    if ((closesocket(sockfd))== SOCKET_ERROR){
        printf("closesocket() failed: %d\n",WSAGetLastError());
        exit(SOCKET_ERROR);
    }

    /* Step 6: Clean up winsocket - this is for Windows only! */
    WSACleanup(); /* in pair with WSAStartup() */
    printf("....Client returned!\n\n");

    return 0;
}
```

13.7.3 Compiling C Programs in Command Prompt

How to compile the serverWin.c and clientWin.c programs in Windows? There are several options, such as using IDEs or command line compilers. IDEs, e.g., Microsoft Visual Studio, can be used for editing, compiling, debugging, and executing C programs. However, using an IDE can be complex and may require a significant amount of time to learn. In this section, a simple method is demonstrated to compile C programs: using a command line compiler.

In the following, let us use the command line compiler `gcc` for Windows as an example first. Then, the `cl` compiler from Microsoft's Visual Studio is introduced.

MinGW gcc Compiler for Windows

Installing `gcc` compiler for Windows takes a few simple steps. First of all, search and download MinGW `gcc` installation file. Then, install it in the default folder `C:\MinGW`. Right after that, a window of **MinGW Installation Manager** is presented. In the **Basic Setup** tab,

- Right click **mingw32-base**, and **Mark it for Installation**,
- Right click **ming32-gcc-g++**, and **Mark it for Installation**, and
- Select other optional packages you like to install.

Now, go to **Installation** → **Apply Changes**, and click on **Apply Changes**. The selected packages will be installed.

After the completion of the MinGW `gcc` installation, the `gcc` compiler is located in the folder `C:\MinGW\bin` if you did not change the installation folder at the beginning of the installation. The full path name of this folder should be added to the user's Path of the Windows Environment Variables. Then, the MinGW `gcc` compiler is ready to use.

Open a command window, go to the folder where the C programs serverWin.c and clientWin.c are stored. Compile them as follows:

```
>gcc -o serverWin.o serverWin.c
>gcc -o clientWin.o clientWin.c
```

Microsoft's Command Line Compiler

Microsoft's Visual Studio comes with a command line compiler `cl`. To use it, set up the necessary environment in two steps. The first step is to add the path of the batch file `vcvarsall.bat` to the Windows environment variables. So, locate the batch file `vcvarsall.bat` through Windows file search. For example, the file may be located at:

C:\Program Files (x86)\Microsoft Visual Studio 14.0\VC

Then, in Windows settings, go into Environment Variables, add this path into the system Path, and click OK to close the settings window. The new path setting becomes effective.

After the environment variable setting is completed, the second step is to open a command line window, e.g., by issuing a `cmd` command. In the command window, execute the batch file `vcvarsall.bat` at the prompt > as follows:

```
>vcvarsall.bat x86
```

If the execution of this batch file has nothing wrong, it runs silently without displaying any information on the monitor. Then, the command line compiler `cl` is ready to use.

To compile the Winsock serverWin.c program, go into your working folder and issue a `cl` command at the prompt followed by the name of the C code to be compiled, i.e., `cl serverWin.c`. On success, an executable file `serverWin.exe` is generated. Execute `serverWin.exe` by typing in its name directly with or without the extension `.exe`. The following shows how to compile and execute the Winsock server program:

```
>cl serverWin.c
>serverWin.exe
```

Similarly, to compile and execute the Winsock client program, the following commands are issued:

```
>cl clientWin.c
>clientWin.exe
```

If both the server and client programs are tested on the same machine by using the loopback IP address, i.e., 127.0.0.1, then open two command line windows, with one for the server and the other for the client.

13.7.4 Further Discussions on Windows Programming

In the examples of the client-server network communication system provided in this chapter, it is not surprising that both the server and client programs will function properly on the same operating system, either Linux or Windows. Since the layered network architecture and standard network protocols have been used in the design and implementation, the system can also be executed across different operating systems. This implies that a Linux socket server can establish communication with a Winsock client. Similarly, a Winsock server can successfully communicate with a Linux client.

In addition to the differences listed in Table 13.8 between Winsock and Linux sockets, a few other aspects also need to be paid attention to for Winsock programming in the examples provided above. They are summarized in the following.

Winsock Library

For Winsock programming, the library `ws2_32.lib` is required but will not be linked automatically. There are two methods to tell the compiler (actually the linker) to link this library:

- Add a preprocessing directive `#pragma comment(lib,"ws2_32.lib")` in the C code as shown previously in our serverWin.c and clientWin.c programs, or
- Link the `win2_32.lib` library explicitly in the compiling command, i.e.,
  - For `cl` compiler: `cl serverWin.c ws2_32.lib`
  - For `gcc` compiler: `gcc -o serverWin.o serverWin.c -lws2_32`

Warning Messages

To suppress warnings or suggestions from the compiler, disable them at the beginning of the C code, before the first `#include` line. The following are a few examples:

```
#define _WINSOCK_DEPRECATED_NO_WARNINGS
#define _CRT_NONSTDC_NO_WARNINGS
#define _CRT_SECURE_NO_WARNINGS
```

By defining these macros, specific warnings or suggestions related to deprecated functions, non-standard functions can be disabled. This can help secure function usage.

kbhit() function

In Windows, there is a '`kbhit()`' function available to detect if a keyboard key is pressed, making it directly usable. However, in Linux, this function is not available by default and needs to be implemented by programmers.

sleep() function

In comparison with the Linux `sleep()` function, which accepts an argument specified in seconds, Windows provides a similar function called `Sleep()`. However, the `Sleep()` function in Windows requires an argument whose value is specified in milliseconds.

getchar() function

With the Linux implementation of `kbhit()` presented previously, the standard library function `getchar()` has been utilized to read a character from the keyboard input buffer without the need to wait for the Enter key to be pressed. However, our tests indicate that employing `getchar()` in the same manner in Windows results in waiting for the Enter key to be pressed. To overcome this issue, the Windows function `getch()` can be employed, although it is not a standard library function.

Socket Disconnection

When the Winsock server closes its socket while the client is still receiving from, or sending data to, the server, the client will experience a socket error known as `WSAECONNRESET` with the error code 10054. This error indicates that an existing connection was forcibly closed by the remote host. It typically occurs when the peer application on the remote host is suddenly stopped, the host is rebooted, the host or remote network interface is disabled, or the remote host performs a hard close. This error can also occur if a connection is broken due to keep-alive activity detecting a failure while one or more operations are in progress. Operations that were in progress will fail with `WSAENETRESET`. Subsequent operations will also fail with `WSAECONNRESET`.

Implementing software handshaking between the server and client for connection and disconnection can help reduce communication disruptions caused by the disconnection of a host that is being communicated with by other hosts. In the example server and client programs provided in this chapter, the `WSAECONNRESET` error can be detected on the client side, and appropriate actions can be taken if this error is detected. This may make the client program more complex. Our tests have shown that if the server waits for a second or two before closing its socket, such an error will not occur on the client if the client behaves passively in the client-server system, as demonstrated in the examples. That being said, how to design and implement a reliable system should be considered on a case-by-case basis. Also, conducting comprehensive testing of a system before its deployment is always important, particularly for critical applications.

13.8 Summary

This chapter has provided a comprehensive guide on building TCP/IP network communication software systems using sockets. Using a top-down approach, it begins with an exploration of the motivation behind socket programming. Then, example client-server systems have been presented to deepen the understanding of the requirements and logic flows involved in socket programming. It has been emphasized that by following the layered network architecture and standard network protocols, the development of TCP/IP network software systems remains consistent for communication within a LAN, across multiple LANs, or over the Internet. This portability and scalability enable TCP/IP network communication programs to be adaptable to various network application scenarios.

To facilitate the practical implementation of sockets, the chapter has thoroughly discussed socket data structures and APIs. The C programming language is used due to its native support on Linux/Unix and Windows platforms. Using these APIs, a client-server system has been built via sockets to illustrate TCP/IP communications between the server and client. This system is extended by incorporating features such as IPv6 networks, keyboard input processing, and multithreading. It is also ported to Windows, taking into account the inherent differences from the Linux environment. All example Linux code programs have been successfully tested on Linux and macOS, while the Windows version of the system implementation has been tested on Windows platforms.

References

1. Jorgensen, B.: Beej's Guide to Network Programming, version 3.0.20. Brian 'Beej Jorgensen' Hall (2016). Published Online https://beej.us/guide/bgnet/. Accessed 6 Sept. 2021
2. Winett, J.M.: The definition of a socket. RFC 147, RFC Editor (1971). https://doi.org/10.17487/RFC0147
3. Gilligan, R., Thomson, S., Bound, J., McCann, J., Stevens, W.: Basic socket interface extensions for IPv6. RFC 3493, RFC Editor (2003). https://doi.org/10.17487/RFC3493
4. Stevens, W., Thomas, M., Nordmark, E., Jinmei, T.: Advanced sockets application program interface (API) for IPv6. RFC 3542, RFC Editor (2003). https://doi.org/10.17487/RFC3542
5. Microsoft: Windows sockets 2. Microsoft Online Documentation (2018). https://docs.microsoft.com/en-au/windows/win32/winsock/windows-sockets-start-page-2. Accessed 5 Sept. 2021
6. Microsoft: Porting socket applications to Winsock. Microsoft Online Documentation (2018). https://docs.microsoft.com/en-au/windows/win32/winsock/porting-socket-applications-to-winsock. Accessed 5 Sept. 2021

Index

A
AAA, 339, 373
AAAA, 394
Access category, 278
Accounting, 374
ACL, 377
Addressing architecture, 141, 161
Addressing requirements, 66
Admission control, 283
AES, 385
AH extension header, 204
Anycast, 186
AODV, 266
AOMDV, 267
APIPA, 164
Application
    mission-critical, 58
    rate-critical, 58
    real-time, 58
Application map, 51, 60
Application requirements, 57
Application virtualization, 451
Architectural model, 15, 124
    component-based, 139
    flow-based, 131
    functioal, 135
    geographical, 127
    hierarchical, 124
Auditing, 374
Authentication, 373
Authorization, 374
Autoconfiguration, 164, 198
    stateful, 202

B
Babel, 269
Backdoor connection, 365
Bandwidth usage pattern, 112
Best-effort service, 40, 79, 88, 90
BGP, 223
Black box, 25
Broadcast behavior, 110
Business goals and constraints, 48

C
C compiler
    cl, 502
    cl in Windows, 536
    gcc, 502
    gcc for windows, 536
CDN, 469
Characterizing networks, 69
CIA, 361
CIDR, 168
Class ID, 165
Client-server communication, 504
Client-server system, 504
Cloud, 15, 447
Cloud auditor, 493
Cloud broker, 492
Cloud carrier, 494
Cloud computing, 447, 477
    architecture, 489
    charateristics, 478
    deployment model, 480
    service model, 483
Cloud consumer, 492
Cloud data center, 408

Y.-C. Tian and J. Gao, *Network Analysis and Architecture*, Signals and Communication Technology, https://doi.org/10.1007/978-981-99-5648-7

Cloud provider, 491
Cloud security, 494
Cloud service, 477
CMIP, 326, 336
CMIS, 326, 336
CMOT, 326
CNF, 473
Colocation data center, 410
Community cloud, 481
Component-based architecture, 139
  addressing, 141
  management, 142
  performance, 143
  routing, 141
  security, 144
Container, 452
Containerization, 452
Contingency planing, 9
Controlled-load service, 91
Core business, 49
Core/distribution/access architecture, 124
CoS, 276
Critical flow, 83

D
Data center, 15, 405
Data center architecture, 430
Data center category, 408
  cloud data center, 408
  colocation data center, 410
  edge data center, 408
  enterprise data center, 409
  managed data center, 410
Data center design model, 431
  multi-tier model, 431, 432
  server cluster model, 432
Data center function, 408
Data center location, 425
Data center network virtualization, 436
Data center security, 441
Data center standard, 410
Data center tiered reliability, 416
Data virtualization, 451
DCF, 279
Decryption, 382
Default mask, 166
Desktop virtualization, 451
Device requirements, 61
DHCP, 164
DHCPv4, 203
DHCPv6, 202
  stateful, 202
  stateless, 203
Diffie-Hellman key exchange, 385
DiffServ, 90, 276, 294
DiffServ domain, 296
DiffServ region, 296
Distance-vector routing protocol, 224
  EIGRP, 226
  RIPv2, 226
DMZ, 387
DNS round robin, 155
Docker, 452
DSCP, 92, 299
DSCP codepoints, 281
DSDV, 268
DSR, 267
Dual stack, 211
Duplicate address detection, 200

E
ECC, 385
ECN codepoints, 282
EDCA, 279
Edge data center, 408
EGP, 221
  path-vector, 223
EIGRP, 226
Emergent behavior, 21
Encryption, 382
  asymmetric, 383
  symmetric, 383
Enterprise data center, 409
Enterprise edge architecture, 129
  multihomed Internet, 130
  redundant WAN, 129
  secure VPN, 130
ESP extension header, 205
Event notification, 346
Extranet, 137

F
FCAPS, 324, 341
Firewall, 375
  cloud firewall, 375
  hardware firewall, 375
  next-generation firewall, 379
  packet-filtering firewall, 376
  proxy firewall, 378
  software firewall, 375
  stateful inspection firewall, 378
First-hop redundancy, 150
Flow, 80

composite, 82
individual, 81
Flow analysis, 15, 79
Flow attributes, 80
Flow-based architecture, 131
client-server, 133
distributed computing, 134
hierarchical client-server, 133
peer-to-peer, 132
Flow boundary, 85
Flow labeling, 209
Flow measurement, 98
architecture, 98
granularity, 101
manager, 99
meter, 99
meter reader, 99
Flow model, 92
client-server, 93
distributed computing, 96
hierarchical client-server, 95
peer-to-peer, 92
Flow performance requirement
best-effort flow, 92, 115
guaranteed flow, 115
predictable flow, 115
Flow prioritization, 91
Flow sink, 84
Flow source, 84
Flowspec, 115, 306
Flowspec algorithm, 116
Flow specification, 114
Flowspec in BGP, 118
n-tuple, 118
Flowspec in IntServ, 117
RSPEC, 118
TSPEC, 118
Flowspec type
multi-part, 116
one-part, 115
two-part, 116
Frame preemption, 277, 288
Functional architecture, 135
application-driven, 136
core/distribution/access, 135
end-to-end service, 137
intranet/extranet, 137
service-provider, 138
Future changes, 52

G
Gcc compiler, 502
Gcc for Windows, 536
Generic network analysis model, 19, 29
Geographical architecture
LAN/MAN/WAN, 127
GLBP, 150
Global address, 187
Guaranteed service, 79, 91

H
HCCA, 279
HCF, 279
Holism, 21
Host ID, 165
HSRP, 147
HSRP datagram, 149
Hybrid cloud, 482
Hypervisor, 448, 449

I
IaaS, 484
ICMP, 213
ICMPv4, 213
ICMPv6, 201
IDPS, 379
IDS, 379
application protocol-based, 381
host-based, 381
hybrid, 381
network-based, 381
protocol-based, 381
IEEE 802.11, 278
IEEE 802.11e, 278
IEEE 802.1p, 278
IGMP, 191
IGP, 221
distance-vector, 224
link-state, 229
IKE, 386
In-band management, 349
Instrumentation, 345
Interface ID, 185
Intranet, 137
Intrusion detection, 379, 381
anomaly-based, 382
signature-based, 382
stateful protocol analysis, 382
IntServ, 276, 300
IntServ architecture, 305
IntServ implementation framework, 309
IntServ service model, 303
IntServ signaling, 307

IPFIX, 334
IPS, 379, 380
  host-based, 381
  network-based, 381
  network behavior-based, 381
  wireless network, 381
IPsec, 203, 397
IPv4 address/addressing, 161
  assignment, 177
  classful, 165
  classless, 168
  mechanisms, 164
  strategies, 177
IPv6 address/addressing, 184
  autoconfiguration, 198
  built-in QoS, 208
  built-in security, 203
  header structure, 193
IPv6 network planning, 214
IPv6 network security, 216
IPv6 socket, 519
IRDP, 147
IS-IS, 235

K
Key applications, 51
Keyboard input processing, 525

L
L2TP, 395
LAN/MAN/WAN architecture, 127
LEACH, 263
Link local address, 187
Link-state routing protocol, 229
  IS-IS, 235
  OSPF, 234
Location dependency, 48, 51, 60, 63, 66, 69, 73
Loopback address, 187

M
MAC, 279, 288
MAC address, 278
Managed data center, 410
Management architectural model, 348
Management architecture, 142, 321
Management data model, 339
Management framework, 323
Management mechanism, 344
Management protocol, 326
Management requirements, 67
Management traffic, 355
Memory virtualization, 451
MIB, 329
MIB-II, 329
MLD, 192
Model-based network management, 343
Multicast address, 188
Multicast flooding, 191
Multicast routing, 250
Multicast tree, 252
  multiple multicast trees, 261
Multihoming, 130
Multithreading, 527

N
NAT, 167
NAT64, 211
Neighbor advertisement, 202
Neighbor solicitation, 202
NETCONF, 335
Network analysis, 15
Network architecture, 15, 123
Network design, 123
Network efficiency, 110
Network ID, 165
Network management, 322
  accounting management, 325
  configuration management, 325
  fault management, 325
  performance management, 325
  security management, 326
Network planning, 3
  as an art, 12
  benefits, 6
  best practice, 13
  deliverables, 6
  motivation, 4
Network planning activity, 9
Network planning approach, 15, 19
Network planning process, 9, 12, 15
Network policy, 290
Network prefix, 185
Network requirements, 64
Network service, 35, 79, 215
Network virtualization, 452
NFV, 447, 459
NFV architecture, 470
NFV challenges, 475
NFV objectives, 460
NFV standards, 461
NFV use cases, 461

O
OLSR, 268
Operational planning, 8
Organization structure, 51
OS-level virtualization, 452
OSPF, 234
Out-of-band management, 349

P
PaaS, 485
Path-vector routing protocol, 223
  BGP, 223
PCF, 279
PCP, 92
Performance architecture, 143, 275
Performance metrics, 40
Performance requirements, 67
PHB, 299
Physical Machine (PM), 448
Physical security, 372
PKI, 386
PPP, 395
Predictable service, 79
Prefix discovery, 201
Prioritization, 277
Private address, 167
Private cloud, 480
Private key, 383
Proactive routing, 265
Project scope, 53
PSAMP, 334
Public cloud, 480
Public key, 383
Publish-subscribe network, 250

Q
QoS, 87, 276
QoS at layer 2, 276
QoS at layer 3, 276
QoS objectives, 89
QoS policy, 276
QoS requirement, 87, 90
Queuing, 277, 287

R
Reactive routing, 265
Redundancy, 145
  route and media redundancy, 156
  router redundancy, 146
  server redundancy, 151
  workstation-to-router redundancy, 146
Remote access, 394
Requirements
  application, 57
  device, 61
  network, 64
  user, 54
Requirements analysis, 15, 47
Requirements map, 48, 51, 60, 73
Requirements specifications, 48, 73
RIPv2, 226
RMON, 329
RMON MIB, 329
Route filtering, 243
Route redistribution, 242
Routing architecture, 141, 221, 238
Routing protocol, 221
  EGP, 221
  IGP, 221
Routing requirements, 66
RSA, 385
RSPEC, 307
RSTP, 156
RSVP message, 312
RSVP operation, 313
RSVP protocol, 310

S
SaaS, 486
Satisfactory solution, 23
Scheduling, 277, 286
SDN, 136, 244
SDN architecture, 246
SDN features, 249
SDN interface, 247
Security architecture, 144, 361, 363
Security awareness, 372
Security mechanism, 372
Security plan, 368
Security policy, 370
Security procedure, 370
Security recommendation, 361
Security requirements, 68
Security threats, 368
Server redundancy, 151
  DHCP server, 152
  DNS server, 152
  file/database server, 153
  mail server, 154
  web server, 153
Server virtualization, 448, 452
Service-based networking, 19, 35
Service offfering, 38

Service request, 38
Site prefix, 185
SLA, 315
SLAAC, 199
SNMP, 326
SNMP architecture, 326
SNMP command, 327
SNMP component, 326
SNMP security, 332
SNMP transport, 333
SNMP trap, 328
SNMPv2, 328
SNMPv3, 332
Socket, 15, 501, 503
Socket API, 505
  bind, 509
  close, 514
  connect, 511
  listen, 511
  send/recv, 512
  socket, 507
Socket programming, 502
  Linux, 505, 515, 531
  Windows, 531
SPIN, 264
SSL, 386
Storage virtualization, 453
STP, 156
Strategic planning, 8
Subnet ID, 185
Subnet mask, 168
Subnetting, 168
Supernetting, 173
Syslog protocol, 334
System, 20
  environment, 20
  subsystem, 20
Systems approach, 19, 20

T
Tactical planning, 8
TCP/IP programming, 501
TCP/IP socket application, 501
TLS, 386
Top-down methodology, 19, 32
Trade-offs, 73
Traffic behavior, 109
Traffic class, 208
Traffic classification, 284, 297
Traffic conditioning, 284, 297
Traffic control, 277
Traffic load, 103
Traffic management, 283
Traffic shaping, 285
Trend analysis, 346
TSPEC, 307
Tunnel, 213
  host-to-host, 213
  host-to-router, 213
  router-to-host, 213
  router-to-router, 213
Tunneling, 211, 213, 395
  layer-2, 395
  layer-3, 397

U
Unicast, 186
Unique local address, 188
User requirements, 54

V
vCDN, 469
Virtual device, 454
Virtualization, 15, 447, 448
Virtualization advantages, 458
Virtualization limitations, 459
Virtualization type, 450
  application virtualization, 451
  data virtualization, 451
  desktop virtualization, 451
  memory virtualization, 451
  network virtualization, 452
  OS-level virtualization, 452
  server virtualization, 452
  storage virtualization, 453
Virtual Machine (VM), 448
Virtual networking, 447
VM networking, 453
  bridge mode, 457
  host-only mode, 455
  NAT mode, 456
VM networking requirements, 455
VPN, 130, 366, 395
  client-to-site, 399
  remote-access, 399
  site-to-site, 398
VRRP, 150

W
Waterfall model, 19, 26
Waterfall networking, 28
Waterfall software development, 28
Wireless routing, 263
  data-centric, 263
  node-centric, 263
Wireless routing protocol, 263

MIX
Papier aus verantwortungsvollen Quellen
Paper from responsible sources
FSC® C105338

If you have any concerns about our products,
you can contact us on
ProductSafety@springernature.com

In case Publisher is established outside the EU,
the EU authorized representative is:
Springer Nature Customer Service Center GmbH
Europaplatz 3, 69115 Heidelberg, Germany

Printed by Libri Plureos GmbH
in Hamburg, Germany